Mining of Microbial Wealth and MetaGenomics

Vipin Chandra Kalia • Yogesh Shouche
Hemant J. Purohit • Praveen Rahi
Editors

Mining of Microbial Wealth and MetaGenomics

Springer

Editors
Vipin Chandra Kalia
Microbial Biotechnology and Genomics
CSIR-Institute of Genomics and Integrative Biology
New Delhi
India

Hemant J. Purohit
Environmental Biotechnology and Genomics Division
National Environmental Engineering Research Institute (CSIR)
Nagpur
India

Yogesh Shouche
Microbial Culture Collection
National Centre for Cell Science
Pune
India

Praveen Rahi
Microbial Culture Collection
National Centre for Cell Science
Pune
India

ISBN 978-981-10-5707-6 ISBN 978-981-10-5708-3 (eBook)
DOI 10.1007/978-981-10-5708-3

Library of Congress Control Number: 2017957066

Printed on acid-free paper

This Springer imprint is published by Springer Nature
The registered company is Springer Nature Singapore Pte Ltd.
The registered company address is: 152 Beach Road, #21-01/04 Gateway East, Singapore 189721, Singapore

Dedicated to our mentors

Preface

Microbes, the tiny little "invisible" organisms, exist along with all living beings. Though difficult to be seen by the naked eye, their effects provide strong evidences of their existence. Microbe–human associations are viewed as dangerous, leading to unhealthy scenes. However, the best part of the living world is that each of them is susceptible to attack by others. Hence, all of them have developed a strong defense mechanism to protect themselves. All living beings have unique genetic background and almost perfect metabolic pathways. The realization that human beings carry microbes on their skin and gut throughout their lives has provoked us to look deeper into these relationships. Recent studies have revealed that microbes produce metabolites which are essential for our health status. The research works of the last century have seen the rise and fall of antibacterials, the most important secondary metabolites. The process of elucidating the identity of organisms has gathered momentum by the advent of novel molecular biology techniques, the latest being the next-generation sequencing technologies. It has led to the discovery of the presence of organisms including those which are extremely low in abundance. Oceans, rivers, mountains, and the gut are unique ecosystems with great potential for harvesting secondary metabolites, i.e., natural products and bioactive molecules, which can find applications in the fields such as agriculture, food, medicine, water, bioremediation, etc. Oceans including ecological habitats such as coral reefs, hydrothermal vents, sponge reefs, sea grass beds, mangroves, and soft sediments are the largest reservoir of unknown living beings. Quite a few organisms are difficult to cultivate under laboratory conditions especially because of variations in temperature and pressure at different depths of the ocean. Metagenomics allows elucidation of the maximum biodiversity within an ecosystem, without the need to actually grow and culture the organisms. Microbiomes are associated with plants (leaves and roots), animals including humans (skin and gut), and marine metazoans (especially sponges). These multidimensional interactions need to be understood for developing sustainable ecosystems, agriculture, and healthy human beings.

Society strongly supports scientific adventures, which are constantly striving hard to improve human lives. In order to ensure that there is strong interest and continuity in scientific pursuits, it is almost imperative that the next generation is prepared to meet the future challenges. We need to train the young minds and impart scientific skills in them. We wish to present the status of the diverse possibilities and our views and opinions to finally provide mankind with novel, innovative, and long-lasting strategies, in the book entitled: *Mining of Microbial Wealth and Metagenomics*. This book has reached this stage only because of the fact that dedicated and academically accomplished members of the scientific community had agreed to share their vision and wisdom, which can be acquired only through decades of sincere and dedicated efforts put in to better understand the living world. This book has been presented in a manner such that all human beings can take advantage of the latent features of the world around us. Our sincere thanks are to all those whose invaluable contributions enabled us to bring out this book. We are indebted to all of them. This acknowledgment may not be sufficient to justify the worthiness of their efforts.

New Delhi Vipin Chandra Kalia

Contents

1 **Mining Metagenomes for Novel Bioactive Molecules** 1
Vipin Chandra Kalia

2 **Rhizosphere Microbiome Metagenomics: Elucidating the Abditive Microflora** 11
Asifa Mushtaq and Seema Rawat

3 **Rhizosphere Microbiome and Its Role in Plant Growth Promotion** ... 29
Rashmi Sharma, Minakshi, and Anjali Chauhan

4 **Microbial Community Dynamics During Soil Ecosystem Development** 57
Divya Deonalli, Rohit Sharma, and Kamlesh Jangid

5 **Bacterial Diversity in Cold Environments of Indian Himalayas** 83
Ramesh Chand Kasana

6 **The Microbiome of the Himalayan Ecosystem** 101
Divya Joshi, Saurabh Kumar, Deep Chandra Suyal, and Reeta Goel

7 **Phytomicrobiome: A Reservoir for Sustainable Agriculture** 117
Praveen Rahi

8 **Rare Biosphere in Human Gut: A Less Explored Component of Human Gut Microbiota and Its Association with Human Health** 133
Shrikant S. Bhute, Saroj S. Ghaskadbi, and Yogesh S. Shouche

9 **Metagenomic Insights into Microbial Diversity and Metabolic Potential of Hot Spring Ecosystems** 143
Jasleen Saini, Ramanpreet Kaur, and Pushpender Kumar Sharma

10 **Bioprospecting Gastrointestinal Microflora of Common Fishes for Disease Control in Aquaculture** 161
Jiun Yan Loh and Adeline Su Yien Ting

11 **Marine Sponge-Associated Microbiome: Reservoir of Novel Bioactive Compounds** 183
Uttara Lele-Rahalkar and Shrikant Pawar

12 **Metagenomic Insights into Herbivore Gut: An Application-Based Perspective** . . . 201
Birbal Singh, Gorakh Mal, Dixit Sharma, Rinku Sharma, Philem S. Kapoor, Gagandeep Singh, and Sarbjit S. Kanwar

13 **Soil Metagenomics: A Tool for Sustainable Agriculture** . . . 217
Reeta Goel, Deep Chandra Suyal, Narayan, Biplab Dash, and Ravindra Soni

14 **Soil Microbiome for Enhanced Crop Productivity** . . . 227
Sankalp Misra, Swapnil Pandey, Vijaykant Dixit, Shashank Kumar Mishra, Mohammad Haneef Khan, Lalit Agarwal, and Puneet Singh Chauhan

15 **Plant-Associated Microbial Endophytes: Promising Source for Bioprospecting** . . . 249
Shipra Singh and Anita Pandey

16 **Bifidobacterial Probiotics Through Fermented Foods** . . . 267
Sajan C. Achi and Prakash M. Halami

17 **Enterococci Prevalent in Processed Food Products: From Probiotics to Food Safety** . . . 287
S.R. Joshi and Koel Biswas

18 **Fermented Foods, Microbiota and Human Health** . . . 301
Ann Catherine Archer and Prakash M. Halami

19 **Metagenomics of Fermented Foods: Implications on Probiotic Development** . . . 333
Shiwani Guleria, Anil Kumar, Samriti Sharma, Saurabh Kulshrestha, and Anjali Chauhan

20 **Probiotics from Fermented Foods** . . . 357
Sundru Manjulata Devi and Prakash M. Halami

21 **Dietary Impacts on the Composition of Microbiota in Human Health and Disease** . . . 377
Anil Kumar Verma, Reena Kumari, Alok Bhattacharya, and Jaishree Paul

22 **Microfungi for the Removal of Toxic Triphenylmethane Dyes** . . . 405
Si Hui Chen and Adeline Su Yien Ting

23 **Applied Aspect of Microalgae in Monitoring of Heavy Metals** . . . 431
Santlal Jaiswar and Prakram Singh Chauhan

24 **Mining Bacterial Diversity for Biosurfactants** . . . 443
Tayebeh Soltanighias, Athoiba Elangbam Singh, and Praveen Rahi

Contributors

Sajan C. Achi Academy of Scientific and Innovative Research (AcSIR), CSIR-Central Food Technological Research Institute, Mysore, India

Microbiology and Fermentation Technology Department, CSIR-Central Food Technological Research Institute, Mysore, India

Lalit Agarwal CSIR-National Botanical Research Institute, Lucknow, Uttar Pradesh, India

Ann Catherine Archer Microbiology and Fermentation Technology, CSIR-Central Food Technological Research Institute, Mysore, India

Alok Bhattacharya School of Life Sciences, Jawaharlal Nehru University, New Delhi, India

Shrikant S. Bhute Department of Zoology, Savitribai Phule Pune University, Pune, India

Koel Biswas Department of Biotechnology and Bioinformatics, North-Eastern Hill University, Shillong, Meghalaya, India

Anjali Chauhan Department of Soil Science and Water Management, College of Forestry, DR YS Parmar University of Horticulture and Forestry, Solan, Himachal Pradesh, India

Prakram Singh Chauhan Marine Biotechnology and Ecology Division, CSIR-Central Salt and Marine Chemicals Research Institute, Bhavnagar, Gujarat, India

Puneet Singh Chauhan CSIR-National Botanical Research Institute, Lucknow, Uttar Pradesh, India

Si Hui Chen School of Science, Monash University Malaysia, Bandar Sunway, Selangor, Malaysia

Biplab Dash Department of Agricultural Microbiology, College of Agriculture, Indira Gandhi Krishi Vishwavidyalaya, Raipur, Chhattisgarh, India

Divya Deonalli Department of Biotechnology, Mithibai College of Arts, Chauhan Institute of Science and Amruthben Jivanlal College of Commerce and Economics, Mumbai, Maharashtra, India

Sundru Manjulata Devi Microbiology and Fermentation Technology Department, CSIR-Central Food Technological Research Institute, Mysore, Karnataka, India

Vijaykant Dixit CSIR-National Botanical Research Institute, Lucknow, Uttar Pradesh, India

Saroj S. Ghaskadbi Department of Zoology, Savitribai Phule Pune University, Pune, India

Reeta Goel Department of Microbiology, CBSH, G. B. Pant University of Agriculture and Technology, Pantnagar, Uttarakhand, India

Shiwani Guleria Department of Microbiology, Lovely Professional University, Phagwara, Punjab, India

DR YS Parmar University of Horticulture and Forestry, Solan, Himachal Pradesh, India

Prakash M. Halami Academy of Scientific and Innovative Research (AcSIR), CSIR-Central Food Technological Research Institute, Mysore, India

Microbiology and Fermentation Technology Department, CSIR-Central Food Technological Research Institute, Mysore, India

Santlal Jaiswar Marine Biotechnology and Ecology Division, CSIR-Central Salt and Marine Chemicals Research Institute, Bhavnagar, Gujarat, India

Kamlesh Jangid Microbial Culture Collection, National Centre for Cell Science, Pune, Maharashtra, India

Divya Joshi Department of Microbiology, Govind Ballabh Pant University of Agriculture and Technology, Pantnagar, Uttarakhand, India

S.R. Joshi Department of Biotechnology and Bioinformatics, North-Eastern Hill University, Shillong, Meghalaya, India

Vipin Chandra Kalia Microbial Biotechnology and Genomics, CSIR-Institute of Genomics and Integrative Biology (IGIB), New Delhi, India

Sarbjit S. Kanwar CSK HPKVV, Palampur, India

Philem S. Kapoor ICAR-Indian Veterinary Research Institute, Palampur, India

Ramesh Chand Kasana ICAR-Central Arid Zone Research Institute, Jodhpur, Rajasthan, India

Ramanpreet Kaur Department of Biotechnology, Sri Guru Granth Sahib World University, Fatehgarh Sahib, Punjab, India

Mohammad Haneef Khan CSIR-National Botanical Research Institute, Lucknow, Uttar Pradesh, India

Saurabh Kulshrestha Faculty of Applied Sciences and Biotechnology, Shoolini University, Solan, Himachal Pradesh, India

Anil Kumar Jaypee University of Information Technology, Solan, Himachal Pradesh, India

DR YS Parmar University of Horticulture and Forestry, Solan, Himachal Pradesh, India

Saurabh Kumar Department of Microbiology, Govind Ballabh Pant University of Agriculture and Technology, Pantnagar, Uttarakhand, India

Reena Kumari Department of Hematology and Bone Marrow Transplantation, University of Kentucky, Lexington, KY, USA

Uttara Lele-Rahalkar Indian Institute of Science Education and Research, Pune, India

Jiun Yan Loh Faculty of Applied Sciences, UCSI University, Kuala Lumpur, Malaysia

School of Science, Monash University Malaysia, Bandar Sunway, Selangor, Malaysia

Gorakh Mal ICAR-Indian Veterinary Research Institute, Palampur, India

Minakshi Department of Basic Sciences, College of Forestry, DR YS Parmar University of Horticulture and Forestry, Solan, Himachal Pradesh, India

Shashank Kumar Mishra CSIR-National Botanical Research Institute, Lucknow, Uttar Pradesh, India

Sankalp Misra CSIR-National Botanical Research Institute, Lucknow, Uttar Pradesh, India

Asifa Mushtaq Department of Botany and Microbiology, H.N.B.G.U. (A Central University), Srinagar Garhwal, India

Narayan Department of Agricultural Microbiology, College of Agriculture, Indira Gandhi Krishi Vishwavidyalaya, Raipur C.G., India

Anita Pandey Biotechnological Applications, G.B. Pant National Institute of Himalayan Environment and Sustainable Development, Almora, Uttarakhand, India

Swapnil Pandey CSIR-National Botanical Research Institute, Lucknow, Uttar Pradesh, India

Jaishree Paul School of Life Sciences, Jawaharlal Nehru University, New Delhi, India

Shrikant Pawar Microbial Culture Collection, National Centre for Cell Science, Central Tower, Pashan Pune, Maharashtra, India

Praveen Rahi National Centre for Microbial Resource, National Centre for Cell Science, Pune, Maharashtra, India

Seema Rawat Department of Botany and Microbiology, H.N.B.G.U. (A Central University), Srinagar Garhwal, India

Jasleen Saini Department of Biotechnology, Sri Guru Granth Sahib World University, Fatehgarh Sahib, Punjab, India

Dixit Sharma Centre for Computational Biology and Bioinformatics, School of Life Sciences, Central University of Himachal Pradesh, Shahpur, India

Pushpender Kumar Sharma, Ph.D. Department of Biotechnology, Sri Guru Granth Sahib World University, Fatehgarh Sahib, Punjab, India

Rashmi Sharma Department of Basic Sciences, College of Forestry, DR YS Parmar University of Horticulture and Forestry, Solan, Himachal Pradesh, India

Rinku Sharma ICAR-Indian Veterinary Research Institute, Palampur, India

Rohit Sharma Microbial Culture Collection, National Centre for Cell Science, Pune, Maharashtra, India

Samriti Sharma DR YS Parmar University of Horticulture and Forestry, Solan, Himachal Pradesh, India

Yogesh S. Shouche Microbial Culture Collection, National Centre for Cell Science, Pune, India

Athoiba Elangbam Singh Department of Environmental Biotechnology, Rajiv Gandhi Institute of IT and Biotechnology, Bharti Vidyapeeth Deemed University, Pune, India

Birbal Singh ICAR-Indian Veterinary Research Institute, Palampur, India

Gagandeep Singh Centre for Computational Biology and Bioinformatics, School of Life Sciences, Central University of Himachal Pradesh, Shahpur, India

Shipra Singh Biotechnological Applications, G.B. Pant National Institute of Himalayan Environment and Sustainable Development, Almora, Uttarakhand, India

Tayebeh Soltanighias Department of Environmental Biotechnology, Rajiv Gandhi Institute of IT and Biotechnology, Bharti Vidyapeeth Deemed University, Pune, India

National Centre for Microbial Resource, National Centre for Cell Science, Pune, Maharashtra, India

Ravindra Soni Department of Agricultural Microbiology, College of Agriculture, Indira Gandhi Krishi Vishwavidyalaya, Raipur C.G., India

Deep Chandra Suyal Department of Microbiology, CBSH, G. B. Pant University of Agriculture and Technology, Pantnagar, Uttaranchal, India

Adeline Su Yien Ting School of Science, Monash University Malaysia, Bandar Sunway, Selangor, Malaysia

Anil Kumar Verma School of Life Sciences, Jawaharlal Nehru University, New Delhi, India

About the Editors

Vipin Chandra Kalia is an emeritus scientist of the CSIR Institute of Genomics and Integrative Biology, Delhi. He holds an M.Sc. and Ph.D. in genetics, both from the Indian Agricultural Research Institute, New Delhi. Further, he is a fellow of the National Academy of Sciences, National Academy of Agricultural Sciences, and Association of Microbiologists of India.

His main areas of research are microbial biodiversity, bioenergy, biopolymers, quorum sensing and quenching, and drug discovery. He has published more than 100 papers in scientific journals including *Nature Biotechnology*, *Biotechnology Advances*, and *Bioresource Technology*.

He has authored 14 book chapters. He has edited seven books and is currently the editor in chief of the *Indian Journal of Microbiology* (since 2013) and editor of several journals, including the *Journal of Microbiology and Biotechnology* (Korea) and *Dataset Papers in Microbiology*. He is a member of numerous professional associations, including the Society of Biological Chemists of India and the Biotech Research Society of India (BRSI). He can be contacted at vckalia@igib.res.in and vc_kalia@yahoo.co.in.

Hemant J. Purohit is the head of the Environmental Biotechnology and Genomics Division, National Environmental Engineering Research Institute (CSIR), Nagpur. He is also a professor at AcSIR (Academy of Scientific and Innovative Research), New Delhi. He completed his Ph.D. at Nagpur University in 1986 and has been involved in designing a strategy for capturing microbial diversity by interfacing culturable and DNA fingerprinting tools; developing genomics-based monitoring tools for EIA and bioremediation process; studying stress-dependent microbial response using dynamic gene expression and its application in bioprocess optimization; and developing better insights into microbial capacities for utilization of organics through genome sequence analysis. He has been project coordinator for a number of high-value projects and has 225 publications to his credit. His Google Scholar Citations is 4711 (as of June 5, 2017), and he has an h-index of 38 and i10-index of 111. He has supervised 25 Ph.D. students and more than 100 M.Sc. student dissertations. He is a recipient of a number of prestigious awards, including the Fogarty International Exchange Program Fellowship; Commonwealth Fellowship, Department of Biochemistry, University of Hull, UK; and CSIR Research Fellowships (JRF and SRF).

Praveen Rahi is a scientist at the National Centre for Microbial Resource, National Centre for Cell Science, Pune. His main areas of research are microbial ecology, plant–symbiont interactions, microbial taxonomy, and microbial proteomics. He has published 25 papers in scientific journals such as *Molecular Ecology*, *Frontiers in Microbiology*, *Microbial Ecology*, *Applied Soil Ecology*, and *Fungal Biology*. He has reviewed publications for various international journals like *BMC Microbiology*, *Current Science*, *Frontiers in Plant Science*, and *Soil Biology and Biochemistry*. He is presently hosting a research topic "MALDI-TOF MS Application in Microbial Ecology Studies" as a guest editor for *Frontiers in Microbiology*. He was awarded the Young Scientist Award in Agriculture Microbiology (November 2013) by the Association of Microbiologists of India and selected as Young Associate (Life Science 2015) by the Maharashtra Academy of Science (MAcS), Pune, India.

Yogesh Shouche is a scientist at the National Centre for Microbial Resource, National Centre for Cell Science, Pune. After working on different aspects of the microbial, especially bacterial, diversity of various ecological niches, his research is now focused on understanding the succession of microbial communities in the human gut during development and their role in health and diseases. He has worked for 25 years in the fields of microbial ecology, microbial molecular taxonomy, and biodiversity; published more than 200 publications in journals of relevant area; and reviewed publications in international journals like *FEMS Ecology*, the *International Journal of Systematic and Evolutionary Microbiology*, and *Microbial Ecology*. His works have been cited 5662 times with an h-index of 37 and an i10-index of 147. He is also on the editorial board of prestigious journals such as *Current Science*, the *European Journal of Soil Biology*, *PLOS ONE*, and *Scientific Reports*. In 2009, he was given responsibility for the establishment of the National Centre for Microbial Resource (NCMR). Dr. Yogesh now heads a large group of researchers at NCMR including 30 scientists and technicians.

1 Mining Metagenomes for Novel Bioactive Molecules

Vipin Chandra Kalia

Abstract

Living organisms especially a wide range of microbes and plants produce secondary metabolites, which prove beneficial in improving the efficiency of metabolic processes. These metabolites with unique properties are categorized as bioactive molecules (BAMs). The utility of these BAMs has found its way to almost all biological processes, such that there has been a dramatic surge in their demand. The unique characteristics can be assigned to the chemical structures and associated groups. Conventional methods for searching novel BAMs are proving counterproductive. Modern molecular biological techniques in association with bioinformatic tools have provided the much necessary boost to the morale of the scientific community. The most effective genomic tool for searching these novel BAMs are (1) sequencing of whole genomes and (2) culture-independent (metagenomic) analysis. These approaches allow generation of huge amount of information, which can be easily analyzed through bioinformatic tools.

Keywords

Bioactive molecules • Metagenomics • Antibacterials • Therapeutics • Anticancer

1.1 Introduction

A wide range of prokaryotic and eukaryotic organisms produce secondary metabolites. These have been quite valuable and categorized as bioactive molecules (BAMs) (Debbab et al. 2010; Guaadaoui et al. 2014; Bandyopadhyay et al. 2015; Trindade

V.C. Kalia
Microbial Biotechnology and Genomics, CSIR – Institute of Genomics and Integrative Biology (IGIB), Delhi University Campus, Mall Road, New Delhi 110 007, India
e-mail: vckalia@igib.res.in

V.C. Kalia et al. (eds.), *Mining of Microbial Wealth and MetaGenomics*,
DOI 10.1007/978-981-10-5708-3_1

et al. 2015; Hernández-Saldaña et al. 2016; Kalia et al. 2016). However, the rate of discovering novel BAMs has declined dramatically. The basic reasons are (1) the conventional methods have been proving quite inefficient and uneconomical, (2) poor access to high-throughput techniques, and (3) inability to predict whether the organisms are genetically and metabolically active to yield the final product. With the advent of molecular techniques coupled with bioinformatics, there has been a dramatic rise in the possibilities of predicting the presence of genes for novel BAMs (Kalia 2013; Karumuri et al. 2015; Yin et al. 2015; Ambardar et al. 2016; Jeyanthi and Velusamy 2016; Parmar et al. 2017; Sharma and Lal 2017). Here, the exploitation of whole genome sequences either of a single organism or a mixture of related or unrelated organisms ranging from prokaryotes to eukaryotes is the most lucrative (Charlop-Powers et al. 2014). The most advantageous feature of the gene/genome sequencing technique is its extrapolation to metagenomics, which eliminates the need to culture the organisms (Daniel 2004). This culture-independent technique allows preparation of gene-based libraries and screening at least a few million genes in a single stroke. These genetic and metabolic reservoirs can be finally exploited for isolating novel BAMs. Metagenomics is motivating enough to substantially influence industrial production of BAMs and their applications in daily life (Torsvik et al. 2002; Handelsman 2004).

The most commonly looked for BAMs are industrial enzymes (detergents, food applications, agriculture, textile processing, leather industry, etc.), antibiotics, and related pharmaceutical products (Singh et al. 2009, 2013, 2015; Arasu et al. 2015; Azman et al. 2017). In spite of high intensity of efforts, the culture-dependent techniques have not proved very productive in the last few decades. The primary reason is that we are not able to provide the culture conditions to the microbes and 99.9% of the genetic potential remains untapped. The alternative approach to out-beat this limitation was to use replace nutrient-rich media with oligotrophic media, which allows slow-growing and nutrient-selective organisms to grow. Other approaches employed were simulated environments, cell encapsulation, etc. (Janssen et al. 2002; Joseph et al. 2003). Though quite impressive, these techniques may not hold good for all the organisms, which are yet to be explored (Daniel 2004).

The course of biological research took a dramatic turn through the development of a few innovative genomic approaches, toward the end of the last century. These opened the ways for dramatic and rapid progress. These genomic tools include (1) genome sequencing, (2) cultivation-independent protocols, and (3) bioinformatics. (Courtois et al. 2003; Robbel et al. 2010; Kalia 2013; Yin et al. 2015; Pooja et al. 2015; Ambardar et al. 2016; Parmar et al. 2017; Sharma and Lal 2017).

1.2 The Potential BAMs

1.2.1 From Terrestrial Microbes

Microbes have been used as active platforms for obtaining BAMs (Lorenz and Eck 2005; Peña-Yam et al. 2016; Radivojevic et al. 2016; Pessione et al. 2017; Saini and

Keum 2017; Sanchart et al. 2017). Microbes isolated from soil have resulted in extraction of BAMs, which act as antifungal and anticancer and can inhibit aminopeptidase activity (Go et al. 2015; Begum et al. 2016; Varsha et al. 2016; Thakur et al. 2017).

In contrast to culturing techniques, and designing the microbes for a mediocre molecule, screening metagenomes for BAMs has been found to hold more promises (Lorenz and Eck 2005). Metagenomic studies have been done largely by employing *Escherichia coli* as host. In order to improve the possibility of finding novel genes, other host organisms such as *Streptomyces*, *Bacillus*, and *Pseudomonas* spp. were employed (Wang et al. 2000; Courtois et al. 2003; Lorenz and Eck 2005). DNA libraries screened for genes encoding the enzyme alcohol oxidoreductases were based on polyol-fermenting microorganisms. For this purpose, 1.2 and 2.1 million clones from enriched and non-enriched samples, respectively, allowed recovery of 20 positive clones (Knietsch et al. 2003a, b). Antimicrobial metabolites from metagenomic libraries (bacterial artificial chromosome, fosmid and cosmid types) yielded positive hits from 1 to 10 at the rate of 1 hit per 100–900 Mb (Lorenz and Eck 2005). Metagenomic soil libraries have been found to yield interesting results on industrial enzymes and BAMs: agarases, amidase, antibacterials, amylases, biotin production, DNase, dehydratases, dehydrogenases, lipases, β-lactamase, indirubin, oxidation of polyols, oxidoreductases, lipase, polyketide synthase, terragine, violacein, turbomycin, and proteases (Richardson et al. 2002; Knietsch et al. 2003c; Daniel 2004; Lorenz and Eck 2005).

1.2.2 From Marine Microbes

Most of the BAMs available so far have been obtained from terrestrial sources. It was realized that the bulk of natural products obtained from marine sources were unique. The ecological niches in the oceans are home to more microbes than in any other environment (Montaser and Luesch 2011; Balakrishnan et al. 2015; Blunt et al. 2015; Tao et al. 2015). The efficiency of marine products is expected to be very high since they get diluted on release. Among marine microbes, algae, sponges, and soft corals have been the focus for quite some time; however, there has been a shift toward bacteria and fungi (Teasdale et al. 2009; Shiva Krishna et al. 2015). Marine sources are rich in antimicrobial, antitumor, antifungal, antiparasitic, anti-nematodal, anti-pathogenic, and anti-inflammatory molecules (Solanki et al. 2008; Teasdale et al. 2009; Trindade et al. 2015).

Marine ecological niches as a source for searching BAMs through metagenomic approaches are yet to make significant contributions to the pharmaceutical industry (Trindade et al. 2015). In spite of these limitations, metagenomic screenings based on prior knowledge on chemical structure and biological function have yielded a few BAMs: (1) bryostatins (type I polyketide) show cytotoxicity against carcinomas and are being tested for its activity as anti-Alzheimer's drug, which were linked to "*Candidatus* Endobugula sertula" (Sudek et al. 2007); (2) ecteinascidin 743 (ET-743) with anticancer activity was shown through metagenomic strategy to be linked to "*Candidatus* Endoecteinascidia frumentensis" (Rath et al. 2011; Schofield et al. 2015); (3) patellazoles-polyketides with antifungal potential and cytotoxicity were proposed to be produced by "*Candidatus* Endolissoclinum faulkneri" (Kwan et al. 2012; Schmidt et al. 2012); (4) psymberin, an antitumor polyketide (Fisch et al. 2009); and (5) polytheonamides, toxins like Calyculin A (Freeman et al. 2012).

1.2.3 From Plant Endophytes

BAMs of plant origin have great potential. Among these endophytes have been recognized as the most promising (Kusari and Spiteller 2011). During the last few years, the following secondary metabolites have been identified as BAMs: paclitaxel, podophyllotoxin, hypericin, camptothecin, and emodin. Their production and utilization of a large scale has yet not been achieved. Whole genome sequencing and metagenomics hold the potential to fish out novel BAMs. Mining of genome sequence of *Streptomyces coelicolor* has revealed *cch* cluster encoding for natural product biosynthetic systems: nonribosomal peptide synthetases (NRPSs). The BAM had a potential role in ferric iron acquisition. Similarly, mining of genomes resulted in BAMs, (1) new peptides: *Pseudomonas fluorescens* yielded orfamide A, a NRP antibiotic where as *Stigmatella aurantiaca* and Myxococcus virescens also produced novel peptides, and (2) terpenoid from *Arabidopsis thaliana* (Van Lanen and Shen 2006; Challis 2008).

Fungi are well known to produce BAMs. Although metagenomic techniques have seen a surge, however, metagenomic analysis of fungal genetic material has been quite scarce. Endophytic fungal DNA extracted from the leaves of *Rhododendron tomentosum* has the potential to produce antibacterial and antioxidant metabolites (Tejesvi et al. 2011). However, metagenomic analysis did not reveal any clear-cut information with respect to fungal genomes. Mining of genomes of *Aspergillus nidulans* has resulted in identification of genes encoding for anthranilate synthases (Scherlach and Hertweck 2006).

1.2.4 From Human Microbiome

Association of microbes with human beings has been reported from their skin and gut. These microbes have been shown to greatly influence human health. The ability of these microbes to produce secondary metabolites can be easily envisaged; however, the current knowledge on these aspects is very limited. Metagenomic studies of human microbiome have revealed a diversity of secondary metabolites (Donia et al. 2014; Joice et al. 2014; Sharon et al. 2014; Donia and Fischbach 2015; Koppel and Balskus 2016). In spite of the enormous quantum of effort, little is known about their identity and functions (Wilson et al. 2017). A few BAMs revealed through metagenomic studies include (1) antibiotic, lactocillin from *Lactobacillus gasseri* (Wieland Brown et al. 2009; Milshteyn et al. 2014); (2) commendamide (Cohen et al. 2015), a signaling molecule exhibiting activities such as antibacterial, activation of calcium channel and agonist; and (3) colibactin (Wilson et al. 2017).

1.3 The Future Prospects

The scope to use BAMs for biotechnological applications is widening with time (Kumar et al. 2014, 2015a, b; Ray and Kalia 2017a, b, c). This potential usage in health-related areas is sufficient to fuel the urge for searching novel and more efficient BAMs. The areas which make this search lucrative are cancer, therapeutics, nutrition, etc. (Milshteyn et al. 2014; Bose and Chatterjee 2015; Dobrucka and Długaszewska 2015; Park et al. 2015; Szweda et al. 2015; Wadhwani et al. 2016; Ahiwale et al. 2017).

1.4 Opinion

The present status of bioactive compounds is quite encouraging in terms of their applications. A large number of such molecules have been elucidated. However, we need to take advantage of recent discoveries and exploit them for accelerating the rate of discovery and development. Expression of genes in multiple host systems and prescreening strategies should be developed. This obviously also demands improved and innovative techniques to detect novel and beneficial BAMs.

Acknowledgments The author wishes to thank the director of CSIR-Institute of Genomics and Integrative Biology, Delhi, India, OLP1126 (CSIR-HRD ES21(1022)/16/EMR-II) for providing the necessary funds, facilities, and moral support.

References

Ahiwale SS, Bankar AV, Tagunde S, Kapadnis BP (2017) A bacteriophage mediated gold nanoparticle synthesis and their anti-biofilm activity. Indian J Microbiol 57:188–194. doi:10.1007/s12088-017-0640-x

Ambardar S, Gupta R, Trakroo D, Lal R, Vakhlu J (2016) High throughput sequencing: an overview of sequencing chemistry. Indian J Microbiol 56:394–404. doi:10.1007/s12088-016-0606-4

Arasu MV, Al-Dhabi NA, Rejiniemon TS, Lee KD, Huxley VAJ, Kim DH, Duraipandiyan V, Karuppiah P, Choi KC (2015) Identification and characterization of *Lactobacillus brevis* P68 with antifungal, antioxidant and probiotic functional properties. Indian J Microbiol 55:19–28. doi:10.1007/s12088-014-0495-3

Azman CA-S, Othman I, Fang C-M, Chan K-G, Goh B-H, Lee L-H (2017) Antibacterial, anticancer and neuroprotective activities of rare actinobacteria from mangrove forest soils. Indian J Microbiol. doi:10.1007/s12088-016-0627-z

Balakrishnan D, Bibiana AS, Vijayakumar A, Santhosh RS, Dhevendaran K, Nithyanand P (2015) Antioxidant activity of bacteria associated with the marine sponge *Tedania anhelans*. Indian J Microbiol 55:13–18. doi:10.1007/s12088-014-0490-8

Bandyopadhyay P, Mishra S, Sarkar B, Swain SK, Pal A, Tripathy PP, Ojha SK (2015) Dietary *Saccharomyces cerevisiae* boosts growth and immunity of IMC *Labeo rohita* (Ham.) juveniles. Indian J Microbiol 55:81–87. doi:10.1007/s12088-014-0500-x

Begum IF, Mohankumar R, Jeevan M, Ramani K (2016) GC–MS analysis of bioactive molecules derived from *Paracoccus pantotrophus* FMR19 and the antimicrobial activity against bacterial pathogens and MDROs. Indian J Microbiol 56:426–432. doi:10.1007/s12088-016-0609-1

Blunt JW, Copp BR, Keyzers RA, Munro MH, Prinsep MR (2015) Marine natural products. Nat Prod Rep 32:116–211. doi:10.1039/c4np00144c
Bose D, Chatterjee S (2015) Antibacterial activity of green synthesized silver nanoparticles using Vasaka (*Justicia adhatoda* L.) leaf extract. Indian J Microbiol 55:163–167. doi:10.1007/s12088-015-0512-1
Challis GL (2008) Genome mining for novel natural product discovery. J Med Chem 51:2618–2628. doi:10.1021/jm700948z
Charlop-Powers Z, Milshteyn A, Brady SF (2014) Metagenomic small molecule discovery methods. Curr Opin Microbiol 19:70–75. doi:10.1016/j.mib.2014.05.021
Cohen LJ, Kang HS, Chu J, Huang YH, Gordon EA, Reddy BV, Ternei MA, Craig JW, Brady SF (2015) Functional metagenomic discovery of bacterial effectors in the human microbiome and isolation of commendamide, a GPCR G2A/132 agonist. Proc Natl Acad Sci U S A 112:E4825–E4834. doi:10.1073/pnas.1508737112
Courtois S, Cappellano CM, Ball M, Francou F-X, Normand P, Helynck G, Martinez A, Kolvek SJ, Hopke J, Osburne MS, August PR, Nalin R, Guérineau M, Jeannin P, Simonet P, Pernodet JL (2003) Recombinant environmental libraries provide access to microbial diversity for drug discovery from natural products. Appl Environ Microbiol 69:49–55. doi:10.1128/AEM.69.1.49-55.2003
Daniel R (2004) The soil metagenome – a rich resource for the discovery of novel natural products. Curr Opin Biotechnol 15:199–204. doi:10.1016/j.copbio.2004.04.005
Debbab A, Aly AH, Lin WH, Proksch P (2010) Bioactive compounds from marine bacteria and fungi. Microb Biotechnol 3:544–563. doi:10.1111/j.1751-7915.2010.00179.x
Dobrucka R, Długaszewska J (2015) Antimicrobial activities of silver nanoparticles synthesized by using water extract of *Arnicae anthodium*. Indian J Microbiol 55:168–174. doi:10.1007/s12088-015-0516-x
Donia MS, Fischbach MA (2015) Small molecules from the human microbiota. Science 349:1254766. doi:10.1126/science.1254766
Donia MS, Cimermancic P, Schulze CJ, Wieland Brown LC, Martin J, Mitreva M, Clardy J, Linington RG, Fischbach MA (2014) A systematic analysis of biosynthetic gene clusters in the human microbiome reveals a common family of antibiotics. Cell 158:1402–1414. doi:10.1016/j.cell.2014.08.032
Fisch K, Gurgui C, Heycke N, van der Sar S, Anderson S, Webb V, Taudien S, Platzer M, Rubio BK, Robinson SJ, Crews P, Piel J (2009) Polyketide assembly lines of uncultivated sponge symbionts from structure-based gene targeting. Nat Chem Biol 5:494–501. doi:10.1038/nchembio.176
Freeman M, Gurgui C, Helf M, Morinaka B, Uria A, Oldham N, Sahl HG, Matsunaga S, Piel J (2012) Metagenome mining reveals polytheonamides as posttranslationally modified ribosomal peptides. Science 338:387–390. doi:10.1126/science.1226121
Go T-H, Cho K-S, Lee S-M, Lee O-M, Son H-J (2015) Simultaneous production of antifungal and keratinolytic activities by feather-degrading *Bacillus subtilis* S8. Indian J Microbiol 55:66–73. doi:10.1007/s12088-014-0502-8
Guaadaoui A, Benaicha S, Elmajdoub N, Bellaoumi M, Hamal A (2014) What is a bioactive compound? A combined definition for a preliminary consensus. Int J Nutr Food Sci 3:174–179. doi:10.11648/j.ijnfs.20140303.16
Handelsman J (2004) Metagenomics: application of genomics to uncultured microorganisms. Microbiol Mol Biol Rev 68:669–685. doi:10.1128/MMBR.68.4.669-685.2004
Hernández-Saldaña OF, Valencia-Posadas M, de la Fuente-Salcido NM, Bideshi DK, Barboza-Corona JE (2016) Bacteriocinogenic bacteria isolated from raw goat milk and goat cheese produced in the Center of México. Indian J Microbiol 56:301–308. doi:10.1007/s12088-016-0587-3
Janssen PH, Yates PS, Grinton BE, Taylor PM, Sait M (2002) Improved culturability of soil bacteria and isolation in pure culture of novel members of the divisions Acidobacteria, Actinobacteria, Proteobacteria, and Verrucomicrobia. Appl Environ Microbiol 68:2391–2396. doi:10.1128/AEM.68.5.2391-2396.2002
Jeyanthi V, Velusamy P (2016) Anti-methicillin resistant *Staphylococcus aureus* compound isolation from halophilic *Bacillus amyloliquefaciens* MHB1 and determination of its mode of

action using electron microscope and flow cytometry analysis. Indian J Microbiol 56:148–157. doi:10.1007/s12088-016-0566-8

Joice R, Yasuda K, Shafquat A, Morgan XC, Huttenhower C (2014) Determining microbial products and identifying molecular targets in the human microbiome. Cell Metab 20:731–741. doi:10.1016/j.cmet.2014.10.003

Joseph SJ, Hugenholtz P, Sangwan P, Osborne CA, Janssen PH (2003) Laboratory cultivation of widespread and previously uncultured soil bacteria. Appl Environ Microbiol 69:7210–7215. doi:10.1128/AEM.69.12.7210-7215.2003

Kalia VC (2013) The visionary: Prof. Carl R. Woese. Indian J Microbiol 53:245–246. doi:10.1007/s12088-013-0417-9

Kalia VC, Prakash J, Koul S (2016) Biorefinery for glycerol rich biodiesel industry waste. Indian J Microbiol 56:113–125. doi:10.1007/s12088-016-0583-7

Karumuri S, Singh PK, Shukla P (2015) *In silico* analog design for terbinafine against *Trichophyton rubrum*: a preliminary study. Indian J Microbiol 55:333–340. doi:10.1007/s12088-015-0524-x

Knietsch A, Waschkowitz T, Bowien S, Henne A, Daniel R (2003a) Construction and screening of metagenomic libraries derived from enrichment cultures: generation of a gene bank for genes conferring alcohol oxidoreductase activity on *Escherichia coli*. Appl Environ Microbiol 69:1408–1416. doi:10.1128/AEM.69.3.1408-1416.2003

Knietsch A, Waschkowitz T, Bowien S, Henne A, Daniel R (2003b) Metagenomes of complex microbial consortia derived from different soils as sources for novel genes conferring formation of carbonyls from short-chain polyols on *Escherichia coli*. J Mol Microbiol Biotechnol 5:46–56. doi:10.1159/000068724

Knietsch A, Bowien S, Whited G, Gottschalk G, Daniel R (2003c) Identification and characterization of genes encoding coenzyme B12-dependent glycerol and diol dehydratases from metagenomic DNA libraries derived from enrichment cultures. Appl Environ Microbiol 69:3048–3060. doi:10.1128/AEM.69.6.3048-3060.2003

Koppel N, Balskus EP (2016) Exploring and understanding the biochemical diversity of the human microbiota. Cell Chem Biol 23:18–30. doi:10.1016/j.chembiol.2015.12.008

Kumar P, Singh M, Mehariya S, Patel SKS, Lee JK, Kalia VC (2014) Ecobiotechnological approach for exploiting the abilities of *Bacillus* to produce co-polymer of polyhydroxyalkanoate. Indian J Microbiol 54:151–157. doi:10.1007/s12088-014-0457-9

Kumar P, Mehariya S, Ray S, Mishra A, Kalia VC (2015a) Biodiesel industry waste: a potential source of bioenergy and biopolymers. Indian J Microbiol 55:1–7. doi:10.1007/s12088-014-0509-1

Kumar P, Mehariya S, Ray S, Mishra A, Kalia VC (2015b) Biotechnology in aid of biodiesel industry effluent (glycerol): biofuels and bioplastics. In: Kalia VC (ed) Microbial factories. Springer, New Delhi, pp 105–119. doi:10.1007/978-81-322-2598-0

Kusari S, Spiteller M (2011) Are we ready for industrial production of bioactive plant secondary metabolites utilizing endophytes? Nat Prod Rep 28:1203. doi:10.1039/c1np00030f

Kwan J, Donia M, Han A, Hirose E, Haygood M, Schmidt E (2012) Genome streamlining and chemical defense in a coral reef symbiosis. Proc Natl Acad Sci U S A 109:20655–20660. doi:10.1073/pnas.1213820109

Lorenz P, Eck J (2005) Metagenomics and industrial applications. Nat Rev Microbiol 3:510–516. doi:10.1038/nrmicro1161

Milshteyn A, Schneider JS, Brady SF (2014) Mining the metabiome: identifying novel natural products from microbial communities. Chem Biol 21:1211–1223. doi:10.1016/j.chembiol.2014.08.006

Montaser R, Luesch H (2011) Marine natural products: a wave of new drugs? Future Med Chem 3:1475–1489. doi:10.4155/fmc.11.118

Park J-M, Radhakrishnan R, Kang S-M, Lee I-J (2015) IAA producing *Enterobacter* sp. I-3 as a potent bio-herbicide candidate for weed control: a special reference with lettuce growth inhibition. Indian J Microbiol 55:207–211. doi:10.1007/s12088-015-0515-y

Parmar NR, Pandit PD, Purohit HJ, Nirmal Kumar JI, Joshi CG (2017) Influence of diet composition on cattle rumen methanogenesis: a comparative metagenomic analysis in Indian and exotic cattle. Indian J Microbiol 57:226–234. doi:10.1007/s12088-016-0635-z

Peña-Yam LP, Esaú Ruíz-Sánchez E, Barboza-Corona JE, Reyes-Ramírez A (2016) Isolation of Mexican *Bacillus* species and their effects in promoting growth of chili pepper (*Capsicum annuum* L. cv Jalapeño). Indian J Microbiol 56:375–378. doi:10.1007/s12088-016-0582-8

Pessione E, Maxxoli R, Riedel K (2017) Bioactive compounds from microbes. Front Microbiol. doi:10.3389/fmicb.2017.00392

Pooja S, Pushpananthan M, Jayashree S, Gunasekaran P, Rajendhran J (2015) Identification of periplasmic a-amlyase from cow dung metagenome by product induced gene expression profiling (Pigex). Indian J Microbiol 55:57–65. doi:10.1007/s12088-014-0487-3

Radivojevic J, Skaro S, Senerovic L, Vasiljevic B, Guzik M, Kenny ST, Maslak V, Nikodinovic-Runic J, O'Connor KE (2016) Polyhydroxyalkanoate-based 3-hydroxyoctanoic acid and its derivatives as a platform of bioactive compounds. Appl Microbiol Biotechnol 100:161–172. doi:10.1007/s00253-015-6984-4

Rath C, Janto B, Earl J, Ahmed A, Hu FZ, Hiller L, Dahlgren M, Kreft R, Yu F, Wolff JJ, Kweon HK, Christiansen MA, Håkansson K, Williams RM, Ehrlich GD, Sherman DH (2011) Meta-omic characterization of the marine invertebrate microbial consortium that produces the chemotherapeutic natural product ET-743. ACS Chem Biol 6:1244–1256. doi:10.1021/cb200244t

Ray S, Kalia VC (2017a) Biomedical applications of polyhydroxyalkanaotes. Indian J Microbiol 57. doi:10.1007/s12088-017-0651-7

Ray S, Kalia VC (2017b) Biological significance of degradation of polyhydroxyalkanoates. In: Kalia VC, Kumar P (eds) Microbial applications, vol 1 Bioremediation and bioenergy. SpringerNature. ISBN 978–3–319-52665-2

Ray S, Kalia VC (2017c) Microbial cometabolism and polyhydroxyalkanaote co-polymers. Indian J Microbiol 57:39–47. doi:10.1007/s12088-016-0622-4

Richardson TH, Tan X, Frey G, Callen W, Cabell M, Lam D, Macomber J, Short JM, Robertson DE, Miller C (2002) A novel, high performance enzyme for starch liquefaction. Discovery and optimization of a low pH, thermostable a-amylase. J Biol Chem 277:26501–26507. doi:10.1074/jbc.M203183200

Robbel L, Knappe TA, Linne U, Xie X, Marahiel MA (2010) Erythrochelin–a hydroxamate-type siderophore predicted from the genome of *Saccharopolyspora erythraea*. FEBS J 277:663–676. doi:10.1111/j.1742-4658.2009.07512.x

Saini RK, Keum Y-S (2017) Progress in microbial carotenoids production. Indian J Microbiol 57:129–130. doi:10.1007/s12088-016-0637-x

Sanchart C, Rattanaporn O, Haltrich D, Phukpattaranont P, Maneerat S (2017) *Lactobacillus futsaii* CS3, a new GABA-producing strain isolated from Thai fermented shrimp (*Kung–Som*). Indian J Microbiol 57:211–217. doi:10.1007/s12088-016-0632-2

Scherlach K, Hertweck C (2006) Discovery of aspoquinolones A-D, prenylated quinoline-2-one alkaloids from Aspergillus nidulans, motivated by genome mining. Org Biomol Chem 4:3517–3520. doi:10.1039/B607011F

Schmidt E, Donia M, McIntosh J, Fricke W, Ravel J (2012) Origin and variation of tunicate secondary metabolites. J Nat Prod 75:295–304. doi:10.1021/np200665k

Schofield MM, Jain S, Porat D, Dick G, Sherman D (2015) Identification and analysis of the bacterial endosymbiont specialized for production of the chemotherapeutic natural product ET-743. Environ Microbiol 17:3964–3975. doi:10.1111/1462-2920.1290

Sharma A, Lal R (2017) Survey of (Meta)genomic approaches for understanding microbial community dynamics. Indian J Microbiol 57:23–38. doi:10.1007/s12088-016-0629-x

Sharon G, Garg N, Debelius J, Knight R, Dorrestein PC, Mazmanian SK (2014) Specialized metabolites from the microbiome in health and disease. Cell Metab 20:719–730. doi:10.1016/j.cmet.2014.10.016

Shiva Krishna P, Sudheer Kumar B, Raju P, Murty MSR, Prabhakar Rao T, Singara Charya MA, Prakasham RS (2015) Fermentative production of pyranone derivate I from marine *Vibrio* sp. SKMARSP9: isolation, characterization and bioactivity evaluation. Indian J Microbiol 55:292–301. doi:10.1007/s12088-015-0521-0

Singh M, Patel SKS, Kalia VC (2009) *Bacillus subtilis* as potential producer for polyhydroxyalkanoates. Microb Cell Factories 8:38. doi:10.1186/1475-2859-8-38

Singh M, Kumar P, Patel SKS, Kalia VC (2013) Production of polyhydroxyalkanoate co-polymer by *Bacillus thuringiensis*. Indian J Microbiol 53:77–83. doi:10.1007/s12088-012-0294-7

Singh M, Kumar P, Ray S, Kalia VC (2015) Challenges and opportunities for customizing polyhydroxyalkanoates. Indian J Microbiol 55:235–249. doi:10.1007/s12088-015-0528-6

Solanki R, Khanna M, Lal R (2008) Bioactive compounds from marine actinomycetes. Indian J Microbiol 48:410–431. doi:10.1007/s12088-0080052-z

Sudek S, Lopanik N, Waggoner L, Hildebrand M, Anderson C, Liu H, Patel A, Sherman DH, Haygood MG (2007) Identification of the putative bryostatin polyketide synthase gene cluster from "Candidatus Endobugulasertula", the uncultivated microbial symbiont of the marine bryozoan *Bugula neritina*. J Nat Prod 70:67–74. doi:10.1021/np060361d

Szweda P, Gucwa K, Kurzyk E, Romanowska E, Dzierzanowska-Fangrat K, Jurek AZ, Kuś PM, Milewski M (2015) Essential oils, silver nanoparticles and Propolis as alternative agents against fluconazole resistant *Candida albicans, Candida glabrata* and *Candida krusei* clinical isolates. Indian J Microbiol 55:175–183. doi:10.1007/s12088-014-0508-2

Tao L, Zhu F, Qin C, Zhang C, Chen S, Zhang P, , Zhang C, Tan C, Gao C, Chen Z, Jiang Y, Chen YZ (2015) Clustered distribution of natural product leads of drugs in the chemical space as influenced by the privileged target-sites. Sci Rep 5:9325. doi: 10.1038/srep09325

Teasdale M, Liu J, Wallace J, Akhlaghi F, Rowley D (2009) Secondary metabolites produced by the marine bacterium *Halobacillus salinus* that inhibit quorum sensing-controlled phenotypes in gram-negative bacteria. Appl Environ Microbiol 75:567–572. doi:10.1128/AEM.00632-08

Tejesvi MV, Kajula M, Mattila S, Pirttilä AM (2011) Bioactivity and genetic diversity of endophytic fungi in *Rhododendron tomentosum* Harmaja. Fungal Divers 47:97. doi:10.1007/s13225-010-0087-4

Thakur R, Sharma KC, Gulati A, Sud RK, Gulati A (2017) Stress-tolerant *Viridibacillus arenosi* strain IHB B 7171 from tea rhizosphere as a potential broad-spectrum microbial inoculant. Indian J Microbiol 57:195–200. doi:10.1007/s12088-017-0642-8

Torsvik V, Øvreås L, Thingstad TF (2002) Prokaryotic diversity-magnitude, dynamics, and controlling factors. Science 296:1064–1066. doi:10.1126/science.1071698

Trindade M, van Zyl LJ, Navarro-Fernández J, Abd Elrazak A (2015) Targeted metagenomics as a tool to tap into marine natural product diversity for the discovery and production of drug candidates. Front Microbiol 6:890. doi:10.3389/fmicb.2015.00890

Van Lanen SG, Shen B (2006) Microbial genomics for the improvement of natural product discovery. Curr Opin Microbiol 9:252–260. doi:10.1016/j.mib.2006.04.002

Varsha KK, Nishant G, Sneha SM, Shilpa G, Devendra L, Priya S, Nampoothiri KM (2016) Antifungal, anticancer and aminopeptidase inhibitory potential of a phenazine compound produced by *Lactococcus* BSN307. Indian J Microbiol 56:411–416. doi:10.1007/s12088-016-097-1

Wadhwani SA, Shedbalkar UU, Singh R, Vashisth P, Pruthi V, Chopade BA (2016) Kinetics of synthesis of gold nanoparticles by *Acinetobacter* sp. SW30 isolated from environment. Indian J Microbiol 56:439–444. doi:10.1007/s12088-016-0598-0

Wang GY, Graziani E, Waters B, Pan W, Li X, McDermott J, Meurer G, Saxena G, Andersen RJ, Davies J (2000) Novel natural products from soil DNA libraries in a streptomycete host. Org Lett 2:2401–2404. doi:10.1021/ol005860z

Wieland Brown LC, Acker MG, Clardy J, Walsh CT, Fischbach MA (2009) Thirteen posttranslational modifications convert a 14-residue peptide into the antibiotic thiocillin. Proc Natl Acad Sci U S A 106:2549–2553. doi:10.1073/pnas.0900008106

Wilson MR, Zha L, Balskus EP (2017) Natural product discovery from the human microbiome. J Biol Chem. doi:10.1074/jbc.R116.762906

Yin X, Ma L, Pei X, Du P, Li C, Xie T, Yu L, Yu L, Wang Q (2015) Creation of functionally diverse chimerical α-glucosidase enzymes by swapping homologous gene fragments retrieved from soil DNA. Indian J Microbiol 55:114–117. doi:10.1007/s12088-014-0493-5

Rhizosphere Microbiome Metagenomics: Elucidating the Abditive Microflora

2

Asifa Mushtaq and Seema Rawat

Abstract

The rhizosphere is a zone of biological activity between plant roots and soil harboring a plethora of microorganisms. The key interactions among a multitude of microorganisms in the rhizosphere have a direct or indirect effect on the plant. Being versatile and intriguingly complex, a comprehension regarding the elementary principles of microbial ecology and functioning is significant to enhance the plant productivity and agroecosystem working. The interplay between plant roots and the associated microbes is regulated by profound chemical signaling. Most of the known facts about these interactions till recently have been derived through the studies based on culturing the microbes; however, it is an established fact that majority of the microbes are uncultivable. Novel insights into enhancing our ability to unravel the quintessential factors determining the rhizosphere microbiome could offer the progress towards the development of sustainable agriculture. We now have the opportunity to utilize the advanced culture independent techniques to have an insight into the intriguing plant-microbe interplay. Metagenomic studies present a strong mandate to understand the enormous richness and diversity of rhizosphere microbiome as well as the key biological processes.

Keywords

Rhizosphere • Microbiome • Interplay • Signaling • Metagenomics

A. Mushtaq • S. Rawat (✉)
Department of Botany and Microbiology, H.N.B.G.U. (A Central University), Srinagar Garhwal, Uttarakhand, India
e-mail: aasifa.peers@gmail.com; seemamillenium@gmail.com

V.C. Kalia et al. (eds.), *Mining of Microbial Wealth and MetaGenomics*,
DOI 10.1007/978-981-10-5708-3_2

2.1 Introduction

The rhizosphere, a microecological zone of soil surrounded and influenced by root system, is a dynamic site for diverse organisms (Dobbelaere et al. 2003; Hartmann et al. 2008; Prashar et al. 2014). This zone, in comparison to the bulk soil, is rich in nutrients due to the accumulation of various plant exudates such as amino acids, aromatic compounds, and sugars, providing a nutrient-rich environment for the colonization of diverse microbes (Gray and Smith 2005; Beneduzi et al. 2012). This zone of influence withholds higher microbial population densities, approximately 10- to 100-fold greater, as compared to bulk soil (Berg et al. 2006; Costa et al. 2006; Hein et al. 2008) suggestive of intense race between microorganisms for the availability of nutrients and for the sustenance of species showing functional diversity and metabolic versatility (Dube and Yeole 1999; Sinha et al. 2001). The rhizosphere itself can be differentiated into three zones: (a) endorhizosphere, the internal root area extending generally to the cortical region which harbors diverse population of microlife with versatile functions; (b) rhizoplane, zone adjacent to the root; and (c) ectorhizosphere, extends from the rhizoplane out to what is called bulk soil (reviewed by Johri et al. 2003). The significance of the rhizosphere as an environment copious in varied microbial populations playing an imperative role in plant health and soil fertility has already been perceived during the late nineteenth century (Bais et al. 2006; Nautiyal et al. 2008).

Rhizosphere microbiota, directly or indirectly, promote plant productivity by a range of varied mechanisms (Buckling et al. 2007; Lugtenberg and Kamilova 2009; Hider and Kong 2010; Leveau et al. 2010; Mapelli et al. 2012; Pii et al. 2015). The diversity of microbes present has a pivotal role in maintaining the soil fertility and, in turn, the plant growth because these microorganisms are involved in many significant biological processes such as soil formation, facilitating the uptake of specific nutrients by the plants, toxin removal, biogeochemical cycling, and many more (Nautiyal et al. 2010). The interplay between plants and the associated microorganisms promotes a physicochemical heterogeneity in the local soil microhabitat making a major contribution to the biotic components of soil, one of the most intricate ecosystems on Earth. Advancement in agricultural sustainability can be achieved by making optimal and systematic use of biotic and abiotic components of soil both of which rely on soil biodiversity and biological processes.

Analyzing bacterial communities traditionally began with culturing microorganisms from an ecological niche. This technique has limitations as the largest proportion of soil microbes cannot be cultured efficiently in the laboratory, due to the complexity of their growth conditions and the presence of cells which are in a viable but noncultivable state. As a result, only 1% of the microbial diversity present has been explored until now using classical cultivation techniques. Various molecular techniques have been put into e to assess microbial communities from diverse environments (Table 2.1). Over the past decade, molecular genetics has opened our eyes to several of the complex host-microbe interactions. One of the rapidly growing techniques used for elucidating the uncultured microorganisms, their functions, cooperation, and evolution from various environments is metagenomics.

Table 2.1 Molecular techniques applied to analyze microbial diversity

Techniques	Applications	Shortcomings	References
FISH (fluorescent in situ hybridization)	Direct visualization of bacteria in the environment Detects active cells by targeting rRNA	Can mislabel cells when probe is not universal Visualization can become difficult with background fluorescence Smaller fraction of the community can be hard to detect	Zwirglmaier (2005), Li et al. (2008), Caracciolo et al. (2010), Lundberg et al. (2012), Schmidt and Eickhorst (2014)
PCR based: DGGE, RFLP, RISA, sequencing of amplified genes, etc.	Easy to implement Detailed picture of rhizosphere diversity Amenable to high throughput analysis	Subject to PCR bias Labor intensive More than one microbial species may be represented by a single band	Acinas et al. (2005), Spiegelman et al. (2005), Thompson et al. (2005), Hong et al. (2006), Bentley et al. (2008), Rawat and Johri (2014), Pascual et al. (2016)
Metagenomics	Whole-community-level genome characterization Characterization of genomes of unculturable microbes	High cost Data analysis is challenging and time consuming Can miss lesser abundant members in the microbial community	Daniel (2005), Uroz et al. (2010), Myrold and Nannipieri (2014), Pascual et al. (2016)
Metatranscriptomics	Gene expression Identification of active community members, correlating them with their metabolic activities	Short mRNA half-life Limiting RNA amounts Presence of enzyme activity inhibitors in soil	Jones and Dangl (2006), Bastida et al. (2012), De Vleesschauwer and Hofte (2009), Simon and Daniel (2011), Carvalhais et al. (2012)
Metaproteomics	Gene activities Metabolic functions Gives an understanding of microbial interactions	Niche discipline restricted Superior separation and measurement protocols Nonannotated and unassembled metagenomic data can be a major hurdle	Wang et al. (2011), Becher et al. (2013), Bao et al. (2014)

Metagenomics aids in evaluating the richness, distribution, and activity of microbial communities in any environmental sample even without the need of culturing. Furthermore, it provides an easy access to the diverse microorganisms present in an environment thus aiding in the discovery of new groups of microbes (Amorim et al. 2008). In this chapter, we will try to explain the potential of metagenomic approach towards the understanding of diverse root-associated microbes.

2.2 Plant-Microbe Interactions

Plants develop an intriguing relationship with the microbial diversity present within the surrounding soil. There are diverse microbial communities present in the soil, and these microbial communities, so far, are considered to be an epitome of biological diversity present on Earth (Gams 2007; Buee et al. 2009; Singh et al. 2009). Rhizosphere, a narrow zone of soil, is home to an overwhelming population of microbes and can harbor up to 10^9 prokaryotic cells per gram (Wegley et al. 2006; Egamberdieva et al. 2008) containing more than 30,000 species (Mendes et al. 2011). As the collective genome of the inhabiting microbial community is much larger than that of the plant itself, it is usually ascribed as plant's second genome (Berendsen et al. 2012). Subsequently, the significance of root microbiome, consisting of the entire rhizosphere-associated microbes, their genetic elements, and their interactions, in determining the plant health (Berendsen et al. 2012), has been demonstrated by various in-depth studies from different parts of the world.

Microbial colonization in the rhizosphere is governed by various biotic and abiotic factors. The heterogeneous physicochemical characteristics of soil are paramount in shaping the rhizosphere microbiome as they affect root exudation patterns by affecting the plant physiology. Moreover, these root exudates are known to vary in accordance with plant species and environmental conditions (Hogberg et al. 2006; Lesuffleur et al. 2007; Micallef et al. 2009) influencing the microbial diversity in rhizosphere (Grayston et al. 1997; Kuklinsky-Sobral et al. 2004; Salles et al. 2004; Somers et al. 2004). The majority of the root exudates are believed to be comprised of amino acids, fatty acids, hormones, organic acids, sugars, and vitamins as well as antimicrobial compounds (Bertin et al. 2003; Jones et al. 2004; Badri and Vivanco 2009). Plant roots exert pronounced effect on the surrounding soil through these "rhizodepositions" (sloughing of root-cap cells along with mucilage secretion and controlled root exudate dispersion), thus render appropriate ecological niche for microbial colonization (Bais et al. 2006). The rhizosphere microbes establish a synergistic relationship with the host plant facilitating nutrient uptake as well as suppressing soilborne phytopathogens, thus helping in improving the plant productivity (Berendsen et al. 2012). (Fig. 2.1).

Rhizosphere microbes perform various ecological functions including maintenance of soil structure and water relationships, organic matter decomposition, biogeochemical cycling, and, above all, growth of the inhabited plant. Various biological processes that are carried out by the microbes in rhizosphere include symbiosis, nutrient uptake, plant protection as well as antibiotic production, geochemical

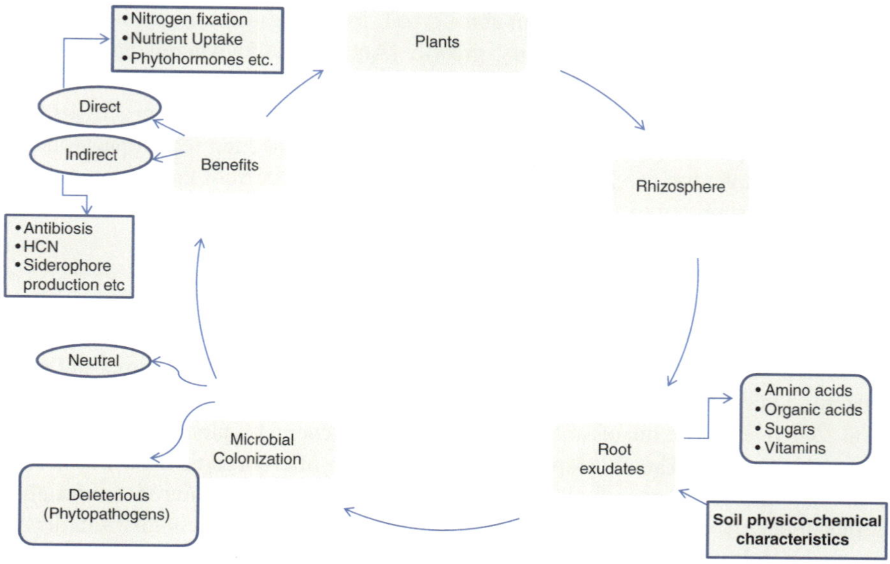

Fig. 2.1 Factors governing plant-microbe interactions (Mushtaq & Rawat, HNBGU)

cycling of minerals, etc. (Kent and Triplett 2002; Bulgarelli et al. 2013; Philippot et al. 2013; Gianfreda 2015). Even though substantial improvement has been made in our understanding of the microbial ecology, more comprehensive studies in soil microbiology must be undertaken to unravel plant-microbe interplay.

2.3 Diversity of Rhizosphere Microbiome

The diversity of root-associated microbes is immense, in the order of tens of thousands of different species. There is a clear-cut difference between the microbial populations with larger number of microbes harboring the rhizosphere as compared to those residing in the bulk soil. The enhanced microbial activity in rhizosphere as compared to bulk soil can be attributed to the availability of large amount of nutrients in the form of root exudates. In general, the organisms found in rhizosphere include bacteria, fungi, oomycetes, nematodes, protozoa, algae, viruses, archaea, and arthropods (Raaijmakers and Weller 2001; Nautiyal et al. 2008; Raaijmakers et al. 2009; Chaudhary et al. 2012). Various studies that have been undertaken to explore the rhizosphere-associated microbes of different plant species from different parts of the world suggest *Proteobacteria* and *Actinobacteria* to be the dominating populations of rhizobacteria (reviewed by Singh et al. 2007; Dokic et al. 2010; Lopes et al. 2016). Other major groups include *Firmicutes*, *Bacteroidetes*, *Planctomycetes*, *Verrucomicrobia*, and *Acidobacteria*. Although extensive studies have been done on decoding the diversity of root-associated microbes, we only know a minor portion of this diversity, the reason being that the majority of the groups of microbial populations inhabiting soil, including rhizosphere microbiota, is still unmanageable in laboratory growth (Amann 1995; Amann et al. 1995; Torsvik et al. 2002; Doornbos et al. 2012). However, over the past decade, a significant progress has been made to comprehend the key players and the processes that are operated in the rhizosphere by applying advanced molecular techniques, one of which is metagenomics. Such studies have extended our current knowledge of rhizosphere microbiome to a greater extent.

2.4 Elucidating the Hidden Root-Associated Microflora

Plants are very well considered as the pivotal primary producers in terrestrial ecosystems which use complex root systems to access resources in soil (Carvalhais et al. 2013). The volume of soil adjacent to and influenced by plant roots is referred to as rhizosphere. The rhizosphere harbors diverse microorganisms which affect plant health directly or indirectly. Plant roots interact with soil microbes through a series of chemical signaling. The current understanding of the mechanisms underlying these multitrophic interactions taking place in rhizosphere is still in its infancy. Paucity of effective methodologies that facilitate the extensive profiling of these extremely diverse communities, largely comprising of noncultivable microbes, poses a major challenge in completely understanding the plant-microbe interplay (Simons et al. 1997; Walker et al. 2003; de Weert et al. 2006). Molecular methods for answering intriguing questions in microbial ecology have pushed the researchers beyond the boundaries of conventional agar plate and culturing broth to bulk environmental sampling, making the analyses much closer to in situ microbial associations engaging interactions of various biotic and abiotic factors in the environment. Application of various advanced culture-independent molecular techniques has provided an advanced understanding of microbial ecology.

2.4.1 Metagenomics: A Classic Tool

One of the widely adopted techniques to assess unculturable microbial population is metagenomics, administering an array of molecular and bioinformatic tools to target the entire genetic composition of an environmental sample. Metagenomics facilitates the characterization of whole-community-level genome. Apart from giving information on the taxonomic diversity, it also gives an insight into the physiological behavior of the microbial communities within an environmental sample by accessing their functional gene composition. The main goals that can be achieved through metagenomics include studying environmental microbes without the need to culture them, bioprospecting (finding new genes with desired bioactivity), and linking function with the phylogeny of a given sample as well as evolutionary contour of the community structure and function (Torsten et al. 2012).

The metagenomic approach for studying phylogeny of microbial communities is based on cloning the collective genome extracted from environmental samples into

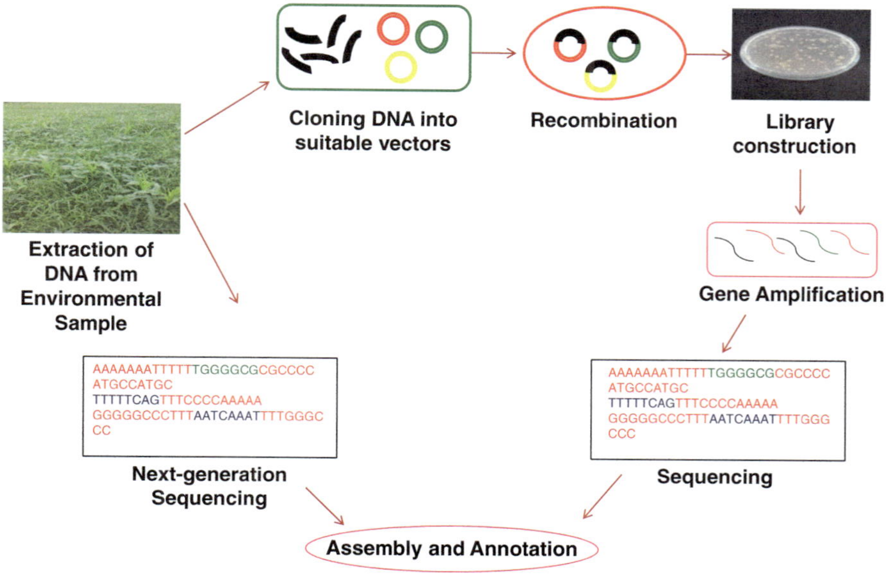

Fig. 2.2 An overview of the steps involved in metagenomic analysis (Mushtaq & Rawat, HNBGU)

suitable vectors, analyzing PCR-amplified 16S rDNA using universal primers, followed by RFLP and sequencing (Venter et al. 2004; Tyson et al. 2004; Guazzaroni et al. 2010; Aakvik et al. 2011) (Fig. 2.2). Although this tool gives an extraordinary advantage over conventional techniques applied for assessing microbial diversity of an environment, it has a major drawback that it might miss lesser abundant members of a microbial community performing indispensable functions.

Metagenomic studies, on the basis of sequencing strategies, can be broadly divided into two categories, viz., unselective (random sequencing) and targeted (directed sequencing) metagenomics (Hikaru 2012). Furthermore, on the basis of screening methods operated, metagenomics can be classified into (1) shotgun analysis, employing mass genome sequencing; (2) activity-driven studies, arranged to examine various microbial functions; and (3) sequence-driven analysis, linking genome information to phylogeny and functional potential of community (Riesenfeld et al. 2004; Hikaru 2012). More recently, the introduction of next-generation sequencing (NGS) technologies (Shendure and Ji 2008; Harismendy et al. 2009) have given a boost to the metagenomic research by enabling the production of large volumes of sequenced data from an environmental sample as compared to traditional sequencing methods (Edwards et al. 2006; Palenik et al. 2009). However, being a simple and economical method, unselective metagenomics has become an increasingly common strategy in DNA sequencing.

Tringe et al. (2005) have reported the first attempt to produce a comprehensive metagenome of Minnesota farm soil applying Sanger sequencing of random lengths of DNA cloned in a phage library. While the assembly of DNA sequences into larger contiguous segments could not be achieved significantly as a result of insufficient depth of sequencing, however, the individual reads were sufficient to draw a comparison of Minnesota farm soil genome with that of various environments. Consequently, various approaches have been put into use to generate metagenomes of different environmental samples with a significant success rate. One of the approaches essentially used to generate soil metagenomes is shotgun sequencing which directly sequences the extracted DNA eliminating the need of constructing a library (David et al. 2013).

2.4.2 Exploring Plant-Microbe Interplay Through Metagenomics

Although metagenomics has not yet been directly applied to explore specific plant-microbe interactions, however, this approach serves as a plausible molecular tool to explore the plenitude of genes involved in any metabolic process, thus linking phylogeny to the functioning of rhizosphere microbiota. The approach should include targeted deep phylogenetic structure determination using short random sequence tags of 16S hypervariable regions (Neufeld et al. 2008), high-throughput sequence analysis using post-Sanger sequencing methods such as pyrosequencing, and other advanced methods along with functional screening and selection strategies for key target genes. The combinatorial strategy will lead to the thorough understanding of the plant-microbial interactions at the community level.

In order to enhance the understanding of mechanisms that influence the plant-microbe interactions, functional metagenomics can be put into use. Functional genes that code for specific enzymes in key metabolic pathways can be used as markers to characterize novel species that may be present in the metagenome. For instance, one of the key enzymes involved in the process of denitrification (nitrogen fixation) is nitrite reductase (*nir* genes) that converts nitrite to nitric oxide. This nitrite reductase is encoded by two structurally different but functionally equivalent genes: *nir*K and *nir*S. Identification of species from the metagenome based on the presence of these genes could be utilized to explore the functional diversity of metagenome. One such study was carried out by Braker et al. (2000), where they deciphered the diversity of denitrifying bacteria present in two soil samples. Amplification of the functional genes from all the species present in the metagenome was carried out using specific primers, followed by construction of clone libraries. These libraries were subjected to restriction digestion, generating a RFLP profile, thus differentiating the species present. Final phylogenetic characterization was done by sequencing. The availability of libraries in plasmid vectors that support replication in the appropriate genomic background will allow phenotypic complementation experiments to be carried out for known biological processes such as N_2 fixation; phosphate solubilization; siderophore, IAA, gibberellin, and antibiotic production; and quorum sensing. A study carried out by Williamson et al. (2005) to screen Alaskan soil metagenomic library for production of quorum sensing relevant activities yielded novel quorum sensing genes both the inducer and agonist type.

Most of the studies for rhizosphere till date have focused on assessing heterogeneity of bacterial taxa in comparison to other rhizosphere dwellers (Table 2.2). Various operational taxonomic units (OTUs) have been reported from different rhizosphere

Table 2.2 A list of dominant bacterial phyla from the explored rhizosphere microbiota of some of the plant species in recent years

Plants	Recovered bacterial phyla	References
Glycine max	Acidobacteria, Actinobacteria, Chloroflexi, Cyanobacteria, Tenericutes, Verrucomicrobia	Mendes et al. (2014)
Lycopersicon esculentum	Acidobacteria, Actinobacteria, Bacteroidetes, Proteobacteria	Lee et al. (2016)
Oryza sativa	Actinobacteria, Proteobacteria	Knief et al. (2012)
Populus deltoides	Actinobacteria, Bacillus, Flavobacterium, Pseudomonads, Sphingobacterium	Brown et al. (2012)
Ramonda nathaliae	Acidobacteria, Actinobacteria, Actinomycetes, Proteobacteria	Dokic et al. (2010)
Saccharum officinarum	Actinobacteria, Nitrospirae, Tenericutes, Verrucomicrobia, Proteobacteria	Lopes et al. (2016)
Thymus zygis L.	Acidobacteria, Actinobacteria, Gemmatimonadetes, Proteobacteria	Pascual et al. (2016)
Zea mays L.	Actinobacteria, Proteobacteria	Chauhan et al. (2011)

studies based on different methods employed. The number ranges from approximately more than 100 to more than 55,000 OTUs (Mendes et al. 2013). Reviewed by Mendes et al. (2013), the data summarized by Hawkes et al. (2007) from 13 papers on meta-analysis of 19 clone libraries generated from the rhizosphere of 14 plant species revealed approximately 1200 diverse bacterial taxa belonging to 35 distinct taxonomic orders, *Proteobacteria* as the most dominant phylum, gram-positive members of the *Cytophaga-Flavobacterium-Bacteroides* (CFB) group, and *Actinobacteria* among others. Several studies have been going on to decipher the rhizosphere microbiota of some important crops across the world. For instance, metagenomic analysis of bacterial diversity of rice rhizosphere by Jaya and Kumarapillai (2011) from the paddy fields of Kerala, India, using 16S rRNA clone library generation along with RFLP, sequencing, and phylogenetic analysis, has revealed that the majority of microbes sequenced showed close relatedness to the *Proteobacteria* with a smaller portion of the 16S rRNA sequences showing a high similarity to rRNA sequences from *Acidobacteria*, *Firmicutes*, and *Bacteriodetes* groups. A similar study on rhizosphere of wheat plants from the city of Zamora Michoacan by Velazquez-Sepulveda et al. (2012) revealed the bacterial diversity associated with the wheat rhizosphere. A total of 30 OTUs were discovered including the classes Alfaproteobacteria, Betaproteobacteria, Deltaproteobacteria, Gammaproteobacteria, Actinobacteria, Bacilli, Clostridia, and uncultivable bacteria. The study also revealed that the genera *Pseudomonas*, *Stenotrophomonas*, and *Bacillus* were most abundant within the Gammaproteobacteria class. A recent study by Lagos et al. (2014) on the bacterial community composition of *Lolium perenne* rhizosphere microsites using pyrosequencing of 16s rRNA genes showed the presence of Proteobacteria, Actinobacteria, and Acidobacteria as dominating phyla.

Subsequently, efforts have been made to explore other microbial taxa as well. With the discovery of *Archaea* representing a major group of ammonia oxidizers in soils (Leininger et al. 2006), there has been a significant increase in the number of studies undertaken to explore the potential of this group across the globe. For instance, Bates et al. (2011) conducted a global survey on 146 soil samples using a set of universal primers for approximately all bacterial and archaeal taxa present. The study showed that nearly 2% of the total 16S rRNA gene sequences recovered from these soils comprised of archaeal domain with its relative abundance in the soils having lower C:N ratios. A similar study by Chelius and Triplett (2001) on maize rhizosphere has reported six unique archaeal sequences inhabiting the maize roots. Furthermore, a study on mangrove rhizosphere by Pires et al. (2012) showed the presence of about 300 archaeal OTUs distributed over four classes (*Halobacteria*, *Methanobacteria*, *Methanomicrobia*, and *Thermoprotei*) associated with *Rhizophora mangle* and *Laguncularia racemosa*.

In order to unravel the unknown mechanisms that drive plant-microbe interactions, it is imperative to implement high-throughput screens to detect novel genes within the microbial population that contribute to improved plant performance. It will be of particular interest to carry out direct functional screens for both plant-promoting and antagonistic effects, which in turn could open up new areas of research investigation.

2.5 Conclusions

Microorganisms are almost an inexhaustible source of metabolic capabilities ready to be exploited for diverse purposes. The rhizosphere, a narrow region of soil influenced by plant root secretions, is home to a myriad of microorganisms where several astonishing plant-microbe interactions take place. Even though the significance of microbes inhabiting rhizosphere has already been acknowledged, classical approaches to elucidate microbial community composition and their functioning are insufficient, and for vast diversity of root-associated microflora, only miniscule knowledge exists. Much needs yet to be understood regarding the key plant-microbe interactions. Taking into account the constraints, there is a need to undertake a systematic approach in order to explore different aspects of rhizosphere microflora and holistic consideration of advanced techniques that can be applied to assess and manage rhizosphere microbiota.

Metagenomics is reshaping the landscape of microbiology. It facilitates the study of rhizosphere microbial communities based on their component genes. While traditional microbial genomics relies upon cultivated cloned cultures, metagenomics combined with other advanced techniques gives a detailed picture of community-level genomics of an environmental sample without the need of culturing and has thus divulged the previously hidden microscopic life. This powerful tool, coupled with other techniques, has the potential to overcome the constraints which researchers have long been facing in understanding the mechanisms that regulate plant-microbe interactions at a much greater scale. Although considerable progress in characterization of microbial communities by metagenomic approach has been made, further improvement of sequencing technologies and bioinformatics tools will enhance our knowledge of rhizosphere biology and thus help in formulating strategies for sustainable agriculture.

2.6 Perspectives

Despite the recent advancement of high-throughput genomics technologies, we are still at the beginning stage of understanding the mechanisms lying beneath these complex yet quintessential plant-microbe interactions. As the measurements of phylogenetic diversity of a community are still inadequate, equating the data obtained to anticipate community structure and functionality seems an uphill battle. Presence of sequences that represent genes of known or yet to be known microbes in larger amounts within the microbial community lays a major setback to explore these communities through metagenomic studies. Assessing the information on genomic linkages between function and phylogeny will remain an important endeavor. The data obtained through metagenomics coupled with that of metatranscriptomics, metaproteomics, and metabolomics will enhance our insights into the major drivers that shape the rhizosphere community structure along with the ecophysiological behavior of resident microbiota, which can ultimately lead to the successful application of these microbes for the sustainable agriculture.

Acknowledgment The authors acknowledge the financial assistance provided by Uttarakhand State Biotechnology Department.

References

Aakvik T, Lale R, Liles M, Valla S (2011) Metagenomic libraries for functional screening. In: de Brujin FJ (ed) Handbook of molecular microbial ecology I: metagenomics and complementary approaches. Wiley, Hoboken, pp 171–181. isbn 978-11-180-1051-8. doi: 10.1002/9781118010518.ch22

Acinas SG, Sarma-Rupavtarm R, Klepac-Ceraj V, Polz MF (2005) PCR-induced sequence artifacts and bias: insights from comparison of two 16S rRNA clone libraries constructed from the same sample. Appl Environ Microbiol 71:8966–8966. doi:10.1128/AEM.71.12.8966-8969.2005

Amann RI (1995) Fluorescently labeled, rRNA-targeted oligonucleotide probes in the study of microbial ecology. Mol Ecol 4:543–553

Amann RI, Ludwig W, Schleifer KH (1995) Phylogenetic identification and in situ detection of individual microbial cells without cultivation. Microbiol Rev 59:143–169

Amorim JH, Macena TN, Lacerda-Junior GV, Rezende RP, Dias JC, Brendel M, Cascardo JC (2008) An improved extraction protocol for metagenomic DNA from a soil of the Brazilian Atlantic Rainforest. Genet Mol Res 7:1226–1232. doi:10.4238/vol7-4gmr509

Badri DV, Vivanco JM (2009) Regulation and function of root exudates. Plant Cell Environ 32:666–681. doi:10.1111/j.1365-3040.2009.01926.x

Bais HP, Weir TL, Perry LG, Gilroy S, Vivanco JM (2006) The role of root exudates in rhizosphere interactions with plants and other organisms. Annu Rev Plant Biol 57:233–266. doi:10.1146/annurev.arplant.57.032905.105159

Bao Z, Okubo T, Kubota K, Kasahara Y, Tsurumaru H, Anda M, Ikeda S, Minamisawa K (2014) Metaproteomic identification of diazotrophic methanotrophs and their localization in root tissues of field-grown rice plants. Appl Environ Microbiol 80:5043–5052. doi:10.1128/AEM.00969-14

Bastida F, Algora C, Hernandez T, Garcia C (2012) Feasibility of a cell separation–proteomic based method for soils with different edaphic properties and microbial biomass. Soil. Biol Biochem 45:136–138. doi:10.1016/j.soilbio.2011.10.017

Bates ST, Berg-Lyons D, Caporaso JG, Walters WA, Knight R, Fierer N (2011) Examining the global distribution of dominant archaeal populations in soil. ISME J 5:908–917. doi:10.1038/ismej.2010.171

Becher D, Bernhardt J, Fuchs S, Riedel K (2013) Metaproteomics to unravel major microbial players in leaf litter and soil environments: challenges and perspectives. Proteomics 13:2895–2909. doi:10.1002/pmic.201300095

Beneduzi A, Ambrosini A, Passaglia LM (2012) Plant growth-promoting rhizobacteria (PGPR): Their potential as antagonists and biocontrol agents. Genet Mol Biol 35:1044–1051. doi:10.1590/S1415-47572012000600020. PMC: PMC3571425

Bentley DR, Balasubramanian S, Swerdlow HP, Smith GP, Milton J, Brown OG, Hall KP, Evers DJ, Barnes CL, Bignell HR, Boutell JM et al (2008) Accurate whole genome human sequencing using reversible terminator chemistry. Nature 456:53–59. doi:10.1038/nature07517

Berendsen RL, Pieterse CMJ, Bakker P (2012) The rhizosphere microbiome and plant health. Trends Plant Sci 17:478–486. doi:10.1016/j.tplants.2012.04.001

Berg G, Opelt K, Zachow C, Lottmann J, Gotz M, Costa R, Smalla K (2006) The rhizosphere effect on bacteria antagonistic towards the pathogenic fungus *Verticillium* differs depending on plant species and site. FEMS Microbiol Ecol 56:250–261. doi:10.1111/j.1574-6941.2005.00025.x

Bertin C, Yang X, Weston LA (2003) The role of root exudates and allelochemicals in the rhizosphere. Plant Soil 256:67–83. doi:10.1023/A:1026290508166

Braker G, Zhou J, Wu L, Devol AH, Tiedje JM (2000) Nitrite reductase genes (*nir*K and *nir*S) as functional markers to investigate diversity of denitrifying bacteria in Pacific Northwest marine sediment communities. Appl Environ Microbiol 66:2096–2104. doi:10.1128/AEM.66.5.2096-2104.2000

Brown SD, Utturkar SM, Klingeman DM, Johnson CM, Martin SL, Land ML, Lu TS, Schadt CW, Doktycz MJ, Pelletier DA (2012) Twenty-one genome sequences from *Pseudomonas* species and 19 genome sequences from diverse bacteria isolated from the rhizosphere and endosphere of *Populus deltoides*. J Bacteriol 194:5991–5993. doi:10.1128/JB.01243-12

Buckling A, Harrison F, Vos M, Brockhurst MA, Gardner A, West SA, Griffin A (2007) Siderophore-mediated cooperation and virulence in *Pseudomonas aeruginosa*. FEMS Microbiol Ecol 62:135–141. doi:10.1111/j.1574-6941.2007.00388.x

Buee M, Reich M, Murat C, Martin FM (2009) 454 Pyrosequencing analyses of forest soils reveal an unexpectedly high fungal diversity. New Phytol 184:449–456. doi:10.1111/j.1469-8137.2009.03003.x

Bulgarelli D, Spaepen SS, Themaat EVL, Shulze-Lefert P (2013) Structure and functions of the bacterial microbiota of plants. Annu Rev Plant Biol 64:807–838. doi:10.1146/annurev-arplant-050312-120106

Caracciolo AB, Bottoni P, Grenni P (2010) Fluorescence in situ hybridization in soil and water: a useful method for studying the effect of xenobiotics on bacterial community structure. Toxicol Environ Chem 92:567–579. doi:10.1080/02772241003620244

Carvalhais LC, Dennis PG, Tyson GW, Schenk PM (2012) Application of metatranscriptomics to soil environments. J Microbiol Methods 91:246–251. doi:10.1016/j.mimet.2012.08.011

Carvalhais, LC, Dennis PG, Tyson GW, Schenk PM (2013) Rhizosphere metatranscriptomics: challenges and opportunities. In: de Brujin FJ (ed) Molecular microbial ecology of the rhizosphere. Wiley, Honboken, pp 1137–1144. doi: 10.1002/9781118297674.ch109. isbn (Print) 978-11-182-9617-2; (Online) 978-11-182-9767-4

Chaudhary V, Rehman A, Mishra A, Chauhan PS, Nautiyal CS (2012) Changes in bacterial community structure of agricultural land due to long-term organic and chemical amendments. Microb Ecol 64:450–460. doi:10.1007/s00248-012-0025-y

Chauhan PS, Chaudhry V, Mishra S, Nautiyal CS (2011) Uncultured bacterial diversity in tropical maize (*Zea mays* L.) rhizosphere. J Basic Microbiol 51(1):15–32. doi:10.1002/jobm.20100017

Chelius MK, Triplett EW (2001) The diversity of archaea and bacteria in association with the roots of *Zea mays* L. Microb Ecol 41:252–263. doi:10.1007/s002480000087

Costa R, Gotz M, Mrotzek N, Lottmann J, Berg G, Smalla K (2006) Effects of site and plant species on rhizosphere community structure as revealed by molecular analysis of microbial guilds. FEMS Microbiol Ecol 56:236–249. doi:10.1111/j.1574-6941.2005.00026.x

Daniel R (2005) The metagenomics of soil. Nat Rev Microbiol 3:470–478. doi:10.1038/nrmicro1160

David DM, Lydia HZ, Janet KJ (2013) The potential of metagenomic approaches for understanding soil microbial processes. Soil Sci Soc Am J 78:3–10. doi:10.2136/sssaj2013.07.0287dgs

De Vleesschauwer D, Hofte M (2009) Rhizobacteria-induced systemic resistance. Adv Bot Res 51:223–281. doi:10.1016/S0065-2296(09)51006-3

de Weert S, Dekkers LC, Bitter W, Tuinman S, Wijfjes AH, Boxtel RV, Lugtenberg BJ (2006) The two-component colR/S system of *Pseudomonas fluorescensWCS365* plays a role in rhizosphere competence through maintaining the structure and function of the outermembrane. FEMS Microbiol Ecol 58:205–213. doi:10.1111/j.1574-6941.2006.00158.x

Dobbelaere S, Vanderleyden J, Okon Y (2003) Plant growth-promoting effects of diazotrophs in the rhizosphere. CRC Crit Rev Plant Sci 22:107–149. doi:10.1080/713610853

Dokic L, Savic M, Narancic T, Vasiljevic B (2010) Metagenomic analysis of soil microbial communities. Arch Biol Sci 62:559–564. doi:10.2298/ABS1003559D

Doornbos RF, Van Loon LC, Bakker PA (2012) Impact of root exudates and plant defense signaling on bacterial communities in the rhizosphere. Agron Sustain Dev 32:227–243. doi:10.1007/s13593-011-0028-y

Dube HC, Yeole RD (1999) Modern approaches and innovations in soil management. Rastogi Publications, Meerut, 33p

Edwards RA, Rodriguez-Brito B, Wegley L, Haynes M, Breitbart M, Peterson DM, Saar M, Alexander S, Alexander IE, Rohwer F (2006) Using pyrosequencing to shed light on deep mine microbial ecology. BMC Genomics 7:57–70. doi:10.1186/1471-2164-7-57

Egamberdieva D, Kamilova F, Validov S, Gafurova L, Kucharova Z, Lugtenberg B (2008) High incidence of plant growth stimulating bacteria associated with the rhizosphere of wheat grown on salinated soil in Uzbekistan. Environ Microbiol 10:1–9. doi:10.1111/j.1462-2920.2007.01424.x

Gams W (2007) Biodiversity of soil-inhabiting fungi. Biodivers Conserv 16:69–72. doi:10.1007/s10531-006-9121-y

Gianfreda L (2015) Enzymes of importance to rhizosphere processes. J Soil Sci Plant Nutr 15:283–306. doi:10.4067/S0718-95162015005000022

Gray EJ, Smith DL (2005) Intracellular and extracellular PGPR: commonalities and distinctions in the plant-bacterium signaling processes. Soil Biol Biochem 37:395–412. doi:10.1016/j.soilbio.2004.08.030

Grayston SJ, Vaughan D, Jones D (1997) Rhizosphere carbon flow in trees, in comparison with annual plants: the importance of root exudation and its impact on microbial activity and nutrient availability. Appl Soil Ecol 5:29–56. doi:10.1016/s0929-1393(96)00126-6

Guazzaroni ME, Golyshin PN, Ferrer M (2010) Analysis of complex microbial community through metagenomic survey. In: Marco D (ed) Metagenomics: theory, methods and application. Caister Academic Press, Argentina, pp 55–77. isbn 978-1-904455-54-7

Harismendy O, Ng PC, Strausberg RL, Wang X, Stockwell TB, Beeson KY, Schork NJ, Murray SS, Topol EJ, Levy S, Frazer KA (2009) Evaluation of next generation sequencing platforms for population targeted sequencing studies. Genome Biol 10:R32. doi:10.1186/gb-2009-10-3-r32

Hartmann A, Rothballer M, Schmid M (2008) Lorenz Hiltner, a pioneer in rhizosphere microbial ecology and soil bacteriology research. Plant Soil 312:7–14. doi:10.1007/s11104-007-9514-z

Hawkes CV, DeAngelis KM, Firestone MK (2007) Root interactions with soil microbial communities and processes. In: Cardon Z, Whitbeck J (ed) The rhizosphere. Elsevier, pp 1–29. doi:10.1016/B978-012088775-0/50003-3. isbn 978-0-12-088775-0

Hein JW, Wolfe GV, Blee KA (2008) Comparison of rhizosphere bacterial communities in *Arabidopsis thaliana* mutants for systemic acquired resistance. Microb Ecol 55:333–343. doi:10.1007/s00248-007-9279-1

Hider RC, Kong X (2010) Chemistry and biology of siderophores. Nat Prod Rep 27:637–657. doi:10.1039/b906679a

Hikaru S (2012) Targeted metagenomics: a high-resolution metagenomics approach for specific gene clusters in complex microbial communities. Environ Microbiol 14:13–22. doi:10.1111/j.1462-2920.2011.02438.x

Hogberg MN, Hogberg P, Myrold DD (2006) Is microbial community composition in boreal forest soils determined by pH, C-to-N ratio, the trees, or all three? Oecologia 150:590–601. doi:10.1007/s00442-006-0562-5

Hong SH, Bunge J, Jeon SO, Epstein SS (2006) Predicting microbial species richness. Proc Natl Acad Sci U S A 103:117–122. doi:10.1073/pnas.0507245102

Jaya KA, Kumarapillai H (2011) Metagenomic analysis of bacterial diversity in the rice rhizosphere soil microbiome. Biotechnol Bioinformatics Bioeng 1:361–367

Johri BN, Sharma A, Virdi JS (2003) Rhizobacterial diversity in India and its influence on soil and plant health. Adv Biochem Eng Biotechnol 84:49–89. doi:10.1007/b11036

Jones JD, Dangl JL (2006) The plant immune system. Nature 444:323–329. doi:10.1038/nature05286

Jones DL, Hodge A, Kuzyakov Y (2004) Plant and mycorrhizal regulation of rhizodeposition. New Phytol 163:459–480. doi:10.1111/j.1469-8137.2004.01130.x

Kent AD, Triplett EW (2002) Microbial communities and their interactions in soil and rhizosphere ecosystems. Annu Rev Microbiol 56:211–236. doi:10.1146/annurev.micro.56.012302.161120

Knief C, Delmotte N, Chaffron S, Stark M, Innerebner G, Wassmann R, von Mering C, Vorholt JA (2012) Metaproteogenomic analysis of microbial communities in the phyllosphere and rhizosphere of rice. ISME J 6:1378–1390. doi:10.1038/ismej.2011.192

Kuklinsky-Sobral J, Araujo WL, Mendes R, Geraldi IO, Pizzirani-Kleiner AA, Azevedo JL (2004) Isolation and characterization of soybean-associated bacteria and their potential for plant growth promotion. Environ Microbiol 6:1244–1251. doi:10.1111/j.1462-2920.2004.00658.x

Lagos L, Navarrete O, Maruyama F (2014) Bacterial community structures in rhizosphere microsites of ryegrass (*Lolium perenne* var. Nui) as revealed by pyrosequencing. Biol Fertil Soils 50:1253–1266. doi:10.1007/s00374-014-0939-2

Lee SA, Park J, Chu B, Kim JM, Joa JH, Sang MK, Song J, Weon HY (2016) Comparative analysis of bacterial diversity in the rhizosphere of tomato by culture-dependent and -independent approaches. J Microbiol 54:823–831. doi:10.1007/s12275-016-6410-3

Leininger S, Urich T, Schloter M, Schwark L, Qi J, Nicol GW, Prosser JI, Schuster SC, Schleper C (2006) Archaea predominate among ammonia-oxidizing prokaryotes in soils. Nature 442:806–809. doi:10.1038/nature04983

Lesuffleur F, Paynel F, Bataille MP, Deunff E, Cliquet JB (2007) Root amino acid exudation: measurement of high efflux rates of glycine and serine from six different plant species. Plant Soil 294:235–246. doi:10.1007/s11104-007-9249-x

Leveau JH, Uroz S, de Boer W (2010) The bacterial genus *Collimonas*: mycophagy, weathering and other adaptive solutions to life in oligotrophic soil environments. Environ Microbiol 12:281–292. doi:10.1111/j.1462-2920.2009.02010.x

Li T, Wu TD, Mazeas L, Toffin L, Guerquin-Kern JL, Leblon G, Bouchez T (2008) Simultaneous analysis of microbial identity and function using NanoSIMS. Environ Microbiol 10:580–588. doi:10.1111/j.1462-2920.2007.01478.x

Lopes LD, Pereira e Silva MC, Andreote FD (2016) Bacterial abilities and adaptation toward the rhizosphere colonization. Front Microbiol 7:1341. doi:10.3389/fmicb.2016.01341

Lugtenberg B, Kamilova F (2009) Plant-growth-promoting rhizobacteria. Annu Rev Microbiol 63:541–556. doi:10.1146/annurev.micro.62.081307.162918

Lundberg DS, Lebeis SL, Paredes SH, Yourstone S, Gehring J, Malfatti S, Tremblay J, Engelbrektson A, Kunin V, del Rio TG, Edgar RC, Eickhorst T, Ley RE, Hugenholtz P, Tringe SG, Dangl JL (2012) Defining the core *Arabidopsis thaliana* root microbiome. Nature 488:86–90. doi:10.1038/nature11237

Mapelli F, Marasco R, Balloi A, Rolli E, Cappitelli F, Daffonchio D, Borin S (2012) Mineral-microbe interactions: biotechnological potential of bioweathering. J Biotechnol 157:473–481. doi:10.1016/j.jbiotec.2011.11.013

Mendes R, Krujit M, de Bruijn I, Raajimakers JM (2011) Deciphering the rhizosphere microbiome for disease-suppressive bacteria. Sci 332:1097–1100. doi:10.1126/science.1203980
Mendes R, Garbeva P, Raaijmakers JM (2013) The rhizosphere microbiome: significance of plant beneficial, plant pathogenic, and human pathogenic microorganisms. FEMS Microbiol Rev 37:634–663. doi:10.1111/1574-6976.12028
Mendes LW, Kuramae EE, Navarrete AA, van Veen JA, Tsai SM (2014) Taxonomical and functional microbial community selection in soybean rhizosphere. ISME J 8:1577–1587. doi:10.1038/ismej.2014.17
Micallef SA, Shiaris MP, Colon-Carmona A (2009) Influence of *Arabidopsis thaliana* accessions on rhizobacterial communities and natural variation in root exudates. J Exp Bot 60:1729–1742. doi:10.1093/jxb/erp053
Myrold DD, Nannipieri P (2014) Classical techniques versus omics approaches. In: Nannipieri P, Pietramellara G, Renella G (eds) Omics in soil science. Caister Academic Press, Haverhill, pp 179–187. isbn 978-1-908230-94-2. doi: 10.1.1.718.6012
Nautiyal CS, Srivastava S, Chauhan PS (2008) Rhizosphere colonization: molecular determinants from plant-microbe coexistence perspective. In: Nautiyal CS, Dion P (eds) Molecular mechanisms of plant and microbe coexistence. Springer, Berlin, pp 99–123. isbn:978-3-540-75575-3
Nautiyal CS, Chauhan PS, Bhatia CR (2010) Changes in soil physicochemical properties and microbial functional diversity due to 14 years of conversion of grassland to organic agriculture in semi-arid agroecosystem. Soil Tillage Res 109:55–60. doi:10.1016/j.still.2010.04.008
Neufeld JD, Chen Y, Dumont MG, Murrell JC (2008) Marine methylotrophs revealed by stable-isotope probing, multiple displacement amplification and metagenomics. Microbiology 10:1526–1535. doi:10.1111/j.1462-2920-2008.01568.x
Palenik B, Ren Q, Tai V, Paulsen IT (2009) Coastal *Synechococcus* metagenome reveals major roles for horizontal gene transfer and plasmids in population diversity. Environ Microbiol 11:349–359. doi:10.1111/j.1462-2920.2008.01772.x
Pascual J, Blanco S, Garcia-Lopez M, Garcia-Salamanca A, Bursakov SA, Genilloud O, Bills GF, Ramos JL, van Dillewjin P (2016) Assessing bacterial diversity in the rhizosphere of *Thymus zygis* growing in the Sierra Nevada National Park (Spain) through culture-dependent and independent approaches. PLoS One 11:e0146558. doi:10.1371/journal.pone.0146558
Philippot L, Raaijmakers JM, Lemanceau P, van der Putten WH (2013) Going back to the roots: the microbial ecology of the rhizosphere. Nat Rev Microbiol 11:789–799. doi:10.1038/nrmicro3109
Pii Y, Mimmo T, Tomasi N, Terzano R, Cesco S, Crecchio C (2015) Microbial interactions in the rhizosphere: beneficial influences of plant growth-promoting rhizobacteria on nutrient acquisition process: a review. Biol Fertil Soils 5:403–415. doi:10.1007/s00374-015-0996-1
Pires AC, Cleary DF, Almeida A, Cunha A, Dealtry S, Mendonca-Hagler LC, Smalla K, Gomes NC (2012) Denaturing gradient gel electrophoresis and barcoded pyrosequencing reveal unprecedented archaeal diversity in mangrove sediment and rhizosphere samples. Appl Environ Microbiol 78:5520–5528. doi:10.1128/AEM.00386-12
Prashar P, Kapoor N, Sachdeva S (2014) Rhizosphere: its structure, bacterial diversity and significance. Rev Environ Sci Biotechnol 13:63–77. doi:10.1007/s11157-013-9317-z
Raaijmakers JM, Weller DM (2001) Exploiting genotypic diversity of 2,4-diacetylphloroglucinol-producing *Pseudomonas* spp.: characterization of superior root-colonizing *P. fluorescens* strain Q8r1-96. Appl Environ Microbiol 67:2545–2554. doi:10.1128/AEM.67.6.2545-2554.2001
Raaijmakers JM, Paulitz TC, Steinberg C, Alabouvette C, Moenne-Loccoz Y (2009) The rhizosphere: a playground and battlefield for soil borne pathogens and beneficial microorganisms. Plant Soil 321:341–361. doi:10.1007/s11104-008-9568-6
Rawat S, Johri BN (2014) Thermophilic fungi: diversity and significance in composting. Kavaka 42:52–68
Riesenfeld CS, Schloss PD, Handelsman J (2004) Metagenomics: genomic analysis of microbial communities. Annu Rev Genet 38:525–552. doi:10.1146/annurev.genet.38.072902.091216
Salles JF, van Veen JA, van Elsas JD (2004) Multivariate analyses of *Burkholderia* species in soil: effect of crop and land use history. Appl Environ Microbiol 70:4012–4020. doi:10.1128/AEM.70.7.4012-4020-2004

Schmidt H, Eickhorst T (2014) Detection and quantification of native microbial populations on soil-grown rice roots by catalyzed reporter deposition–fluorescence in situ hybridization. FEMS Microbiol Ecol 87:390–402. doi:10.1111/1574-6941.12232

Shendure J, Ji H (2008) Next-generation DNA sequencing. Nat Biotechnol 26:1135–1145. doi:10.1038/nbt1486

Simon C, Daniel R (2011) Metagenomic analyses: past and future trends. Appl Environ Microbiol 77:1153–1161. doi:10.1128/AEM.02345-10

Simons M, Permentier HP, de Weger LA, Wijffelman CA, Lugtenberg BJ (1997) Amino acid synthesis is necessary for tomato root colonization by *Pseudomonas fluorescens* strain WCS365. Mol Plant Microbe Interact 10:102–106. doi:10.1094/MPMI.1997.10.1.102

Singh BK, Munro S, Potts JM, Millard P (2007) Influence of grass species and soil type on rhizosphere microbial community structure in grassland soils. Appl Soil Ecol 36:147–155. doi:10.1016/j.apsoil.2007.01.004

Singh BK, Campbell CD, Sorenson SJ, Zhou J (2009) Soil genomics. Nat Rev Microbiol 7:756. doi:10.1038/nrmicro2119-c1

Sinha R, Goel R, Johri BN (2001) Molecular markers in rhizosphere microbiology. In: Maheshwari DK, Dubey RC (eds) Innovative approaches in microbiology. Bishen Singh Mahendra Pal Singh, Dehradun, p 255. isbn:978-81-211-0210-0

Somers E, Vanderleyden J, Srinivasan M (2004) Rhizosphere bacterial signaling: a love parade beneath our feet. Crit Rev Microbiol 30:205–235. doi:10.1080/10408410490468786

Spiegelman D, Whissell G, Greer CW (2005) A survey of the methods for the characterization of microbial consortia and communities. Can J Microbiol 51:355–386. doi:10.1139/w05-003

Thompson JR, Pacocha S, Pharino C, Klepac-Ceraj V, Hunt DE, Benoit J, Sarma-Rupavtarm R, Distel DL, Polz MF (2005) Genotypic diversity within a natural coastal bacterioplankton population. Sci 307:1311–1313. doi:10.1126/science.1106028

Torsten T, Jack G, Folker M (2012) Metagenomics–a guide from sampling to data analysis. Microb Inform Exp 2:3. doi:10.1186/2042-5783-2-3

Torsvik V, Overas L, Thingstad T (2002) Prokaryotic diversity magnitude dynamics, and controlling factors. Science 296:1064–1066. doi:10.1126/science.1071698

Tringe SG, Von Mering C, Kobayashi A, Salamov AA, Chen K, Chang HW, Podar M, Short JM, Mathur EJ, Detter JC, Bork P (2005) Comparative metagenomics of microbial communities. Sci 308:554–557. doi:10.1126/science.1107851

Tyson GW, Chapman J, Hugenholtz P, Allen EE, Ram RJ, Richardson PM, Solovyev VV, Rubin EM, Rokhsar DS, Banfield JF (2004) Community structure and metabolism through reconstruction of microbial genomes from the environment. Nature 428:37–43. doi:10.1038/nature02340

Uroz S, Buee M, Murat C, Frey-Klett P, Martin F (2010) Pyrosequencing reveals a contrasted bacterial diversity between oak rhizosphere and surrounding soil. Environ Microbiol Rep 2:281–288. doi:10.1111/j.1758-2229.2009.00117.x

Velazquez-Sepulveda I, Orozco-Mosqueda MC, Prieto-Barajas CM, Santoyo G (2012) Bacterial diversity associated with the rhizosphere of wheat plants (*Triticum aestivum*): toward a metagenomic analysis. Phyton 81:81–87. doi: 10.1.1.472.577

Venter JC, Remington K, Heidelberg JF, Halpern AL, Rusch D, Eisen JA, Wu D, Paulsen I, Nelson KE, Nelson W, Fouts DE, Levy S, Knap AH, Lomas MW, Nealson K, White O, Peterson J, Hoffman J, Parsons R, Baden-Tillson H, Pfannkoch C, Rogers Y, Smith HO (2004) Environmental genome shotgun sequencing of the Sargasso Sea. Sci 304:66–74. doi:10.1126/science.1093857

Walker TS, Bais HP, Grotewold E, Vivanco JM (2003) Root exudation and rhizosphere biology. Plant Physiol 132:44–51. doi:10.1104/pp.102.019661

Wang HB, Zhang ZX, Li H, Lin WX (2011) Characterization of metaproteomics in crop rhizospheric soil. J Proteome Res 10:932–940. doi:10.1021/pr100981r

Wegley L, Mosier-Boss P, Lieberman S, Andrews J, Graff-Baker A, Rohwer F (2006) Rapid estimation of microbial numbers in water using bulk fluorescence. Environ Microbiol 8:1775–1782. doi:10.1111/j.1462-2920.2006.01062.x

Williamson LL, Borlee BR, Schloss PD, Guan C, Allen HK, Handelsman J (2005) Intracellular screen to identify metagenomic clones that induce or inhibit a quorum-sensing biosensor. Appl Environ Microbiol 71:6335–6344. doi:10.1128/AEM.71.10.6335-6344.2005

Zwirglmaier K (2005) Fluorescence in situ hybridization (FISH)-the next generation. FEMS Microbiol Lett 246:151–158. doi:10.1016/j.femsle.2005.04.015

Rhizosphere Microbiome and Its Role in Plant Growth Promotion

3

Rashmi Sharma, Minakshi, and Anjali Chauhan

Abstract

Microbial communities play a vital role in the growth and development of plants by influencing their physiological processes. The role of rhizodeposits to shape the rhizobacterial community structure is well established. Plant roots release various organic chemicals that attract and choose specific kind of microbes within the rhizosphere. In response, the plants associated with microbes enhance plant growth and productivity via different mechanisms. Therefore, in order to develop sustainable farming approaches such as biofertilizers and biopesticides, the study of host plants and associated microbial interactions in the rhizosphere plays an important role. Although plant growth-promoting microbial communities are abundant in the rhizosphere, many plant pathogens are also present that break through the plant defense mechanisms and cause various diseases. Therefore, to promote growth and productivity of crop plants, it is central to know what types of microorganisms are present and what functions they are performing in the rhizosphere. In this chapter, we have discussed the chief components of rhizosphere microbiome and its role in plant growth and management of various phytopathoens. The rhizospheric plant-microbe interactions and function of rhizosphere

R. Sharma
Assistant Professor, Department of Microbiology, DAV University Jalandhar, Nauni, Solan 173230, Himachal Pradesh, India

Minakshi
Research Scholar, Department of Basic Sciences (Microbiology Section), Dr YS Parmar University of Horticulture and Forestry, Nauni, Solan 173230, Himachal Pradesh, India

A. Chauhan (✉)
Assistant Professor, Department of Soil Science and Water Management, College of Forestry, DR YS Parmar University of Horticulture and Forestry, Nauni, Solan 173230, Himachal Pradesh, India
e-mail: anjali_chauhan22@yahoo.co.in

V.C. Kalia et al. (eds.), *Mining of Microbial Wealth and MetaGenomics*,
DOI 10.1007/978-981-10-5708-3_3

microbiome in agriculture have been described. Finally, we have drawn attention to various approaches to manipulate and redirect the microbial population in rhizosphere to enhance plant growth and crop productivity.

Keywords
Rhizobacteria • Rhizospheric microbiome • Rhizodeposits • Plant growth promoting

3.1 Plant Microbiome: An Introduction

Joshua Lederberg used the term "microbiome" as the microorganisms inhabiting our body as commensals, symbionts, or pathogens (Lynch 1990), and plant microbiome is the dynamic community of microorganisms associated with the plant. Phyllosphere, rhizosphere, and endosphere are the regions of the plant which provide niche for microbial community (Berg et al. 2016).

The aerial portion of plants inhabited by microorganisms is called phyllosphere; the latter is further divided into caulosphere (stems), phylloplane (leaves), anthosphere (flowers), and carposphere (fruits). Conservative estimates indicate that bacteria are the most abundant colonizers of leaf surfaces (Morris and Kinkel 2002). On the basis of degree of sway of diverse leaf exudates or other materials, assorted microorganisms may either die or endure and propagate in contrast to phyllosphere, the belowground portion of plants colonized by microbiome, broadly soil under the sway of roots, is known as **rhizosphere,** whereas the **rhizoplane** encompasses the root surface and its adhering soil. Various compounds are released from plant roots that take part in symbiotic functioning in the soil area under the influence of plant roots (rhizosphere) (Barea et al. 2005). For instance, rhizosphere provides a nutrient-rich environment for diazotrophic bacteria that fix atmospheric nitrogen and made the nitrogen available to plants. Abundance of bacteria, fungi, and archaea is high in the rhizosphere due to the presence of nutrient-rich environment (Egamberdiyeva et al. 2008; Mendes et al. 2011). Based upon the kind of colonization, plant microbiome is classified into epiphytes and endophytes. Microbes living on plant tissues or in close proximity (phyllosphere) and in the rhizosphere are well thought-out as epiphytes, while microbes living within plant tissues (stem, root, leaf tissues) are well considered as endophytes (Turner et al. 2013).

3.1.1 Contribution of Plant Microbiome in Plant Intensification and Health

The combinations as well as concentration of different kind of nutrients in the soil affect plant health and development. Furthermore, owing to immobility of some nutrients, plants frequently face considerable challenges in obtaining an enough provision of these nutrients in order to fulfill the demands of basic cellular processes. Limiting nutrient supply results in decreased plant productivity. Plant roots uptake mineral nutrients from the soil, but several factors influence the effectiveness of nutrient acquisition which includes chemistry as well as composition of certain soils that makes plants difficult to absorb nutrients and that either the nutrients may not be existing in such soils or may be present in forms that the plants cannot use.

As a result, many plant species develop an evolution of the mutually positive symbiotic associations with the soilborne microorganisms termed as plant growth-promoting rhizobacteria. As a consequence of their association, both the plant and the microsymbiont associated with that plant obtain valuable resources that they need for their own productivity and survival.

3.1.1.1 Mechanisms of Improving Plant Growth and Health

1. Direct mechanisms

   Plant microbiome may unswervingly assist the proliferation of their host plant through various methods:

   (a) Fundamental macronutrient for plant growth and health is nitrogen (N). Although, an approximate of 78% by volume of the atmosphere is occupied by nitrogen, but because of its inert form growing plants are incapable of nitrogen uptake. Diazotrophic bacteria have capability to convert atmospheric inert form of N_2 to the plant utilizable form (ammonia) by the action of intricate enzyme: nitrogenase (Kim and Rees 1994).

   (b) Phosphorus next to nitrogen is an essential macronutrient which is recognized as one of the vital elements that limit plant development (Feng et al. 2004). Preponderance of phosphate is fixed in soils, and hence, plant available P is barely accessible even though the copiousness of phosphorus (both organic and inorganic) in soils. Some bacterial species have solubilization potential for inorganic phosphorus through the release of metabolites such as organic acids (Rodríguez et al. 2006; Bianco and Defez 2010; Shahid et al. 2012), the functional groups (hydroxyl and carboxyl)of organic acids are responsible for chelation the phosphate cation and convert it into soluble form (Chen et al. 2006; Vyas and Gulati 2009; Lavania and Nautiyal 2013). *Pseudomonas putida*, *Pantoea agglomerans*, and *Microbacterium laevaniformans* are some common examples of inorganic phosphate-solubilizing bacteria (Park et al. 2011).

       Organic phosphorous is available to plants in mineralized form, and the process of mineralization is carried out by some bacterial genera through the liberation of phosphatase enzyme that catalyzes dephosphorylation of chemical bonds (phosphoester or phosphoanhydride) present in organic phosphorus (soil phytate) (Jorquera et al. 2008).

   (c) Some bacterial genera associated with plants, particularly the rhizobia, are known to produce several plant hormones like indole-3-acetic acid (IAA) (Ghosh et al. 2011), and some of the *Bacillus* spp. produce gibberellins (Gutierrez-Manero et al. 2001). Some strains of *Pseudomonas* produce hormone analogs that induce jasmonate and ethylene signaling within the plants resulting in plant defense responses against different plant pathogens (Melotto et al. 2006). Some bacterial genera have also been documented for hormone precursor degradation or degradation of hormones. For instance, deamination of ACC catalyzed by ACC deaminase of bacterial origin prevents ethylene signaling in plants that results in increased tolerance of plants to environmental stress (Glick 2005).

2. Indirect mechanisms:
   Preclusion from deleterious effects of plant pathogens on plants via increasing natural resistance of host plant or through the synthesis of several inhibitory compounds involves in indirect mechanisms (Nehl et al. 1997). Generally, these mechanisms include niche exclusion and production of antifungal metabolites: HCN, viscosinamide, phenazines, pyrrolnitrin, tensin, and pyoluteorin (Bhattacharyya and Jha 2012). In addition, many rhizobacteria have also been reported for the production of siderophores that prevent the proliferation of plant pathogens by limiting the supply of iron required for their growth (Ali and Vidhale 2013). **Induced systemic resistance** is the outcome of interaction of some microbial strains that induces impedance against several pathogenic microbes to host plant (Lugtenberg and Kamilova 2009).

3.2 Bioprospecting Microbes Along with Metagenome of Plant Rhizosphere

Rhizosphere provides a suitable environment for the growth of microorganisms. Multiple numbers of beneficial microorganisms are also residents of rhizosphere which are responsible for multiple biological as well as ecological processes that are essential for plant health (Kent and Triplett 2002). The abundance of plant growth-promoting microorganisms assumes a natural importance from agriculture point of view. The diversity of such microorganisms in soil is much higher than any other habitat; therefore (Delmont et al. 2011; Janssen 2006), rhizosphere ecosystem is a great pool for discovering novel microbes and their products; the term used to denote the discovery of novel microbes from natural system is "bioprospecting" (Lee and Lee 2013).

Amann et al. (1995) reported that 99% of microbial population in soil cannot be cultured under laboratory conditions. Therefore, utilization of non-culturable microbial assets would provide inimitable prospect to find novel microbial resources (Lee and Lee 2013). A total microbial genome which is directly isolated from microbial habitat known as metagenome is a rich source for bioprospecting (Berry et al. 2003; Zhou et al. 1996; Bertrand et al. 2005). For this metagenome from rhizosphere is cloned in a suitable host to comprise metagenome library (Rondon et al. 2000; Kim et al. 1992). This cloned library is either used for analysis of microbial community through direct sequencing of whole genome or used for selecting novel genetic resources for non-culturable rhizosphere microbiome.

Some common examples of bioprospecting potential of rhizosphere metagenome

1. Novel genus ***Swaminathania*** *salitolerans* from the mangroves
   A variety of bacterial strains belonging to taxa *Swaminathania*, *Vibrio*, *Bacillus*, *Enterobacter*, and *Azospirillum* were recovered from *Porteresia coarctata* Tateoka. Among these bacterial genera, *Swaminathania salitolerans*—a novel salt-tolerant strain—possess nitrogen-fixing activity along with phosphate solubilization. Likewise *Mangroveibacter plantisponsor*, a novel diazotrophic strain, was recognized as new genus in *Enterobacteriaceae* (Loganathan and Nair 2004).

2. Novel *Vibrio* associated with the mangroves
   Four new species (*Vibrio rhizosphaerae* sp. nov., *Vibrio porteresiae* sp. nov. (Rameshkumar et al. 2008), *Vibrio mangrovi* sp. nov (Rameshkumar et al. 2010), and *Vibrio plantisponsor* sp. nov (Rameshkumar et al. 2011)) isolated from the rhizosphere of mangrove define innovative ecological function of *Vibrio* as a rhizosphere-associated heterotrophic nitrogen-fixing bacteria.

3.3 The Rhizosphere Microbiome

The soil fraction under the sway of root secretions is termed as rhizosphere, this zone of soil can hold approximately 10^{11} microbial populace per gram of soil sample (Egamberdiyeva et al. 2008) as well as more than 30,000 species of prokaryotes (Mendes et al. 2011). A diverse array of compounds are accumulated and secreted through plant roots that will attract a diverse group of microorganisms that are metabolically active (Lugtenberg and Kamilova 2009). All this activity makes the rhizosphere the most dynamic environment in the soil (Walker et al. 2003). Rhizosphere consists of plant beneficial as well as plant pathogenic microbial species. Beneficial microbial community includes nitrogen fixers, mycorrhizae, plant growth-promoting rhizobacteria (PGPR), antagonistic microorganisms, as well as protozoa (Bonkowski et al. 2009; Buée et al. 2009; Raaijmakers et al. 2009). Pathogenic fungi, some bacterial species, and nematodes are deleterious to plant health.

3.4 Composition, Abundance, and Diversity of Rhizospheric Microbiome

It has been recognized that microbial life is present in a trifling area of soil which is localized in hot spots like rhizosphere, where microorganisms have continuous access to the flow of various plant root-derived organic substrates (Nannipieri et al. 2003). Flow of such nutrients together with physicochemical and biological factors can influence microbial community structure and function of rhizosphere (Sorensen 1997; Brimecombe et al. 2001). Microbial community and its abundance present in rhizosphere are represented in Fig. 3.1 as follows:

3.4.1 Bacteria

Wide variety of bacterial genera are present in rhizosphere whose composition differ among different plant species, root zone as well as plant phonological phase (Rovira 1965; Hinsinger et al. 2009; Marschner et al. 2011). Mendes et al. (2011), Weinert et al. (2011), and Yang et al. (2012) reported that the most dominant bacterial groups present in rhizosphere of sugarcane, pea native hardwood forest, and conifer plantations belong to *Proteobacteria*, *Firmicutes*, *Actinobacteria*, *Bacteroidetes*, and *Acidobacteria*. Among them, *Proteobacteria* are considered to be the most abundant bacterial group in

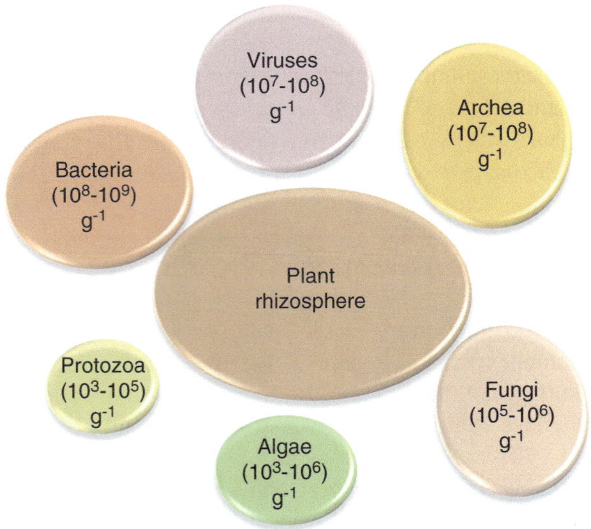

Fig. 3.1 Generalized diagram showing abundance of microorganisms present in the rhizosphere. The size of the circle is a measure of abundance of group of particular microbial community

the rhizosphere due to their ability to respond to labile C sources, showing fast growth and adaptation to the diverse plant rhizospheres. *Proteobacteria* is followed by *Acidobacteria*, which have been attributed to carbon cycle in soils because of their cellulose and lignin degradation capabilities (Ward et al. 2009). In relation to *Actinobacteria*, they have been linked with disease stifling soils and increasing root nodulation in plants thereby contributing in plant growth promotion (Mendes et al. 2011; Tokala et al. 2002). Due to fast growth and response to labile carbon sources, *Proteobacteria* is known to be the most abundant group of bacteria adapted to diverse plant rhizospheres (Bulgarelli et al. 2013; Chaparro et al. 2012; Peiffer et al. 2013). Due to the ability of *Acidobacteria* to degrade cellulose and lignin (Ward et al. 2009) which has also been attributed to carbon cycle, *Acidobacteria* is dominant in soil next to *Proteobacteria*. *Actinobacteria* has been found dominant in disease-suppressive soils, and these bacteria have also been found to increase root nodulation and hence plant growth promotion (Mendes et al. 2011; Tokala et al. 2002). *Rhizobium*, *Azospirillum*, *Burkholderia*, and *Pseudomonas* are the rhizobacterial genera acknowledged from the GenBank database contributing to plant growth promotion. Moreover, a less dominant group of rhizobacteria which do not play a role in plant growth and development includes *Verrucomicrobia*, *Sphingobacteria*, *Flavobacteriia*, *Deinococcus*, and *Epsilonproteobacteria*.

3.4.2 Archaea

Archaea are common but not major inhabitants of rhizosphere; also less information is acknowledged for Archaea from soil. On the basis of earlier 16S rRNA gene amplification, Crenarchaeota has been isolated from tomato rhizosphere which constituted of about 4–16% (relative to bacteria) (Bintrim et al. 1997; Borneman and Triplett 1997). It has also been reported that archaeal abundance is decreased by root exudates due to lower competitiveness than bacteria as well as slower growth rate (Karlsson et al. 2012).

3.4.3 Fungi in the Rhizosphere

The presence of organotropic fungi in the rhizosphere has been shown by both cultivable and non-cultivable techniques (Smit et al. 1999; Viebahn et al. 2005; De Boer et al. 2008; Zachow et al. 2008). Fungi as a member of the rhizosphere microbiome are termed as "mycorrhiza." The representatives of fungi in the rhizosphere include both yeast and filamentous fungi (*Ascomycota* as well as *Basidiomycota*) (Renker et al. 2004; Berg et al. 2005; Vujanovic et al. 2007). Joergensen (2000) reported that grassland plant rhizosphere consists of large amount of fungal biomass, i.e., average of 39% and range of 20–60%. The chief role of mycorrhiza is to increase nutrient uptake by extending the reach of plant root systems as well as the decomposition of root exudates containing simple or complex organic compounds (Butler et al. 2003; Treonis et al. 2004).

3.4.4 Other Rhizosphere Inhabitants

Organisms whose populations tend not to react to influxes of readily decomposable organic matter are usually not affected by root growth. This grouping includes actinomycetes, protozoa, and algal populations.

The actinomycetes generally derive their energy supply from decomposition of less readily decomposable soil organic matter components, whereas algal population uses solar energy. Protozoan populations are limited by the distribution and density of prey required to support increases in protozoan cell numbers (Robert and Tate 1994).

3.5 Interactions Between Plants and Microbes in the Rhizosphere

The rhizosphere, a zone in the close vicinity of plant roots is a hot spot for potentially important microbes and copious organisms. Various unicellular and multicellular organisms such as bacteria, archaea, algae, fungi, protozoans, and arthropods together with plant roots form the most intricate ecosystem on earth (Raaijmakers et al. 2009). Plants release ample amount of nutrients in the form of rhizodeposits which determine the type and composition of rhizosphere microbiome. Various categories of compounds are exuded from plant roots including sugars, organic acids, nucleotides, peptides, enzymes, and other secondary metabolites which together regulate the microbial diversity and activity inside the rhizosphere. The plants may exert selective pressure by releasing unique rhizodeposits to stimulate the growth of beneficial microorganisms for their growth and development (Cook et al. 1995). Rhizospheric microorganisms thus impart ecological fitness to their host plant and vice versa. These plant-microbe interactions thus can be categorized as neutral, positive, and negative interactions, depending upon the type of microorganisms, host plants, as well as existing environmental conditions.

3.5.1 Negative Interactions in the Rhizosphere

Beneficial microorganisms colonize plant roots in response to root exudates, but they can attract pathogenic population as well. Plant diseases are directly involved in damaging crop plants and destructing agricultural economy. Soilborne pathogens cause significant damage to the crops, and among these fungi are the most devastating. Their damaging effects include mild as well as severe symptoms causing inconsiderate crop losses. Thus, they are the foremost chronic threat to food production and economic stability worldwide. The most common fungal pathogens include members of genus *Aspergillus*, *Fusarium*, *Pythium*, *Phytophthora*, *Mucor*, *Rhizopus*, and *Verticillium* (Tournas and Katsoudas 2005) and the common forest fungi, viz., *Armillaria* and *Poria* (Asiegbu et al. 2005). The common and most widely studied bacterial pathogens belong to the genus *Pseudomonas*, *Erwinia*, *Ralstonia*, and *Xanthomonas* (Tournas and Katsoudas 2005).

3.5.2 Positive Interactions in the Rhizosphere

In the rhizosphere, plant-microbe interactions are involved in imperative ecosystem functioning processes, such as nutrient mineralization and immobilization in biogeochemical cycles. Microorganisms form a number of symbiotic associations with plants such as colonization of rhizosphere by plant growth-promoting rhizobacteria (PGPR), mycorrhizae, and legume-rhizobium association. These interactions impart several benefits to plants and are of three types: First are the type of microorganisms that increase availability of the nutrients to plants and are referred as biofertilizers. They either directly interact with plants or are involved in soil biotic and abiotic processes of plant growth promotion. The second type are the group of microbes that increase the plant growth indirectly by protecting plants from pathogen attack. These are referred to as biocontrol agents. The third group includes microorganisms that stimulate the plant growth by secreting growth-promoting hormones and growth regulators such as auxins, cytokinins, gibberellins, etc. They are known as biostimulants. Literature has described the importance of rhizospheric microorganisms in stimulating plant growth and maintaining soil health (Welbaum et al. 2004), whereas plant roots exude various metabolites in rhizosphere that are used as nutrients and signaling molecules by the bacteria (Bais et al. 2004).

3.5.3 Root Exudates-Mediators of Plant-Microbe Interactions

The "rhizosphere," a term coined by Hiltner (1904), was later redefined by Pinton as the zone in soil influenced by the plant roots along with the root tissues colonized by microorganisms (Pinton et al. 2001). Here, the soil plant-microbe interactions alter soil physical and chemical properties which further determine the soil microbial population (Nihorimbere et al. 2011). Approximately 5–20% of total photosynthetically fixed carbon is released in the form of root exudates by plants which are further used as nutrients by microbes in the rhizosphere (Chaparro et al. 2013). Furthermore, these rhizodeposits determine the plant-microbe interactions in the rhizosphere (Chaparro et al. 2013). An array of distinct signature compounds are released by plant roots which determine the microbial diversity in the rhizosphere, so if diverse is the plant community above ground, diverse will be the microbial population in the rhizosphere.

Plant root exudates can be subdivided into two main categories: molecules with low molecular weight, viz., sugars, phenolic compounds, other secondary metabolites, and hormones, and compounds with high molecular weight, viz., proteins and polysaccharides (Badri and Vivanco 2009). The amount and composition of rhizodeposits depend upon host plant, cultivar, growth stage, and a range of environmental factors, such as soil type, temperature, pH, microbial activities, and soil type (Uren 2000). These differences create the type of rhizobacterial communities that have a certain level of specificity for the host plant.

Certain compounds imitating bacterial quorum-sensing (QS) signals are also released by plant roots that either repress or stimulate QS responses of related bacterial species. In plants, root-microbe associations are governed by QS signals, whether they are beneficial, antagonistic, or symbiotic (Gao et al. 2003). Identifying these QS imitating compounds may lead to the development of a new antimicrobial compound or discovery of novel molecules. For example, different molecules that mimic the activity of *N*-acyl homoserine lactones and pose specific effects on bacterial quorum-sensing-mediated activities have been found in *Coronilla varia* L. (crown vetch), *Pisum sativum* L. (pea), *Oryza sativa* L. (rice), and *Solanum lycopersicum* (L.) Karst. (tomato) and also in *Chlamydomonas reinhardtii* (Teplitski et al. 2000, 2004; Daniels et al. 2002).

3.5.4 Impact of Root Exudates on Rhizospheric Microbiome

Plants roots pose selective pressure by releasing their unique signature molecules for attracting distinct microbial population, therefore, amend the diversity and composition of rhizospheric microbial communities. For example, root exudates from mutant *Arabidopsis* plant had more concentration of phenolic compounds than sugars when compared to its wild type, thus causing considerable changes in native microbial community structure (Badri et al. 2009). This change in the rhizodeposits composition can be further linked to the development of beneficial microbial population composed of PGPR, nitrogen fixers, and bioremediating bacterial population. Previous studies have also described that plants can develop a unique rhizobacterial community structure by releasing their unique root exudates profile (Berendsen et al. 2012; Bakker et al. 2012). For example, application of a root exudate compound, *p*-coumaric acid to cucumber seedlings, increased the native microbial population, thus alters the composition and organization of rhizobacterial communities, and increased the population of a soilborne fungal pathogen (*F. oxysporum* f.sp. *cucumerinum*) (Zhou and Wu 2012).

3.5.5 Rhizosphere Microbiome Influences Root Exudation Process

The rhizosphere colonizing microbes such as bacteria and fungi, also influence root exudation process (Matilla et al. 2010). Colonization of plant roots by arbuscular mycorrhizal (AM) fungi qualitatively changed the rhizodeposition, e.g., increasing the secretions of amino acids, gibberellins, and phenolics and decreasing the secretions of potassium, phosphorus, and sugars (Jones et al. 2004). Studies have described that various ectomycorrhizae have profound effects on composition as well as abundance of rhizodeposits on plants (Rosling et al. 2004). Moreover, certain compounds such as, oxalic acid and phytoalexins, are exuded by plant roots in response to pathogenic attack (Steinkellner et al. 2007). In addition to fungi, bacteria also influence root exudation profile. For example, an auxin secreting strain of *Bacillus amyloliquefaciens* FZB42 stimulates root exudation but reduce the uptake of phosphorus in *Triticum aestivum* (Talboys et al. 2014).

3.6 Rhizospheric Microbiome in Agriculture

Agricultural productivity lies on the foundation of microbial activities taking place in soil. The soil harbors an enormous microbial diversity, and existing extensive research has reinforced this fact (Roesch et al. 2007). The genetic potential and functional importance of the soil microbiome is gaining appreciation due to its direct role in plant productivity. Within a given soil type, native plants exert a selective pressure on this vast pool of biodiversity, thus restructuring and shaping rhizospheric microbial communities (Berg and Smalla 2009). At the same time, plants are also responsive to microbial activity, and may show either improved or reduced performance depending on the microbial activities in the rhizosphere. This active, two-way interaction between soil microbes and plants is significant in agricultural ecosystem, and enhancing our ability to redirect these interactions could offer progress toward agricultural sustainability through development of crop varieties that enhance beneficial functions within the rhizosphere microbiome.

There are many mechanisms by which rhizosphere microorganisms may positively impact plant performance. Whenever host plants can capture services provided by the soil microbiome, agricultural productivity may be enhanced by fully exploiting beneficial microbial activities. Beneficial microbial activities in the rhizosphere microbiome include:

- Production of phytohormones
- Provision of nutrients
- Rhizoremediation
- Improvement of plant resistance to pathogen infection

3.6.1 Production of Phytohormones (Phytostimulation)

Rhizobacteria are known to produce various phytohormones such as auxins, cytokinins, and gibberellins which are involved in plant growth promotion process. These growth regulators are either synthesized by plant itself or they can be released by various rhizobacterial species such as *Bacillus* and *Pseudomonas* (Steenhoudt and Vanderleyden 2000). Rhizobacteria belonging to genus *Bacillus* and *Pseudomonas* are known to produce different growth regulators that make plants to develop a number of fine roots, thus increasing the total surface area for nutrient and water absorption. Various growth hormones released are auxins, mainly indole-3-acetic acid, cytokinins, gibberellins, and ethylene inhibitors. Indole-3-acetic acid is known to stimulate root initiation, cell division, and elongation and is known to be produced by 70% of the rhizobacterial species (Barazani and Friedman 2001).

Generally indole-3-acetic acid increases root extension, cell division process, germination of seeds and tubers, flow rate of water and nutrients through vascular tissues, and secondary root development, mediates of geotropic and phototropic response, and provides resistance to plant stress. In addition to this, bacterial IAA alters the root exudation profile by loosening the plant cell wall, thus enhancing the amount of rhizodeposits and providing more nutrients to enhance rhizobacterial growth (Glick 2012). Due to this, rhizobacterial IAA plays an important role in phytostimulation as well as in pathogenesis and is regarded as a central molecule in plant-microbe interactions (Spaepen et al. 2007). Production of other hormones by rhizobacteria such as cytokinins and gibberellins is also known to be responsible for plant growth and development (Ullah et al. 2014).

In addition, rhizobacteria can also manipulate hormonal balance in plants. For example, ethylene is regarded as a senescence hormone which is known to inhibit plant growth in normal conditions, but at low concentration, it stimulates growth in many plants including *Arabidopsis thaliana*.

3.6.2 Provision of Nutrients (Biofertilization)

A number of plant growth-promoting rhizobacteria have been commercialized already which are known to promote plant growth through a variety of mechanisms including repression of plant diseases (bioprotectants), growth hormone production (biostimulants), and increased nutrient availability (biofertilizers). Biofertilization is one of the most extensively studied mechanisms which involve increasing the availability of plant deficient nutrients, viz., nitrogen, phosphorus, and iron. In India, most of the cultivable land lack satisfactory amount of one or more of these nutrients which resulted in suboptimal plant growth. To preclude this nutrient deficiency and to get high plant yield, farmers are depending upon more and more chemical fertilizers which, besides being pricey, has resulted in depletion of nonrenewable sources of energy, used for their synthesis. Chemical fertilizers are also known to adversely affect the human health and environment; therefore, it would be advantageous to fulfill nutrient demand of crop plants by certain biological means that could replace a part if not the full demand of these chemical fertilizers (Glick 2012).

3.6.2.1 Nitrogen Fixation

About 65% of the total nitrogen demand of crop plants is fulfilled by biological nitrogen fixation (BNF) (Bloemberg and Lugtenberg 2001). Nitrogen (N) is the most crucial primary mineral element required for plant growth and development. Although nitrogen is abundant (78%) in the atmosphere, it remains in the fixed form which cannot be utilized by the plants. The atmospheric nitrogen is converted to plant-usable forms through the process of BNF which involves conversion of nitrogen to ammonia by a complex enzyme system called nitrogenase present in nitrogen-fixing microorganisms (Kim and Rees 1994). About 2/3rd of the total nitrogen fixed is through BNF, whereas rest of the nitrogen is industrially synthesized by Haber and Bosch process (Rubio and Ludden 2008). The BNF constitutes a cost-effective and environment-friendly substitute for chemical fertilizers (Ladha et al. 1997).

Nitrogen fixers can be categorized as (1) symbiotic nitrogen fixers including bacteria belonging to family *Rhizobiaceae* (e.g., *Rhizobium*) that form symbiosis with legume plants and actinomycetes *Frankia* which form symbiotic association with non-legume trees such as *Alnus* and *Casuarina* and (2) free-living (*Azotobacter*, *Derxia*) or associative (*Azospirillum*) non-symbiotic nitrogen-fixing bacteria or cyanobacteria such as *Anabaena* and *Nostoc* (Bhattacharyya and Jha 2012). Free-living nitrogen-fixing plant growth-promoting rhizobacteria are also known as diazotrophs and form a nonobligatory association with the host plant (Glick et al. 1999). A few examples of nitrogen-fixing rhizobacterial species along with their host plants have been illustrated in Table 3.1.

3.6.2.2 Phosphorus Solubilization

Phosphorus (P), is the second most crucial mineral element required by plants, and is found in abundance in soil in both organic as well as inorganic forms. Despite of its abundance in soil, the majority of P is present in fixed insoluble forms which are unavailable to plants. The plants take up P in only two soluble forms, the monobasic ($H_2PO_4^-$) and the diabasic (HPO_4^{2-}) forms (Bhattacharyya and Jha 2012). The insoluble forms of soil phosphorus include inorganic mineral phosphates such as apatite or organic forms such as phosphomonesters, phosphotriesters, and inositol phosphate (soil phytate) (Glick 2012). To conquer phosphorus deficiency, farmers are frequently applying phosphatic chemical fertilizers in fields of which plants absorb a relatively small amount and the rest is quickly converted into insoluble forms of mineral phosphates in the soil. The regular application of phosphatic fertilizers is expensive as well as damaging to the environment. So this has led to the search for an environment-friendly and economic method to improving crop yield in P-deficient soil. In this context, the phosphate-solubilizing microorganisms (PSMs) may make available the unavailable forms of phosphorus to the plants and may act as a good replacement to chemical phosphatic fertilizers (Khan et al. 2006).

Table 3.1 Rhizobacterial species and their ability to fix atmospheric N_2 in certain plants

Rhizobacteria	Host crops	References
Burkholderia sp.	Rice	Baldani and Dobereiner (2000)
Azotobacter sp.	Wheat	Mrkovacki and Milic (2001)
Gluconacetobacter sp.	Sugarcane	Boddey et al. (2001)
Herbaspirillum sp.	Rice	James et al. (2002)
Pseudomonas stutzeri	Green pepper	Yan et al. (2008)
Bacillus subtilis	Tomato	Walia et al. (2014)
Bacillus methylotrophicus	Apple	Mehta et al. (2014)
Bacillus subtilis	Tomato	Sharma et al. (2015)

Among a variety of PSMs found in the rhizosphere, phosphate-solubilizing bacteria (PSB) are most widely recognized biofertilizers that supply plants with P from insoluble or unavailable sources of phosphorus. Bacteria belonging to genus *Bacillus*, *Pseudomonas*, *Rhizobium*, *Enterobacter*, *Erwinia*, *Azotobacter*, *Thiobacillus*, and *Serratia* are the most widely recognized phosphate-solubilizing bacteria (Bhattacharyya and Jha 2012).

Mechanism Involved in Phosphate Solubilization

Phosphorus exists in two forms in soil: first as inorganic phosphorus which includes insoluble mineral compounds which are mostly formed after application of chemical fertilizers and second type as organic phosphate compounds which form major pool of bound P constituting 20–80% of total P in soil (Richardson 1994).

(a) Mineral phosphate solubilization
Mineral phosphate solubilization involves action of various organic acids secreted by soil microorganisms. The organic acids acidify the surrounding medium resulting in low pH, as a result, phosphorus may be released from mineral phosphates by proton substitution for Ca^{2+} (Goldstein et al. 1993). Various organic acids are known to be produced by various phosphate-solubilizing microorganisms, and among them, gluconic acid has been documented to be the most significant agent of mineral phosphate solubilization (Rodriguez and Fraga 1999). Another organic acid identified in PSB strains is 2-ketogluconic acid, found in *Rhizobium leguminosarum*, *Rhizobium meliloti*, and *Bacillus firmus* (Banik and Dey 1982; Halder et al. 1990; Halder and Chakrabartty 1993). Various other organic acids, such as citric, glycolic, oxalic, and succinic acid, have also been identified in phosphate-solubilizing *Bacillus* strains (Mehta et al. 2014). Other mechanisms have also been considered, such as inorganic acid production, like sulphydric, carbonic, and nitric acid, and synthesis of various phosphorus chelating agents by rhizobacteria (Rodriguez and Fraga 1999).
(b) Organic phosphate solubilization
The mechanism of mineralization of organic phosphorus involves action of various phosphatases or phosphohydrolases. Soil organic matter is decomposed by saprophytic microorganisms, and the saprobes containing various phosphatases cause the release of radical orthophosphate from carbon skeleton of organic molecule in a hydrolysis reaction. The release of phosphorus involves breaking down of phosphoester bonds. Phosphatases are of two types depending upon their pH requirement for optimum catalytic activity: alkaline phosphatases work in an alkaline environment and acid phosphatases show optimum catalysis at acidic pH. This organic phosphate solubilization by microbes is greatly influenced by different environmental factors; more particularly, slightly alkaline conditions favor the solubilization of organic phosphorus (Paul and Clark 1988).

3.6.2.3 Sequestering Iron by Rhizobacteria

Iron is one more essential nutrient for plants besides nitrogen and phosphorus. It serves as a cofactor in enzymes involved in various physiological processes such as nitrogen fixation, respiration, and photosynthesis. Iron is taken up by plants as ferric iron (Fe^{+3}), which readily reacts to form various oxides and hydroxides that cannot be utilized by the plants. Plants absorb iron for soil in two ways:

- Release low molecular weight organic compounds called siderophores which can chelate iron, and make it available to the plants.
- Plants absorb the complex formed between organic compound and ferric iron, where the iron is reduced and easily absorbed by the plants.

Iron can bind reversibly to various functional groups present in the siderophores. These iron transport siderophores are normally either hydroxamates or phenolates-catecholates. In these types of siderophores, the distance between various functional groups is optimal to bind iron. *Bacillus* and *Pseudomonads* are widely known to produce the siderophores, and among these pyoverdine and pyochelin are the most commonly produced siderophores by *Pseudomonads*. Besides iron nutrition, siderophores also provide protection from fungal pathogens (Glick 1995). They hamper the growth of fungal pathogens by limiting the iron availability to pathogen (generally fungi), since bacterial siderophores have more affinity to chelate iron than fungal siderophores.

3.6.3 Rhizobacteria as Rhizoremediators

Rhizobacterial communities are sensitive and can sequester heavy metals due to presence of various functional groups; therefore, they can be used in bioremediation of soil (Umrania 2006). Various communities of microbes have been known to treat metal-polluted soil, but the composition of microbes inhabiting these heavy metal-polluted soil is exactly not known. The rhizosphere, with high concentration of root exudates, is known to attract more bacterial genera compared to the bulk soil (Penrose and Glick 2001). The root exudates as well as metal pollutants in the rhizosphere are utilized as nutrients by rhizoremediating bacteria and in reverse, they facilitate the plant growth by various mechanisms. The bioremediating PGPRs treating various heavy metals are given in Table 3.2.

Table 3.2 Bioremediation of heavy metals by PGPR

Bacteria	Plant	Heavy metal	Function	References
Kluyvera ascorbata	Indian mustard, tomato, canola	Lead, nickel, zinc	Decreased plant growth inhibition by heavy metals	Burd et al. (2000)
Azotobacter chroococcum	*Brassica juncea*	Lead, zinc	Stimulated plant growth and protected from metal toxicity	Wu et al. (2006)
Bacillus subtilis	*Brassica juncea*	Nickel	Facilitated nickel accumulation	Zaidi et al. (2006)
Aeromonas aquarium, *Pseudomonas composti*, and *Bacillus* sp.	*Spartina densiflora*	Different heavy metals	High PGP activities and resistance to heavy metals	Moreno et al. (2014)

The main problem associated with bacterial soil bioremediation system is that the process is effective under in situ conditions, but fails under ex situ remediation of the bulk soil, where their mechanism of action involves metabolism of soil pollutants. The other bottleneck of this technology is that the microorganisms starve very soon, and thus they are inept for bioremediation in the long run (Bottiglieri and Keel 2006). This can be overcome by separating the energy required in microbial metabolism from that required to remediate the pollutants. For this, a method has been developed in which rhizobacteria involved in rhizoremediation utilize the root exudates as primary source of nutrients. One of these methods includes the use of *P. putida* PCL1444 in rhizoremediation by Kuiper et al. (2001). The strain depends upon root exudates as primary source of nutrients and simultaneously degrades naphthalene in the rhizosphere.

3.6.4 Improvement of Plant Resistance to Pathogen Infection

Improvement of plant resistance to pathogens through one or more mechanisms of biocontrol is one of the aspects of indirect plant growth promotion by rhizobacteria. The use of microorganisms as biocontrol agents is an environment-friendly approach.

3.6.4.1 Mechanisms of Biocontrol

PGPR provide resistance to plant diseases through various mechanisms (Fig. 3.2), such as production of different antibiotics, viz., pyocyanin, phenazine, 2,4-diacetylphloroglucinol, iturin, surfactin, and fengycins; production of siderophores, HCN, and hydrolytic enzymes; and competition for nutrients and space (Elad and Chet 1987; Defago et al. 1990; Pierson and Thomashow 1992; Velazhahan et al. 1999). The brief accounts of important mechanisms involved in biocontrol are given in Table 3.3.

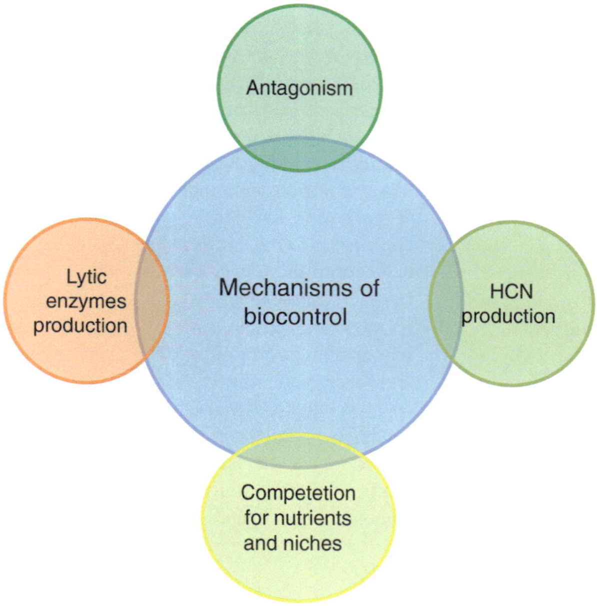

Fig. 3.2 Mechanisms of biocontrol by rhizobacteria

Table 3.3 Mechanisms involved in biocontrol of plant diseases

Method	Mechanism	Examples
Antagonism by rhizobacteria	The rhizobacteria produce various antibiotics which can kill pathogens. Some of these either act upon cell membrane or interfere with metabolic processes. To be a suitable biocontrol agent, the bacteria must release the antibiotics, in the right place around the root surface at the right time	Phenazines, 2,4-diacetylphloroglucinol, and pyoluteorin by *Pseudomonas* spp. (Mavrodi et al. 2006; Thomashow and Weller 1996; Nowak-Thompson et al. 1999); zwittermycin A by *Bacillus* (Emmert et al. 2004); fengycin, iturin, and surfactins by *Bacillus subtilis* (Kim et al. 2010)
HCN production	It is a secondary metabolite produced by the rhizobacteria and negatively influences the plant growth and root development. HCN acts as an inhibitor of electron transport chain at cytochrome oxidase complex and acts as an environment-friendly mechanism of biocontrol of weeds. Besides this, HCN is synthesized by various organisms such as bacteria, fungi, algae, insects, and plants as a mean to avoid predation	HCN production by *Pseudomonas fluorescens* resulted in inhibition of fungal pathogens such as *Pythium ultimum* and *Rhizoctonia solani* in sugar beet rhizosphere (Nelson et al. 2002); HCN-producing *Pseudomonas fluorescens* strain resulted in reduced root and shoot growth of weeds in rye, barley, and wheat rhizosphere (Ramette et al. 2003)
Lytic enzyme production	Enzymes are the biocatalysts produced generally by microorganisms, differing from other catalysts, and comprise the tools which determine the course of the multitude of life processes. Various hydrolytic enzymes are produced by rhizospheric microorganisms such as chitinases, proteases, cellulases, and β-1,3-glucanases that lyse various components of fungal cell wall which contributed to the antagonistic behavior of these microbes	A strain of *Pseudomonas* produced chitinase against pathogenic *Rhizoctonia solani* which hydrolyze the cell wall and reduced the disease development (Radjacommare et al. 2004); *Paenibacillus* strains releasing cell wall-degrading enzymes such as chitinases, cellulases, and β-1,3-glucanases were found to inhibit various bacterial and fungal pathogens (Von der Weid et al. 2000)
Competition for nutrient and space	Competition of beneficial rhizobacterial strains with phytopathogens for nutrition and space in the rhizosphere has been accounted as a biocontrol mechanism, and these competitively advanced strains can be effectively utilized to combat the pathogens	A mixture of competitive root-colonizing strains was applied to the seedlings and showed better growth in response to aggressive root colonization as compared to moderate colonizer *P. fluorescens* WCS365 (Spaink et al. 1998)

3.7 Manipulation of Rhizosphere Microbiome

Microbes form the most important module in the rhizosphere, and the composition of rhizobacterial communities greatly influences the plant-soil environment. The rhizobacterial community structure and its distribution greatly affect the plant growth. Hence, in order to enhance the population of helpful native or foreign soil microbes that enhance plant growth directly or indirectly, there is a need to develop efficient methods to alter the rhizosphere. Since the rhizosphere is a complex habitat, so efforts can be made to generate suitable environment for maximum plant growth by soil amendments, engineering improved plants and redirecting plant-microbe interactions. Rhizosphere microbiome is greatly affected by soil type and plant genotype, and plants are known to employ their unique members in rhizosphere microbiome. For example, malic acid release in root exudates stimulates the growth of beneficial *Bacillus subtilis* in plant rhizosphere (Rudrappa et al. 2008). Nonetheless, several metabolites acting as chemoattractants for useful microbes can also stimulate seed germination and growth of phytopathogens. For example, isoflavones exuded from soybean roots not only attract the symbiotic bacterium *Bradyrhizobium japonicum* but also stimulate the growth of phytopathogen *Phytophthora sojae* (Morris et al. 1998).

Soil is a multifaceted and diverse environment, which regulates physiology and metabolism in plants, accumulation of rhizodeposits, and community structure of rhizosphere microbiome in tandem. Various techniques have been advised to manipulate the rhizobacterial community structure and redirecting their metabolic activities in soil.

3.7.1 Manipulation by Introducing or Stimulating Microorganisms

The most direct methods to manipulate the rhizosphere microbiome include (1) introduction of one or more useful microbes in soil, on seeds or plant materials, and (2) stimulating indigenous beneficial rhizospheric population by soil and plant management practices. In spite of the fact that root colonization by beneficial microbes is still poorly understood, soil microbiologists and agronomists have been trying to alter the rhizosphere microflora by introducing selected beneficial microbial strains, either by coating seeds with inoculum or by placing the inoculum into the soil in close proximity to the seeds and seedlings. Numerous plant growth-promoting rhizobacterial strains have been inoculated into bulk soil to boost up the growth of crop plants (Bhattacharyya and Jha 2012). Introduction of other beneficial strains in order to protect the host plant from various phytopathogens can also result in alteration of rhizospheric microbiome. Small-scale inoculation of a biocontrol agent for short time interval is not sufficient to completely remove the pathogen as if the treatment is stopped very early, the pathogen can recover and reestablish again in the environment. Hence, the inoculated strain should multiply and aggressively colonize the rhizosphere to achieve a cell density above than the threshold value in a particular time to give maximum benefits to the host plant. Instead of single-strain inoculum, the use of multi-strain consortia with synergistic plant growth-promoting potential may prove more beneficial to overall plant growth and disease resistance (Bakker et al. 2012).

3.7.2 Soil Sterilization and Application of Specific Compounds

In soil sterilization, heating, drying, and irradiation are used. Sterilization is also achieved by fumigation using certain chemicals like chloroform, chloropicrin, methyl bromide, or carbon sulfide. These treatments resulted in improved plant growth even in the absence of phytopathogens (Rovira 1976). These beneficial effects can be attributed to chemical modifications like increase in ammonia content; organic matter decomposition, including dead microorganisms; recolonization of soil by non-pathogenic microbes, especially *Bacillus* and *Pseudomonas* which are known to stimulate plant growth (Ridge 1976); and elimination of nitrifying bacteria, which are particularly susceptible to soil fumigation (Jenkinson and Powlson 1976).

Among the different types of inhibitors that have been known, nitrification inhibitors are considered most important due to their possible use in the field. Inhibitors such as 2-chloro-6-(trichloromethyl)-pyridine have been successfully used to inhibit nitrification, thereby increasing the use efficiency of nitrogen fertilizers by reducing denitrification and leaching loss of nitrate ion. Unfortunately, in tropical areas, the inhibitors are readily decomposed by the soil microbes so that nitrification occurs even before plant requirements for nitrogen are at their peak. Another reason for the limited use of nitrification inhibitors is their high price. Although some low-cost substitutes have been proposed, such as neem cake, but these are not much effective as 2-chloro-6-(trichloromethyl)-pyridine (Prasad and De Datta 1978). The stimulation of a particular component of the microflora can also be achieved by adding their specific substrate to the soil. A classic example is that of the selective multiplication of amylase-producing bacteria in a soil amended by starch (Madhav et al. 2011).

3.7.3 Soil Management and Fertilization

Inoculation of soil even with specific microorganisms, like *Rhizobium*, is unsuccessful when one or more limiting environmental factors are still operating in the soil microenvironment. Therefore, improving environmental conditions is a prerequisite that can be successfully achieved by various soil management techniques, such as applying of organic inputs, irrigation, liming, or slow release from mineral fertilizers.

The whole rhizospheric environment is a function of various interactions as well as competing processes that are defined by soil type, moisture level, and metabolic and physiological activities of root-associated microorganisms and host plant.

Farmers manipulate physical and chemical environment around the roots of their crop plants during the irrigation or at the time of application of organic and chemical fertilizers. Nitrate-based fertilizers increase the alkalinity in rhizosphere, whereas ammonium-based fertilizers tend to acidify the rhizosphere. These changes in the soil pH alter the soil chemical environment which affect the plant growth and can alter soil chemistry around plant roots and influence the growth as well as composition of rhizosphere microbiome.

3.7.4 Manipulating the Rhizosphere Microbiome Using Biotechnological Approaches of Plant Breeding and Genetic Engineering

Numerous approaches have been developed to manipulate and redirect the composition and activities of rhizospheric microorganisms. Various root exudates are involved in attracting phytopathogens and activation of their virulence factors. Therefore, it is important to alter the quality as well as quantity of root exudates via genetic engineering which alters the structure of rhizosphere microbiome. More long-standing alterations in the rhizosphere that carry along the plant growth cycle can be generated using biotechnological approaches. In this type of approach, the plants with superior rhizospheric traits are selected, and these traits are further included into the breeding line that resulted in significant alterations in the rhizosphere. The identification of useful, heritable, and easily detectable traits is a prerequisite for successful breeding program to alter the rhizosphere. Although our knowledge about rhizospheric interactions and communications is increasing day by day, studies regarding engineering of rhizosphere through breeding program for improvement of rhizosphere-associated characters are still lacking (Wissuwa et al. 2009). At present, no breeding plan is available for evaluation of multidimensional interactions between plant and rhizosphere microbiome (Bakker et al. 2012). Gene loci linked to the resistance against *Pythium torulosum* were identified among various phenotypic variants of recombinant inbred lines of tomato (Smith et al. 1999). The study revealed that genetic variations within host plants can be utilized for the improvement of positive interactions between rhizosphere, microorganisms, and plants.

Genetic engineering is a more efficient process to manipulate the rhizosphere compared to conventional breeding programs. Previously, the process of genetic engineering has been employed to alter various rhizospheric factors such as pH and organic and inorganic ion effluxes (Gevaudant et al. 2007; Li et al. 2005). Besides this, plants are engineered to secrete specific signal compounds that attract a unique group of microbes, thus revealing that plants communicate with rhizosphere microbiome. For example, potato plant engineered to show high expression of lactonase gene interferes with bacterial quorum-sensing signal and showed increased resistance to pathogenic bacteria *Pectobacterium carotovorum*. Genetically engineered potato plants showing higher production of 5-*O*-glucosyltransferase and pectate lyase also showed increased resistance toward *Pectobacterium carotovorum* (Dong et al. 2001; Wegener 2001; Lorenc-Kukula et al. 2005).

3.8 Conclusion and Future Perspectives

The role of rhizosphere microbiome to execute and maintain the plant ecosystem is well established, but the conventional techniques used to understand their function in the rhizosphere are still not sufficient. Furthermore, for the enormous majority of microorganisms in the rhizosphere, no knowledge exists. Therefore, combining conventional methods with highly advanced next-generation sequencing techniques will strengthen our understanding about microbial community structure and function in the rhizosphere environment. Unraveling new plant signal molecules and root exudates in the root environment will make available biochemical as well as microbial markers to reveal that how beneficial microbes are being recruited and stimulated by the plants in rhizosphere. Exploring the rhizosphere microbiome also holds great potential to discover plentiful but previously unidentified soil microbes, their functioning, and mining of genes for various applications.

Various effective strategies should be designed to alter the rhizosphere microbiome in a way to favor the growth of antagonistic microorganisms that prevent the growth of devastating phytopathogens in soil. Among few possible approaches are plant breeding and genetic engineering programs; those are directed to investigate the molecular mechanism involved in plant-microbe interactions in the rhizosphere. Breeding of tomato using QTL mapping to incorporate characters supporting growth of favorable microbes in soil (Smith et al. 1999) and genetic modification of potato for expression of lactonase gene product interfering with bacterial quorum-sensing phenomenon and showing increased resistance toward *Pectobacterium carotovorum* (Dong et al. 2001) have given a remarkable success in this.

A lot of work is still required to be done in the future, to completely understand the structure and function of rhizosphere microbiome. Less than 5% of total soil bacterial and fungal population is culturable, and the rest of the vast population is still not culturable. Therefore, it is challenging to understand the functioning of non-cultivable microbial population in rhizosphere. Their function to alter the rhizosphere and response to external environment is poorly understood. Moreover, there are many unexplored species of rhizospheric bacteria and fungi having explicit role in biofertilization, biostimulation, and phytoprotection, but they until now are still unidentified. Lastly, global climate change has also an impact on structure and function of microbial species in rhizosphere microbiome. To decipher the extent of climate change on rhizosphere microbiome is still to be explored.

References

Ali SS, Vidhale NN (2013) Bacterial siderophore and their application: a review. Int J Curr Microbiol App Sci 2:303–312

Amann RI, Ludwig W, Schleifer KH (1995) Phylogenetic identification and in situ detection of individual microbial cells without cultivation. Microbiol Rev 59:143–169

Asiegbu FO, Nahalkova J, Li G (2005) Pathogen-inducible cDNAs from the interaction of the root rot fungus *Heterobasidion annosum* with Scots pine (*Pinus sylvestris* L.) Plant Sci 168:365–372. doi:10.1016/j.plantsci.2004.08.010

Badri DV, Vivanco JM (2009) Regulation and function of root exudates. Plant Cell Environ 32:666–681. doi:10.1111/j.1365-3040.2009.01926.x
Badri DV, Quintana N, El Kassis EG, Kim HK, Choi YH, Sugiyama A, Verpoorte R, Martinoia E, Manter DK, Vivanco JM (2009) An ABC transporter mutation alters root exudation of phytochemicals that provoke an overhaul of natural soil microbiota. Plant Physiol 151:2006–2017. doi:10.1104/pp.109.147462
Bais HP, Park SW, Weir TL, Callaway RM, Vivanco JM (2004) How plants communicate using the underground information superhighway. Trends Plant Sci 9:26–32
Bakker M, Manter D, Sheflin A, Weir T, Vivanco J (2012) Harnessing the rhizosphere microbiome through plant breeding and agricultural management. Plant Soil 360:1–13. doi:10.1007/s11104-012-1361-x
Baldani VLD, Dobereiner J (2000) Inoculation of rice plants with the endophytic diazatrophs *Herbaspirillum seropedicae* and *Burkholderia* spp. Biol Fertil Soils 30:485–491
Banik S, Dey BK (1982) Available phosphate content of an alluvial soil is influenced by inoculation of some isolated phosphate-solubilizing microorganisms. Plant Soil 69:353–364
Barazani O, Friedman J (2001) Allelopathic bacteria and their impact on higher plants. Crit Rev Microbiol 27:41–55. doi:10.1080/20014091096693
Barea JM, Pozo MJ, Azcon R, Aguilar CA (2005) Microbial co-operation in the rhizosphere. J Exp Bot 56:1761–1778
Berendsen RL, Pieterse CM, Bakker PA (2012) The rhizosphere microbiome and plant health. Trends Plant Sci 17:478–486. doi:10.1016/j.tplants. 2012.04.001
Berg G, Smalla K (2009) Plant species and soil type cooperatively shape the structure and function of microbial communities in the rhizosphere. FEMS Microbiol Ecol 68:1–13
Berg G, Zachow C, Lottmann J, Götz M, Costa R, Smalla K (2005) Impact of plant species and site on rhizosphere-associated fungi antagonistic to *Verticillium dahliae* Kleb. Appl Environ Microbiol 71:4203–4213
Berg G, Rybakova D, Grube M, Koberl M (2016) The plant microbiome explored: implications for experimental botany. J Exp Bot. doi:10.1093/jxb/erv466
Berry AE, Chiocchini C, Selby T, Sosio M, Wellington EM (2003) Isolation of high molecular weight DNA from soil for cloning into BAC vectors. FEMS Microbiol Lett 223:15–20
Bertrand H, Poly F, Van VT, Lombard N, Nalin R, Vogel TM (2005) High molecular weight DNA recovery from soils prerequisite for biotechnological metagenomic library construction. J Microbiol Methods 62:1–11
Bhattacharyya PN, Jha DK (2012) Plant growth promoting rhizobacteria (PGPR): emergence in agriculture. World J Microbiol Biotechnol 28:1327–1350
Bianco C, Defez R (2010) Improvement of phosphate solubilization and Medicago plant yield by an indole-3-acetic acid-overproducing strain of Sinorhizobium meliloti. Appl Environ Microbiol 76:4626–4632
Bintrim SB, Donohue TJ, Handelsman J, Roberts GP, Goodman RM (1997) Molecular phylogeny of Archaea from soil. PNAS 94:277–282
Bloemberg GV, Lugtenberg BJJ (2001) Molecular basis of plant growth promotion and biocontrol by rhizobacteria. Curr Opin Plant Biol 4:343–350. doi:10.1016/S1369-5266(00)00183-7
Boddey RM, Polidoro JC, Resende AS, Alves BJR, Urquiaga S (2001) Use of the ^{15}N natural abundance technique for the quantification of the contribution of N_2 fixation to sugarcane and other grasses. Aust J Plant Physiol 28:889–895
Bonkowski M, Villenave C, Griffiths B (2009) Rhizosphere fauna: the functional and structural diversity of intimate interactions of soil fauna with plant roots. Plant Soil 321:213–233
Borneman J, Triplett EW (1997) Molecular microbial diversity in soils from Eastern Amazonia: evidence for unusual microorganism and microbial population shifts associated with deforestation. Appl Environ Microbiol 63:2647–2653
Bottiglieri M, Keel C (2006) Characterization of PhlG, a hydrolase that specifically degrades the antifungal compound 2,4-diacetylphloroglucinol in the biocontrol agent *Pseudomonas fluorescens* CHA0. Appl Environ Microbiol 72:418–427
Brimecombe MJ, De Lelj FA, Lynch JM (2001) The rhizosphere–the effect of root exudates on rhizosphere microbial populations. In: Pinton R, Varanini Z, Nannipieri P (eds) The rhizosphere

biochemistry and organic substances at the soil-plant interface. Marcel Dekker, New York, pp 95–140
Buée M, De Boer W, Martin F, van Overbeek L, Jurkevitch E (2009) The rhizosphere zoo: an overview of plant-associated communities of microorganisms, including phages, bacteria, archaea, and fungi, and of some of their structuring factors. Plant Soil 321:189–212
Bulgarelli D, Schlaeppi K, Spaepen S, Van Themaat EVL, Schulze-Lefert P (2013) Structure and functions of the bacterial microbiota of plants. Annu Rev Plant Biol 64:807–838
Burd GI, Dixon DG, Glick BR (2000) Plant growth-promoting bacteria that decrease heavy metal toxicity in plants. Can J Microbiol 46:237–245
Butler JL, Williams MA, Bottomley PJ, Myrold DD (2003) Microbial community dynamics associated with rhizosphere carbon flow. Appl Environ Microbiol 69:6793–6800
Chaparro JM, Sheflin AM, Manter DK, Vivanco JM (2012) Manipulating the soil microbiome to increase soil health and plant fertility. Biol Fertil Soils 48:489–499
Chaparro JM, Badri DV, Bakker MG, Sugiyama A, Manter DK, Vivanco JM (2013) Root exudation of phytochemicals in *Arabidopsis* follows specific patterns that are developmentally programmed and correlate with soil microbial functions. PLoS One 8:e55731. doi:10.1371/journal.pone. 0055731
Chen YP, Rekha PD, Arun AB, Shen FT, Lai WA, Young CC (2006) Phosphate solubilizing bacteria from subtropical soil and their tricalcium phosphate solubilizing abilities. Appl Soil Ecol 34:33–41. doi:10.1016/j.apsoil.2005.12.002
Cook RJ, Thomashow LS, Weller DM, Fujimoto D, Mazzola M, Bangera G, Kim DS (1995) Molecular mechanisms of defense by rhizobacteria against root disease. Proc Natl Acad Sci U S A 92:4197
Daniels R, De Vos DE, Desair J, Raedschelders G, Luyten E, Rosemeyer V, Verreth C, Schoeters E, Vanderleyden J, Michiels J (2002) The *cin* quorum sensing locus of *Rhizobium etli* CNPAF512 affects growth and symbiotic nitrogen fixation. J Biol Chem 277:462–468. doi:10.1074/jbc.M106655200
De Boer W, de Ridder-Duine AS, Klein Gunnewiek PJA, Smant W, van Veen JA (2008) Rhizosphere bacteria from sites with higher fungal densities exhibit greater levels of potential antifungal properties. Soil Biol Biochem 40:1542–1544
Defago G, Berling CH, Burger V, Hass D, Hhar G, Keel C, Voisard C, Wirthner P, Wuthrich B (1990) Suppression of black rot of tobacco and other root diseases by strains of *Pseudomonas* fluorescens: potential applications and mechanisms. In: Hormby D (ed) Biological control of soil-borne plant pathogens. CAB International, Wellingford, pp 93–108
Delmont TO, Robe P, Cecillon S, Clark IM, Constancias F, Simonet P (2011) Accessing the soil metagenome for studies of microbial diversity. Appl Environ Microbiol 77:1315–1324
Dong YH, Wang LH, Xu JL et al (2001) Quenching quorum-sensing-dependent bacterial infection by an N-acyl homoserine lactones. Nature 411:813–817
Egamberdiyeva D, Kamiliva F, Validov S, Gafurova L, Kucharova Z, Lugentenberg B (2008) High incidence of plant growth stimulating bacteria associated with the rhizosphere of wheat grown in salinated soil in Uzbekistan. Environ Microbiol 10:1–9
Elad Y, Chet I (1987) Possible role of competition for nutrition in biocontrol of *Pythium*-damping-off by bacteria. Phytopathology 77:190–195
Emmert EA, Klimowicz AK, Thomas MG, Handelsman J (2004) Genetics of zwittermicin A production by *Bacillus cereus*. Appl Environ Microbiol 70:104–113
Feng K, Lu HM, Sheng HJ, Wang XL, Mao J (2004) Effect of organic ligands on biological availability of inorganic phosphorus in soils. Pedosphere 14:85–92
Gao M, Teplitski M, Robinson JB, Bauer WD (2003) Production of substances by *Medicago truncatula* that affect bacterial quorum sensing. Mol Plant-Microbe Interact 16:827–834. doi:10.1094/MPMI.2003.16.9.827
Gevaudant F, Duby G, von Stedingk E, Zhao R, Morsomme P, Boutry M (2007) Expression of a constitutively activated plasma membrane H^+-ATPase alters plant development and increases salt tolerance. Plant Physiol 144:1763–1776
Ghosh S, Ghosh P, Maiti TK (2011) Production and metabolism of indole acetic acid (IAA) by root nodule bacteria (Rhizobium): a review. J Pure Appl Microbiol 5:523–540

Glick BR (1995) The enhancement of plant growth by free-living bacteria. Can J Microbiol 41:109–117
Glick BR (2005) Modulation of plant ethylene levels by the bacterial enzyme ACC deaminase. FEMS Microbiol Lett 251:1–7
Glick BR (2012) Plant growth-promoting bacteria: mechanisms and applications. Scientifica. doi:10.6064/2012/963401
Glick BR, Patten CL, Holguin G, Penrose DM (1999) Biochemical and genetic mechanisms used by plant growth promoting bacteria. Imperial College Press, London, UK, pp 111–118
Goldstein AH, Rogers RD, Mead G (1993) Separating phosphate from ores via bioprocessing. Biotechnology 11:1250–1254
Gutierrez-Manero FJ, Ramos-Solano B, Probanza A, Mehouachi J, Tadeo FR, Talon M (2001) The plant-growth-promoting rhizobacteria *Bacillus pumilus* and *Bacillus licheniformis* produce high amounts of physiologically active gibberellins. Physiol Plant 111:206–211
Halder AK, Chakrabartty PK (1993) Solubilization of inorganic phosphate by *Rhizobium*. Folia Microbiol 38:325–330
Halder AK, Mishra AK, Bhattacharyya P, Chakrabartty PK (1990) Solubilization of rock phosphate by *Rhizobium* and *Bradyrhizobium*. J Gen Appl Microbiol 36:81–92
Hiltner L (1904) Uber neuere Erfahrungen und probleme auf dem gebiet der bodenbakteriologie und unter besonderer berucksichtigung der grundungung und brache. Arbeiten der Deutschen Landwirtschafts Gesellschaft 98:59–78
Hinsinger P, Bengough AG, Vetterlein D, Young IM (2009) Rhizosphere: biophysics, biogeochemistry, and ecological relevance. Plant Soil 321:117–152
James EK, Gyaneshwar P, Mathan N et al (2002) Infection and colonization of rice seedlings by the plant growth-promoting bacterium *Herbaspirillum seropedicae* Z67. Mol Plant Microbe Interact 15:894–906
Janssen PH (2006) Identifying the dominant soil bacterial taxa in libraries of 16S rRNA and 16S rRNA genes. Appl Environ Microbiol 72:1719–1728
Jenkinson DS, Powlson DS (1976) The effects of biocidal treatments on metabolism in soil in fumigation with chloroform. Soil Biol Biochem 8:167–177
Joergensen RG (2000) Ergosterol and microbial biomass in the rhizosphere of grassland soils. Soil Biol Biochem 32:647–652
Jones DL, Hodge A, Kuzyakov Y (2004) Plant and mycorrhizal regulation of rhizodeposition. New Phytol 163:459–480. doi:10.1111/j.1469-8137.2004.01130.x
Jorquera MA, Hernández MT, Rengel Z, Marschner P, Luz Mora M (2008) Isolation of culturable phosphobacteria with both phytate-mineralization and phosphate-solubilization activity from the rhizosphere of plants grown in a volcanic soil. Biol Fertil Soils 44:1025–1034. doi:10.1007/s00374-008-0288-0
Karlsson AE, Johansson T, Bengtson P (2012) Archaeal abundance in relation to root and fungal exudation rates. FEMS Microbiol Ecol 80:305–311
Kent AD, Triplett EW (2002) Microbial communities and their interactions in soil and rhizosphere ecosystems. Annu Rev Microbiol 56:211–236
Khan MS, Zaidi A, Wani PA (2006) Role of phosphate solubilizing microorganisms in sustainable agriculture-a review. Agron Sustain Dev 27:29–43
Kim J, Rees DC (1994) Nitrogenase and biological nitrogen fixation. Biochemistry 33:389–397
Kim UJ, Shizuya H, de Jong PJ, Birren B, Simon MI (1992) Stable propagation of cosmid sized human DNA inserts in an F factor based vector. Nucleic Acid Res 20:1083–1085
Kim P, Ryu J, Kim YH, Chi YT (2010) Production of biosurfactant lipopeptides iturin A, fengycin, and surfactin A from *Bacillus subtilis* CMB32 for control of *Colletotrichum gloeosporioides*. J Microb Biotechnol 20:138–145
Kuiper I, Bloemberg GV, Lugtenberg BJJ (2001) Selection of a plant-bacterium pair as a novel tool for rhizostimulation of polycyclic aromatic hydrocarbon-degrading bacteria. Mol Plant Microbe Interact 14:1197–1205
Ladha JK, de Bruijn FJ, Malik KA (1997) Introduction: assessing opportunities for nitrogen fixation in rice-a frontier project. Plant Soil 124:1–10

Lavania M, Nautiyal C (2013) Solubilization of tricalcium phosphate by temperature and salt tolerant *Serratia marcescens* NBRI1213 isolated from alkaline soils. Afr J Microbiol Res 7:4403–4413. doi:10.5897/AJMR2013.5773

Lee MH, Lee SW (2013) Bioprospecting potential of the soil metagenome: novel enzymes and bioactivities. Genomics Inform 11:114–120. doi:10.5808/GI.2013.11.3.114

Li JY, Xu RK, Xiao SC, Ji GL (2005) Effect of low-molecular-weight organic anions on exchangeable aluminum capacity of variable charge soils. J Colloid Interface Sci 284:393–399

Loganathan P, Nair S (2004) *Swaminathania salitolerans* gen. nov., sp. nov., a salt-tolerant, nitrogen-fixing and phosphate-solubilizing bacterium from wild rice (*Porteresia coarctata Tateoka*) P. Int J Syst Evol Microbiol 54:1185–1190

Lorenc-Kukula K, Jafra S, Oszmianski J, Szopa J (2005) Ectopic expression of anthocyanin 5-O-glucosyltransferase in potato tuber causes increased resistance to bacteria. J Agric Food Chem 53:272–281

Lugtenberg B, Kamilova F (2009) Plant-growth-promoting rhizobacteria. Annu Rev Microbiol 63:541–556

Lynch JM (1990) Soil biotechnology: beneficial interactions between microorganisms and roots. Biotechnol Adv. Blackwell, Oxford, pp 335–346

Madhav K, Verma S, Tanta R (2011) Isolation of amylase producing *Bacillus* species, from soil sample of different regions in Dehradun and to check the effect of pH and temperature on their amylase activity. J Pharm Biomed Sci 12:1–8

Marschner P, Crowley D, Rengel Z (2011) Rhizosphere interaction between microorganisms and plants govern iron and phosphorus acquisition along the root axis model and research methods. Soil Biol Biochem 43:883–894

Matilla MA, Ramos JL, Bakker PAHM, Doornbos R, Badri DV, Vivanco JM, Ramos-Gonzalez MI (2010) *Pseudomonas putida* KT2440 causes induced systemic resistance and changes in Arabidopsis root exudation. Environ Microbiol Rep 2:381–388. doi:10.1111/j.1758-2229.2009.00091.x

Mavrodi DV, Blankenfeldt W, Thomashow LS (2006) Phenazine compounds in fluorescent *Pseudomonas* spp.: biosynthesis and regulation. Annu Rev Phytopathol 44:417–445

Mehta P, Walia A, Kakkar N, Shirkot CK (2014) Tricalcium phosphate solubilization by new endophyte *Bacillus methylotrophicus* CKAM isolated from apple root endosphere and its plant growth-promoting activities. Acta Physiol Plant. doi:10.1007/s1173801415811

Melotto M, Underwood W, Koczan J, Nomura K, He SY (2006) Plant stomata function in innate immunity against bacterial invasion. Cell 126:969–980

Mendes R, Kruijit M, de Bruijin I, Dekker E, der Voort V, Schneider M (2011) Deciphering the rhizosphere microbiome for disease-suppressive bacteria. Science 332:1097–1100

Moreno AL, Castillo ID, Parra R, Doukkali B, Redondo-Gómez S, Pérez-Palacios P, Caviedes M, Pajuelo E, Rodríguez-Llorente ID (2014) Prospecting metal-resistant plant-growth promoting rhizobacteria for rhizoremediation of metal contaminated estuaries using *Spartina densiflora*. Environ Sci Pollut Res 21:3713–3721

Morris CE, Kinkel LL (2002) Fifty years of phyllosphere microbiology: significant contribution to research in related fields. In: SE L, Hicht-Poinar EI, Elliott V (eds) Phyllosphere microbiology. APS Press, St. Paul, pp 365–375

Morris PF, Bone E, Tyler BM (1998) Chemotropic and contact responses of Phytophthora sojae hyphae to soybean isoflavonoids and artificial substrates. Plant Physiol 117:1171–1178

Mrkovacki N, Milic V (2001) Use of *Azotobacter chroococcum* as potentially useful in agricultural application. Ann Microbiol 51:145–158

Nannipieri P, Ascher J, Ceccherini MT, Landi L, Pietramellara G, Renella G (2003) Microbial diversity and soil functions. Eur J Soil Sci 54:655–670

Nehl DB, Allen SJ, Brown JF (1997) Deleterious rhizosphere bacteria: as integrating perspective. Appl Soil Ecol 5:1–20

Nelson TH, Sorensen D, Tobiasen C, Anderson JB, Christophersen C, Givskov M, Sorensen J (2002) Antibiotic and lipopeptides produced by fluorescent *Pseudomonas* sp. from sugar beet rhizosphere. Appl Environ Microbiol 68:3416–3424

Nihorimbere V, Ongena M, Smargiassi M, Thonart P (2011) Beneficial effect of the rhizosphere microbial community for plant growth and health. Biotechnol Agron Soc Environ 15:327–337
Nowak-Thompson B, Chaney N, Wing JS, Gould SJ, Loper JE (1999) Characterization of the pyoluteorin biosynthetic gene cluster of *Pseudomonas fluorescens* Pf-5. J Bacteriol 181:2166–2174
Park JH, Bolan N, Megharaj M, Naidu R (2011) Isolation of phosphate solubilizing bacteria and their potential for lead immobilization in soil. J Hazard Mater 185:829–836
Paul EA, Clark FE (1988) Soil microbiology and biochemistry. Academic, San Diego
Peiffer JA, Spor A, Koren O, Jin Z, Tringe SG, Dangl JL, Buckler ES, Ley RE (2013) Diversity and heritability of the maize rhizosphere microbiome under field conditions. Proc Natl Acad Sci U S A 110:6548–6553
Penrose DM, Glick BR (2001) Levels of 1-aminocyclopropane-1-carboxylic acid (ACC) in exudates and extracts of canola seeds treated with plant growth promoting bacteria. Can J Microbiol 47:368–372
Pierson LS, Thomashow LS (1992) Cloning and heterologus expression of phenazine biosynthetic locus from *Pseudomonas aureofaciens*. Mol Plant Microbe Interact 5:330–339
Pinton R, Varanini Z, Nannipieri P (2001) The rhizosphere: biochemistry and organic substances at the soil–plant interface. Marcel Dekker, New York
Prasad R, De Datta SK (1978) Increasing fertilizer nitrogen efficiency In: Nitrogen and rice symposium, Los Banos, Sept 1978
Raaijmakers JM, Paulitz TC, Steinberg C, Alabouvette C, Moenne-Loccoz Y (2009) The rhizosphere: a playground and battlefield for soilborne pathogens and beneficial microorganisms. Plant Soil 321:341–361
Radjacommare R, Kandan A, Nanda KR, Samiyappan R (2004) Association of the hydrolytic enzyme chitinase against *Rhizoctonia solani* in rhizobacteria treated rice plants. J Phytopathol 152:365–370
Rameshkumar N, Fukui Y, Sawabe T, Nair S (2008) *Vibrio porteresiae* sp. nov., a diazotrophic bacterium isolated from a mangrove-associated wild rice (*Porteresia coarctata Tateoka*). Int J Syst Evol Microbiol 58:1608–1615. doi:10.1099/ijs.0.65604-0
Rameshkumar N, Sproer C, Lang E, Nair S (2010) *Vibrio mangrovi* sp. nov., a diazotrophic bacterium isolated from mangrove-associated wild rice (*Poteresia coarctata Tateoka*). FEMS Microbiol Lett 307:35–40. doi:10.1111/j.1574-6968.2010.01958.x
Rameshkumar N, Gomez-Gil B, Sproer C, Lang E, Kumar ND, Krishnamurthi S, Nair RA (2011) *Vibrio plantisponsor* sp. nov., a diazotrophic bacterium isolated from a mangrove associated wild rice (*Porteresia coarctata Tateoka*). Syst Appl Microbiol 34:487–493
Ramette A, Frapolli M, Défago G, Moënne-Loccoz Y (2003) Phylogeny of HCN synthase encoding *hcnBC* genes in biocontrol fluorescent Pseudomonads and its relationship with host plant species and HCN synthesis ability. Mol Plant Microbe Interact 16:525–535
Renker C, Blanke V, Börstler B, Heinrichs J, Buscot F (2004) Diversity of Cryptococcus and Dioszegia yeasts (Basidiomycota) inhabiting arbuscular mycorrhizal roots or spores. FEMS Yeast Res 4:597–603
Richardson AE (1994) Soil microorganisms and phosphorus availability. In: Pankhurst CE, Doubeand BM, Gupta VVSR (eds) Soil biota: management in sustainable farming systems. CSIRO, Victoria, pp 50–62
Ridge EH (1976) Studies on soil fumigation: effects on bacteria. Soil Biol Biochem 3:249–253
Robert L, Tate III (1994) Soil microbiology. Wiley, New York, p 506
Rodriguez H, Fraga R (1999) Phosphate solubilizing bacteria and their role in plant growth promotion. Biotechnol Adv 17:319–339
Rodríguez H, Fraga R, Gonzalez T, Bashan Y (2006) Genetics of phosphate solubilization and its potential applications for improving plant growth-promoting bacteria. Plant Soil 287:15–21. doi:10.1007/s11104-006-9056-9
Roesch LF, Fulthorpe RR, Riva A, Casella G, Hadwin AKM, Kent AD et al (2007) Pyrosequencing enumerates and contrasts soil microbial diversity. ISME J 1:283–290
Rondon MR, August PR, Bettermann AD, Brady SF, Grossman TH, Liles MR (2000) Cloning the soil metagenome: a strategy for accessing the genetic and functional diversity of uncultured microorganisms. Appl Environ Microbiol 66:2541–2547

Rosling A, Lindahl BD, Taylor AFS, Finlay RD (2004) Mycelial growth and substrate acidification of ectomycorrhizal fungi in response to different minerals. FEMS Microbiol Ecol 47:31–37. doi:10.1016/S0168-6496(03)00222-8

Rovira A (1965) Interactions between plant roots and soil microorganisms. Annu Rev Microbiol 19:241–266

Rovira AD (1976) Studies on soil fumigation: effects on ammonium, nitrate and phosphate in soil and on the growth, nutrition and yield of wheat. Soil Biol Biochem 8:241–247

Rubio LM, Ludden PW (2008) Biosynthesis of the iron-molybdenum cofactor of nitrogenase. Annu Rev Microbiol 62:93–111

Rudrappa T, Czymmek KJ, Pare PW, Bais HP (2008) Root-secreted malic acid recruits beneficial soil bacteria. Plant Physiol 148:1547–1556. doi:10.1104/pp.108.127613

Shahid M, Hameed S, Imran A, Ali S, van Elsas JD (2012) Root colonization and growth promotion of sunflower (*Helianthus annuus* L.) by phosphate solubilizing Enterobacter sp. Fs-11. World J Microbiol Biotechnol 28:2749–2758

Sharma R, Walia A, Chauhan A, Shirkot CK (2015) Multi-trait plant growth promoting bacteria from tomato rhizosphere and evaluation of their potential as bioinoculant. Appl Biol Res 17:1–12

Smit E, Leeflang P, Glandorf B, van Elzas JD, Wernars K (1999) Analysis of fungal diversity in the wheat rhizosphere by sequencing of cloned PCR-amplified genes encoding 18 S rRNA and temperature gradient gel electrophoresis. Appl Environ Microbiol 65:2614–2621

Smith KP, Handelsman J, Goodman RM (1999) Genetic basis in plants for interactions with disease-suppressive bacteria. Proc Natl Acad Sci U S A 96:4786–4790

Sorensen J (1997) The rhizosphere as a habitat for soil microorganisms. In: Van Elsas JB, Trevors JT, Wellington MH (eds.). Modern Soil Microbiology. Marcel Dekker pp 21–45

Spaepen S, Vanderleyden J, Remans R (2007) Indole-3-acetic acid in microbial and microorganism-plant signaling. FEMS Microbiol Rev 31:425–448

Spaink HP, Kondorosi A, Hooykaas PJJ (1998) The Rhizobiaeceae. Kluwer Academic, Dordrecht

Steenhoudt O, Vanderleyden J (2000) *Azospirillum*, a free-living nitrogen-fixing bacterium closely associated with grasses: genetic, biochemical and ecological aspects. FEMS Microbiol Rev 24:487–506. doi:10.1016/S0168-6445(00)00036-X

Steinkellner S, Lendzemo V, Langer I, Schweiger P, Khaosaad T, Toussaint JP, Vierheilig H (2007) Flavonoids and strigolactones in root exudates as signals in symbiotic and pathogenic plant-fungus interactions. Molecules 12(7):1290–1306

Talboys PJ, Owen DW, Healey JR, Withers PJA, Jones DL (2014) Auxin secretion by *Bacillus amyloliquefaciens* FZB42 both stimulates root exudation and limits phosphorus uptake in Triticum aestivum. BMC Plant Biol 14(51):2–9

Teplitski M, Robinson JB, Bauer WD (2000) Plants secrete substances that mimic bacterial *N*-acyl homoserine lactone signal activities and affect population density-dependent behaviors in associated bacteria. Mol Plant-Microbe Interact 13:637–648. doi:10.1094/MPMI.2000.13.6.637

Teplitski M, Chen H, Rajamani S, Gao M, Merighi M, Sayre RT, Robinson JB, Rolfe BG, Bauer WD (2004) *Chlamydomonas reinhardtii* secretes compounds that mimic bacterial signals and interfere with quorum sensing regulation in bacteria. Plant Physiol 134:137–146. doi:10.1104/pp.103.029918

Thomashow LS, Weller DM (1996) Current concepts in the use of introduced bacteria for biological disease control: mechanisms and antifungal metabolites. In: Stacey G, Keen NT (eds) Plant-microbe interaction, vol 1. Chapman & Hall, New York, pp 187–235

Tokala RK, Strap JL, Jung CM, Crawford DL, Salove MH, Deobald LA (2002) Novel plant-microbe rhizosphere interaction involving Streptomyces lydicus WYEC108 and the pea plant (*Pisum sativum*). Appl Environ Microbiol 68:2161–2171

Tournas VH, Katsoudas E (2005) Mould and yeast flora in fresh berries, grapes and citrus fruits. Int J Food Microbiol 105:11–17. doi:10.1016/j.ijfoodmicro.2005.05.002

Treonis AM, Ostleb NJ, Stotth AW, Primrosea R, Graystona SJ, Ineson P (2004) Identification of groups of metabolically active rhizosphere microorganisms by stable isotope probing of PLFAs. Soil Biol Biochem 36:533–537

Turner TR, James EK, Poole PS (2013) The plant microbiome-a review. Genome Biol 14:209
Ullah I, Khan AR, Jung BK, Khan AL, Lee IJ, Shin JH (2014) Gibberellins synthesized by the entomopathogenic bacterium, *Photorhabdus temperata* M1021 as one of the factors of rice plant growth promotion. Plant Micro Interact 9:775–782
Umrania VV (2006) Bioremediation of toxic heavy metals using acidothermophilic autotrophes. Bioresour Technol 97:1237–1242
Uren NC (2000) Types, amounts, and possible functions of compounds released into the rhizosphere by soil-grown plants. In: Pinton R, Varanini Z, Nannipieri P (eds) The rhizosphere: biochemistry and organic substances at the soil-plant interface. Marcel Dekker, New York, pp 19–40
Velazhahan R, Samiyappan R, Vidhasekaran P (1999) Relationship between antagonistic activities of *Pseudomonas fuorescens* isolates against *Rhizoctonia solani* and their production of lytic enzymes. J Plant Dis Protect 106:244–250
Viebahn M, Veenman C, Wernars K, van Loon LC, Smit E, Bakker PAHM (2005) Assessment of differences in ascomycete communities in the rhizosphere of field-grown wheat and potato. FEMS Microbiol Ecol 53:245–253
Von der Weid IA, Paiva E, Nobrega A, Van Elsas JD, Seldin L (2000) Diversity of *Paenibacillus polymyxa* strains isolated from the rhizosphere of maize planted in Cerrado soil. Res Microbiol 151:369–381
Vujanovic V, Hamelin RC, Bernier L, Vujanovic G, St-Arnaud M (2007) Fungal diversity, dominance, and community structure in the rhizosphere of clonal Picea marina plants throughout nursery production chronosequences. Microbiol Ecol 54:672–684
Vyas P, Gulati A (2009) Organic acid production in vitro and plant growth promotion in maize under controlled environment by phosphate-solubilizing fluorescent *Pseudomonas*. BMC Microbiol 22:174
Walia A, Mehta P, Chauhan A, Shirkot CK (2014) Effect of *Bacillus subtilis* strain CKT1 as inoculum on growth of tomato seedlings under net house conditions. Proc Natl Acad Sci India Sect B Biol Sci 84:145–155
Walker TS, Bais HP, Grotewold E, Vivanco JM (2003) Root exudation and rhizosphere biology. Plant Physiol 132:44–51. doi:10.1104/pp.102.019661
Ward NL, Challacombe JF, Janssen PH, Henrissat B, Coutinho PM, Wu M (2009) Three genomes from the phylum Acidobacteria provide insight into the lifestyles of these microorganisms in soils. Appl Environ Microbiol 75:2046–2056
Wegener CB (2001) Transgenic potatoes expressing an *Erwinia* pectate lyase gene-results of a 4-year field experiment. Potato Res 44:401–410
Weinert N, Piceno Y, Ding GC, Meincke R, Heuer H, Berg G, Schloter M, Andersen G, Smalla K (2011) Phylochip hybridization uncovered an enormous bacterial diversity in the rhizosphere of different potato cultivars: many common and few cultivar-dependent taxa. FEMS Microbiol Ecol 75:497–506
Welbaum GE, Sturz AV, Dong Z, Nowak J (2004) Fertilizing soil microorganisms to improve productivity of agroecosystems. Crit Rev Plant Sci 23:175–193. doi:10.1080/07352680490433295
Wissuwa M, Mazzola M, Picard C (2009) Novel approaches in plant breeding for rhizosphere-related traits. Plant Soil 321:409–430
Wu SC, Cheung KC, Luo YM, Wong MH (2006) Effects of inoculation of plant growth-promoting rhizobacteria on metal uptake by *Brassica juncea*. Environ Pollut 140:124–135
Yan Y, Yang J, Dou Y et al (2008) Nitrogen fixation island and rhizosphere competence traits in the genome of root-associated *Pseudomonas stutzeri* A1501. Proc Natl Acad Sci U S A 105:7564–7569
Yang C, Hamel C, Gan Y, Vujanovic V (2012) Pyrosequencing reveals how pulses influence rhizobacterial communities with feedback on wheat growth in the semiarid Prairie. Plant Soil 367:493–505
Zachow C, Tilcher R, Berg G (2008) Sugar beet-associated bacterial and fungal communities show a high indigenous antagonistic potential against plant pathogens. Microb Ecol 55:119–129

Zaidi S, Usmani S, Singh BR, Musarrat J (2006) Significance of *Bacillus subtilis* strain SJ 101 as a bioinoculant for concurrent plant growth promotion and nickel accumulation in *Brassica juncea*. Chemosphere 64:991–997

Zhou X, Wu F (2012) *p*-Coumaric acid influenced cucumber rhizosphere soil microbial communities and the growth of *Fusarium oxysporum* f.sp. *cucumerinum* Owen. PloS One **7**:e48288. doi:10.1371/journal.pone.0048288

Zhou J, Bruns MA, Tiedje JM (1996) DNA recovery from soils of diverse composition. Appl Environ Microbiol 62:316–322

Microbial Community Dynamics During Soil Ecosystem Development

4

Divya Deonalli, Rohit Sharma, and Kamlesh Jangid

Abstract

Ecological succession is a gradual change in community structure with time, which is important for proper functioning of an ecosystem. For a long time, plants have been in the limelight of community succession. In contrast, the significance of microbial succession in ecosystem development and various functions has only been emphasized relatively recently. Owing to the development of molecular methods over the last decade, microbial communities can now be investigated in much greater depths eventually leading to a better understanding of their contribution in an ecosystem. With the advent of these technologies, it is now possible to study primary succession in diverse ecosystems, such as wildfires, impact regions, paddy fields, etc. In this chapter, using case studies, we discuss the scenarios and successional paths that microbial communities follow in various ecosystems and show that the patterns are very similar in many. However, generalizing these to all communities at this time is not recommended, and the correlations observed from this analyses must be used with caution while applying them to a specific ecosystem.

Keywords

Microbial ecology • Primary succession • Ecosystem development

D. Deonalli
Department of Biotechnology, Mithibai College of Arts, Chauhan Institute of Science & Amruthben Jivanlal College of Commerce and Economics,
Vile Parle, Mumbai 400 056, Maharashtra, India

R. Sharma • K. Jangid (✉)
National Centre for Microbial Resource, National Centre for Cell Science,
SP Pune University Campus, Ganeshkhind, Pune 411 007, Maharashtra, India
e-mail: jangidk@nccs.res.in

V.C. Kalia et al. (eds.), *Mining of Microbial Wealth and MetaGenomics*,
DOI 10.1007/978-981-10-5708-3_4

4.1 Introduction

Primary succession is an important ecological phase that refers to an orderly and predictable process of establishment of biological activity in an area where none occurs (Fierer et al. 2010). The study of soil ecosystems undergoing primary succession is crucial as it provides a significant futuristic link regarding the conditions likely to occur in similar new habitats or during habitat rehabilitation.

Patterns of microbial community succession in different soil ecosystems at various stages of development have been a subject of great interest for years. Soil, being an important component in the development of varied ecosystems, is largely influenced by variables like climate, diversity of organisms, topography, parent material, and time, which are collectively known as "soil-forming factors" (Stevens and Walker 1970). This suggests that the underlying ecosystem in the primary successive stages has a significant impact on the soil properties being developed, and vice versa. The rate and type of soil and the ecosystem under development have a close dependence on one another. Primary succession studies therefore provide a significant link to the soil development history and future perspectives (Ohtonen et al. 1999; Walker and del Moral 2003; Bardgett et al. 2005; Siddique et al. 2012).

Primary succession can be studied during the development of numerous soil types. The absence of soil, newly exposed surfaces, and barren surfaces are some crucial entities in determining the nature of primary succession that would lead to formation of complex soil ecosystems with the available biotic and abiotic components. Earth movements expose barren surfaces for primary succession. Various natural calamities create conditions for primary succession. Volcanic eruptions, earthquakes, landslides, uplifts, avalanches, and erosion are some of the extreme disturbances that trigger primary succession by either creating new surfaces or exposing bedrock (Merila et al. 2002; Besemer et al. 2005; Cutler et al. 2014). Similarly, winds in the form of hurricanes, tornadoes, and severe storms can initiate primary succession in newly created dunes, eroded soils, and reshaped coastlines. Floods have been widely known for creation of erosional and depositional plains (Tarlera et al. 2008; Langhans et al. 2009). Deglaciation and disturbances in the permafrost provide new substrates for primary succession (Jumpponen 2003; Wagner et al. 2005; Lazzaro et al. 2010). Though fire primarily leads to secondary succession, uncontrolled prolonged fires can sometimes result in complete mortality of biological life, thus leading to primary succession (Ferrenberg et al. 2013). Primary succession at impact sites is also of great interest (Cockell and Lee 2002). Similarly, human activities, such as deforestation, urbanization, extensive farming, agriculture, mining, and other activities severely damage ecosystems (Calderon et al. 2001; Muller et al. 2001; Bossio et al. 2005; Wang et al. 2011; Banning et al. 2011). These activities majorly alter the soil structure, and succession at such damaged sites plays an important role in their recovery.

All these years, lack of techniques to study microbial succession has been a major concern. Even though microbes play an important role in shaping and proper functioning of an ecosystem, research on their succession has received little attention in the past due to these technological limitations. Along with primary producers and herbivorous insects, microbial decomposers are also important drivers of ecosystem functioning (Soliveres et al. 2016). However, with the development of methods, the contribution of microbial succession to the development of any ecosystem is now well recognized. In this chapter, we present how microbial succession acts as a proxy of ecosystem development, discuss the various stages in succession, and then study the patterns of succession in microbial communities during the development of different soil ecosystems.

4.2 Microbial Succession: A Proxy of Ecosystem Development

Microbes form an essential component in the development of newly exposed surfaces that haven't been colonized before. They are among the initial colonizers of such ecosystems and help higher organisms in establishment. Microbial succession in itself is strongly influenced by interaction between biotic and abiotic factors like nutrients, pH, temperature, time intervals, vegetation, species diversity, and many more (Bardgett et al. 2005; Besemer et al. 2005; Nemergut et al. 2006; Baldrian et al. 2008). As microbial succession can be studied from the beginning, such bare surfaces serve as a template for the study of soil ecosystem development and community dynamics (Fierer et al. 2010). They help in understanding the trajectories undertaken based on the existing community structure and factors which affect such processes. Although microbial succession may be initiated by a set of interacting disturbances, not strictly restricted to a single cause, it occurs over a wide range of scales. Thus, there are countless ways to study microbial succession patterns in soil ecosystems. The types of microbes colonizing a given area influence the underlying successional trajectories (Walker et al. 2010). This makes it possible to study and compare numerous globally distributed environments to reveal the intricate successional patterns.

The major function of soil microflora is decomposition, mineralization, and energy transfer, thus contributing to long-term stability to soil properties. The actual functional processes of microbial community succession in soil ecosystem development are still unclear. Previous theories state that succession proceeds in a single direction and has a predictable endpoint. After a certain time interval, climax communities exist (Fierer et al. 2010; Williams et al. 2013). However, the more common view held now is that succession isn't strictly linear and has multiple trajectories with many possibilities, such as cyclic, convergent, divergent, parallel, or reticulate (for details on these trajectories, please refer to Walker and del Moral 2003). The directionality and predictability attributed to succession indicate the change in species diversity and not in the sense of heading toward an endpoint in a specific period. Microbial communities are subject to changes over different periods. These periods vary from days to years to even decades (Crews et al. 1995; Nemergut et al. 2006). The phenomenon of climax communities is highly uncertain as the microbial communities are constantly subject to change by major or minor alterations (Sigler et al. 2002; Fukami and Morin 2003; Fierer et al. 2010). According to various studies, microbial species composition changes between different ages of a forest tree in its complete life span (forest ecosystem), different time of a crop season (agricultural field ecosystem), different time of a year (creator ecosystem), etc. Hence, it is a continuous process wherein community change keeps on occurring with change in age, time, season, etc.

Microbes play an indispensable role in carbon fixation and decomposition of fixed carbon to accumulate soil organic matter. This incorporates a higher proportion of C in the microbial biomass, and its availability is likely to be the preliminary driver of microbial succession. It determines the proportion of autotrophs and heterotrophs colonizing the ecosystem. Both communities are likely to co-occur; however, in the initial stages of succession, autotrophs may be dominant in distinct environments where heterotrophs rely completely on fixed carbon by autotrophs due to limited resources (Ohtonen et al. 1999; Blaalid et al. 2012). Microbes interact with the minerals and organic carbon to bring about weathering that mobilizes inaccessible nutrients essential for succession and development of other ecosystem components. In most ecosystems, the common observation suggests that due to weathering the proportion of soil phosphorus decreases, whereas carbon, nitrogen, and sulfur accumulate. To have a good balance of nutrients, the existing microbial communities produce enzymes that play a role in cycling and acquisition of nutrients that are lost due to weathering (Allison et al. 2005; Baldrian et al. 2008). At early stages, there is significant loss of C due to respiration; overtime the loss decreases resulting in higher C accumulation. Due to increased allocation of C in the microbial biomass with succession, there is a shift in microbial communities from energy-inefficient to energy-efficient state, thus maintaining stability of developing soils (Bekku et al. 1999; Fierer et al. 2010; Coomes et al. 2013). For example, in a portion of forest affected with forest fire, we see small herbs growing after the rains. They may be the primary plant species colonizers but before that microbial community have colonized and had already began the successional process. The successional changes and function contributed by microbes to an ecosystem have not been studied in detail and need more attention. The recent incorporation of molecular techniques has helped gather more information on microbial succession. Whether microbes decide the aboveground plant ecosystem composition or vice versa during succession has been a contentious issue for botanists and microbiologists alike. In our opinion and as explained below, it is a direct influence of either component on the other in a developing ecosystem, which decides the succession trajectory.

4.3 Stages in Microbial Succession

There is significant change in the population of various taxa, from bacteria and archaea to fungi to *Actinobacteria*. Among microbes, bacteria and fungi are the first colonizers during ecosystem succession. A relative proportion of archaea stably colonize early successional stages (Bardgett et al. 2005; Ohtonen et al. 1999). Along with physical and chemical parameters, they contribute to decomposition, biogeochemical cycles, enzymatic activity, and subsequent release of nutrients, thus contributing to the proper functioning of the ecosystem. The presence of specific taxa such as photosynthetic cyanobacteria, nitrogen fixers, and sulfide oxidizers leads to pronounced changes in the successional patterns (Kastovska et al. 2005; Schmidt et al. 2008). Microbial biomass is often highest at the earlier stages, tapering off as succession progresses primarily depending on environment (climate, season, etc.) and availability of organic biomass. Similarly, microbial diversity is significantly high at the initial and intermediate stages of succession that may significantly decline in the later stages. It also indicates the degree of stability of a community. The species richness and evenness greatly vary through succession and most significant in dynamic ecosystem. Over succession, numerous environmental and nutritional parameters determine the abundance of specific taxa. There is a shift in the biomass balance of microbial autotrophs, copiotrophs, and oligotrophs depending on the presence or absence of environmental stress during succession, thereby helping us gather ecological information on the ecosystem (Bardgett et al. 2005; Okabe et al. 2007; Fierer et al. 2010).

In the initial stages of succession, bacterial communities dominate which are the primary colonizers within microbial communities (Fig. 4.1). However, as succession proceeds, there is a shift from bacterial-dominated to fungal-dominated communities commonly observed (Cherif and Loreau 2007). This shift is mainly due to the fact that fungi are more efficient in uptake of limited nutrients by secreting enzymes and metabolites to the surrounding. They can tolerate environmental fluctuations better and can bring about recycling of recalcitrant and complex compounds. Fungi are also more effective in energy use. Another crucial function of fungi is their association with plant roots. The mycorrhizal association plays a vital role in the establishment of pioneering plant species by enhancing uptake of nutrients, modifying the pH and toxicity of soils (Kaye and Hart 1997; Bardgett et al. 2005; Kikvidze et al. 2010). Most of the mycorrhizal fungi are non-specific, i.e., single fungal mycorrhizal species can form in association with many tree plant species. At the same time, they help the plants in colonizing a habitat by providing water and nutrition from a long distance especially in a nutrient-deficit initial successional stage of an ecosystem. They also help by protecting the initial plant species from pathogenic fungi, thus helping them to survive the initial successional stage which is also crucial for the development of an ecosystem because it will decide which species will finally colonize the ecosystem.

During microbial succession in soil ecosystems, the interaction between plants and microbes is most important. Plant root, exudates, and leaf litter have a significant impact on the colonizing microflora. Microbial diversity is patchy with greatest concentration in vegetated areas typically exhibiting the rhizosphere effect. There is a significant increase in the soil organic matter, microbial biomass, and microbial enzymatic activity in vegetated areas, symbolizing a synergistic relationship between plants and microbes. However, the exact nature of this relationship is not clear. A forest tree, apart from yearly change in microbial composition, is known to have different microbial associations during its entire life span (herb, shrub, and tree stage). However, studies also show that changes in microbial community pattern in

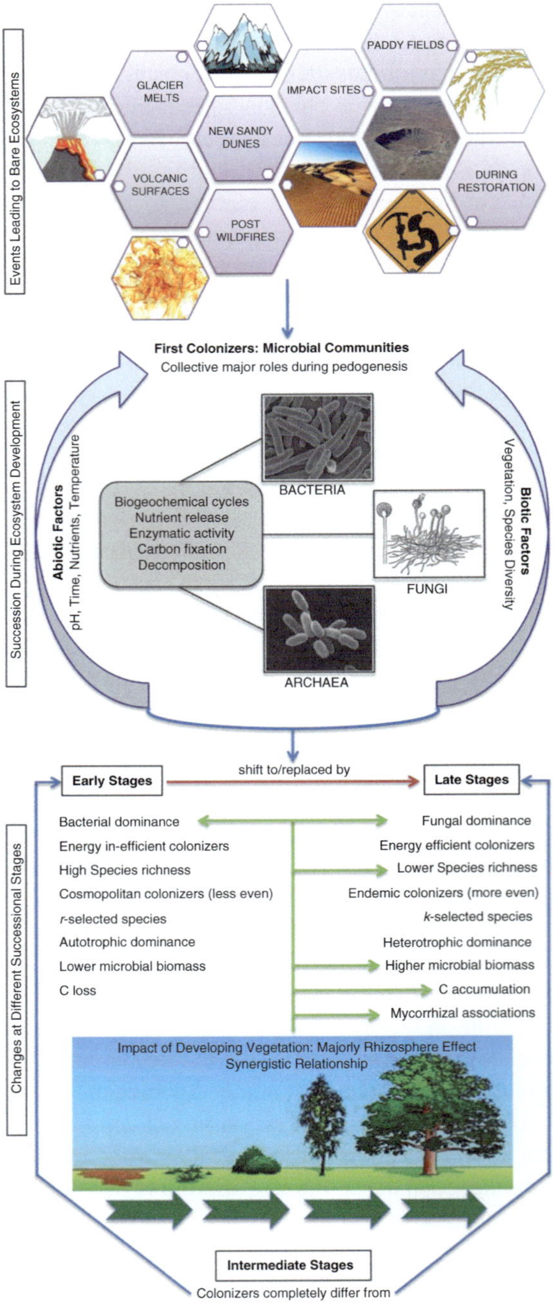

Fig. 4.1 Microbial community succession across different ecosystems. Different natural and human-induced events lead to the creation of "near-pristine" ecosystems on which primary succession proceeds. Among the first colonizers of such "bare" ecosystems are the microbial communities that develop concomitantly with the development of the ecosystem. At the same time, the interactions with the abiotic and biotic factors directly impact the development of microbial communities that are responsible for many functions within the ecosystem for it to develop further. During such successional events, many changes take place within the ecosystem, e.g., the early stages primarily dominated by bacteria and late stages characterized by fungal dominance. However, there is an intermediate stage during this transition which significantly differs from the other two

association with vegetation reflect increasing competition between the two for acquisition of nutrients (Kaye and Hart 1997; Pennanen et al. 2001; Bardgett et al. 2005; Kikvidze et al. 2010). Because a tree plant may release different amounts and kinds of exudates which may be preferred by different microbes, it changes the microbial community composition over a successional period. Moreover, a litter rich with organic matter will show more changes in their successional community composition than an ecosystem with less nutrition which may show more or less uniform community composition. Further, a mixed forest will be more dynamic than pure evergreen forest. For instance, in the forests of *Shorea robusta* (Sal), both litter and organic matter is in good quantity due to the evergreen forest, which helps in good microbial biomass and diversity. In ectomycorrhizal association between a forest tree plant and fungi, a single tree species may form mycorrhiza with several fungal species. Thus, collectively all the fungal species may support a single forest tree species resulting in monoculture forest. In contrast, *Tectona grandis* (teak) forests have less microbial diversity due to lack of litter and organic matter. These soils are likely to harbor selected microbes as they have affiliation to the teak leaf litter, can survive intense sunlight on the soil surface, and assimilate the specific root secretions. Regardless of these two scenarios, there is likely a core community whose composition does not change at any stage. It is this community which is likely to impose the most significant effect on the aboveground plant community and possibly the most resilient to the top-down effect of the diverse plant community in such forests.

4.4 Chronosequences to Study Succession

Microbial succession during soil pedogenesis can encompass a period from years to decades. Indirect measures to analyze succession are essential to determine age of subsequent stages and derive the soil conditions over long-term scales, and chronosequences have been historically used to study succession (Huggett 1998). A chronosequence is a set of sites or habitats of related soils with similar properties but differs only on temporal scale. This is useful to study the succession of an ecosystem or effect of the environment or climate change on ecosystem by collecting data in a short duration of time and make predictions on a large scale of time (decades or centuries). It uses the concept of substituting space for time in order to understand ecosystem establishment and evolution. Soil chronosequences are indicative of the rate and direction of pedogenic change providing critical information about soil development and associated change in the microflora. They are suitable for measuring soil community characteristics, species richness, and soil organic matter accumulation. Well-dated chronosequences with knowledge of subsequent history permit observation of ecological progression or retrogression overtime. However, in studies of chronosequence it is presumed that no other variable changes except time during the studies on chronosequence. The linkage between stages, type of trajectories, predictability, and time span involved determine how reliably chronosequences may be utilized in successional studies (Kirk et al. 2004; Walker et al. 2010; Bernasconi et al. 2011; Nyberg et al. 2012). However, chronosequences can provide misleading information also. They prove to be least appropriate when observing succession with missing linkages, short span, divergent trajectories, or highly descriptive patterns. Moreover, it is difficult to find and correlate sites that have evolved similarly. Errors are further added by stochastic colonization, recurrent disturbances, and varying climate (Johnson and Miyanishi 2008; Walker et al. 2010). To avoid these limitations, study of microbial succession should preferably be analyzed in real time for precision. However, this is impractical.

Chronosequences have been historically used to study successional patterns in different ecosystems, such as glacial moraines, surfaces after lava flow, sand dunes, landslide scars, flood plains, old pastures, mining sites and restoration (Huggett 1998; Kirk et al. 2004). In further sections, various ecosystem-specific chronosequence studies will be discussed as case studies.

4.5 Case Studies: Widely Studied Ecosystems for Microbial Succession During Soil Pedogenesis

4.5.1 Primary Succession in Receding Glaciers

It is now a well-known fact that global warming has led to glacier retreat in several mountainous areas of the world. As glacier melts, underlying bare rock and gravel gets exposed which becomes a potential site for primary succession to proceed, thus helping in the study of community construction. Chronosequences of these deglaciated or uplifted areas are pivotal in understanding primary succession and ecosystem development (Schmidt et al. 2008). During soil pedogenesis, exposed rock is subjected to weathering. The soil-forming process is significantly affected by various biotic and abiotic factors to give soil a characteristic structure. Since soils had been under cover for a long period with minimal microbial activity, these soils lack essential nutrients like carbon, nitrogen, phosphorus, and sulfur (Lazzaro et al. 2010). Among the biotic factors affecting soil pedogenesis, colonization of deglaciated areas by microbes is the most essential one. Colonization of exposed bedrock, by microbes that can adapt to low-nutrient conditions and other fluctuating factors, is the pioneering step (Jumpponen 2003; Nicol et al. 2008; Schmidt et al. 2008). Pioneering microbes play essential roles in biogeochemical cycling, mineral weathering, and subsequent release of nutrient in order to provide substrates for colonization of other complex microbial and plant communities (Schutte et al. 2009).

In any deglacier soil, primary colonizers of deglaciated areas have been mainly bacteria, fungi, and some species of archaea (Nicol et al. 2003; Zumsteg et al. 2012). Then, as the soil development takes place, these are taken over by some metabolically versatile bacteria and psychrophilic yeasts which in turn give way to bacteria degrading complex organic matter and higher fungi. *Cyanobacteria*, *Proteobacteria*, and *Actinobacteria* are commonly found in the early stages of succession in the deglaciated regions of the Arctic and Antarctic (Bradley et al. 2016), whereas

McCann et al. (2016) observed that in Arctic polar desert, *Proteobacteria*, *Actinobacteria*, and *Chloroflexi* dominated the microbial community composition. *Actinobacteria*, being an early colonizer, are known to produce several secondary metabolites and thus play important role in ecosystem development and biogeochemical cycling. The microbial diversity, including that of *Actinobacteria*, has been found to be different in different glaciers (Zhang et al. 2016). With almost similar nutritional status of the glacier, there is a difference in diversity emphasizing the spatial effect. The microbial communities undergo several shifts as succession proceeds (Knelman et al. 2012). Early microbial colonizers are more cosmopolitan and replaced by more endemic species along the soil gradient. Studies indicate that there is a shift from r-selected to k-selected species in glacier forefields (Nemergut et al. 2006). In the early stages, the species richness is high till an intermediate stage; however, species evenness increases in the later stages with decreasing overall diversity (Blaalid et al. 2012; Zumsteg et al. 2012). Colonizing communities show greatest variations in the early stages which later diminishes with ecosystem establishment as has been observed at the Franz Josef glacier (Jangid et al. 2013). Similarly, in a 10-year deglaciation chronosequence on the Tibetan Plateau, Liu et al. (2014) showed that the microbial communities rapidly developed and remained stable in old soils due to increase in total organic carbon and nitrogen resulting from autotrophs. Cyanobacterial diversity also increases and plays a key role in such ecosystems. Studies by Schutte et al. (2009) indicate that microbial communities colonizing the youngest, intermediate, and oldest site in the high Arctic glacier forelands and similar deglaciated regions completely differ from one another, suggestive of the fact that there are major shifts in the richness and evenness of colonizing communities with succession. These parameters also suggest that microbial succession in deglaciated regions may not follow a linear or predictable trajectory (Walker and del Moral 2003). Even the depth also plays a significant role in the microbial community development in deglacier soils (Edwards and Cook 2015; Rime et al. 2015).

In the initial stages, the nutrients are low in the soil. Hence cyanobacteria, microalgae, form major component at that time and thus contribute to the autotrophic activity of the soil. Microbial biomass and activity increases with the progress of succession. The increase in microbial biomass is positively correlated with increasing carbon and nitrogen content (Bekku et al. 1999; Ohtonen et al. 1999; Jangid et al. 2013; Zhang et al. 2016). With succession, microbes accumulate C into its biomass indicating that the pioneering colonizer microbes are major contributors on the accumulating organic matter. The large ratio between microbial biomass and organic matter indicates greater input of liable C of microbial origin (Ohtonen et al. 1999). As succession proceeds, the colonizing microbes shift is based on the availability of C. The core taxon frequency shift with successional age suggests niche distinction based on input patterns of autochthonous and allochthonous origin (Brown and Jumpponen 2014). In recently deglaciated soils, ancient inputs of C may fuel the initial stages, but microbial species especially fungi and plant communities developing in later stages are significantly dependent on C inputs by pioneering microbes (Schmidt et al. 2008). Moreover, as vegetation develops in these deglaciated areas, it has a significant impact on the kind and amount of resources available for microbes (Bowman et al. 2004; Bardgett et al. 2005). Microbial communities produce various enzymes during successional stages. These enzymatic patterns are correlated with regulation of carbon and nitrogen contents of soils (Saleh-Lakha et al. 2005). Ecosystem productivity is dependent on mineralization by colonizing microbes. Allison and Martiny (2007) studied the chronosequence of the retreating Franz Josef Glacier; it was revealed that extracellular enzymes regulate the process of mineralization in successional soils. There is a negative relationship between nutrient concentration and enzyme accumulation in order to maintain a balance during succession. When the required nutrient is abundant or there is a shift in the microbial community, which requires low level of the nutrient, the enzyme production is downregulated (Allison and Martiny 2007).

It was often previously implied that primary succession in soil ecosystem begins with plant establishment. Over the time research has proved that microbial succession across postglacial chronosequence occurs well before plant colonization (Nemergut et al. 2006; Schmidt et al. 2008). Therefore, beneath the soil, the changes are much more dynamic and start much earlier than aboveground changes. In other words, microbes (bacteria, archaea, and fungi) set the platform or arrange for nutrition for plant colonization. Microbial colonization in un-vegetated deglaciated regions represent a critical first step in setting ecological, biochemical, and pedological trajectories of developing ecosystem (Knelman et al. 2012). Microbes colonizing the initial ecosystem may have been dormant beneath glaciers or can be brought to the deglaciated site by air currents, glacial runoff, snowmelts, and deposition of particulate matter (Kastovska et al. 2005; Bhatia et al. 2006). The bacteria colonizing the recently deglaciated regions range from autotrophs, free-living nitrogen fixers, and sulfide oxidizers to heterotrophs. The major dominant active groups in glacier forefields during primary colonization are photosynthetic diazotrophs and cyanobacteria (Lazzaro et al. 2010; Bradley et al. 2016). These bacteria carry out essential functions of photosynthesis and nitrogen fixation at early stages, therefore replenishing the carbon and nitrogen pools to stimulate activity of heterotrophs. Heterotrophs like *Proteobacteria*, *Actinobacteria*, and *Acidobacteria* play essential roles in further biogeochemical cycles in young soils (Schmidt et al. 2008; Lazzaro et al. 2010; Knelman et al. 2012). Schmidt et al. (2008), based on research on receding glaciers at Andes, suggested that the diversity of *Cyanobacteria* along succession shows variation during the development of ecosystem. There was a transition from ice-dwelling *Cyanobacteria* to biological crust-like *Cyanobacteria.* The phylotypes and major groups of *Cyanobacteria* varied in significant number. However, as plants colonize the role of autotrophic, *Cyanobacteria* are likely to be diminished with the increase in availability of fixed plant carbon for heterotrophs (Knelman et al. 2012).

Early stages of succession are characterized by a large bacterial biomass as compared to fungal communities, which are scattered through the stages of soil development (Blaalid et al. 2012). The fungal biomass increases over time and is attributed to decrease in pH due to plant litter and exudates in the soil as plants develop (Schutte et al. 2009). Fungi are known to better adapt to acidic environment than bacteria, are more efficient in energy use, and therefore can build up large inactive biomass. They are also known to be effective in utilization of recalcitrant compounds (Ohtonen et al. 1999; Welc et al. 2012; Jangid et al. 2013). Fungi have a special association with developing plant communities. Studies have demonstrated that mycorrhizal fungi are present and are potentially important players in glacial forefront succession (Blaalid et al. 2012; Brown and Jumpponen 2014). Some fungi are pioneers but most are late colonizers. A shift in diversity from *Ascomycota* dominated in young soils to more *Basidiomycota* dominated in old soils in receding Damma glacier forefield (Zumsteg et al. 2012). Archaeal communities have been seen to increase along the successional gradient and then remain stable for long periods (Nemergut et al. 2006). Young soils are predominantly colonized by Euryarchaeota, whereas Crenarchaeota mainly colonize mature soils (Nicol et al. 2003). However, little is known about the role of archaea in primary succession; it is believed that they are involved in subglacial methanogenesis (Lazzaro et al. 2010).

There is a significant link between the aboveground and belowground communities during succession (Bardgett et al. 2005; Jangid et al. 2013). It seems plant colonization dramatically alters soil microbial community composition and function but actually microbes change the soil in such a way that it helps the plants to colonize. They may provide carbon inputs in the form of root exudates and litter (Knelman et al. 2012). As plants become abundant in later stages of succession, they become hotspots for microbial activity and may guide their development in the ecosystem as per their benefits (Lazzaro et al. 2010). Microbial biomass is greatest under plant canopies and much different from non-vegetated soils (Ohtonen et al. 1999). In later stage, the composition of the litter and exudates determine the colonizing microbes. Bardgett et al. (2005) revealed that phenolic compounds in the litter enhance fungi, whereas high carbohydrate content enhances bacteria. Richness of species tends to decrease with vegetation cover, suggesting that niches available are reduced, thus specifying a negative relationship (Alfredsen and Høiland 2001; Jangid et al. 2013).

With the increase in global warming, the change in temperature is likely to determine the development of ecosystem in deglaciated soils. The development of microbes will also depend on soil properties, spatial location of glacier, and nutritional availability in the soil which in turn will govern the aboveground successional development of the ecosystem. Moreover, within microbes, apart from general bacteria, *Actinobacteria* and *Cyanobacteria* will play important roles which are primarily carbon- and nitrogen-fixing autotrophs.

4.5.2 Primary Succession on Volcanic Substrates

Volcanic eruptions and lava flow are one of the ideal sites for studying microbial colonization and succession since they are one of the most deficient sites in available nutrition and water. They are also unique due to their extreme environment. After volcanic eruption, virgin surfaces such as lava, tephra, and volcanic ash are available for primary succession (Sato et al. 2004). Microbial succession on these substrates is known to follow similar patterns as that of succession on glacier forefields (Gomez-Alvarez et al. 2007; Hopkins et al. 2007). The source of microbes in these areas is mainly aerial dust (Fujimura et al. 2012). Similar to succession in deglaciated regions, the pioneering colonization of volcanic substrates is by carbon- and nitrogen-fixing autotrophs and fewer heterotrophs dependent on existing Aeolian inputs, trace gases, and organic matter (King 2003). The pioneering microbes help in colonization of other microbial communities and establishment of vegetation (Van der Heijden et al. 2008; Dennis et al. 2010). There is a gradual increase in the microbial biomass, soil organic matter, diversity, and energy efficiency with succession (King 2003) along with a shift in microbial communities that bring about more efficient soil pedogenesis (Crews et al. 2001). While the succession stages may show similarity with deglaciated surfaces, volcanic eruptions are a much drastic occurrence and hence may differently influence smaller stages within the common trajectories.

Volcanic material imposes severe constraint on microbial colonization and activity. Substrates formed following lava flows are severely deficient in exogenous nutrients, organic matter, fixed nitrogen, and other autochthonous inputs (Cutler et al. 2014). Allochthonous inputs carried by the wind are likely to participate in soil development on lava flows (Walker and del Moral 2003). Several abiotic processes such as varying pH and cation exchange add further difficulties in colonization of microbial species. Further, solidified lava is impermeable leading to major water unavailability (Gomez-Alvarez et al. 2007). Microbes, which successfully overcome these odds, initiate the series of events leading to primary succession in volcanic substrates. The harsh conditions may lead to formation of specific confined microsites that may offer physical protection for enhanced utilization of resources. In such stressful conditions, the colonizing species may plateau for several years with slow increment in biomass. It has been shown that microbial colonization and succession in volcanic substrates take a shorter time period than plant succession. The succession of vegetation is slow as established plant species significantly lack dispersal capabilities in such environments (Sigler et al. 2002). Plant species are highly dependent on weathering, nutrient accumulation, and regulation of biogeochemical cycles by pioneering microbial communities (Van der Heijden et al. 2008). The plant litter composition and species are also known to impact the colonizing microbial species during alter stages (King 2003; Hopkins et al. 2007). Overall, the direction and pace of ecosystem development may hinge on microbial utilization of nutrients and energy resources (King 2003).

Bacterial diversity in the early stages is extensive with divergent phylotypes (Gomez-Alvarez et al. 2007). The initial stages are dominated by autotrophic and phototrophic bacteria like *Cyanobacteria* and *Chlorobacteria*, which bring about accumulation of organic matter, photosynthesis, and nitrogen fixation (Cutler 2010). Microbial analysis shows the presence of heterotrophic thermophiles, acidophiles, chemolithotrophs, and iron-oxidizing bacteria (King 2003; Fujimura et al. 2012). Common representatives at early stages during succession are *Acidobacteria*, *Actinobacteria*, and *Proteobacteria* (Gomez-Alvarez et al. 2007). *Proteobacteria* are the most abundant class detected in volcanic deposits as it has traits such as phototrophy, photoheterotrophy, and chemolithotrophy. However, there is a shift in the species as succession proceeds. *Betaproteobacteria* found in early stages are replaced by *Alphaproteobacteria* in the later stages (Fujimura et al. 2012). In volcanic deposits there is a special significance to carbon monoxide and hydrogen-oxidizing bacteria. These bacteria have numerous physiological and functional traits that significantly contribute to nutrient cycling, nitrogen fixation, and utilization of organic substrates. Bacterial communities stabilize over a period of time and show minor alterations with plant succession in later stages (Cutler et al. 2014). Fungal communities dominate the later stages of succession for reasons similar to that observed at glacier forefronts. Establishment and shift in fungal species are highly dependent on vegetation (Xu et al. 2006; Cutler et al. 2014). Studies have shown significant differences in fungal species colonizing the youngest lava flow from those on older lava flow. Evidence suggests that mycorrhizal associations mainly exist at later stages of succession (Hopkins et al. 2007; Cutler et al. 2014). Further, only a negligible change has been observed in the functional capacity of microbial communities during later stages of succession on such volcanic sites (Xu et al. 2006).

4.5.3 Primary Succession on Newly Formed Sandy Dunes

Microbial successional patterns are less studied in dune ecosystems as compared to glaciers and volcanic substrates. Dunes are primarily formed by deposition of sand by wind, tornadoes, hurricanes, and severe storms (Walker and del Moral 2003; Tarlera et al. 2008; Kikvidze et al. 2010). The sandy nature of dunes signifies extremely low water-holding capacity. This water stress makes the soils arid and prone to droughts. Moreover, dunes are also characterized by extreme temperature and of regions exposed to radiation. In spite of the harsh environments, biological succession in these environments has been observed (Johansen et al. 1997; Tarlera et al. 2008; Langhans et al. 2009). It has been noticed that colonizing microbes in dune ecosystems differ with succession stages as every other ecosystem. Microbes in dune ecosystems exhibit high diversity and richness in the young soils, which often reverses at the terminal stages (Tarlera et al. 2008), suggesting that there is a transition to a new community with succession.

Primary succession in dune ecosystems is divided into many stages. First stage is necessarily the colonization by microbes that can survive in harsh nutrient-deficient conditions, followed by the development of dependent microbes and vegetation that replaces most of the primary colonizing species (Grootjans et al. 1997). The predominant colonizing microbes are bacteria and fungi. Bacteria play essential roles in stabilizing the dune soils by producing sticky polysaccharide that adheres to the soils, therefore stabilizing them. Similarly, fungi help in trapping sand particles by their hyphal network. Blue-green algae play an indispensable role in nitrogen fixation and soil aggregation, increasing the chlorophyll content and producing mucilaginous sheaths to aggregate soils (Forster and Nicolson 1981; Jangid et al. 2013). Soil aggregation helps in stabilizing dunes, prevents drought and erosion, increases nutrient content, and helps sandy soils to retain moisture (Forster and Nicolson 1981). Vascular plants, which find it difficult to develop in such harsh conditions, can form substantial coverage after aggregation of sand by the primary microflora (Forster and Nicolson 1981; Langhans et al. 2009). The underlying bacterial community structure is highly related to the aboveground plant structure once vegetation develops on dunes (Jangid et al. 2013; Williams et al. 2013). It is presumed that the aboveground vegetation impacts the bacterial communities through an indirect mechanism in the later stages of succession.

Acidobacteria is one of the most abundant taxa in dune soils, which undergo shifts in species with succession (Barns et al. 1999). Blue-green algae like *Microcoleus*, *Nostoc*, *Scytonema*, *Calothrix*, and *Gloeocapsa* are found to colonize in the early stages along with Chlorophyceae *and* Bacillariophyceae (Lukesova 2001). In the later stages, there is an abundance of *Chloroflexi* and fungal communities similar to succession in glaciers and other ecosystems (Ohtonen et al. 1999; Tarlera et al. 2008). According to Liu et al. (2014), *Actinobacteria* and *Proteobacteria* (especially *Alphaproteobacteria*) are the dominant group of bacteria in sand dunes. *Planctomycetes* abundance during succession is an indication of native conditions inherent to dune soils, as the taxa shows specialized characters restricted to dune succession (Buckley et al. 2007). The general trend observed is that large filamentous *Cyanobacteria* are first colonizers, followed by smaller *Cyanobacteria* and eukaryotic algae, and lastly by lichens and bryophytes (Langhans et al. 2009).

Free-living and plant-associated microbes are both essential in development of the soil. Microbial consortium, such as mycorrhizal fungi associated with specific colonizing plants, is thought to be essential in establishing a model for primary succession on sand dunes (Grootjans et al. 1997; Jangid et al. 2013). Mycorrhizal fungi have a control on plant populations and community dynamics in arid ecosystems (Pugnaire et al. 2006; Armas et al. 2010; Kikvidze et al. 2010). In a recent study, Rao et al. (2016) while studying the microbial diversity in Indian monsoon desert showed fungi dominated the sand dunes. It is found that bacterial communities help in the stabilization of succession in sand dunes. However, mineral and nutrient status of the sand dunes are also important in ecosystem stability (Ronca et al. 2015).

4.6 Case Studies: Less Studied, Unique Ecosystems for Microbial Succession During Soil Pedogenesis

4.6.1 Primary Succession in Paddy fields

Among the less studied wetland ecosystems, microbial succession in paddy fields is best studied. Understanding microbial dynamics in paddy field ecosystems allows for the characterization of pathways that lead to soil pedogenesis under its unique water-logged conditions (Conrad and Frenzel 2002; Kirk et al. 2004). Flooding leads to formation of an oxygen gradient with an oxic-anoxic upper interface (Noll et al. 2005). The interface formation triggers essential reactions like methane oxidation, nitrification, denitrification, sulfur-formation, and many more. It has been observed that microbial communities in the upper oxic layer differ completely from that of the lower anoxic layer and that the rate of succession differs in both the layers. Microbial colonization changes rapidly in the oxic layer, whereas no major difference is observed in the lower layer for long periods (Murase et al. 2006).

Another significant feature of paddy soils is methanogenesis. Paddy is one of the most cultivated crops, and paddy fields contribute approximately 10% of global methane emissions. Methanogenesis and methane oxidation are important activities of a paddy field, which are affected by pH, temperature, seasonal variation, variety of rice, etc. Methanotrophs can easily survive harsh, unfavorable, nutrient-limiting conditions (Eller et al. 2005). Anoxic conditions favor methanogenic archaea and methanotrophic bacteria to persist widely in such wetland ecosystems. Methanotrophs are absent or present in small counts under oxic conditions. However, in paddy ecosystems it has been revealed that methanotrophs are extremely crucial in maintaining functions of the dynamic ecosystem (Krause et al. 2010). According to Ma et al. (2016), increase in soil phosphorus is the key in the change of microbial community structure in paddy field. However, bacterial and archaeal population are ubiquitously distributed in paddy fields (Mueller-Niggemann et al. 2016) along with cyanobacterial population.

Studies have demonstrated that soil properties and components undergo constant change at different rates under paddy cultivation conditions primarily due to interaction between rice plants, soil biochemistry, and microbes (Bodelier 2003; Murase et al. 2006). Microbial colonization in paddy fields depends on two main factors, first efficiently using limiting resources and next based on survival under anoxic conditions (Sugano et al. 2005; Murase et al. 2006; Shrestha et al. 2007). As studied in most ecosystems, similar trend is followed in paddy soils wherein the primary colonization is by opportunistic microbes, which are replaced when favorable conditions are achieved (Sigler et al. 2002). At the later stages, greater species diversity may be achieved (Shrestha et al. 2007). Similar to succession in volcanic deposits, *Betaproteobacteria* and *Gammaproteobacteria* are major colonizers in the primary stages later replaced by *Alphaproteobacteria*. *Actinobacteria*, CFB group, and spirochetes are also detected in the later stages of succession. The fungal succession is known to play a minor role in microbial succession in paddy soils and is not clearly studied yet (Nakamura et al. 2003; Sugano et al. 2005; Shrestha et al. 2007).

Studies on the microbial succession of paddy fields will allow typing of microbial communities based on location and cultivar. This information will be helpful in mapping the paddy field communities and changes in nutrient cycling associated with it. At the same time, the complete successional study in paddy fields can be completed in a few months time, thereby allowing for a much faster assessment of the entire successional process.

4.6.2 Primary Succession on Impact Sites

Impact craters have received little attention over the years. Over the years, 184 impact craters have been recorded over the world in diverse biomes like desserts, grasslands, forests, underwater, and the polar region (source: earth impact database). Impact events lead to the formation of sterile habitats due to the intense effect of heat and pressure, exposing substrates for superior microbial colonization at the impacted sites. Allogenic succession is mainly observed at impacted sites. The colonizing microflora shows a considerable difference after the impact. At Haughton impact site, the colonization by *Chroococcidiopsis* species indicated that impact sites could potentially be invaded and colonized by pioneering microorganisms in the earliest stages of primary succession (Grieves 1988). The colonizing microbes brought to the site depend upon the geological characteristics of the crater and are mainly transported to the site from the outlying region primarily through water and wind (Gronlund et al. 1990; Cockell and Lee 2002). Asteroids and comets that impact the target structure impose extreme physical conditions leading to development of fractures and increase the pore space. These fractures provide suitable conditions for microbial colonization by protecting them from external physical stressors such as the UV radiations. Increased *Cyanobacteria* colonization has been observed at impact sites. However, there is an irregularity in the colonization due to the heterogenous nature of available substrates postimpact (Cockell and Lee 2002). To reach climax communities at impact sites, it takes multiple episodes and numerous changes, much more as compared to other ecosystems, thus making impact sites much unique for studying microbial succession. It has been observed that vegetation cover is usually low at these impact sites and is mainly dominated by microbial processes as extreme physical conditions make it difficult for establishment of higher life (Cockell and Lee 2002).

4.6.3 Primary Succession Post Wildfires

Soil microorganisms are extremely sensitive to environmental changes; therefore, fires can majorly alter the colonizing microflora at such locations. Both heat and ash are important parameters involved in alteration of the soil ecosystems (Ferrenberg et al. 2013). Partial and total sterilization of soil microflora have been reported after forest, grassland, and chaparral fires, which makes it an entity for the study of primary succession (Vazquez et al. 1998). The total wipeout of existing microbes depends largely on the intensity of the fire. Slow burning, extremely hot fires can destroy most organic matter leading to initiation of primary succession on new substrates (Walker and del Moral 2003). The severity of the fire determines the extent of biogeochemical heterogeneity and environmental variation likely to occur and the rate at which recolonization will commence (Ferrenberg et al. 2013). Post fires the nutrient content is altered in a way to allow better colonization of bacteria over those of fungi and algae. High density of aerobic heterotrophic bacteria, mainly *Acidophiles*, is followed by spore bearers, but a significant decrease in fungal and cyanobacterial population has been observed. This shift may be due to the qualitative difference of the organic substrates post fires (Vazquez et al. 1998; Ferrenberg et al. 2013). In addition, fires result in ephemeral pulse of ammonia, creation of reactive charcoal layer, formation of metal oxides, and significant alterations in the pH, moisture, and temperature that allow bacteria to proliferate better (Ferrenberg et al. 2013). However, in the later stages of succession, there is a gradual equilibration due to leaching of ash minerals to the lower soil layers, decreased aeration, and soil compaction, which eventually lead to recreate soil conditions similar to the ones before fire. With the soil structure returning to its original stage over time, the effects of burning gradually disappear with soil age (Vazquez et al. 1998; Cadotte 2007).

4.7 Case Studies: Microbial Dynamics at Restoration Sites

Numerous land ecosystems have been severely damaged by human activities. Such severe loss of valuable natural ecosystems has emphasized on the importance of ecological restoration. Timely and effective restoration measures need to be considered in order to retrieve the ecosystem (Harris 2009). Soil microbes play critical roles in the development and maintenance of soil ecosystems; hence, they are crucial for reestablishing functioning and biodiversity of damaged ecosystems. Microbial characteristics now enable the categorization of degradation state and the effects of management practices in these ecosystems (Harris 2003). During restoration, microbial succession observed at post-mining sites and in species-rich grasslands is one of the main targets under study for ecosystem conservation bodies (Chodak et al. 2009; Harris 2009).

4.7.1 Primary Succession Following Restoration of Mining Sites

Physical and biological properties of soil are severely hampered by activities such as mining. Piling of large amounts of excavated soil material at mining sites leads to major destruction of plant cover and deficiency of organic matter and nutrients and makes the soil prone to erosion. This material typically exhibits very low biological activity and a degradation of stable soil structure. Mined sites typically remain barren for numerous years due to these conditions. Hence, the restoration of soil structure is essential for restoring ecosystem functions at these excavated sites. Reclamation measures involving microbial communities are therefore applied to accelerate ecosystem development and soil formation in post-mining sites. Mining sites are usually reclaimed for forestry in order to restore stable and productive forest ecosystems (Chodak et al. 2009). These reclaimed sites provide an ideal opportunity to study primary succession in terms of ecosystem reconstruction.

Soil microbial communities play a crucial role in development of functional ecosystems in post-mine soils and mine tailings (Chodak et al. 2009). The initial belowground colonizing population is indispensably involved in nutrient cycling and turnover of organic matter to mediate development of stable soil structure and quality (Jeffries et al. 2003). It has been studied that mining leads to high irregularity in the surface, therefore leading to high spatial and temporal heterogeneity in the colonizing microflora. Taxonomically diverse species of nitrogen-fixing diazotrophs assist in necessary reclamation for long-term stability of soils. *Proteobacteria*, *Cyanobacteria*, and *Firmicutes* are widely detected in the early stages as every other successional pattern observed in different ecosystems. Early fungal populations show a shorter mycelium than normal native colonizers. Microbial enzymes act as early indicators for reflecting the degree of rehabilitation process (Caravaca et al. 2003). Microbial enzymes are highly sensitive and specific to any alteration in nutrient and organic matter content as soils are reclaimed (Garcia et al. 1997; Kandeler et al. 1999). Presence of hydrolases observed on reclamation is an indication of improvement of soil fertility as hydrolases contribute to mineralization of nitrogen, phosphorus, and carbon in mined soils. Similarly, dehydrogenases are indicative of the increase in viable microbial population. Fungal populations significantly contribute in increasing the phosphatase activity of reclaimed soils. A similar trend of succession is observed wherein the microbial biomass, respiration, and metabolic quotient increase with soil age and are found to be the highest at the intermediate stages. The ratio of microbial biomass to the content of organic carbon diminishes in the later stages due to progressive accumulation of recalcitrant humic material from other sources (Chodak et al. 2009).

The diminishing microbial activity with age is highly linked with development of vegetation. In order to rehabilitate soils, post-mining, fast-growing, and exotic species are planted which contribute to nitrogen fixation and organic matter through exudates and litter, hence lowering the pressure on microbes. The vegetative cover determines the colonizing microbes. There is an increase in fungal to bacterial ratio after development of vegetation since fungi are dominant decomposers of plant litter. Moreover, the symbiotic relationship between arbuscular mycorrhizal fungi and plants has shown to increase the stability of soil microaggregates in post-mined site (Bearden and Petersen 2000; Baldrian et al. 2008). The results on the chronosequence on coal mine spoils (reclamation and revegetation sites) in China show that there is positive impact of reclamation and vegetation restoration on soil microbial diversity. Comparing young and old reclamation sites, Li et al. (2014) showed that microbial community recovery occurred between 15 and 20 years after reclamation. Such positive effects have also been observed in mine soils, grasslands, etc. (Chodak et al. 2009; Esperschutz et al. 2011).

In spite of progressive patterns observed in post-mining areas, complete restoration of metabolic abilities of microbial community is slow (Chodak et al. 2009) as it takes time to recreate the native structure. It has been observed that soil microbial communities are more similar to the surrounding non-mined soils during restoration. In contrast, fungal communities do not resemble the nearby surrounding communities due to their differential dispersal and colonizing abilities during restoration (Banning et al. 2011). Regardless, reclamation measures enhance gross microbial properties and promote development of metabolic capabilities typical as that for native soils (Baldrian et al. 2008; Chodak et al. 2009).

4.7.2 Primary Succession During Grassland Restoration

Grasslands and forest ecosystems have been altered for agricultural and development purposes. Agriculture mainly involves use of conventional tillage and the use of herbicides and weedicides that have numerous negative impacts on the soil.

These include disruption of the soil structure, lowering the organic matter, decreased microbial activity, altered nutrient supply and soil moisture, increased soil temperature fluctuations, and causing extreme wet and dry cycles in the soil (Calderon et al. 2001; Samson et al. 2004; Allison et al. 2005; Mckinley et al. 2005; Bach et al. 2010). The change in nutrient dynamics is largely due to the loss of nitrogen through denitrification, nitrate leaching, and carbon loss due to high carbon dioxide flux, which hampers the existing biological ecosystems (Calderon et al. 2001). The realization about conservation of endangered grasslands has led to an effort to restore these ecosystems to their original healthy state. However, the restoration of agricultural land is highly dependent on land use history. Cultivation has a lasting impact on soil communities, and it might take decades for them to recover. However, it has been noticed that discontinuation of cultivation and deliberate efforts to restore the ecosystem bring a positive change in the degraded soil parameters which makes reclamation and restoration of such ecosystems necessary (Knops and Tilman 2000; Mckinley et al. 2005).

It is a widely known fact that soil microbes play critical roles in development and maintenance of soil structures during restoration. They play a role in aggregate formation, carbon sequestration, nutrient turnover, pore space, and soil hydrology during restoration (Six et al. 2000; Bach et al. 2010). Microbial biomass increases over time during restoration with change in various variables like cessation of tillage, vegetation development, soil texture, and the composition of adjacent undisturbed communities (Baer et al. 2002; Allison et al. 2005; Mckinley et al. 2005; Bach et al. 2010; Jangid et al. 2011). Restoration studies have shown that agricultural practices majorly affect only the uppermost layers of soil microbial communities. Immigration and interaction between microbial groups at early stages significantly contribute to restoration (Potthoff et al. 2006; Bach et al. 2010; Jangid et al. 2011). Cessation of agricultural practices show an abundance of Gram-positive bacteria such as *Bacillus* species. Gram-negative bacteria and fungi are seen to increase in the later stages of succession as vegetation develops. *Actinomycetes* are seen to dominate restored grassland soils. They can be classified as a part of autochthonous soil microbial communities that can survive and maintain themselves in lower carbon availability, anaerobic conditions, and environmental fluctuations, therefore having an advantage over Gram-negative bacterial and fungal communities (Mckinley et al. 2005; Potthoff et al. 2006). The bacteria/fungi ratio remains high during the initial restoration as tillage severely damages fungal hyphae and mycorrhizal communities. However, the fungal biomass improves with age and development of vegetation (Calderon et al. 2001; Mckinley et al. 2005).

Vegetation facilitates the recovery of soil environment and microbial communities. After increase in plant litter and exudates, greater microbial biomass and decomposition rates are observed. Plant species and their products have variedly been observed in contributing and determining soil development microbes in diverse ecosystems (Bardgett and Shine 1999; Mckinley et al. 2005; Potthoff et al. 2006). However, restored soils show lesser dependence of colonizing microbes on aboveground communities (Bach et al. 2010; Jangid et al. 2010; Marshall et al. 2011).

In addition, soil texture plays a very important role in restoration of disturbed agricultural soils. It has been observed that clayey soil remnants of post-agricultural practices are more effective in increasing microbial biomass on restoration than sandy remnants. This is because clayey soils have greater available carbon pools, and they provide protection to microbes from predation, desiccation, heat, and pH fluctuations, allowing for a greater microbial recovery during grassland restoration (Mc Lauchlan 2006; Bach et al. 2010). Other factors such as increase in microbial enzymes and decrease in the emission of greenhouse gases with time are an indication of progressing restoration (Drijber et al. 2000). The changes in the due course of restoration suggest that microbial communities tend to become more like native soils with time. Restoration generally shifts them in the direction of native soils but may not be able to recreate native conditions for a long span (Mckinley et al. 2005; Bach et al. 2010).

Conclusion

The importance of studying microbial ecology in ecosystem succession is unquestionable. The existing lacunae in the study of microbial ecology due to technological barrier are now overcome with the advent of newer methods of DNA isolation and innovations of various kits along with the breakthrough in the sequencing technology. It has helped in yielding new insights on the previously hidden microbial flora. Presently, the low cost of sequencing per read has afforded in-depth analysis of microbes, even rare ones, in an ecosystem. In successional ecosystems, such as paddy field or with extreme environment like deglaciated land and impact sites, the ecosystem shows changes from low-nutrient to high-nutrient status concomitantly leading to a change in microbial communities from archaea and bacteria to fungi followed by plants, from low microbial biomass to high microbial biomass, and from low microbial diversity to high microbial diversity. The ability to sequence more and statistically analyze metagenomic data has also helped in carrying out continental scale studies. While attributing function to microbes has also helped in functional genomics of an ecosystem, it is still limited due to the non-availability of cultured representatives of these microbes. If experiments leading to cultivation of the interesting microbes go hand in hand, these are likely to offer unequivocal evidence about the role of microbial communities during development of an ecosystem. This does not undermine the role of computational programs, which are helping us to analyze the data with reasonable amount of precision. It has helped to compare the data from various ecosystems which are geographically separated at continental scale but share very common climate systems or ecotypes. Such systematic analyses carried out at continental scales are likely to significantly contribute in developing models and theories of microbial community succession and possibly predict the future course of communities that are under continuous threat due to the changing climate.

Acknowledgments This work was supported by the Department of Biotechnology (DBT; Grant no. BT/PR/0054/NDB/52/94/2007), Government of India, under the project "Establishment of microbial culture collection." KJ acknowledges the critical inputs from Profs. David C. Coleman, William "Barny" Whitman, and Mark A. Williams throughout his research that helped decipher the successional patterns shown by soil microbial communities during restoration and/or development of several ecosystems.

References

Alfredsen G, Høiland K (2001) Succession of terrestrial macrofungi along a deglaciation gradient at Glacier Bisen, South Norway. Nor J Bot 21:19–37. doi:10.1111/j.1756-1051.2001.tb01336.x

Allison SD, Martiny JBH (2007) Resistance, resilience and redundancy in microbial communities. Proc Natl Acad Sci U S A 105:11512–11519. doi:10.1073/pnas.0801925105

Allison V, Miller M, Jastrow J, Matamala R, Zak D (2005) Changes in soil microbial community structure in a Tallgrass Prairie chronosequence. Soil Sci Soc Am J 69:1412–1421. doi:10.2136/sssaj2004.0252

Armas C, Padilla FM, Pugnaire FI, Jackson RB (2010) Hydraulic lift and tolerance to salinity of semiarid species: consequences for species interactions. Oecologia 162:11–21. doi:10.1007/s00442-009-1447-1

Bach E, Baer S, Meyer C, Six J (2010) Soil texture affects soil microbial and structural recovery during grassland restoration. Soil Biol Biochem 42:2182–2191. doi:10.1016/j.soilbio.2010.08.014

Baer SG, Kitchen DJ, Blair JM, Rice CW (2002) Changes in ecosystem structure and function along a chronosequence of restored grasslands. Ecol Appl 12:1688–1701. doi:10.1890/1051-0761(2002)012[1688:CIESAF]2.0.CO;2

Baldrian P, Trogl J, Frouz J, Snajdr J, Valaskova V, Merhautova V, Cajthaml T, Herinkova J (2008) Enzyme activities and microbial biomass in topsoil layer during spontaneous succession in spoil heaps after brown coal mining. Soil Biol Biochem 40:2107–2115. doi:10.1016/j.soilbio.2008.02.019

Banning NC, Gleeson DB, Grigg AH, Grant CD, Andersen GL, Brodie EL, Murph DV (2011) Soil microbial community successional patterns during forest ecosystem Restoration. Appl Environ Microbiol 77:6158–6164. doi:10.1128/AEM.00764-11

Bardgett RD, Shine A (1999) Linkages between plant litter diversity, soil microbial biomass and ecosystem function in temperate grasslands. Soil Biol Biochem 31:317–321

Bardgett RD, Bowman WD, Kaufmann R, Schmidt SK (2005) A temporal approach to linking aboveground and belowground ecology. Trends Ecol Evol 20:634–641. doi:10.1016/j.tree.2005.08.005

Barns SM, Takala SL, Kuske CR (1999) Wide distribution and diversity of members of the bacterial kingdom *Acidobacterium* in the environment. Appl Environ Microbiol 65:1731–1737

Bearden BN, Petersen L (2000) Influence of arbuscular mycorrhizal fungi on soil structure and aggregate stability of vertisols. Plant Soil 218:173–183. doi:10.1023/A:1014923911324

Bekku Y, Kum A, Nakatsubo T, Masuzawa T, Kanda H, Koizu H (1999) Microbial biomass in relation to primary succession on Artic deglaciated moraines. Polar Biosci 12:47–53

Bernasconi SM, Bauder A, Bourdon B et al (2011) Chemical and biological gradients along the Damma glacier soil chronosequence, Switzerland. Vadose Zone J 10:867–883. doi:10.2136/vzj2010.0129

Besemer K, Moeseneder MM, Arrieta JM, Herndl GJ, Peduzzi P (2005) Complexity of bacterial communities in a river-floodplain system (Danube, Austria). Appl Environ Microbiol 71:609–620. doi:10.1128/AEM.71.2.609-620.2005

Bhatia M, Sharp M, Foght J (2006) Distinct communities exist beneath a High Arctic polythermal glacier. Appl Environ Microbiol 72:5838–5845. doi:10.1128/AEM.00595-06

Blaalid R, Carlsen T, Kumar S, Halvorsen R, Ugland K, Fontana G, Kauserud S (2012) Changes in the root-associated fungal communities along a primary succession gradient analysed by 454 pyrosequencing. Mol Ecol 21:1897–1908. doi:10.1111/j.1365-294X.2011.05214.x

Bodelier PLE (2003) Interactions between oxygen-releasing roots and microbial processes in flooded soils and sediments. Root Ecol 168:331–362. doi:10.1007/978-3-662-09784-7_13

Bossio DA, Girvan MS, Verchot L, Bullimore J, Borelli T, Albrecht A, Scow KM, Ball AS, Pretty JN, Osborn AM (2005) Soil microbial community response to land use change in an agricultural landscape of Western Kenya. Microb Ecol 49:50–62. doi:10.1007/s00248-003-0209-6

Bowman WD, Steltzer H, Rosenstiel TN, Cleveland CC, Meier CL (2004) Litter effects of two co-occurring alpine species on plant growth, microbial activity and immobilization of nitrogen. Oikos 104:336–344

Bradley JA, Arndt S, Šabacká M, Benning LG, Barker GL, Blacker JJ, Yallop ML, Wright KE, Bellas CM, Telling J, Tranter M, Anesio A (2016) Microbial dynamics in a High-Arctic glacier forefield: a combined field, laboratory, and modelling approach. Biogeosciences 13:5677–5696. doi:10.5194/bg-13-5677-2016

Brown S, Jumpponen A (2014) Contrasting primary successional trajectories of fungi and bacteria in retreating glacier soils. Mol Ecol 23:481–497. doi:10.1111/mec.12487

Buckley DH, Huangyutitham V, Hsu SF, Nelson TA (2007) Stable isotope probing with $^{15}N_2$ reveals novel noncultivated diazotrophs in soil. Appl Environ Microbiol 73:3196–3204. doi:10.1128/AEM.02610-06

Cadotte MW (2007) Competition-colonization trade-offs and disturbance effects at multiple scales. Ecology 88:823–829. doi:10.1890/06-1117

Calderon FJ, Jackson LE, Scow KM, Rolston DE (2001) Short-term dynamics of nitrogen, microbial activity, and phospholipid fatty acids after tillage. Soil Sci Soc Am J 65:118–126. doi:10.2136/sssaj2001.651118x

Caravaca F, Alguacil MM, Figueroa D, Barea JM, Roldan A (2003) Re-establishment of *Retama sphaerocarpa* as a target species for reclamation of soil physical and biological properties in a semiarid Mediterranean area. For Ecol Manag 182:49–58. doi:10.1016/S0378-1127(03)00067-7

Cherif M, Loreau M (2007) Stoichiometric constraints on resource use, competitive interactions, and elemental cycling in microbial decomposers. Am Nat 169:709–724. doi:10.1086/516844

Chodak M, Pietrzykowski M, Niklinska M (2009) Development of microbial properties in a chronosequence of sandy mine soils. Appl Soil Ecol 41:259–268. doi:10.1016/j.apsoil.2008.11.009

Cockell CS, Lee P (2002) The biology of impact craters–a review. Biol Rev 77:279–310. doi:10.1017/S146479310100584X

Conrad R, Frenzel P (2002) Flooded soils. In: Bitton G (ed) The encyclopedia of environmental microbiology. Wiley, New York, pp 1316–1333. doi:10.1002/0471263397.env034

Coomes D, Bentley W, Tanentzap A, Burrows L (2013) Soil drainage and phosphorus depletion contribute to retrogressive succession along a New Zealand chronosequence. Plant Soil 367:77–91. doi:10.1007/s11104-013-1649-5

Crews TE, Kitayama K, Fownes JH, Riley RH, Herbert DA, Mueller-Dombois D, Vitousek PM (1995) Changes in soil phosphorus fractions and ecosystem dynamics across a long chronosequence in Hawaii. Ecology 76:1407–1424

Crews TE, Kurina LM, Vitousek PM (2001) Organic matter and nitrogen accumulation and nitrogen fixation during early ecosystem development in Hawaii. Biogeochemistry 52:259–279. doi:10.1023/A:1006441726650

Cutler N (2010) Long-term primary succession: a comparison of non-spatial and spatially explicit inferential techniques. Plant Ecol 208:123–136. doi:10.1007/s11258-009-9692-2

Cutler NA, Chaput DL, van der Gast CJ (2014) Long-term changes in soil microbial communities during primary succession. Soil Biol Biochem 69:359–370. doi:10.1016/j.soilbio.2013.11.022

Dennis PG, Miller AJ, Hirsch PR (2010) Are root exudates more important than other sources of rhizodeposits in structuring rhizosphere bacterial communities? FEMS Microbiol Ecol 72:313–327. doi:10.1111/j.1574-6941.2010.00860.x

Drijber RA, Doran JW, Parkhurst AM, Lyon DJ (2000) Changes in soil microbial community structure with tillage under long-term wheat-fallow management. Soil Biol Biochem 32:1419–1430

Edwards A, Cook S (2015) Microbial dynamics in glacier forefield soils show succession is not just skin deep. Mol Ecol 24:963–966. doi:10.1111/mec.13098

Eller G, Kruger M, Frenzel P (2005) Comparing field and microcosm experiments: a case study on methano- and methylotrophic bacteria in paddy soil. FEMS Microbiol Ecol 51:279–291. doi:10.1016/j.femsec.2004.09.007

Esperschutz J, Perez-de-Mora A, Schreiner K, Welzl G, Buegger F, Zeyer J, Hagedorn F, Munch JC, Schloter M (2011) Microbial food web dynamics along a soil chronosequence of a glacier forefield. Biogeosciences 8:3283–3294. doi:10.5194/bg-8-3283-2011

Ferrenberg S, ONeill SP, Knelman JE, Todd B, Duggan S, Bradley D, Robinson T, Schmidt SK, Townsend AR, Williams MW, Cleveland CC, Melbourne BA, Jiang L, Nemergut DR (2013) Changes in assembly processes in soil bacterial communities following a wildfire disturbance. ISME J 7:1102–1111. doi:10.1038/ismej.2013.11

Fierer N, Nemergut D, Knight R, Craine JM (2010) Changes through time: integrating microorganisms into the study of succession. Res Microbiol 161:635–642. doi:10.1016/j.resmic.2010.06.002
Forster S, Nicolson T (1981) Microbial aggregation of sand in a Maritime Dune Succession. Soil Biol Biochem 13:205–208
Fujimura R, Sato Y, Nishizawa T, Nanba K, Oshima K, Hattori M, Kamijo T, Ohta H (2012) Analysis of early bacterial communities on volcanic deposits on the Island of Miyake (Miyakejima), Japan: a 6-year study at a fixed site. Microbes Environ 27:19–29. doi:10.1264/jsme2.ME11207
Fukami T, Morin PJ (2003) Productivity-biodiversity relationships depend on the history of community assembly. Nature 424:423–426. doi:10.1038/nature01785
Garcia C, Hernandez T, Costa F (1997) Potential use of dehydrogenase activity as an index of microbial activity in degraded soils. Commun Soil Sci Plant Anal 28:123–134
Gomez-Alvarez V, King GM, Nusslein K (2007) Comparative bacterial diversity in recent Hawaiian volcanic deposits of different ages. FEMS Microbiol Ecol 60:60–73. doi:10.1111/j.1574-6941.2006.00253.x
Grieves RAF (1988) The Haughton impact structure: summary and synthesis of the results of the HISS project. Meteorit Planet Sci 23:249–254
Gronlund T, Lortie G, Guibault J, Bouchard M, Sanissto M (1990) Diatoms and arcellaceans from Lac du Cratere du Nouveau-Quebec, Ungava, Quebec, Canada. Can J Bot 68:1187–1200
Grootjans AP, van den Ende FP, Walsweer AF (1997) The role of microbial mats during primary succession in calcareous dune slacks: an experimental approach. J Coast Conserv 3:95–102
Harris JA (2003) Measurements of the soil microbial community for estimating the success of restoration. Eur J Soil Sci 54:801–808. doi:10.1046/j.1351-0754.2003.0559.x
Harris J (2009) Ecology: facilitators or followers? Soil microbial communities and restoration. Science 325:573–574. doi:10.1126/science.1172975
Hopkins DW, Badalucco L, English LC, Meli SM, Chudek JA, Ioppolo A (2007) Plant litter decomposition and microbial characteristics in volcanic soils (Mt Etna, Sicily) at different stages of development. Biol Fertil Soils 43:461–469. doi:10.1007/s00374-006-0124-3
Huggett R (1998) Soil chronosequences, soil development, and soil evolution: a critical review. Catena 32:155–172
Jangid K, Williams MA, Franzluebbers AJ, Blair JM, Coleman DC, Whitman WB (2010) Development of soil microbial communities during tall grass prairie restoration. Soil Biol Biochem 42:302–312. doi:10.1016/j.soilbio.2009.11.008
Jangid K, Williams MA, Franzluebbers AJ, Schmidt TM, Coleman DC, Whitman WB (2011) Land-use history has a stronger impact on soil microbial community composition than aboveground vegetation and soil properties. Soil Biol Biochem 43:2184–2193. doi:10.1016/j.soilbio.2011.06.022
Jangid K, Whitman W, Condron L, Turner B, Williams M (2013) Progressive and retrogressive ecosystem development coincide with soil bacterial community change in a dune system under lowland temperate rainforest in New Zealand. Plant Soil 367:235–247. doi:10.1007/s11104-013-1720-2Jeffries P, Gianinazzi S, Perotto S, Turnau K, Barea JM (2003) The contribution of arbuscular mycorrhizal fungi in sustainable maintenance of plant health and soil fertility. Biol Fertil Soils 37:1–16. doi:10.1007/s00374-002-0546-5
Johansen A, Finlay RD, Olsson PA (1997) Nitrogen metabolism of external hyphae of the arbuscular mycorrhizal fungus *Glomus intraradices*. New Phytol 133:705–712
Johnson EA, Miyanishi K (2008) Testing the assumptions of chronosequences in succession. Ecol Lett 11:419–431. doi:10.1111/j.1461-0248.2008.01173.x
Jumpponen A (2003) Soil fungal community assembly in a primary successional glacier forefront ecosystem as inferred from rDNA sequence analyses. New Phytol 158:569–578. doi:10.1046/j.1469-8137.2003.00767.x
Kandeler E, Tscherko D, Spiegel H (1999) Long-term monitoring of microbial biomass. Biol Fertil Soils 28:343–351
Kastovska K, Elster J, Stibal M, Santruckova H (2005) Microbial assemblages in soil microbial succession after glacial retreat in Svalbard (high Arctic). Microb Ecol 50:396–407. doi:10.1007/s00248-005-0246-4

Kaye JP, Hart SC (1997) Competition for nitrogen between plants and soil microorganisms. Trends Ecol Evol 12:139–143

Kikvidze Z, Armas C, Fukuda K, Martinez-Garcia L, Miyata M, Oda-Tanaka A, Pugnaire F, Wu B (2010) The role of arbuscular mycorrhizae in primary succession: differences and similarities across habitats. Web Ecol 10:50–57. doi:10.5194/we-10-50-2010

King GM (2003) Contributions of atmospheric CO and hydrogen uptake to microbial dynamics on recent Hawaiian volcanic deposits. Appl Environ Microbiol 69:4067–4075. doi:10.1128/AEM.69.7.4067-4075.2003

Kirk J, Beaudette L, Hart M, Moutoglis P, Klironomos J, Lee H, Trevors J (2004) Methods of studying soil microbial diversity. J Microbiol Methods 58:169–188. doi:10.1016/j.mimet.2004.04.006

Knelman JE, Legg TM, O'Neill SP, Washenberger CL, González A, Cleveland CC, Nemergut DR (2012) Bacterial community structure and function change in association with colonizer plants during early primary succession in a glacier forefields. Soil Biol Biochem 46:172–180. doi:10.1016/j.soilbio.2011.12.001

Knops JMH, Tilman D (2000) Dynamics of soil nitrogen and carbon accumulation for 61 years after agricultural abandonment. Ecology 81:88–98. doi:10.1890/0012-9658(2000)081[0088:DOSNAC]2.0.CO;2

Krause S, Luke C, Frenzel P (2010) Succession of methanotrophs in oxygen-methane counter-gradients of flooded rice paddies. ISME J 4:1603–1607. doi:10.1038/ismej.2010.82

Langhans TM, Storm C, Schwabe A (2009) Community assembly of biological soil crusts of different successional stages in a temperate sand ecosystem, as assessed by direct determination and enrichment techniques. Microb Ecol 58:394–407. doi:10.1007/s00248-009-9532-x

Lazzaro A, Franchini AG, Brankatschk R, Zeyer J (2010) Pioneer communities in the forefields of retreating glaciers: how microbes adapt to a challenging. In: Current research, technology and education topics in applied microbiology and microbial biotechnology. Formatex, Badajoz, pp 43–52

Li Y, Wen H, Chen L, Yin T (2014) Succession of bacterial community structure and diversity in soil along a chronosequence of reclamation and re-vegetation on coal mine spoils in China. PLoS One 9:e115024. doi:10.1371/journal.pone.0115024

Liu R, Li K, Zhang H, Zhu J, Joshi D (2014) Spatial distribution of microbial communities associated with dune landform in the Gurbantunggut Desert, China. J Microbiol 52:898–907. doi:10.1007/s12275-014-4075-3

Lukesova A (2001) Soil algae in brown coal and lignite post-mining areas in Central Europe (Czech Republic and Germany). Restor Ecol 9:341–350. doi:10.1046/j.1526-100X.2001.94002.x

Ma X, Liu M, Li Z (2016) Shifts in microbial biomass and community composition in subtropical paddy soils under a gradient of manure amendment. Biol Fertil Soils 52:775–787. doi:10.1007/s00374-016-1118-4

Marshall CB, McLaren JR, Turkington R (2011) Soil microbial communities resistant to changes inplant functional group composition. Soil Biol Biochem 43:78–85

Mc Lauchlan K (2006) The nature and longevity of agricultural impacts on soil carbon and nutrients: a review. Ecosystems 9:1364–1382. doi:10.1007/s10021-005-0135-1

McCann CM, Wade MJ, Gray ND, Roberts JA, Hubert CRJ, Graham DW (2016) Microbial communities in a high arctic polar desert landscape. Front Microbiol 7:419. doi:10.3389/fmicb.2016.00419

McKinley VL, Peacock AD, White DC (2005) Microbial community PLFA and PHB responses to ecosystem restoration in tallgrass prairie soils. Soil Biol Biochem 37:1946–1958. doi:10.1016/j.soilbio.2010.09.016

Merila P, Strommerb R, Fritzc H (2002) Soil microbial activity and community structure along a primary succession transect on the land-uplift coast in western Finland. Soil Biol Biochem 34:1647–1654. doi:10.1016/S0038-0717(02)00148-7

Mueller-Niggemann C, Utami SR, Marxen A, Mangelsdorf K, Bauersachs T, Schwark L (2016) Distribution of tetraether lipids in agricultural soils–differentiation between paddy and upland management. Biogeosciences 13:1647–1666. doi:10.5194/bg-13-1647-2016

Muller AK, Westergaard K, Christensen S, Sorensen SJ (2001) The effect of long-term mercury pollution on the soil microbial community. FEMS Microbiol Ecol 36:11–19. doi:10.1111/j.1574-6941.2001.tb00821.x

Murase J, Noll M, Frenzel P (2006) Impact of protists on the activity and structure of the bacterial community in a rice field soil. Appl Environ Microbiol 72:5536–5444. doi:10.1128/AEM.00207-06

Nakamura A, Tun CC, Asakawa S, Kimura M (2003) Microbial community responsible for the decomposition of rice straw in a paddy field: estimation by phospholipids fatty acid analysis. Biol Fertil Soils 38:288–295. doi:10.1007/s00374-003-0658-6

Nemergut DR, Anderson SP, Cleveland CC, Martin AP, Miller AE, Seimon A, Schmidt SK (2006) Microbial community succession in an unvegetated, recently deglaciated soil. Microb Ecol 53:110–122. doi:10.1007/s00248-006-9144-7

Nicol GW, Glover LA, Prosser JI (2003) Spatial analysis of archaeal community structure in grassland soil. Appl Environ Microbiol 69:7420–7429. doi:10.1128/AEM.69.12.7420-7429.2003

Nicol GW, Leininger S, Schleper C, Prosser JI (2008) The influence of soil pH on the diversity, abundance and transcriptional activity of ammonia oxidizing archaea and bacteria. Environ Microbiol 10:2966–2978. doi:10.1111/j.1462-2920.2008.01701.x

Noll M, Matthies D, Frenzel P, Derakshani M, Liesack W (2005) Succession of bacterial community structure and diversity in a paddy soil oxygen gradient. Environ Microbiol 7:382–395. doi:10.1111/j.1462-2920.2005.00700.x

Nyberg G, Tobella B, Kinyangi J, Ilstedt U (2012) Soil property changes over a 120-yr chronosequence from forest to agriculture in western Kenya. Hydrol Earth Syst Sci 16:2085–2094. doi:10.5194/hess-16-2085-2012

Ohtonen R, Fritze H, Pennanen T, Jumpponen A, Trappe J (1999) Ecosystem properties and microbial community changes in primary succession on a glacier forefront. Oecologia 119:239–246

Okabe S, Odagiri M, Ito T, Satoh H (2007) Succession of sulfur-oxidizing bacteria in the microbial community on corroding concrete in sewer systems. Appl Environ Microbiol 73:971–980. doi:10.1128/AEM.02054-06

Pennanen T, Stroemmer R, Markkola A, Fritze H (2001) Microbial and plant community structure across a primary succession gradient. Scand J For Res 16:37–43. doi:10.1080/028275801300004398

Potthoff M, Steenwerth KL, Jackson LE, Drenovsky RE, Scow KM, Joergensen RG (2006) Soil microbial community composition as affected by restoration practices in California grassland. Soil Biol Biochem 38:1851–1860. doi:10.1016/j.soilbio.2005.12.009

Pugnaire FI, Luque MT, Armas C, Gutierrez L (2006) Colonization processes in semi-arid Mediterranean old-fields. J Arid Environ 65:591–603. doi:10.1016/j.jaridenv.2005.10.002

Rao S, Chan Y, Bugler-Lacap DC, Bhatnagar A, Bhatnagar M, Pointing SB (2016) Microbial diversity in soil, Sand Dune and Rock Substrates of the Thar Monsoon Desert, India. Indian J Microbiol 56:35–45. doi:10.1007/s12088-015-0549-1

Rime T, Hartmann M, Brunner I et al (2015) Vertical distribution of the soil microbiota along a successional gradient in a glacier forefield. Mol Ecol 24:1091–1108. doi:10.1111/mec.13051

Ronca S, Ramond J-B, Jones BE, Seely Mand Cowan DA (2015) Namib Desert dune/inter dune transects exhibit habitat-specific edaphic bacterial communities. Front Microbiol 6:845. doi:10.3389/fmicb.2015.00845

Saleh-Lakha S, Miller M, Campbell RG, Schneider K, Elahimanesh P, Hart MM, Trevors JT (2005) Microbial gene expression in soil: methods, applications and challenges. J Microbiol Methods 63:1–19. doi:10.1016/j.mimet.2005.03.007

Samson F, Knopf F, Ostlie W (2004) Great Plains ecosystems: past, present, and future. Wildl Soc Bull 32:6–15. doi:10.2193/0091-7648(2004)32[6:GPEPPA]2.0.CO;2

Sato Y, Nishihara H, Yoshida M, Watanabe M, Rondal JD, Ohta H (2004) Occurrence of hydrogen-oxidizing Ralstonia species as primary microorganisms in the Mt. Pinatubo volcanic mudflow deposits. Soil Sci Plant Nutr 50:855–861. doi:10.1080/00380768.2004.10408546

Schmidt K, Reed S, Nemergut D, Grandy A, Cleveland C, Weintraub M, Hill A, Costello E, Meyer A, Neff J, Martin A (2008) The earliest stages of ecosystem succession in high-elevation (5000 metres above sea level), recently deglaciated soils. Proc R Soc 275:2793–2802. doi:10.1098/rspb.2008.0808

Schutte MEU, Abdo Z, Bent S, Williams C, Schneider M, Solheim B, Forney L (2009) Bacterial succession in a glacier foreland of the High Arctic. ISME J 3:1258–1268. doi:10.1038/ismej.2009.71

Shrestha P, Noll M, Liesack W (2007) Phylogenetic identity, growth-response time and rRNA operon copy number of soil bacteria indicate different stages of community succession. Environ Microbiol 9:2464–2474. doi:10.1111/j.1462-2920.2007.01364.x

Siddique MAM, Barua P, Ghani MH (2012) Comparative study of physico-chemical properties of soil according to the age of aquaculture pond of Bangladesh. Mesopot J Mar Sci 27:29–38
Sigler WV, Crivii S, Zeyer J (2002) Bacterial succession in glacial forefield soils characterized by community structure, activity and opportunistic growth dynamics. Microb Ecol 44:306–316. doi:10.1007/s00248-002-2025-9
Six J, Elliott ET, Paustian K (2000) Soil macroaggregate turnover and microaggregate formation: a mechanism for C sequestration under no-tillage agriculture. Soil Biol Biochem 32:2099–2103. doi:10.1016/S0038-0717(00)00179-6
Soliveres S, van der Plas F, Manning P et al (2016) Biodiversity at multiple trophic levels is needed for ecosystem multifunctionality. Nature 536:456–459. doi:10.1038/nature19092
Stevens PR, Walker TW (1970) The chronosequence concept and soil formation. Q Rev Biol 45:333–350
Sugano A, Tsuchimoto H, Tun CC, Asakawa S, Kimura M (2005) Succession and phylogenetic profile of eubacterial communities in rice straw incorporated into a rice field: estimation by PCR-DGGE analysis. Soil Sci Plant Nutr 51:51–60. doi:10.1111/j.1747-0765.2005.tb00006.x
Tarlera S, Jangid K, Ivester AH, Whitman WB, Williams MA (2008) Microbial community succession and bacterial diversity in soils during 77,000 years of ecosystem development. FEMS Microbiol Ecol 64:129–140. doi:10.1111/j.1574-6941.2008.00444.x
Van der Heijden MGA, Bardgett RD, Van Straalen NM (2008) The unseen majority: soil microbes as drivers of plant diversity and productivity in terrestrial ecosystems. Ecol Lett 11:296–310. doi:10.1111/j.1461-0248.2007.01139.x
Vazquez G, Moreno-Casasola P, Barrera O (1998) Interaction between algae and seed germination in tropical dune slack species: a facilitation process. Aquat Bot 60:409–416. doi:10.1016/S0304-3770(97)00087-9
Wagner D, Lipski A, Embacher A, Gattinger A (2005) Methane fluxes in permafrost habitats of the Lena Delta: effects of microbial community structure and organic matter quality. Environ Microbiol 7:1582–1592. doi:10.1111/j.1462-2920.2005.00849.x
Walker L, del Moral R (2003) Primary succession and ecosystem rehabilitation. Cambridge University Press, Cambridge
Walker LR, Wardle DA, Bardgett RD, Clarkson BD (2010) The use of chronosequences in studies of ecological succession and soil development. J Ecol 98:725–736. doi:10.1111/j.1365-2745.2010.01664.x
Wang B, Liu GB, Xue S, Zhu B (2011) Changes in soil physico-chemical and microbiological properties during natural succession on abandoned farmland in the Loess Plateau. Environ Earth Sci 62:915–925. doi:10.1007/s12665-010-0577-4
Welc M, Banemann E, Fliebbach A, Frossard E, Jansa J (2012) Soil bacterial and fungal communities along a soil chronosequence assessed by fatty acid profiling. Soil Biol Biochem 49:184–192. doi:10.1016/j.soilbio.2012.01.032
Williams MA, Jangid K, Shanmugam SG, Whitman WB (2013) Bacterial communities in soil mimic patterns of vegetative succession and ecosystem climax but are resilient to change between seasons. Soil Biol Biochem 57:749–757. doi:10.1016/j.soilbio.2012.08.023
Xu X, Inubushi K, Sakamoto K (2006) Effect of vegetations and temperature on microbial biomass carbon and metabolic quotients of temperate volcanic forest soils. Geoderma 136:310–319. doi:10.1016/j.geoderma.2006.03.045
Zhang B, Wu X, Zhang G, Li S, Zhang W, Chen X, Sun L, Zhang B, Liu G, Chen T (2016) The diversity and biogeography of the communities of *Actinobacteria* in the forelands of glaciers at a continental scale. Environ Res Lett 11:054012. doi:10.1088/1748-9326/11/5/054012
Zumsteg A, Luster J, Goransson H, Smittenberg R, Brunner I, Bernasconi S, Zeyer J, Frey B (2012) Bacterial, archaeal and fungal succession in the forefield of a receding glacier. Microb Ecol 63:552–564. doi:10.1007/s00248-011-9991-8

Bacterial Diversity in Cold Environments of Indian Himalayas

5

Ramesh Chand Kasana

Abstract

The remote cold environments of Indian Himalayas are witnessed by extreme situations with a lot of variations in temperatures, low availability of water and nutrients and exposure to a lot of radiations. These extreme environments generally considered unfavourable to growth and survival of plants and animals are usually colonized by the microorganisms capable of growth and survivability under the prevailing severe conditions. Because of the extremophilic enzymes, proteins and biomolecules possessed by cold-adapted microorganism, they are of importance for industry, agriculture and biotechnology. In this chapter, (1) diversity of bacteria present in cold environments based on culturing and metagenomics approaches, (2) microorganisms from cold environments in agriculture, (3) novel bacteria from cold environments and (4) genome sequencing of bacteria from cold environments have been discussed. Bacteria affiliated to various phyla like *Proteobacteria*, *Firmicutes*, *Actinobacteria*, *Bacteroidetes*, *Acidobacteria*, *Gemmatimonadetes*, *Planctomycetes*, *Chlamydiae*, *Chlorobi*, *Chloroflexi*, *Dictyoglomi*, *Fibrobacteres*, *Nitrospirae* and *Verrucomicrobia* have been reported from the Indian Himalayas. Microorganisms belonging to various genera for improving agriculture production under cold environment have been isolated and identified. Twenty-one novel species of bacteria have been isolated from different locations in the cold environments of the Himalayas. A genome of 18 strains isolated from these cold environments has been sequenced and published.

Keywords

Low-temperature • Psychrotrophs • Genome sequencing • Novel bacteria • Metagenomics

R.C. Kasana
ICAR-Central Arid Zone Research Institute, Jodhpur 342 003, Rajasthan, India
e-mail: rameshkasana@yahoo.co.in

V.C. Kalia et al. (eds.), *Mining of Microbial Wealth and MetaGenomics*,
DOI 10.1007/978-981-10-5708-3_5

5.1 Introduction

In the world, most biodiversity does not exist as animals or plants, but as microorganisms mainly bacteria and archaea. The environments which are too harsh for normal life to exist such as hot springs, cold areas and deserts, acidic springs, saline and/or alkaline lakes and the ocean beds are also prevalent in nature and are known as extreme environments. Species diversity in these extreme environments is mainly limited to microorganisms. Among the extreme environments, cold areas are major one with exceptionally low-temperature areas of Arctic as well as Antarctic and low-temperature areas such as the high reaches in the mountains (Rashid et al. 1999). More than 80% of the biosphere has temperature below 5 °C (Brenchley 1996). The psychrophilic/psychrotrophic bacteria are virtually present in all these areas. These organisms play an important role in biological and biochemical process, and functioning of all the ecosystems, but we know very little about the identity and characteristics of microorganisms in the natural cold environments. The cold environments/deserts presents extreme variation in climate due to dropping of temperature in winters much below 0 °C and a lot of difference in diurnal temperature, very little rain and high intensity of solar radiations. In these areas generally there is very little rainfall and humidity is also very low throughout the year. Microorganisms present in the cold environments particularly in cold deserts and glaciers are generally exposed to high light and ultraviolet radiations and seasonal fluctuations in temperature. These microorganisms inhabiting those extreme environments can serve as potential bioresource for producing the commercially and industrially important enzymes, secondary metabolites and environmental bioremediations (Kasana 2010). This article describes the culturable and non-culturable bacterial diversity, use of cold-adapted microorganisms in agriculture and novel bacteria and genome sequences of microorganisms from the cold environments of Indian Himalayas. A pictorial view of few locations in the cold environments/deserts in Lahaul-Spiti, Himachal Pradesh, India, is given (Fig. 5.1).

5.2 Bacterial Diversity in Cold Environments by Culture-Dependent Approach

As assumed earlier, the cold environments are no longer barrier to microbial life, because microorganisms belonging to various groups survive and grow in these environments. Microorganisms thriving in the cold environments are known as psychrotolerant/psychrotroph, meaning that although they are capable of growth at low temperature (5 °C), they grow better at high temperatures, usually 25–35 °C, while psychrophiles are defined as the microorganisms which grow optimally at 15 °C or below and cannot grow at temperature higher

Fig. 5.1 Cold environments/deserts in Lahaul-Spiti, Himachal Pradesh, India

than 20 °C. Exploration of bacterial diversity in the cold environments has become very important due to the presence of novel microorganisms with academic, commercial and biotechnological potentials. This section discusses the microbial diversity studies by the culture-dependent approach carried out in India from the samples collected from the cold environments and glaciers of Indian Himalayas. While exploring the sediment samples from Leh, for anaerobic bacteria, 15 anaerobic bacteria were isolated on casein-containing agar plates (Dube et al. 2000). Five protease-producing isolates showing a unique protein on further characterization were found to produce the volatile fatty acids eliminating the need of acidogenic and up to some extent acetogenic population required for biomethanation. Processing of soil samples collected from Gangotri Glacier on skim milk agar plates resulted in selection of eleven bacterial isolates (Baghel et al. 2005). Among these, four isolates which showed protease production at alkaline pH and different temperatures were identified as *Pseudomonas aeruginosa*, *Bacillus subtilis* (two isolates) and *Bacillus licheniformis*. An anaerobic psychrophilic *Clostridium* species obtained from sediment sample of lake showed highest biomass production in the temperature range of 10–20 °C and was capable of producing serine type of metalloprotease extracellularly (Alam et al. 2006). A bacterial strain showing solubilization of phosphate as well as having antagonistic activities was isolated from a sample collected from subalpine area in the Indian Himalayas. The strain was characterized and identified as *Pseudomonas putida* (Pandey et al. 2006). Among the nine cold-tolerant bacterial isolates from samples of various low-temperature locations of the Western Himalayan region, one showing largest zone of clearance and capable of growth over a wide temperature and pH range was identified as *Exiguobacterium* sp. through sequencing of 16S rRNA gene (Kasana and Yadav 2007). A psychrotrophic bacterium *Curtobacterium luteum* from the soils of a glacier near Gangotri showed highest protease production at 15 °C in skim milk agar medium, and maximum enzyme activity was reported at 20 °C and pH 7 (Kuddus and Ramteke 2008). Two-hundred sixteen fluorescent pseudomonads having phosphate-solubilizing ability were isolated from Lahaul-Spiti valleys in the Himalayan region. Twelve isolates displaying higher tricalcium phosphate-solubilizing ability belonged to *Pseudomonas fluorescens*, *Pseudomonas poae*, *Pseudomonas trivialis* and *Pseudomonas* spp. (Gulati et al. 2008). Among 43 cold-tolerant bacterial isolates from soils of Chandra River in Lahaul-Spiti, 11 isolates showed lipase production. One of the isolates producing biggest clearance zone in plate and producing lipase over broad temperature and pH range was identified by 16S rRNA gene sequencing as *Acinetobacter* sp. (Kasana et al. 2008). A psychrotrophic bacterium from the soils of cold environments in the Himalayas having plant growth-promoting traits was identified using 16S rRNA gene sequencing as *Pantoea dispersa* (Selvakumar et al. 2008). A bacterium isolated from the rhizosphere of *Amaranth* growing in Himalayas showed growth and expressed plant growth promotion traits at 4 °C and was identified as *Pseudomonas* sp. (Mishra et al. 2009). Fourteen bacterial isolates belonging to various species of genus *Pseudomonas* and showing phosphate solubilization were isolated from the Uttarakhand Himalayas (Selvakumar et al. 2009). Studies conducted on analysis of diversity of bacteria in samples from cold habitats of the Himalayas carried out by culture-dependent approach showed that strains belonging to the phyla *Proteobacteria*, *Firmicutes*, *Actinobacteria* and *Bacteroidetes* were present. Mostly the strains were affiliated with genus *Arthrobacter*, *Bacillus* and *Pseudomonas*. Furthermore, the isolates were found to produce various hydrolytic enzymes and lipase being the most common enzyme secreted by many strains (Gangwar et al. 2009). Processing of the samples of soils obtained from Pindari Glacier in the Himalayas resulted in the isolation of 78 strains of actinomycetes. Based on antagonistic properties, the five efficient isolates were identified as *Streptomyces aurantiacus*, *Streptomyces griseobrunneus*, *Streptomyces griseoluteus* and *Streptomyces sampsonii* (Malviya et al. 2009). Among the various bacteria obtained from the rhizospheric soils of *Hippophae rhamnoides* growing in Lahaul-Spiti, and showing phosphate solubilization, one bacterial isolate showing a sharp phosphate-solubilizing activity with about 22 mm zone of clearance was identified as *Rahnella* sp. (Vyas et al. 2010). *Acinetobacter rhizosphaerae* a bacterium which was isolated from rhizospheric soil of *Hippophae rhamnoides* growing in Lahaul-Spiti produced organic acids and showed phosphate solubilization, resulting in the growth promotion of plants (Gulati et al. 2009, 2010). The diversity of bacteria isolated from soil and glacier in Lahaul-Spiti was investigated and among 217 isolates, 109 reported as protease producers. Among the protease producers, 20 psychrotrophic bacterial isolates showing comparatively higher enzyme activity at lower temperature and under alkaline conditions were identified as belonging to genus *Arthrobacter*, *Pseudomonas*, *Acinetobacter*, *Mycoplana*, *Pseudoxanthomonas*, *Serratia* and *Stenotrophomonas* (Salwan et al. 2010). The studies conducted on bacterial numbers in samples from glacier and lake showed bacteria count in the range of 0.9–4.2 × 10^7 bacteria/g of soil (Pradhan et al. 2010), whereas bacterial number in two samples of soil collected from the Kafni Glacier was found to be 6.25 × 10^8 and 30.71 × 10^8 bacteria/g soil, respectively. Bacterial isolates were grouped into 11 morphotypes and strains belonged to *Acinetobacter*, *Bacillus*, *Viridibacillus*, *Lysinibacillus*, *Pseudomonas* and *Psychrobacter* (Srinivas et al. 2011). In another study on the diversity of bacteria from soil samples of various altitudes in Himalayan region, bacterial isolates belonging to *Arthrobacter methylotrophus*, *Flavobacterium limicola*, *Pseudomonas frederiksbergensis*, *Bacillus soli*, *Pseudomonas fluorescens*, *Arthrobacter agilis* and *Pseudomonas veronii* were identified. All of these isolates showed production of extracellular enzymes, with lipase as one of the most common enzyme being secreted by many of them (Gangwar et al. 2011). Analysis of three samples of soil collected from the Pindari Glacier for bacterial enumeration showed bacterial count ranging from 2.2 × 10^8 to 8.7 × 10^8 cells/g of soil. After isolation of 40 bacterial strains, further studies grouped them into 20 morphotypes, which belonged to 12 genera, *Arthrobacter*, *Advenella*, *Bacillus*, *Brachybacterium*, *Lysinibacillus*, *Sporosarcina*, *Rhodococcus*, *Flavobacterium*, *Microterricola*, *Pseudomonas*, *Paenisporosarcina* and *Viridibacillus* (Shivaji et al. 2011). Evolutionary studies carried on with 120 bacteria from the root nodules of pea grown at various locations of Lahaul-Spiti in the Himalayan region suggested that there are genomic variations in bacteria from root nodules of two valleys, Lahaul and Spiti (Rahi et al. 2012). Among 38 isolates from cold environments of Himalaya, two strains which synthesized hydroxystyrenes from substituted cinnamic acids through biological catalysis were identified as *Pantoea agglomerans* by sequencing of 16S rRNA gene (Sharma et al. 2012). A psychrotolerant *Stenotrophomonas* sp. isolated from the Thajiwas Glacier of Kashmir produced alkaline protease of industrial importance (Saba et al. 2012). A *Bacillus* sp. producing thermostable alkaline L-glutaminase was isolated from Gangotri region of Uttarakhand Himalaya (Kumar et al. 2012a). Studies on bacterial diversity in tropical Eastern Himalaya showed that tropical forests were richer in the diversity of bacteria as compared to the temperate and alpine forests. The 16S rRNA phylogenetic studies of 155 bacterial isolates using 16S rRNA gene sequences revealed that *Firmicutes* represented the most dominant group with *Proteobacteria* and *Bacteroidetes* following it next. Isolates belonged to 27 genera and 77 species with *Bacillus* and *Pseudomonas* as most dominant genera (Lyngwi et al. 2013). Characterization of 232 isolates of bacteria from the cold environments of Northwestern Himalaya using 16S rRNA gene sequencing and phylogenetic analysis revealed the presence of 82 different species belonging to 31 genera, affiliated to four phyla represented by *Firmicutes* (41%), *Proteobacteria* (37%), *Actinobacteria* (19%) and *Bacteroidetes* (3%) (Yadav et al. 2015). These studies show that the microorganisms from cold environments of Indian Himalayas are repository of commercial and industrial enzymes of basic and applied significance, as well as these microorganisms also play a vital role in agriculture in these harsh environments.

5.3 Bacterial Diversity by Culture-Independent Approach

In the past, assessment of microbial diversity in any sample was carried out exclusively by culturing the microorganisms in nutrient medium. Nevertheless, estimation based on observations carried out by microscopic analysis as well as mathematical modelling have suggested that merely a very small portion representing about 0.1–1% of all types of microorganisms existing in any sample can be grown/cultivated by using the traditional/conventional laboratory techniques; hence about 99% of microorganisms are unculturable under the standard conditions employed in the laboratory (Amann et al. 1995; Schloss and Handelsman 2006). To overcome these problems and limitations in analysing the microbial diversity by culturing methods, the genomes of communities representing all the microorganisms prevailing in the given sample/habitat can be sequenced and explored by a new methodology known as metagenomics (Handelsman et al. 1998). The metagenomic approach has been used in microbial diversity and community analysis studies of samples from diverse habitats. It has been proposed that for describing the microbial communities existing in any kind of environmental samples in terms of quantifiable number, the metagenomic approach is one of the best precise techniques (von Mering et al. 2007). In this section, microbial diversity studies by culture-independent approach carried out in India from the samples collected from the cold environments and glaciers of Indian Himalayas are discussed. Bacterial diversity analysis of samples from the cold habitats of the Himalayas carried out by culture-independent approaches showed that the phylotypes belonged to the phyla *Proteobacteria* with the dominance of β-*Proteobacteria* (Gangwar et al. 2009). Exploration of bacterial diversity present in three samples of soil collected from glacier in Himalayan region conducted by employing the 16S rRNA gene clone library technique reported that in all these three libraries, *Actinobacteria* followed by *Firmicutes* and *Proteobacteria* were the dominant group of bacteria. In addition to it, the two samples have common groups represented by *Acidobacteria*, *Bacteroidetes*, *Gemmatimonadetes* and *Planctomycetes*. Furthermore, many other phyla absent in samples one and two were reported only from the third sample (Shivaji et al. 2011). In another study on bacterial diversity in three soil samples from other glacier and glacial lake in Himalayan region by employing the 16S rRNA gene, clone library technique reported 798 clones which belonged to 25 different classes. In two libraries clones affiliated to ten different phyla were reported, while in case of third library, the clones belonged to only two phyla (Pradhan et al. 2010). Bacterial diversity in two soil samples collected from another glacier (Kafni Glacier) by 16S rRNA gene clone libraries generated 648 clones belonging to 199 taxa. There were nine phyla which were found common in both the libraries, while clones belonging to phyla *Acidobacteria*, *Chlamydiae*, *Lentisphaerae*, *Nitrospirae* and candidate phylum TM7 were reported only in one of the libraries. Based on 16S rRNA gene sequence studies, 237 clones generated from the first sample were affiliated to 106 phyla of bacteria, and 411 clones generated from second sample were affiliated to 93 phyla of bacteria (Srinivas et al. 2011). Bacterial diversity present in the cold desert of Drass, using the 16S rDNA amplicon pyrosequencing, showed the presence of bacteria belonging to 15 different phyla (Gupta et al. 2015). This shows the richness of bacterial diversity present in the samples from cold environments and glaciers employing the culture-independent approach.

5.4 Bacteria from Cold Environments in Agriculture

Bacteria isolated from cold environments of Indian Himalayas possessed various growth-promoting attributes and have been used for promoting the plant growth under cold climatic situations prevailing there. In a pot culture experiment conducted in non-sterile soils at suboptimal temperature environments, the bacterization of wheat seedlings with cold-tolerant *Pseudomonas* sp. MTCC-9002 from the Indian Himalayas recorded 19.2% more germination of seed, 30.0 and 22.9% more shoot and root length, respectively, in comparison to the uninoculated control plants (Mishra et al. 2009). Psychrotrophic *Exiguobacterium acetylicum* MTCC 8707 isolated from Uttarakhand Himalaya expressed various plant growth-promoting traits like phosphate-solubilizing ability, production of siderophore, indole acetic acid and hydrogen cyanide at temperatures of 15 and 4 °C, which are below optimal growth temperatures. Inoculation of seed with this species showed a positive effect on the various growth parameters as well as uptake of nutrient by wheat seedlings under the low-temperature conditions in glass house studies (Selvakumar et al. 2010). Inoculation of maize with bacterium *Acinetobacter rhizosphaerae* BIHB 723 which was isolated from cold environments of Himalayan region increased the height of plants, fresh and dry weight of shoots and length and dry weight of roots significantly compared to uninoculated plants. There was also enhancement in the phosphorus contents of root, shoot and soil on inoculation (Gulati et al. 2010). Studies carried out at low temperature of 8 °C in greenhouse on wheat seedling inoculated with various pseudomonad isolates to mitigate the effect of low temperature and growth of wheat demonstrated that there was significant increase in root and shoot weight, uptake of nutrients as well as improvement in the various cellular metabolites compared to non-bacterized control (Mishra et al. 2011). The *Bacillus* sp. BPR7 isolated from common bean growing at Uttarakhand Himalaya was found positive for indole acetic acid production, siderophore production, organic acid production and enzymes; phytase, ACC deaminase, oxalate oxidase and lytic enzymes; and solubilization of various organic phosphates, inorganic phosphates, potassium and zinc. *Bacillus* sp. BPR7 also showed strong inhibitory effect on the growing ability of many plant pathogens such as *Colletotricum* sp., *Fusarium solani*, *F. oxysporum*, *Macrophomina phaseolina*, *Rhizoctonia solani* and *Sclerotinia sclerotiorum* under laboratory conditions (Kumar et al. 2012b). Five bacterial isolates belonging to genera *Arthrobacter*, *Klebsiella* and *Pseudomonas* isolated from Garhwal Himalaya efficiently increased the biomass and phosphate uptake in rice under controlled conditions (Gusain et al. 2015). Among the 82 bacterial species isolated from low-temperature environments in the deserts of Himalayan region, 28 reported positive for possessing many plant growth-promoting attributes. Eight isolates from these which showed phosphate solubilization, production of indole acetic acid, siderophores and ACC deaminase were identified as belonging to eight species of five genera, *Providencia*, *Pseudomonas*, *Psychrobacter*, *Sanguibacter* and *Yersinia*. Furthermore, ten isolates showed antagonism against the phytopathogenic fungi *Macrophomina phaseolina* and *Rhizoctonia solani* (Yadav et al. 2015).

5.5 Novel Bacteria from Cold Environments of India

A large number of bacteria belonging to various classes and genera have been isolated from various locations in the cold environments worldwide, and all of them have been found to be psychrotolerant/psychrophilic in nature. In India the cold environments have not been explored much for isolation of novel bacteria. The first report on novel bacterium isolated from cold environments/glaciers appeared in the year 2005. As of now (2016), there are only 21 new/novel species of Gram −ve and Gram +ve bacteria reported form soil/water/ice samples collected from different locations in cold environments of Indian Himalayas. The list of novel bacterial isolates and the location from which samples were collected is given in Table 5.1. Most of these novel species have been isolated and identified by the scientist working in two institutions of CSIR (Centre for Cellular and Molecular Biology, Hyderabad, and Institute of Microbial Technology, Chandigarh). More efforts are required to be put in by the researches for exploring the cold environments of India for harnessing the unexplored potential of bacterial diversity in terms of new and novel taxa of bacteria.

Table 5.1 Novel bacterial species isolated from cold environments of Indian Himalayas

Bacterial species	Location	Type strain	16S rRNA gene sequence accession number	References
Pedobacter himalayensis	Hamta Glacier, Lahaul-Spiti	MTCC 6384	AJ583425	Shivaji et al. (2005)
Planococcus stackebrandtii	Kibber, Lahaul-Spiti	MTCC 6226	AY437845	Mayilraj et al. (2005)
Dyadobacter hamtensis	Hamta Glacier, Lahaul-Spiti	MTCC 7023	AJ619979	Chaturvedi et al. (2005)
Actinoalloteichus spitiensis	Rangrik, Lahaul-Spiti	MTCC 6194	AY426714	Singla et al. (2005)
Ornithinimicrobium kibberense	Kibber, Lahaul-Spiti	MTCC 6545	AY636111	Mayilraj et al. (2006a)
Dietzia kunjamensis	Kunjam Pass, Lahaul-Spiti	MTCC 7007	AY972480	Mayilraj et al. (2006b)
Agrococcus lahaulensis	Lahaul-Spiti	MTCC 7154	DQ156908	Mayilraj et al. (2006c)
Kocuria himachalensis	Kibber, Lahaul-Spiti	MTCC 7020	AY987383	Mayilraj et al. (2006d)
Rhodococcus kroppenstedtii	Kibber, Lahaul-Spiti	MTCC 6634	AY726605	Mayilraj et al. (2006e)
Exiguobacterium indicum	Hamta Glacier, Lahaul-Spiti	LMG 23471	AJ846291	Chaturvedi and Shivaji (2006)
Rhodobacter changlensis	Changla Pass	DSM 18774	AM399030	Anil Kumar et al. (2007)
Desulfovibrio psychrotolerans	Pangong Lake	DSM 19430	AM418397	Sasi Jyothsna et al. (2008)
Leifsonia pindariensis	Pindari Glacier	MTCC 9128	AM900767	Reddy et al. (2008a)
Bacillus cecembensis	Pindari Glacier	MTCC 9127	AM773821	Reddy et al. (2008b)
Cryobacterium roopkundense	Roopkund Lake	DSM 21065	EF467640	Reddy et al. (2010)
Paenibacillus glacialis	Kafni Glacier	DSM 22343	EU815300	Kishore et al. (2010)
Azospirillum himalayense	Chamba Valley	KCTC 23189	GQ284588	Tyagi and Singh (2014)
Paenisporosarcina indica	Pindari Glacier	LMG 23933	FN397659	Reddy et al. (2013)
Bacillus lehensis	Leh	MTCC 7633	AY793550	Ghosh et al. (2007)

(continued)

Table 5.1 (continued)

Bacterial species	Location	Type strain	16S rRNA gene sequence accession number	References
Exiguobacterium himgiriensis	Lahaul-Spiti	MTCC 7628	JX999056	Singh et al. (2013)
Pseudacidovorax austerolens	Chamba Valley	DSM 24877	FJ581042	Tyagi and Singh (2015)

5.6 Genome Sequencing of Bacteria from Cold Environments of India

In the last 5–6 years, researchers from India have started working on the genome sequencing of bacteria, and genomes of approximately 300 bacteria isolated from various samples have been sequenced. Publications on genome sequencing of bacteria isolated from cold environments of India have also reached to 18 (Table 5.2). The first report on genome sequencing of bacteria from cold environments of India appeared in 2011. Two bacterial strains were isolated from the alkaline brackish water samples collected from Pangong Lake, located at high altitude of Ladakh, India. The first strain showing the highest *rrs* gene sequence identity of 95.2%, with the type strain of *Idiomarina maris*, and second strain showing 98.68% sequence identity with *Rheinheimera soli* were used for genome sequencing (Gupta et al. 2011a, b). Along with the presence of genes required for various basic metabolic processes, the sequenced genomes of bacteria from the cold environments of Indian Himalayas also have gene related to survival under the extreme conditions. Genes for various cold shock proteins like *CspA, CspD, CspG and CspC*; genes for DNA repair system like RecA, UvrABC, RecBCD, RecFOR, UvrD, RecQ, MutS and *MutT*; carotenoid/terpenoids biosynthesis pathway; multienzyme complex UvrABC; and group of chaperone proteins including *ClpB, CopZ, DnaJ, GroEL, HtpG, HtrA, HspR, GrpE, DnaK* and *GroES* have been reported from these bacteria (Mahato et al. 2014; Salwan et al. 2014; Sharma et al. 2014; Gulati et al. 2015; Kumar et al. 2015a, 2016; Dhar et al. 2016; Himanshu et al. 2016). Genes related to plant growth promotion like phosphate solubilization, auxin biosynthesis, siderophore production, HCN production, ACC deaminase and ammonia production were reported from the genome sequence of *Pseudomonas trivialis* (Gulati et al. 2015). The genomes of more bacteria isolated from cold environments should be sequenced which will broaden our knowledge and understanding on the adaptations, survival and growth of these microorganism under these harsh conditions.

Table 5.2 Bacterial genomes sequenced and published from cold environments of Indian Himalayas

	Strain	Isolation source, location	Size in (Mb)	Number of CDS	G + C content	Accession number	References
1.	*Arthrobacter agilis* strain L7	Pangong Lake, Ladakh	3.6	3316	69.79%	JWSU00000000	Singh et al. (2016)
2.	*Paenibacillus* sp. strain IHB B 3084	Sediment, Lahaul-Spiti	5.88	6093	46.83%	CP013203-CP013209	Dhar et al. (2016)
3.	*Arthrobacter alpinus* strain ERGS4:06	Glacier, East Rathong	4.3	4154	60.64%	CP013200 & CP013201	Kumar et al. (2016)
4.	*Microterricola viridarii* strain ERGS5:02	Glacier, East Rathong	3.7	3456	68.70%	CP014145	Himanshu et al. (2016)
5.	*Arthrobacter* sp. strain ERGS1:01	Glacier, East Rathong	4.03	4623	65.37%	CP012477–CP012479	Kumar et al. (2015a)
6.	*Paenibacillus* sp. strain IHBB 10380	Glacier, Kunzum Pass	5.77	5638	41.33%	CP010976 & CP010977	Pal et al. (2015)
7.	*Pseudomonas trivialis* strain IHBB 745	Soil, Lahaul-Spiti	6.45	6032	59.91%	CP011507	Gulati et al. (2015)
8.	*Arthrobacter* sp. strain IHBB 11108	Lake, Chandra Tal	3.6	3454	58.97%	CP011005 & CP011006	Kiran et al. (2015)
9.	*Cellulosimicrobium* sp. strain MM	Microbial mats, Manikaran	3.85	3718	74.4%	JPQW00000000	Sharma et al. (2014)
10.	*Cryobacterium roopkundensis* strain RuGl7	Glacial Lake, Roopkund	4.36	4048	65.3%	JPXF00000000	Sathyanarayana Reddy et al. (2014)
11.	*Chryseobacterium polytrichastri* ERMR1:04	Glacier, East Rathong	5.53	4524	34.07%	LIRF00000000	Kumar et al. (2015b)
12.	*Paenibacillus* strain, IHB B 3415	Soil, Lahaul-Spiti	8.44	7335	50.77%	JUEI00000000	Dhar et al. (2015)

(continued)

Table 5.2 (continued)

	Strain	Isolation source, location	Size in (Mb)	Number of CDS	G + C content	Accession number	References
13.	*Acinetobacter* sp. strain MN12 (MTCC 10786)	Soil, Lahaul-Spiti	4.31	4017	40.75%	JROB00000000	Swarnkar et al. (2014)
14.	*Planomicrobium glaciei* strain CHR43	Chandra River, Lahaul-Spiti	3.9	3934	46.97%	AUYR00000000	Salwan et al. (2014)
15.	*Thermus* sp. strain RL	Water, Manikaran	2.03	1986	68.77%	AIJQ00000000	Dwivedi et al. (2012)
16.	*Deinococcus* sp. strain RL	Water, Manikaran	2.79	2614	69.4%	JMQF00000000	Mahato et al. (2014)
17.	*Idiomarina* sp. strain A28L	Pangong Lake, Ladakh	2.59	2299	45.5%	AFPO00000000	Gupta et al. (2011a, b)
18.	*Rheinheimera* sp. strain A13L	Pangong Lake, Ladakh	4.52	3942	46.23%	AFHI00000000	Gupta et al. (2011a, b)

5.7 Perspective

In conclusion, although many of the microorganisms isolated and identified in the past have been commercially exploited, our knowledge on the microbial diversity and major roles being played by these microorganisms in sustaining global life-supporting systems is still limited. Moreover, cold environments/deserts and glaciers in the Indian Himalayas have been explored for the microbial diversity to very small extent. It is clear from the various studies conducted with the samples collected from cold environments areas that they can serve as source of novel/new microorganisms with potential application in academics, agriculture and industry. So, there is a great need to investigate these hot spots of microbial diversity for exploring the diversity for human benefits.

References

Alam SI, Dube S, Agarwal MK, Singh L (2006) Purification and characterization of an extracellular protease produced by psychrotolerant *Clostridium* sp. LP3 from lake sediment of Leh, India. Can J Microbiol 52:1238–1246. doi:10.1139/w06-089

Amann RI, Ludwig W, Schleifer KH (1995) Phylogenetic identification and in situ detection of individual microbial cells without cultivation. Microbiol Rev 59:143–169. PMID:7535888

Anil Kumar P, Srinivas TN, Sasikala C, Ramana CV (2007) *Rhodobacter changlensis* sp. nov. a psychrotolerant, phototrophic alphaproteobacterium from the Himalayas of India. Int J Syst Evol Microbiol 57:2568–2571. doi:10.1099/ijs.0.65110-0

Baghel VS, Tripathi RD, Ramteke PW, Gopal K, Dwivedi S, Jain RK, Rai UN, Singh SN (2005) Psychrotrophic proteolytic bacteria from cold environment of Gangotri glacier, Western Himalaya, India. Enzym Microb Technol 36:654–659. doi:10.1016/j.enzmictec.2004.09.005

Brenchley JE (1996) Psychrophilic microorganisms and their cold active enzymes. J Ind Microbiol 17:432–437. doi:10.1007/BF01574774

Chaturvedi P, Shivaji S (2006) *Exiguobacterium indicum* sp. nov. a psychrophilic bacterium from the Hamta glacier of the Himalayan mountain ranges of India. Int J Syst Evol Microbiol 56:2765–2770. doi:10.1099/ijs.0.64508-0

Chaturvedi P, Reddy GS, Shivaji S (2005) *Dyadobacter hamtensis* sp. nov. from Hamta glacier, located in the Himalayas, India. Int J Syst Evol Microbiol 55:2113–2117. doi:10.1099/ijs.0.63806-0

Dhar H, Swarnkar MK, Gulati A, Singh AK, Kasana RC (2015) Draft genome sequence of a cellulase-producing psychrotrophic *Paenibacillus* strain, IHB B 3415, isolated from the cold environment of the western Himalayas, India. Genome Announc 3:e01581–e01514. doi:10.1128/genomeA.01581-14

Dhar H, Swarnkar MK, Rana A, Kaushal K, Singh AK, Kasana RC, Gulati A (2016) Complete genome sequence of a low-temperature active and alkaline-stable endoglucanase producing *Paenibacillus* sp. strain IHBB3084 from the Indian Trans-Himalayas. J Biotechnol 230:1–2. doi:10.1016/j.jbiotec.2016.04.037

Dube S, Alam SI, Singh L (2000) Studies on anaerobic proteolytic bacteria of Leh, India. World J Microbiol Biotechnol 16:297–301. doi:10.1023/A:1008935425175

Dwivedi V, Sangwan N, Nigam A, Garg N, Niharika N, Khurana P, Khurana JP, Lal R (2012) Draft genome sequence of *Thermus* sp. strain RL, isolated from a hot water spring located atop the Himalayan ranges at Manikaran, India. J Bacteriol 194:3534. doi:10.1128/JB.00604-12

Gangwar P, Alam SI, Bansod S, Singh L (2009) Bacterial diversity of soil samples from the western Himalayas, India. Can J Microbiol 55:564–577. doi:10.1139/w09-011

Gangwar P, Alam SI, Singh L (2011) Metabolic characterization of cold active *Pseudomonas*, *Arthrobacter*, *Bacillus*, and *Flavobacterium* spp. from Western Himalayas. Indian J Microbiol 51:70–75. doi:10.1007/s12088-011-0092-7

Ghosh A, Bhardwaj M, Satyanarayana T, Khurana M, Mayilraj S, Jain RK (2007) *Bacillus lehensis* sp. nov. an alkalitolerant bacterium isolated from soil. Int J Syst Evol Microbiol 57:238–242. doi:10.1099/ijs.0.64617-0

Gulati A, Rahi P, Vyas P (2008) Characterization of phosphate-solubilizing fluorescent pseudomonads from the rhizosphere of seabuckthorn growing in the cold deserts of Himalayas. Curr Microbiol 56:73–79. doi:10.1007/s00284-007-9042-3

Gulati A, Vyas P, Rahi P, Kasana RC (2009) Plant growth-promoting and rhizosphere-competent *Acinetobacter rhizosphaerae* strain BIHB 723 from the cold deserts of the Himalayas. Curr Microbiol 58:371–377. doi:10.1007/s00284-008-9339-x

Gulati A, Sharma N, Vyas P, Sood S, Rahi P, Pathania V, Prasad R (2010) Organic acid production and plant growth promotion as a function of phosphate solubilization by *Acinetobacter rhizosphaerae* strain BIHB 723 isolated from the cold deserts of the trans-Himalayas. Arch Microbiol 192:975–983. doi:10.1007/s00203-010-0615-3

Gulati A, Swarnkar MK, Vyas P, Rahi P, Thakur R, Thakur N, Singh AK (2015) Complete genome sequence of the rhizobacterium *Pseudomonas trivialis* strain IHBB745 with multiple plant growth-promoting activities and tolerance to desiccation and alkalinity. Genome Announc 3(5):e00943–e00915. doi:10.1128/genomeA.00943-15

Gupta HK, Gupta RD, Singh A, Chauhan NS, Sharma R (2011a) Genome sequence of *Rheinheimera* sp. strain A13L, isolated from Pangong Lake, India. J Bacteriol 193:5873–5874. doi:10.1128/JB.05636-11

Gupta HK, Singh A, Sharma R (2011b) Genome sequence of *Idiomarina* sp. strain A28L, isolated from Pangong Lake, India. J Bacteriol 193:5875–5876. doi:10.1128/JB.05648-11

Gupta P, Sangwan N, Lal R, Vakhlu J (2015) Bacterial diversity of Drass, cold desert in Western Himalaya and its comparison with Antarctic and Arctic. Arch Microbiol 197:851–860. doi:10.1007/s00203-015-1121-1124

Gusain YS, Kamal R, Mehta CM, Singh US, Sharma AK (2015) Phosphate solublilizing and indole-3-acetic acid producing bacteria from the soil of Garhwal Himalaya aimed to improve the growth of rice. J Environ Biol 36:301–307. PMID:26536808

Handelsman J, Rondon MR, Brady SF, Clardy J, Goodman RM (1998) Molecular biological access to the chemistry of unknown soil microbes: a new frontier for natural products. Chem Biol 5:R245–R249. PMID:9818143

Himanshu, Swarnkar MK, Singh D, Kumar R (2016) First complete genome sequence of a species in the genus *Microterricola*, an extremophilic cold active enzyme producing bacterial strain ERGS5:02 isolated from Sikkim Himalaya. J Biotechnol 222:17–18. doi:10.1016/j.jbiotec.2016.02.011

Kasana RC (2010) Protease from psychrotrophs: an overview. Crit Rev Microbiol 36:134–145. doi:10.3109/10408410903485525

Kasana RC, Yadav SK (2007) Isolation of a psychrotrophic *Exiguobacterium* sp. SKPB5 (MTCC 7803) and characterization of its alkaline protease. Curr Microbiol 54:224–229. doi:10.1007/s00284-006-0402-1

Kasana RC, Kaur B, Yadav SK (2008) Isolation and identification of a psychrotrophic *Acinetobacter* sp. CR9 and characterization of its alkaline lipase. J Basic Microbiol 48:207–212. doi:10.1002/jobm.200700160

Kiran S, Swarnkar MK, Pal M, Thakur R, Tewari R, Singh AK, Gulati A (2015) Complete genome sequencing of protease-producing novel *Arthrobacter* sp. strain IHBB 11108 using PacBio single-molecule real-time sequencing technology. Genome Announc 3(2):e00346–e00315. doi:10.1128/genomeA.00346-15

Kishore KH, Begum Z, Pathan AA, Shivaji S (2010) *Paenibacillus glacialis* sp. nov. isolated from the Kafni glacier of the Himalayas, India. Int J Syst Evol Microbiol 60:1909–1913. doi:10.1099/ijs.0.015271-0

Kuddus M, Ramteke PW (2008) A cold–active extracellular metalloprotease from *Curtobacterium luteum*. (MTCC 7529), enzyme production and characterization. J Gen Appl Microbiol 54:385–392. doi:10.2323/jgam.54.385

Kumar L, Singh B, Adhikari DK, Mukherjee J, Ghosh D (2012a) A temperature and salt-tolerant L-glutaminase from Gangotri region of uttarakhand himalaya: enzyme purification and characterization. Appl Biochem Biotechnol 166:1723–1735. doi:10.1007/s12010-012-9576-0

Kumar P, Dubey RC, Maheshwari DK (2012b) *Bacillus* strains isolated from rhizosphere showed plant growth promoting and antagonistic activity against phytopathogens. Microbiological Res 167:493–499. doi:10.1016/j.micres.2012.05.002

Kumar R, Singh D, Swarnkar MK, Singh AK, Kumar S (2015a) Complete genome sequence of Arthrobacter sp. ERGS1:01, a putative novel bacterium with prospective cold active industrial enzymes, isolated from East Rathong glacier in India. J Biotechnol 214:139–140. doi:10.1016/j.jbiotec.2015.09.025

Kumar R, Singh D, Swarnkar MK, Singh AK, Kumar S (2015b) Genome assembly of *Chryseobacterium polytrichastri* ERMR1:04, a psychrotolerant bacterium with cold active proteases, isolated from East Rathong Glacier in India. Genome Announc 3(6):e01305–e01315. doi:10.1128/genomeA.01305-15

Kumar R, Singh D, Swarnkar MK, Singh AK, Kumar S (2016) Complete genome sequence of *Arthrobacter alpinus* ERGS4:06, a yellow pigmented bacterium tolerant to cold and radiations isolated from Sikkim Himalaya. J Biotechnol 20(220):86–87. doi:10.1016/j.jbiotec.2016.01.016

Lyngwi LA, Koijam K, Sharma D, Joshi SR (2013) Cultivable bacterial diversity along the altitudinal zonation and vegetation range of tropical Eastern Himalaya. Rev Biol Trop 61:467–490. PMID:23894996

Mahato NK, Tripathi C, Verma H, Singh N, Lal R (2014) Draft genome sequence of *Deinococcus* sp. strain RL isolated from sediments of a hot water spring. Genome Announc 2(4):e00703–e00714. doi:10.1128/genomeA.00703-14

Malviya MK, Pandey A, Trivedi P, Gupta G, Kumar B (2009) Chitinolytic activity of cold tolerant antagonistic species of streptomyces isolated from glacial sites of Indian Himalaya. Curr Microbiol 59:502–508. doi:10.1007/s00284-009-9466-z

Mayilraj S, Prasad GS, Suresh K, Saini HS, Shivaji S, Chakrabarti T (2005) *Planococcus stackebrandtii* sp. nov. isolated from a cold desert of the Himalayas, India. Int J Syst Evol Microbiol 55:91–94. doi: 10.1099/ijs.0.63290-0

Mayilraj S, Saha P, Suresh K, Saini HS (2006a) *Ornithinimicrobium kibberense* sp. nov. isolated from the Indian Himalayas. Int J Syst Evol Microbiol 56:1657–1661. doi:10.1099/ijs.0.64138-0

Mayilraj S, Suresh K, Kroppenstedt RM, Saini HS (2006b) *Dietzia kunjamensis* sp. nov. isolated from the Indian Himalayas. Int J Syst Evol Microbiol 56:1667–1671. doi:10.1099/ijs.0.64212-0

Mayilraj S, Suresh K, Schumann P, Kroppenstedt RM, Saini HS (2006c) *Agrococcus lahaulensis* sp. nov. isolated from a cold desert of the Indian Himalayas. Int J Syst Evol Microbiol 56:1807–1810. doi:10.1099/ijs.0.64247-0

Mayilraj S, Kroppenstedt RM, Suresh K, Saini HS (2006d) *Kocuria himachalensis* sp. nov. an actinobacterium isolated from the Indian Himalayas. Int J Syst Evol Microbiol 56:1971–1975. doi:10.1099/ijs.0.63915-0

Mayilraj S, Krishnamurthi S, Saha P, Saini HS (2006e) *Rhodococcus kroppenstedtii* sp. nov. a novel actinobacterium isolated from a cold desert of the Himalayas, India. Int J Syst Evol Microbiol 56:979–982. doi:10.1099/ijs.0.63831-0

Mishra PK, Mishra S, Bisht SC, Selvakumar G, Kundu S, Bisht JK, Gupta HS (2009) Isolation, molecular characterization and growth-promotion activities of a cold tolerant bacterium *Pseudomonas* sp. NARs9 (MTCC9002) from the Indian Himalayas. Biol Res 42:305–313. doi: /S0716-97602009000300005

Mishra PK, Bisht SC, Ruwari P, Selvakumar G, Joshi GK, Bisht JK, Bhatt JC, Gupta HS (2011) Alleviation of cold stress in inoculated wheat (*Triticum aestivum* L.) seedlings with psychrotolerant Pseudomonads from NW Himalayas. Arch Microbiol 193:497–513. doi:10.1007/s00203-011-0693-x

Pal M, Swarnkar MK, Thakur R, Kiran S, Chhibber S, Singh AK, Gulati A (2015) Complete genome sequence of *Paenibacillus* sp. strain IHBB 10380 using PacBio single molecule real-time sequencing technology. Genome Announc 3:e00356–e00315. doi:10.1128/genomeA.00356-15

Pandey A, Trivedi P, Kumar B, Palni LM (2006) Characterization of a phosphate solubilizing and antagonistic strain of *Pseudomonas putida* (B0) isolated from a sub-alpine location in the Indian Central Himalaya. Curr Microbiol 53:102–107. doi:10.1007/s00284-006-4590-5

Pradhan S, Srinivas TN, Pindi PK, Kishore KH, Begum Z, Singh PK, Singh AK, Pratibha MS, Yasala AK, Reddy GS, Shivaji S (2010) Bacterial biodiversity from Roopkund Glacier, Himalayan mountain ranges, India. Extremophiles 14:377–395

Rahi P, Kapoor R, Young JP, Gulati A (2012) A genetic discontinuity in root-nodulating bacteria of cultivated pea in the Indian trans-Himalayas. Mol Ecol 21:145–159. doi:10.1111/j.1365-294X.2011.05368.x

Rashid N, Kikuchi H, Ezaki S, Atomi H, Imanaka T (1999) Isolation and characterization of psychrotrophs from subterranean environments. J Biosci Bioeng 87:746–751. PMID:16232548

Reddy GS, Prabagaran SR, Shivaji S (2008a) *Leifsonia pindariensis* sp. nov. isolated from the Pindari glacier of the Indian Himalayas, and emended description of the genus *Leifsonia*. Int J Syst Evol Microbiol 58:2229–2234. doi:10.1099/ijs.0.65715-0

Reddy GS, Uttam A, Shivaji S (2008b) *Bacillus cecembensis* sp. nov. isolated from the Pindari glacier of the Indian Himalayas. Int J Syst Evol Microbiol 58:2330–2335. doi:10.1099/ijs.0.65515-0

Reddy GS, Pradhan S, Manorama R, Shivaji S (2010) *Cryobacterium roopkundense* sp. nov. a psychrophilic bacterium isolated from glacial soil. Int J Syst Evol Microbiol 60:866–870. doi:10.1099/ijs.0.011775-0

Reddy GS, Manasa BP, Singh SK, Shivaji S (2013) Paenisporosarcina indica sp. nov. a psychrophilic bacterium from a glacier, and reclassification of *Sporosarcina antarctica* Yu et al. 2008 as *Paenisporosarcina antarctica* comb. nov. and emended description of the genus *Paenisporosarcina*. Int J Syst Evol Microbiol 63:2927–2933. doi:10.1099/ijs.0.047514-0

Saba I, Qazi PH, Rather SA, Dar RA, Qadri QA, Ahmad N, Johri S, Taneja SC, Shawl S (2012) Purification and characterization of a cold active alkaline protease from *Stenotrophomonas* sp. isolated from Kashmir, India. World J Microbiol Biotechnol 28:1071–1079. doi:10.1007/s11274-011-0905-1

Salwan R, Gulati A, Kasana RC (2010) Phylogenetic diversity of alkaline protease-producing psychrotrophic bacteria from glacier and cold environments of Lahaul and Spiti, India. J Basic Microbiol 50:150–159. doi:10.1002/jobm.200800400

Salwan R, Swarnkar MK, Singh AK, Kasana RC (2014) First draft genome sequence of a member of the genus *Planomicrobium*, isolated from the Chandra River, India. Genome Announc 2(1):e01259–e01213. doi:10.1128/genomeA.01259-13

Sasi Jyothsna TS, Sasikala C, Ramana CV (2008) *Desulfovibrio psychrotolerans* sp. nov. a psychrotolerant and moderately alkaliphilic sulfate-reducing deltaproteobacterium from the Himalayas. Int J Syst Evol Microbiol 58:821–825. doi:10.1099/ijs.0.65402-0

Sathyanarayana Reddy G, Sreenivas A, Shivaji S (2014) Draft genome sequence of *Cryobacterium roopkundensis* strain RuGl7, isolated from a soil sample in the vicinity of Roopkund Lake, Himalayas, India. Genome Announc 2:e01206–e01214. doi:10.1128/genomeA.01206-14

Schloss PD, Handelsman J (2006) Toward a census of bacteria in soil. PLoS Comput Biol 2:786–793. doi:10.1371/journal.pcbi.0020092

Selvakumar G, Kundu S, Joshi P, Nazim S, Gupta AD, Mishra PK, Gupta HS (2008) Characterization of a cold-tolerant plant growth-promoting bacterium *Pantoea dispersa* 1A isolated from a sub-alpine soil in the North Western Indian Himalayas. World J Microbiol Biotechnol 24:955–960. doi:10.1007/s11274-007-9558-5

Selvakumar G, Joshi P, Mishra PK, Bisht JK, Gupta HS (2009) Mountain aspect influences the genetic clustering of psychrotolerant phosphate solubilizing Pseudomonads in the Uttarakhand Himalayas. Curr Microbiol 59(4):432–438. doi:10.1007/s00284-009-9456-1

Selvakumar G, Kundu S, Joshi P, Nazim S, Gupta AD, Gupta HS (2010) Growth promotion of wheat seedlings by *Exiguobacterium acetylicum* 1P (MTCC 8707) a cold tolerant bacterial strain from the Uttarakhand Himalayas. Indian J Microbiol 50:50–56. doi:10.1007/s12088-009-0024-y

Sharma UK, Sharma N, Salwan R, Kumar R, Kasana RC, Sinha AK (2012) Efficient synthesis of hydroxystyrenes via biocatalytic decarboxylation/deacetylation of substituted cinnamic acids by newly isolated *Pantoea agglomerans* strains. J Sci Food Agric 92:610–617. doi:10.1002/jsfa.4616

Sharma A, Hira P, Shakarad M, Lal R (2014) Draft genome sequence of *Cellulosimicrobium* sp. strain MM, isolated from arsenic-rich microbial mats of a Himalayan hot spring. Genome Announc 2(5):e01020–e01014. doi:10.1128/genomeA.01020-14

Shivaji S, Chaturvedi P, Reddy GS, Suresh K (2005) *Pedobacter himalayensis* sp. nov. from the Hamta glacier located in the Himalayan mountain ranges of India. Int J Syst Evol Microbiol 55:1083–1088. doi:10.1099/ijs.0.63532-0

Shivaji S, Pratibha MS, Sailaja B, Hara Kishore K, Singh AK, Begum Z, Anarasi U, Prabagaran SR, Reddy GS, Srinivas TN (2011) Bacterial diversity of soil in the vicinity of Pindari glacier, Himalayan mountain ranges, India, using culturable bacteria and soil 16S rRNA gene clones. Extremophiles 15:1–22. doi:10.1007/s00792-010-0333-4

Singh NK, Raichand R, Kaur I, Kaur C, Pareek S, Mayilraj S (2013) *Exiguobacterium himgiriensis* sp. nov. a novel member of the genus *Exiguobacterium*, isolated from the Indian Himalayas. Antonie Van Leeuwenhoek 103:789–796. doi:10.1007/s10482-012-9861-9865

Singh RN, Gaba S, Yadav AN, Gaur P, Gulati S, Kaushik R, Saxena AK (2016) First high quality draft genome sequence of a plant growth promoting and cold active enzyme producing psychrotrophic *Arthrobacter agilis* strain L77. Stand Genomic Sci 11:54. doi:10.1186/s40793-016-0176-4

Singla AK, Mayilraj S, Kudo T, Krishnamurthi S, Prasad GS, Vohra RM (2005) *Actinoalloteichus spitiensis* sp. nov. a novel actinobacterium isolated from a cold desert of the Indian Himalayas. Int J Syst Evol Microbiol 55:2561–2564. doi: 10.1099/ijs.0.63720-0

Srinivas TN, Singh SM, Pradhan S, Pratibha MS, Kishore KH, Singh AK, Begum Z, Prabagaran SR, Reddy GS, Shivaji S (2011) Comparison of bacterial diversity in proglacial soil from Kafni Glacier, Himalayan Mountain ranges, India, with the bacterial diversity of other glaciers in the world. Extremophiles 15:673–690. doi:10.1007/s00792-011-0398-8

Swarnkar MK, Salwan R, Kasana RC, Singh AK (2014) Draft genome sequence of psychrotrophic *Acinetobacter* sp. strain MN12 (MTCC 10786), which produces a lowtemperature-active and alkaline-stable peptidase. Genome Announc 2:e01167–e01114. doi:10.1128/genomeA.01167-14

Tyagi S, Singh DK (2014) *Azospirillum himalayense* sp. nov. a nifH bacterium isolated from Himalayan valley soil, India. Ann Microbiol 64:259–266. doi:10.1007/s13213-013-0658-1

Tyagi S, Singh DK (2015) *Pseudacidovorax austerolens* sp. nov. *a nifH* bacterium isolated from Himalayan valley soil, India. Ann Microbiol 65:217–223. doi:10.1007/s13213-014-0852-9

von Mering C, Hugenholtz P, Raes J, Tringe SG, Doerks T, Jensen LJ, Ward N, Bork P (2007) Quantitative phylogenetic assessment of microbial communities in diverse environments. Science 315:1126–1130. doi:10.1126/Science.1133420

Vyas P, Joshi R, Sharma KC, Rahi P, Gulati A, Gulati A (2010) Cold-adapted and rhizosphere-competent strain of *Rahnella* sp. with broad-spectrum plant growth-promotion potential. J Microbiol Biotechnol 20:1724–1234. doi:10.4014/jmb.1007.07030

Yadav AN, Sachan SG, Verma P, Saxena AK (2015) Prospecting cold deserts of north western Himalayas for microbial diversity and plant growth promoting attributes. J Biosci Bioeng 119:683–693. doi:10.1016/j.jbiosc.2014.11.006

The Microbiome of the Himalayan Ecosystem

6

Divya Joshi, Saurabh Kumar, Deep Chandra Suyal, and Reeta Goel

Abstract

The Himalayas are referred to as the highest mountain ranges in the world. They are also one of the youngest to be formed. It is approximately 2400 km and consists of more than 40 mountains which exceed 7000 m in height. The entire Himalayan mountain range has its unique flora and fauna. Moreover, traditional agricultural practices in the Himalayan Mountains have always been a storehouse of agro-biodiversity. The bacterial diversity of the Himalayan soil has been widely studied using 16S rRNA gene cloning and sequencing. Clone library analysis revealed the dominance of proteobacteria especially genus *Pseudomonas* and *Rhizobium* in the soil of high-altitude agroecosystems. Moreover, *Bacteroidetes*, *Chloroflexi*, *Acidobacteria*, *Firmicutes*, *Nitrospira*, *Planctomycetes*, *Cyanobacteria*, *Gemmatimonadetes*, *Verrucomicrobia*, *Chlorobi*, *Actinobacteria*, OD1, OP11, and BRC1 were also found in the Himalayan soils. Furthermore, metagenomic studies from these regions revealed the presence of "yet not cultured" microbial communities from these high-altitude niches, thus indicating the need of culture-dependent studies from Himalayan regions.

Keywords

Himalaya • Agro-biodiversity • Micro-diversity conversion • high-altitude niches • metagenomics

D. Joshi • S. Kumar • D.C. Suyal • R. Goel (✉)
Department of Microbiology, Govind Ballabh Pant University of Agriculture and Technology, Pantnagar, Uttarakhand 263 145, India
e-mail: rg55@rediffmail.com

V.C. Kalia et al. (eds.), *Mining of Microbial Wealth and MetaGenomics*,
DOI 10.1007/978-981-10-5708-3_6

6.1 Introduction

Himalaya is one of the "hotspots" of biodiversity. The unusually wide altitudinal range (over 3000 m), rapid change in altitude and high endemism (Chitale et al. 2014) make it an interesting area for studies. The Himalayas are the largest, highest and densely populated mountain range in the world. Extreme variations in altitude, geology and soils over short distances have resulted in a wealth of natural ecosystems. Mountains occupy approximately 24% of the world's land area and maintain 12% of the global population that are living within mountain regions (Sharma et al. 2010). One-fifth of mankind gets a vast array of goods and services from mountains which include water, energy, timber, biodiversity maintenance and opportunities for recreation. They also fulfil the freshwater needs of more than half of humanity and are regarded as one of the important water storage reservoirs. Their ecological, aesthetic and socio-economic significance is not only important and beneficial to the people living in them but also to those living downstream and beyond them. Mountains harbour high level of biological diversity, which is the result of the compression of climatic zones with altitude and the diversified small-scale habitats produced by varied topological climates. However, mountain ecosystems are among the most delicate in the world and are vulnerable to climate change, urbanization, invasive alien species and other anthropogenic changes.

The Indian Himalayan Region (IHR) is blessed with abundant natural resources in the forms of forest, water, pleasant climate, fertile soil and beautiful landscapes. Various water sources in form of glaciers, the highland lakes, glacier-fed perennial rivers, springs and ground water is in surplus amount in this region. The economic viability of these forests is very important. Alpine meadows are the repository of many medicinal plants. The Himalayas mainly comprises of lofty snow-clad mountain peaks, alpine pastures, valleys and thick forest covers. Unsustainable land use practices and negligence to conserve biodiversity and water resources have contributed to various natural hazards such as species extinction and habitat degradation, leading to unsustainable development (Sharma et al. 2010). In the present scenario, biodiversity across all the major systems like terrestrial or freshwater and levels (genetic, species and ecosystem) is undergoing major changes and resulting in altered biodiversity, circulation of various ecosystem services downstream and welfare of people linked to them (Pandit et al. 2014).

Microorganisms are the important life form of the Earth that occurs in all climatic areas, including even those which were once considered impossible to support life, i.e. the cold deserts like Arctic and Antarctic (Boetius et al. 2015), the very high temperature of geysers and oceanic hot vents (Sylvan et al. 2012) and deep rocks inside the Earth's crust. Natural microbial diversity includes a broad range of microorganisms that have a strong impact on global major processes such as the nitrogen, carbon and sulphur biogeochemical cycles (Golubiewsk and McGinley 2010). As the scenario of microbial diversity has become more apparent, the need to preserve them and their gene pools has become equally important.

Cold environments represent a major portion of the Earth's biosphere which has been colonized by cold-adapted microbes, generally termed as "psychrophiles". These microorganisms are not only surviving at low temperatures but they have also maintained their viability. These microorganisms play significant roles in their habitats and include a variety of representatives of all the three domains of life. Temperature has a strong influence on whether a particular organism can survive and/or thrive, which can be both indirect and direct through its impact on water and/or on the organic molecules constituting the living cells, respectively (Maayer et al. 2014). Psychrophiles have evolved a complete set of complex morphological and physiological adaptations for their survival. As a result, there are evidences of diverse metabolic activities in cold ecosystems.

6.2 The Great Himalayan Range

The Himalayas generally refers to a system of parallel ranges of tertiary mountains which stretch over 3000 km, from Myanmar (east) to Afghanistan (west). The Himalayas rises from the Indian plains and stretches to the high plateau of Tibet, in the north, resulting in a complex series of parallel, converging and diverging ranges. The width of the system ranges from 80 km to more than 300 km, resulting in a unique distribution of elevation, temperature gradients, low partial pressure, slope, winds and UV radiation. This creates a great climatic variation. These mountains mainly act as natural barricade to atmospheric circulation for the winter westerlies and the summer monsoon. The summer monsoon is predominant, lasting for 8 months (March–October) in the Eastern Himalayas, 4 months (June–September) in the Central Himalayas and 2 months (July–August) in the Western Himalayas. The dry inner valleys receive less rainfall than the adjacent mountain slopes (Shreshtha et al. 2012). These extreme climatic conditions create an annual freezing and thawing cycles and make the Himalaya a perfect realm for native microorganisms especially psychrotrophs that show optimum growth at near-mesophilic temperatures but can also tolerate cold temperatures.

Physically the Himalayas is divided into parallel zones: (1) The Great Himalaya, (2) the Inner Himalayas or Middle or Lesser Himalayas and (3) the Sub-Himalayas foothills and the adjacent Terai and Duar plain (Fig. 6.1).

6.2.1 The Great Himalayas

It is the highest and consists of a great line of snow-clad peaks with an average height exceeding 20,000 ft. It projects from the Great Himalayas and spurs towards south into the Inner Himalayas irregularly. The Nepal and Sikkim portion of the Great Himalayas has maximum number of highest peaks. Next is the Kumaon section, followed by Punjab and Bhutan section. The Great Himalayas is least high in Assam. The snow line in the southern slopes varies from 14,700 ft. (Nepal and Sikkim) to 17,000 (Punjab). In the north there are several ranges such as Zanskar, Ladakh and the Kailas. The Karakoram Range is situated on the Tibetan side. There are some high valleys that are inhabited by small clustered settlement. The extremely low temperature and short growing season restrict the farmers to one crop per annum.

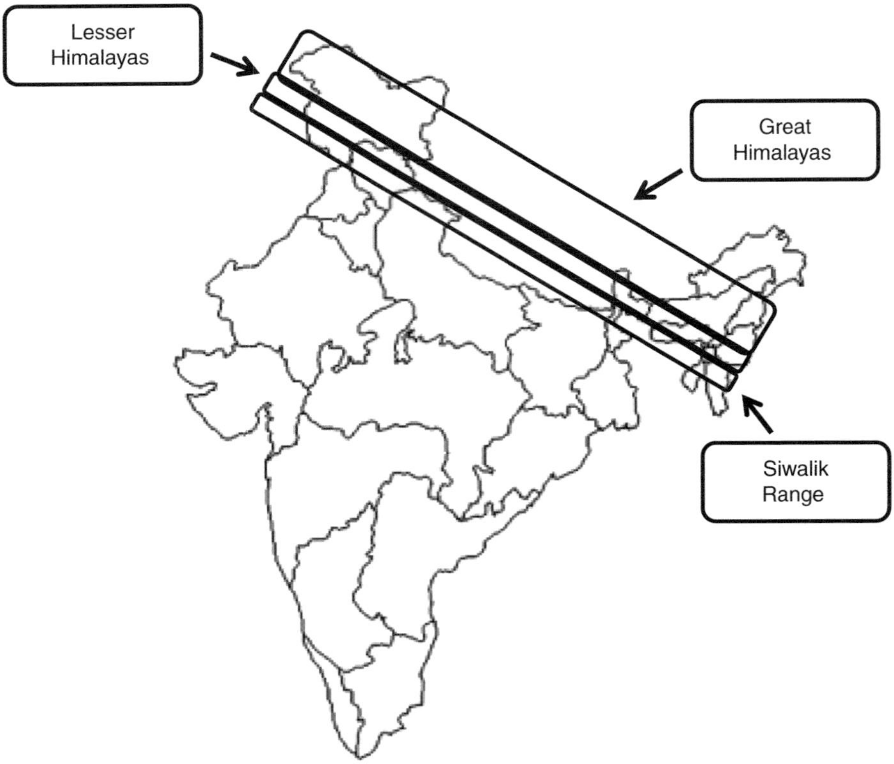

Fig. 6.1 Three parallel zones of the Himalayas

6.2.2 The Inner Himalayas

It is about 50 m wide and located at the borders of the Himalayan range to the south. It principally comprises of lofty peaks rising obliquely from the Great Himalayan range at several points. These primarily comprise Nag Tibba, the Dhauladhar range, the Pir Panjal range and the North Kashmir range. The three main outer parallel ranges are the Mahabharat, the Mussoorie range and the Ratanpir in the southern Kashmir which is separated from the Pir Panjal by the Punch River. In the eastern side of Sikkim, Bhutan and Assam, the Inner Himalayas have been divided into blocks of north-south ranges by rivers which are originating in the Great Himalayan area. It is a beautiful mosaic of dense forest-covered ranges and fertile valleys located in between. It has nonetheless isolated the fertile valleys of the Himalayas from the Gangetic plains.

6.2.3 Sub-Himalayas Foothills

It is the outermost zone which is also the lowest and includes the Siwalik range of tertiary rocks and shares a common border with the plains of Indi. Its width slowly narrows from about 30 miles in the west where it completely disappears in Bhutan and Assam. Considering a gap of about 50 miles opposite the Raidak and Tista basins, the outcrop of the Siwalik zone ranges from the Indus to Brahmaputra. A remarkable feature of this Himalaya is the presence of many numbers of longitudinal valleys with flat bottom which are called as "duns", consisting gravelly alluvium. To the south of the forested foothills lies the Sub-Himalayas piedmont plain called as Terai and Duars which forms a densely cultivated belt among much of its length.

6.3 Microbiology of Himalaya

Microorganisms have been evolving for nearly four billion years and have ability to exploit a vast range of energy sources and thriving in almost every habitat such as extremes of cold, heat, radiation, pressure, salt, acidity and darkness. The Himalayas are rich in microbial diversity. Its climatic conditions and topological characteristics have served to be a home for thousands of psychrophiles or cold-adapted microbes. Extreme environmental conditions are very common in Himalayan regions, and the study of microbial diversity of these regions is important because it provides good opportunity to understand the better adaptability of these microorganisms to extreme environmental conditions. Besides having keen interest to understand the biology of these microorganisms, slow growth rate makes the study of these psychrotolerants difficult, but still, the Himalayan region is extensively explored for the study of microbial biodiversity, various plant growth-promoting rhizobacteria (PGPR) and economically important microorganisms (Suyal et al. 2014; Kumar et al. 2014).

Cold environments of Himalaya are known to harbour the great diversity of microorganisms which includes bacteria, archaebacteria, unicellular algae and fungi. Persistent cold conditions in this habitat leads to reduced bioavailability of nutrients in soil, reduced enzymatic activity, altered soil pH, water activity and soil salinity. Psychrophiles have different adaptation to combat the cold stress and thrive in low-temperature environments. These microorganisms have cold shock proteins (CSPs) which provide protection against cold stress, and the presence of CSPs in these microorganisms has been confirmed by the previously done genomic and proteomic analysis of bacterial isolates from Western Indian Himalayas. Several cold-adapted nitrogen-fixing bacterial species were isolated from Himalayan soil, and proteome of psychrophilic diazotroph *Pseudomonas migulae* S10724 (Suyal et al. 2014) and psychrotroph *Pseudomonas palleroniana* N26 (Soni et al. 2015) was studied to document the protein profile under low-temperature diazotrophy.

6.3.1 Bacterial and Fungal Diversity in Himalayan Soil Ecosystem

Microorganisms living in soil ecosystem control carbon and nitrogen cycle and establish a link between plant diversity and soil ecosystem. Similarly, plant diversity also has its influence on the microbial community structure of the soil around the plant roots. Plants root secrets root exudates which are used by microorganism for their growth and development. Due to rich source of nutrients the region around the vicinity of root harbours rich bacterial diversity. This region of contact between root and soil is called as "rhizosphere". The rhizosphere is the thin zone of soil around the plant roots which is rich in substrates, compared to bulk soil, that affect microbial activity. This rhizosphere effect is caused because 5–21% of carbon fixed by the plant is secreted mainly as root exudate (Lugtenberg and Kamilova 2009). Pala et al. (2011) reported two species of macrofungi from Gulmarg, Doodhpathri, Uri, Yusmarg and Kellar forests of Kashmir. Sharma et al. (2015a, b) had reported a number of species of fungi from different vegetation zones of Arunachal Himalayas. *Ascomycota* was the dominant phylum, followed by *Zygomycota*, *Basidiomycota* and *Heterokontophyta* (Table 6.1).

Suyal et al. (2015) studied soil bacterial diversity in rhizosphere of *Phaseolus vulgaris* from Western Indian Himalaya using unculturable approach and found that

Table 6.1 Fungi reported from Arunachal Himalayas

Phylum	Class	Genus	References
Ascomycota	*Eurotinomycetes*	*Chromocleista, Paecilomyces, Aspergillus, Penicillium, Talaromyces, Thysanophora*	Sharma et al. (2015a, b)
	Sodariomycetes	*Fusarium, Trichoderma, Hypocrea, Emericellopces, Nectria, Chamaeleomyces, Niesslia, Apiospora, Pleurostomophora*	Sharma et al. (2015a, b)
	Dothideomycetes	*Davidiella, Cladosporium*	Sharma et al. (2015a, b)
		Alternaria	Vashisht and Chauhan (2016)
	Pezizomycetes	*Helvella*	Sharma et al. (2015a, b)
		Gyromitra	Pala et al. (2011)
	Lecanoromycetes	*Umbilicaria*	Krzewicka (2010)
Zygomycota	*Zygomycetes*	*Umbelopsis, Absidia, Mucor, Mortierella*	Sharma et al. (2015a, b)
Basidiomycota	*Urediniomycetes*	*Torula, Pseudotorula*	Sharma et al. (2015a, b)
	Agaricomycetes	*Rhizoctonia*	Dar et al. (2011)
		Amanita, Russula	Sharma et al. (2015a, b)
		Suillus	Verma and Reddy (2015)
		Mutinus	Pala et al. (2011)
Heterokontophyta	*Oomycetes*	*Pythium*	Sharma et al. (2015a, b)

Table 6.2 Plant growth-promoting psychrotrophic bacteria

Division	Class	Genus	References
Firmicutes	*Bacilli*	*Bacillus, Paenibacillus, Sporosarcina*	Yadav et al. (2016)
		Exiguobacterium	Singh et al. (2013)
		Staphylococcus	Spergser et al. (2003)
Actinobacteria	*Actinobacteria*	*Arthrobacter, Microbacterium*	Gangwar et al. (2011)
		Rhodococcus	De Mandal et al. (2015)
		Sanguibacter	Yadav et al. (2015)
Proteobacteria	α-protebacteria	*Brevundimonas, Methylobacterium*	Yadav et al. (2015)
	β-proteobacteria	*Janthinobacterium*	Suman et al. (2015)
		Burkholderia	Soni et al. (2016)
	γ-proteobacteria	*Aeromonas, Pseudomonas*	Yadav et al. (2015)
		Yersinia	Sangwan et al. (2015)
		Psychrobacter	Latha et al. (2009)
Bacteroidetes	*Sphingobacteria*	*Sphingobacterium*	Yadav et al. (2015)
	Flavobacteria	*Flavobacterium*	Gangwar et al. (2011)

proteobacteria was the dominant group of bacteria with *Pseudomonas* as predominant genus. Other bacterial groups present were *Chloroflexi*, *Actinobacteria*, *Acidobacteria*, *Gemmatimonadetes*, *Planctomycetes*, *Bacteroidetes*, *Firmicutes*, *Nitrospira*, *Cyanobacteria*, *Verrucomicrobia* and *Chlorobi*. Yadav et al. (2015) reported soil bacterial diversity from different altitudes (3978–4069 m) of Rohtang Pass (Himachal Pradesh) situated in North-western Himalayas. *Proteobacteria* was the major group followed by *Firmicutes* and *Actinobacteria*, respectively. These bacteria exhibited remarkable plant growth promotory properties (Table 6.2). Similarly, bacterial and fungal diversity from Sikkim Himalaya revealed the presence of bacteria, actinomycetes and fungi, which was found to be decreased along the increasing altitude (Rai and Kumar 2015).

Institute of Bioresource Technology, Palampur, studied the rhizospheric microbial diversity in different plants growing in the cold regions of Himalaya and found that the rhizospheres of sea buckthorn (*Hippophae rhamnoides* L.), tea (*Camellia sinensis* L.), black gram (*Vigna mungo* L.) and carnation (*Dianthus caryophyllus* L.) were dominated by various species of *Pseudomonas* (Gulati et al. 2010). Moreover, to conserve the diversity of cold-adapted fluorescent *Pseudomonas*, a repository has been established at Ranichauri Hill Campus of Govind Ballabh Pant University of Agriculture and Technology (GBPUA&T) Pantnagar, Uttarakhand (Negi et al. 2005).

Other dominating group of microorganism inhabiting Himalayan soil other than bacteria are fungi. Fungi are the most diverse group of organisms, which are the largest group of living organism in terms of species richness. Besides this fact fungi also contribute more of soil biomass compared to bacteria and maintain global carbon cycle by decomposing plant-derived polymeric substances like cellulose, hemicelluloses and lignin. The fungal distribution in soil ecosystem is affected by the soil organic content, soil texture, surface vegetation and other physiochemical properties of soil.

Fungal diversity in soil at higher altitudes of Sikkim and Uttarakhand Himalaya has shown that the *Penicillium* is the most abundant and diverse genus present over there. *P. raistrickii*, *P. janthinellum*, *P. pinophillum*, *P. javanicum*, *P. chrysogenum*, *P. oxalicum*, *P. purpurogenum* and *P. aurantiogriseum* are some of the commonly occurring species of *Penicillium*. *Aspergillus*, *Epicoccum*, *Fusarium*, *Myrothecium*, *Cladosporium*, *Paecilomyces*, *Gangronella* and *Trichoderma* were other genera contributing to the fungal flora of this Himalayan region (Rai and Kumar 2015).

Further, Devi et al. (2012) studied fungal diversity in Eastern Himalayan range and found that the major diversity of fungi belonged to phyla *Zygomycota* and *Ascomycota*. *Eurotiales*, *Calosphaeriales*, *Hypocreales*, *Mortierellales*, *Pleosporales*, *Mucorales* and *Capnodiales* were the seven dominant orders corresponding to these phyla. Studies from four altitude ranges [1–500, 500–1000, 1000–1500 and 1500–2000 masl (metres above sea level)] suggested that *Eurotiomycetes* was dominant in all altitudes followed by *Sordariomycetes*, except 500–1000 masl, where *Sordariomycetes* were dominant. Only very few *Dothideomycetes* were observed in the altitudinal range 500–1000 masl. This poor diversity of *Dothideomycetes* could be due to presence of different vegetation types because vegetation type plays an important role in the diversity of organisms residing in that particular habitat. *Penicillium*, *Eurotiales* and *Aspergillus* were the most diverse and evenly distributed genus in Eastern Himalayan range. In this region of Himalaya, the most frequently occurring genus of *Ascomycota* were *Penicillium*, *Talaromyces*, *Chromocleista*, *Paecilomyces*, *Thysanophora*, *Fusarium*, *Trichoderma*, *Hypocrea*, *Aspergillus*, *Emericellopces*, *Nectria*, *Chamaeleomyces* and *Niesslia*, while *Umbelopsis*, *Absidia*, *Mucor* and *Mortierella* were commonly occurring *Zygomycota*. Diversity indexes were highest for genus *Penicillium* followed by *Aspergillus* and *Talaromyces*.

Symbiotic association of fungi with plants root is termed as mycorrhiza. Seven types of mycorrhizal associations are known to be present between fungi and plants root, namely, ectomycorrhizae, ecto-endo mycorrhiza, monotropoid, arbuscular mycorrhizae, ericoid, orchidoid mycorrhizae and arbutoid mycorrhiza (Brundrett 2004). Diversity of fungi varies according to the plant species present in the particular habitat regardless to latitude of habitat; hence, the diversity pattern of mycorrhizal fungi does not increase from poles to tropics, which is mostly observed in the diversity of plants and other organisms. Hence plant community plays an important role in the establishment of mycorrhizal diversity. Singh (2014) described the diversity of ectomycorrhizae in context of Himalayan trees and their plant growth-promoting potential. *Amanita*, *Russula*, *Boletus*, *Lactarius*, *Suillus* and *Hygrophorus* are the common ectomycorrhizal fungal genera associated with oaks and conifers in temperate forest of Western Himalaya (Wang et al. 2015).

6.3.2 Diversity of Microorganism in Aquatic Habitat of Himalaya

Metagenomic analysis from cold climatic water ecosystem of North-western Himalayas revealed the existence of *Actinobacteria*, *Bacteroidetes*, *Firmicutes* and *Proteobacteria* in the Himalayas (Yadav et al. 2015). Moreover, hot springs present in the Himalayan region represent extreme environment for growth of microorganisms because outlet temperature of these hot springs ranges about 45 °C and above.

Microbial diversity of Indian Himalayan hot springs was extensively studied using metagenomics and culture-dependent approaches. Sharma et al. (2015a, b) and Bhardwaj and Tiwari (2010) studied culturable diversity of aquatic microorganism from two hot springs of Soldhar and Ringigad, respectively, which are located in the Chamoli district of the Garhwal Himalaya. Bacteria, filamentous organisms and yeast were the major groups observed in these soils, and spore-forming rods were dominant bacteria among these groups. Study of bacterial diversity of the Soldhar hot spring, using 16S rRNA gene clone library, denaturing gradient gel electrophoresis (DGGE) and sequencing of bands from the DGGE revealed that *Proteobacteria* was the predominant group in this habitat, followed by *Deinococcus*, *Thermus* and *Aquificae*. The only archaea found in this hot spring was the *Pyrobaculum* (Sharma et al. 2015a, b). Moreover, several cyanobacterial species were also reported from Soldhar and Ringigad hot springs, viz. *Chroococcus minimus*, *Chroococcus turgidus*, *Chroococcus tenax*, *Synechococcus elongatus*, *Synechococcus sallensis*, *Gloeocapsa livida*, *Myxosarcina* sp. *Oscillatoria animalis* and *Oscillatoria limnosa* (Kumar et al. 2011; Yadav et al. 2016).

6.3.3 Diversity of Microorganism in Himalayan Glacier

Glaciers and ice sheets are the largest freshwater reservoirs which embody about 10% of the surface of the Earth. They are vital components for the Earth's climate, and due to harmful effect of global warming, melting rate of these freshwater reservoirs is increasing which leads to rise of sea level (Cazenave and Cozannet 2014). Various Himalayan glaciers are explored for the microbial diversity and for the documentation and conservation of psychrophilic and psychrotolerant microorganisms (Řeháková et al. 2011; Shivaji et al. 2011). Studies of these Himalayan glaciers reported that the *Proteobacteria*, *Cytophaga–Flavobacterium–Bacteroides* (CFB) and high G + C gram-positive bacteria are common inhabitant of such cold habitats (Srinivas et al. 2011).

Clone library analysis by 16S rDNA from Himalayan glacier revealed the dominance of *Bacteroidetes* and *Betaproteobacteria*, followed by the *Actinobacteria*. Further, study of dry mountains of Ladakh revealed the dominance of phototrophic microbial communities with wide diversity of soil cyanobacteria and microalga (Řeháková et al. 2011). Furthermore, Srinivas et al. (2011) studied bacterial diversity of Kafni Glacier in Kumoan Himalayas, at an altitude of about 3853 m using 16S rRNA gene clone libraries, and suggested that majority of bacterial genera belonged to phylum *Proteobacteria*. In *Firmicutes*, *Bacilli*, *Clostridia*, *Erysipelotrichia* and *Negativicutes* were the dominant. Other important bacteria were *Cryobacterium roopkundense* (*Actinobacteria*), *Acidobacteria*, *Tenericutes*, *Chloroflexi*, *Chlamydiae* and candidate phylum TM7s TM7a. Nevertheless, Shivaji et al. (2011) studied bacterial diversity in Pindari Glacier of Himalaya, where *Actinobacteria*, *Firmicutes* and *Proteobacteria* were common. Dominant bacterial genera were from *Acidobacteria*, *Fibrobacteres*, *Chloroflexi*, *Nitrospirae* and *Planctomycetes*. *Chlamydiae*, *Verrucomicrobia*, *Bacteroidetes*, *Chlorobi*, *Dictyoglomi*, *Gemmatimonadetes* and candidate TM7s TM7a phylum were also observed in this glacier.

6.4 Challenges for Micro-diversity Conservation

6.4.1 Rising Temperature

Temperature rise observed in the twenty-first century depicts enhanced greenhouse effect observed in the twentieth century (Cook et al. 2014). Over the last 100 years, the average rise in temperature for the Himalayas is significantly large than the global average of 0.74 °C (Joshi et al. 2015). The major problem is its least predictability but the effects are detrimental in the long run. Mountains can be considered as important indicators of climatic change (Singh et al. 2010). Glacial retreat and shrinking snow lines leading to changed flow of rivers along with alteration in water flow may result in various social and environmental issues. In nutshell, climate change has affected the agriculture, forestry, tourism, livestock, etc. as well as human health. These changes ultimately influence the microbial diversity from the Himalayan ranges.

6.4.2 Variation in Precipitation Pattern

The Himalayas form the main natural water resource of the major river systems of the Indian region (Nandargi and Dhar 2011). During the last few decades, variable rainfall trend has been prevalent across Asia. Both increasing and decreasing precipitation patterns were detected in the Himalayan region. Unfortunately, a vast reduction in rainfall has been recorded for the southern and eastern Tibetan Plateau and the Central Himalayas. These variations have imposed the harmful effects on natural ecosystems and, therefore, created great threat to indigenous microflora.

6.4.3 Retreating Glaciers

The rate of receding of Himalayan glaciers is faster than the world average (Kulkarni et al. 2011), thus posing a serious threat. It is generally a combination of precipitation decrease and temperature increase in the Himalayas. According to an estimate, the shrinking of glaciers will speed up if the global warming persists for prolonged period, and it is believable that many glaciers will retreat in the coming years even more (Li et al. 2015), while smaller ones may completely vanish. This has become a serious threat to the microbiota that is found to be the native of these glaciers.

Permafrost constitutes a major and unique ecological niche for psychrophiles with relatively high microbial diversity. In some cases, the community composition is variable across different permafrost features. Permafrosts are highly susceptible to deterioration with climate change. Various studies conducted have shown that the shrinkage of permafrost and the thickness of upper portion of soil have increased which has resulted in altered hydrological cycle, carbon dioxide and vegetation composition (Fort 2015). The increased rate of seasonal thawing over large areas has led the ground to be unstable and easily eroded (Gruber et al. 2016). This will consequently result in loss of biodiversity in the near future.

6.5 Indigenous Approach for Micro-diversity Conservation

Conservation of the biodiversity is the need of the hour. It is clearly evident that the various tribal communities or local groups of the Himalayan region have knowingly or unknowingly contributed a lot in conservation of this resource. These communities are following their traditions for years which have significantly shown some improvement in this direction. There are several ways in which these local groups help in maintaining the diversity (Fig. 6.2). Cultural diversity is in close relation with the biodiversity. These interrelationships need to be studied simply for the reason that culture is ethically essential for development, as well as it provides a condition of its sustainability as they exist in a symbiotic relationship. The association of religion with ecosystem is eternal in the Himalayan traditional communities. Since a long time, the idea of traditional methods for conservation of resources has

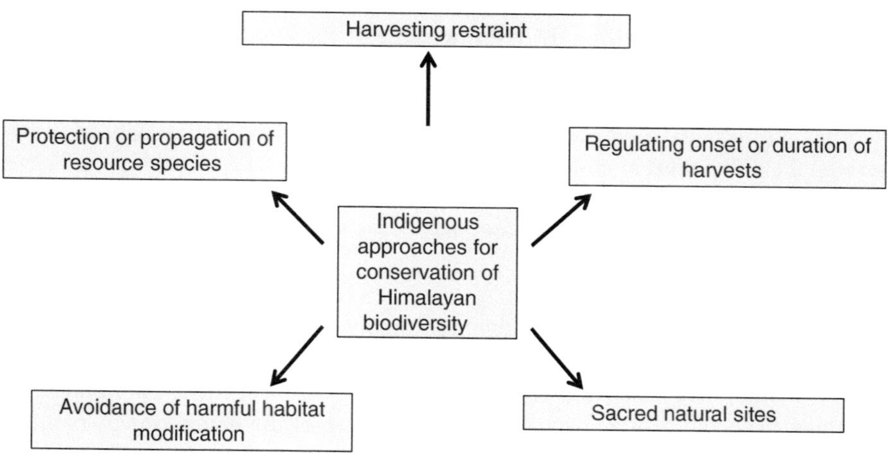

Fig. 6.2 Indigenous approach for the conservation of Himalayan microbial diversity

gained a worldwide recognition. Traditional knowledge is important for sustainability of natural resources that include forests, water and agroecosystems across landscape, farms, village, commons and wildlife. Some of these strategies are discussed as follows:

- Harvesting restraint
- Protecting sacred pastures and landscapes would ultimately lead to protection of microbial diversity residing in these areas.
- Protection of resourceful species
- It involves protection of resource species to conserve associated microflora effectively.
- Regulating harvest
- It also involves the conservation of microorganisms by confining the number of harvest of the sacred plants involved in the rituals.
- Avoidance of harmful habitat modification
- Some types of habitats are more sensitive to the effects of modification than others, and hence avoidance or mitigation of such habitat change can be a form of conservation.
- Example: *Putuk-tu bugyals* in the Vyas Valley
- Sacred natural sites
- One could easily visualize the important role played by these sacred forests in the protection of the biodiversity as these areas strictly prohibit any unnecessary encroachment.

A variety of traditional crops are grown in the Himalayan agroecosystems which are being looked after by local farmers since time immemorial. The rhizosphere of the crops is inhabited by diverse and potential microorganisms (Suyal et al. 2014). These microorganisms are adapted to the local environment and possess the innate qualities to withstand various harsh environmental conditions. This ability to adapt according to the environment has ensured the food and nutritional security of the hill farmers. However, during the last few decades, the area for traditional agriculture has drastically declined (>60%), and thus many of the microbial species are at the verge of extinction (Suyal et al. 2015). In this perspective, the encouragement of marginal farmers and their traditional cropping pattern seems very essential for preserving the indigenous microbial reservoir.

6.6 Conclusion

The Himalayan ranges are the most dynamic mountain systems as well as extremely vulnerable to global warming. It has been a home for a number of organisms since ages. The rate, magnitude and potential outcome of climate change are gradually changing the ecological and socio-economic pattern of the Himalayas. A number of diverse microbes have been successfully residing in the various niches such as soil, water, air and glacial ice, etc. which indicates its natural prosperity. These microbes have skilfully adapted themselves according to the extreme environmental conditions and successfully overcame the two main physical challenges, namely, high viscosity and low thermal energy, both of which slows down the metabolism. Their adaptation may be due to highly flexible vital parts of the molecular structure that reduce or nullify the freezing effect of low temperatures. They are successful inhabitants of the harsh habitats of Earth which otherwise are considered impossible to live. The further studies will lead to the characterization and exploration of potential microorganisms for productive and sustainable agricultural practices and industries.

6.7 Opinion

The Himalayas are one of the most significant mountain systems in the world. It is not only one of the prime topological features regulating the climate and seasonal variations in the surrounding nations, but also it serve as a huge reservoir to a number of animal, plant and microbial species. The conservation of the Himalayan diversity will in turn lead to sustainable development as these ecosystems form complex interrelations among the various entities of nature. Various advance conservation strategies involve preserving the microorganisms in culture collections and conservation of their potential genes in gene or cDNA libraries, but several indigenous approaches have also proven to be effective to maintain the integrity of the Himalayan ecosystem. Thus, integrated approach involving the advance and indigenous strategies would ultimately lead to stable and long-lasting development.

Acknowledgement The work mentioned in this chapter from author group was supported by the National Bureau of Agriculturally Important Microorganisms, India (NBAIM/ICAR), grant to R.G.

References

Boetius A, Anesio AM, Demimg JW, Mikucki JA, Rapp JZ (2015) Microbial ecology of the cryosphere: sea ice and glacial habitats. Nat Rev Microbiol 13:677–690. doi:10.1038/nrmicro3522

Brundrett MC (2004) Diversity and classification of mycorrhizal associations. Biol Rev 79:473–495. doi:10.1017/S1464793103006316

Cazenave A, Cozannet GL (2014) Sea level rise and its coastal impacts. Earth's Future 2:15–34. doi:10.1002/2013EF000188

Chitale VS, Behera MD, Roy PS (2014) Future of endemic flora of biodiversity hotspots in India. PLoS One 9:1–15. doi:10.1371/journal.pone.0115264

Cook BI, Smerdon JE, Seager R, Coats L (2014) Global warming and 21st century drying. Clim Dyn 43:2607–2627. doi:10.1007/s00382-014-2075-y

Dar GH, Beig MA, Ahanger FA, Ganai NA, Ahanger MA (2011) Management of root rot caused by *Rhizoctonia solani* and *Fusarium oxysporum* in blue pine (Pinus walchiana) through use of fungal antagonists. Asian J Plant Pathol 5:62–67. doi:10.3923/ajppaj.2011

De Mandal S, Sanga Z, Kumar NS (2015) Metagenome sequencing reveals Rhodococcus dominance in Farpuk Cave, Mizoram, India, an Eastern Himalayan biodiversity hot spot region. Genome Announc 3:1–2. doi:10.1128/genomeA.00610-15

Devi LS, Khaund P, Nongkhlaw FM, Joshi SR (2012) Diversity of culturable soil Micro-fungi along altitudinal gradients of Eastern Himalayas. Mycobiology 40:151–158. doi:10.5941/MYCO.2012.40.3.151

Fort M (2015) Permafrost in the Himalayas: specific characteristics, evolution vs. climate change and impacts on potential natural hazards. Geophys Res Abstr, Conference paper, 17, EGU General Assembly

Gangwar P, Alam SI, Singh L (2011) Metabolic characterization of cold active *Pseudomonas*, *Arthrobacter*, *Bacillus*, and *Flavobacterium* spp. from Western Himalayas. Indian J Microbiol 51:70–75. doi:10.1007/s12088-011-0092-7

Golubiewsk N, McGinley M (2010) Species influences upon ecosystem function. In: Cleveland CJ (ed) Encyclopedia of earth. Environmental Information Coalition, National Council for Science and the Environment, Washington, DC. http://www.eoearth.org/article/Species_influences_upon_ecosystem_function

Gruber S, Fleiner F, Guegan E, Panday P, Schmid MO, Stumm D, Wester P, Zhang Y, Zhao L (2016) Inferring permafrost and permafrost thaw in the mountains of the Hindu-Kush Himalaya region. Cryosphere Discuss, pp 1–29. doi: 10.5194/tc-2016-104

Gulati A, Sharma N, Vyas P, Sood S, Rahi P, Pathania V, Prasad R (2010) Organic acid production and plant growth promotion as a function of phosphate solubilization by *Acinetobacter rhizosphaerae* strain BIHB 723 isolated from the cold deserts of the trans-Himalayas. Arch Microbiol 192:975–983. doi:10.1007/s00203-010-0615-3

Joshi R, Kumar K, Palni LMS (eds) (2015) Dynamics of climate change and water resources of north-western Himalaya. In: Society of earth scientists series. Springer, Switzerland. ISBN: 978-3-319-13742-1

Krzewicka B (2010) *Umbilicaria rhizinata* comb. nov. (Lichenized ascomycota). Lichenologist 42:491–493. doi:10.1017/S0024282910000010

Kulkarni AV, Rathore BP, Singh SK, Bahuguna IM (2011) Understanding changes in the Himalayan cryosphere using remote sensing techniques. Int J Remote Sens 32:601–615. doi:10.1080/01431161.2010.517802

Kumar M, Gupta RK, Bhatt AB, Tiwari SC (2011) Epiphytic cyanobacterial diversity in the sub-Himalayan belt of Garhwal region of Uttarakhand, India. Botan Orientalis J Plant Sci 8:77–89. doi:10.3126/botor.v8i0.5955

Kumar S, Suyal DC, Dhauni N, Bhoriyal M, Goel R (2014) Relative plant growth promoting potential of Himalayan psychrotolerant *Pseudomonas jesenii* strain MP1 against native *Cicer arietinum* (L.). *Vigna mungo* (L.) Hepper; *Vigna radiata* (L.) Wilczek. *Cajanus cajan* (L.) Millsp. and *Eleusine coracana* (L.) Gaertn. Afr J Microbiol Res 8:3931–3943. doi:10.5897/AJMR2014.7035

Latha PK, Soni R, Khan M, Marla SS, Goel R (2009) Exploration of *csp* genes from temperate and glacier soils of the Indian Himalayas and *in silico* analysis of encoding proteins. Curr Microbiol 58:343–348. doi:10.1007/s00284-008-9344-0

Li Z, Fan K, Tian L, Zhang J (2015) Response of lake and glacier dynamics in four inland basins to climate change at the transition zone between Karakoram and Himalayas. PLoS One 10:1–12. doi:10.1371/journal.pone.0144696

Lugtenberg B, Kamilova F (2009) Plant-growth-promoting rhizobacteria. Annu Rev Microbiol 63:541–556. doi:10.1146/annurev.micro.62.081307.162918

Maayer PD, Anderson D, Cary C, Cowan DA (2014) Some like it cold: understanding the survival strategies of psychrophiles. EMBO Rep 15:508–517. doi:10.1002/embr.201338170

Nandargi S, Dhar ON (2011) Extreme rainfall events over the Himalayas between 1871 and 2007. Hydrol Sci J 56:930–945. doi:10.1080/02626667.2011.595373

Negi YK, Garg SK, Kumar J (2005) Cold tolerant fluorescent *Pseudomonas* isolates from Garhwal Himalayas as potential plant growth promoting and biocontrol agents in pea. Curr Sci 89:2151–2156

Pala SA, Wani AH, Bodha RH, Wani BA, Bhat MY, Mir RA, un-Nisa T (2011) Two hitherto unreported macro fungi from Kashmir; Himlaya. Biores Bull 2:130–134

Pandit MK, Kumar M, Koh LP (2014) Dancing on the roof of the world: ecological transformation of the Himalayan landscape. Bioscience 64:980–992. doi:10.1093/biosci/biu152

Rai AK, Kumar R (2015) Potential of microbial bio-resources of Sikkim Himalayan region. ENVIS Bull Himalayan Ecol 23:99–105

Řeháková K, Chlumská Z, Doležal J (2011) Soil cyanobacterial and microalgal diversity in dry mountains of Ladakh, NW Himalaya, as related to site, altitude, and vegetation. Microb Ecol 62:337–346. doi:10.1007/s00248-011-9878-8

Sangwan N, Lambert C, Sharma A, Gupta V, Khurana P, Khurana JP, Sockett RE, Gilbert JE, Lal R (2015) Arsenic rich Himalayan hot spring metagenomics reveal genetically novel predator-prey genotypes. Environ Microbiol Rep 7:812–823. doi:10.1111/1758-2229.12297

Sharma E, Chettri N, Oli KP (2010) Mountain biodiversity conservation and management: a paradigm shift in policies and practices in the Hindu Kush-Himalayas. Ecol Res 25:909–923. doi:10.1007/s11284-010-0747-6

Sharma D, Gosai K, Dutta J, Arunachalam A, Shukla AK (2015a) Fungal diversity of twelve major vegetational zones of Arunachal Himalaya, India. Curr Res Environ Appl Mycol 5:101–119. doi:10.5943/cream/5/2/4

Sharma A, Jani K, Shouche YS, Pandey A (2015b) Microbial diversity of the Soldhar hot spring, India, assessed by analyzing 16S rRNA and protein-coding genes. Ann Microbiol 65:1323–1332. doi:10.1007/s13213-014-0970-4

Shivaji S, Pratibha MS, Sailaja B, Kishore KH, Singh AK, Begum Z, Srinivas TNR (2011) Bacterial diversity of soil in the vicinity of Pindari glacier, Himalayan mountain ranges, India, using culturable bacteria and soil 16S rRNA gene clones. Extremophiles 15:1–22. doi:10.1007/s00792-010-0333-4

Shreshtha UB, Gautam S, Bawa KS (2012) Widespread climate change in the Himalayas and associated changes in local ecosystems. PLoS One 7:1–10. doi:10.1371/journal.pone.0036741

Singh SP (2014) Attributes of Himalayan forest ecosystems: they are not temperate forests. Proc Indian Natl Sci Acad 80:221–233. doi:10.16943/ptinsa/2014/v80i2/55103

Singh SP, Singh V, Skutsch M (2010) Rapid warming in the Himalayas: ecosystem responses and development options. Clim Dev 2:1–13. doi:10.3763/cdev.2010.0048

Singh NK, Raichand R, Kaur I, Kaur C, Pareek S, Mayilraj S (2013) *Exiguobacterium himgiriensis* sp. nov. a novel member of the genus *Exiguobacterium,* isolated from the Indian Himalayas. A Van Leeuw J Microb 103:789–796. doi:10.1007/s10482-012-9861-5

Soni R, Suyal DC, Agrawal K, Yadav A, Shouche Y, Goel R (2015) Differential proteomic expression of Himalayan psychrotrophic diazotroph *Pseudomonas palleroniana* N26 under low temperature diazotrophic conditions. Cryo-Letters 36:74–82

Soni R, Suyal DC, Sai S, Goel R (2016) Exploration of *nifH* gene through soil metagenomes of the western Indian Himalayas. 3Biotech 6:1–4. doi:10.1007/s13205-015-0324-3

Spergser J, Wieser M, Taubel M, Rosselló-Mora RA, Rosengarten R, Busse HJ (2003) Staphylococcus nepalensis sp. nov. isolated from goats of the Himalayan region. Int J Syst Evol Microbiol 53:2007–2011. doi:10.1099/ijs.0.02646-0

Srinivas TNR, Singh SM, Pradhan S, Pratibha MS, Kishore KH, Singh AK, Shivaji S (2011) Comparison of bacterial diversity in proglacial soil from Kafni Glacier, Himalayan Mountain ranges, India, with the bacterial diversity of other glaciers in the world. Extremophiles 15:673–690. doi:10.1007/s00792-011-0398-8

Suman R, Sharma P, Gupta S, Sourirajan A, Dev K (2015) A Novel Psychrophilic *Janthinobacterium lividum* MMPP4 Isolated from Manimahesh Lake of Chamba District of Himachal Pradesh, India. J Biochem Technol 6:846–851

Suyal DC, Yadav A, Shouche Y, Goel R (2014) Differential proteomics in response to low temperature diazotrophy of Himalayan psychrophilic nitrogen fixing *Pseudomonas migulae* S10724 strain. Curr Microbiol 68:543–550. doi:10.1007/s00284-013-0508-1

Suyal DC, Yadav A, Shouche Y, Goel R (2015) Bacterial diversity and community structure of Western Indian Himalayan red kidney bean (*Phaseolus vulgaris*) rhizosphere as revealed by 16S rRNA gene sequences. Biologia 70:305–313. doi:10.1515/biolog-2015-004

Sylvan JB, Toner BM, Edwards KJ (2012) Life and death of deep-sea vents: bacterial diversity and ecosystem succession on inactive hydrothermal sulfides. mBio 3:1–10. doi:10.1128/mBio.00279-11

Vashisht I, Chauhan RS (2016) First report of leaf spot on medicinal herb *Picrorhiza kurroa* (Royle ex. Benth) caused by *Alternaria tenuissima* in India. Plant Dis 100:647. doi:10.1094/PDIS-07-15-0809-PDN

Verma B, Reddy MS (2015) *Suillus indicus* sp. nov. (Boletales, Basidiomycota), a new boletoid fungus from northwestern Himalayas, India. Mycology 6:35–41. doi:10.1080/21501203.2014.988770

Wang P, Zhang Y, Mi F, Tang X, He X, Cao Y, Liu C, Yang D, Dong J, Zhang K, Xu J (2015) Recent advances in population genetics of ectomycorrhizal mushrooms *Rusulla* spp. Mycology 6:110–120. doi:10.1080/21501203.2015.1062810

Yadav AN, Sachan SG, Verma P, Saxena AK (2015) Prospecting cold deserts of north western Himalayas for microbial diversity and plant growth promoting attributes. J Biosci Bioeng 119:683–693. doi:10.1016/j.jbiosc.2014.11.006

Yadav AN, Sachan SG, Verma P, Saxena AK (2016) Bioprospecting of plant growth promoting psychrotrophic Bacilli from the cold desert of north western Indian Himalayas. Indian J Exp Biol 54:142–150

Phytomicrobiome: A Reservoir for Sustainable Agriculture

7

Praveen Rahi

Abstract

Global demands for agricultural produce will increase up to 70% by the next two decades. To meet this goal, the productivity of crops needs to be increased under harsher climate conditions and with declining soil and water quality. In addition to this, widespread emergence of new pathogens and pests is posing a serious threat to agricultural produce. Plant-microbe interactions may have a beneficial, neutral, or negative effect on one or both partners. The whole world needs to address climate change scenario, by enabling agriculture to adopt climate-friendly practices. Plants host a rich microbiota, which supports plants in nutrition uptake and health management. Phytomicrobiome represents a great reservoir, which can be harnessed to improve the productivity of crops and food quality. A systematic research focused on the understanding of structure and function of phytomicrobiome driven by crop management and environmental factors is critical for devising reliable strategies to improve crop health and productivity in a sustainable way, by reducing chemical inputs and emissions of greenhouse gases. Microbial diversity associated with plants is highly underestimated, as the data generated by next-generation sequencing technologies have exhibited that only a small portion of microorganisms could be cultivated until now. Recently, the high-throughput microbial cultivation approach "culturomics" in conjunction with high-speed and low-cost identification technique based on MALDI-TOF MS has offered an opportunity to cultivate the not-yet-cultivated microorganisms and harness them to manipulate the crop phytomicrobiome. In this chapter, we discussed various aspects of phytomicrobiome, like beneficial and harmful effects of its members and colonization in rhizosphere and phyllosphere.

P. Rahi
National Centre for Microbial Resource, National Centre for Cell Science, Pune, Maharashtra 411 007, India
e-mail: praveen@nccs.res.in

V.C. Kalia et al. (eds.), *Mining of Microbial Wealth and MetaGenomics*,
DOI 10.1007/978-981-10-5708-3_7

Keywords
Beneficial microbes • PGPR • Pathogenic microbes • Plant pathogens • Human pathogens • NGS • Culturomics

7.1 Introduction

Plants are surrounded with diverse microorganisms, and release of specific compounds leads to the development of niche-specific microbial communities. These microorganisms are in generally recruited from surrounding environments and many times from the parent plant through seeds. Plants invest a huge portion of photosynthates in the form of carbonaceous compounds to nurture their microbiota (Kuzyakov and Domanski 2000; Bulgarelli et al. 2013). These carbonaceous compounds differ among plants species, varieties of species, and plant growth and development phases (Akiyama et al. 2005; Singh et al. 2007; Bascom-Slack et al. 2012; Neal et al. 2012; Chaparro et al. 2014). The involvement of plants in selecting beneficial soil microbial communities is not conclusively known, but still there are many examples on selection of specific group of microorganisms by plants (see review by Reinhold-Hurek et al. 2015). Legume-rhizobia symbiosis is the best-understood plant-microbe interaction reflecting highly specific relationship and dependence of both partners on to each other (Rahi et al. 2012; Oldroyd 2013; Vuong et al. 2017).

Microorganisms occupy almost every part of plant. As the plant grows healthy and strong, its microbiome is also benefited. The term "phytomicrobiome" represents the collection of microorganisms that colonize in and on different parts of plant (Fig. 7.1). Most members of phytomicrobiome assist plants by a variety of mechanisms like nutrient uptake, growth hormone production, and combating biotic and abiotic stress (Berendsen et al. 2012; Wallenstein 2017). The nitrogen-fixing bacteria, mycorrhizal fungi, plant growth-promoting rhizobacteria, and biocontrol microorganisms have been well studied for their beneficial effects on plant health and growth (Mendes et al. 2013). In addition to their benefits to plants, several members of phytomicrobiome are pathogenic to plants and may pose negative effect on the plant growth by causing various diseases (Mendes et al. 2013; Burgess et al. 2017). Plant pathogens like *Fusarium*, *Phytophthora*, *Rhizoctonia*, *Pseudomonas*, *Ralstonia*, *Xanthomonas*, and *Erwinia* can infect a wide range of trees, woody shrubs, and herbs (Mansfield et al. 2012; Lecomte et al. 2016; Burgess et al. 2017). Control methods for such plant pathogens are very few and that too mostly based on harmful chemical pesticides, thus leading to severe losses to the crops by pathogenic microorganisms. The application of chemicals to control plant pathogens can cause serious environmental problems. Once the disease has broken out, it influences the quality of the plant produce, which makes it unsuitable for commercialization (Lecomte et al. 2016). The other important aspect of phytomicrobiome is the presence of human pathogenic microorganisms in and on plants. Several human pathogens from soil or irrigation water may enter inside the plant tissues through the

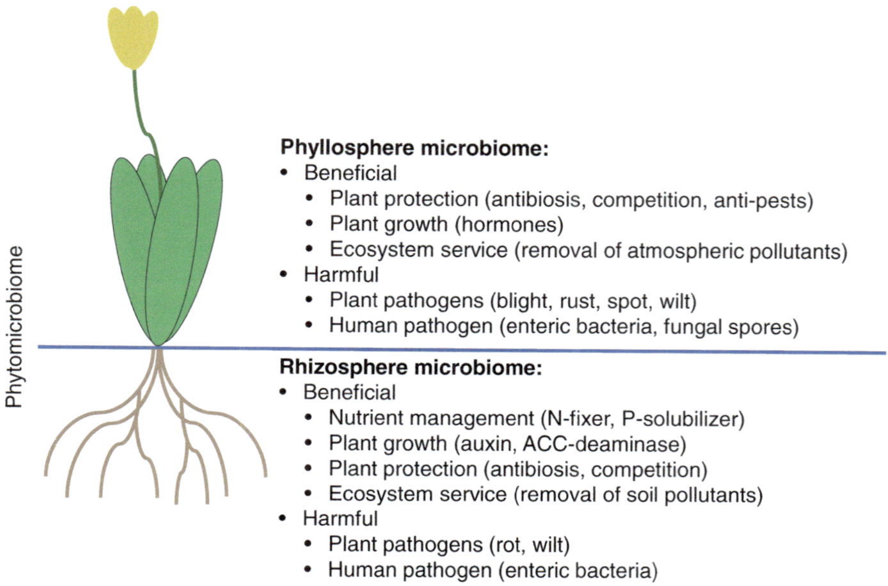

Fig. 7.1 The phytomicrobiome and roles of microbial communities

shoot as well as the root and translocate, colonize, and persist inside the plants (Hofmann et al. 2014). Amendment agriculture fields with organic manures and irrigation with treated or untreated effluents are the most common source for human pathogenic microorganisms (Hofmann et al. 2014; Ongeng et al. 2014; Ximenes et al. 2017). Plant-microbe interactions are important aspect for both agriculture and horticulture, but a relatively small portion of it is already explored and exploited. However, a large portion remain unexplored, which can be exploited for optimizing plant production. A better understanding of phytomicrobiome could help find novel ways to engineer the microbiome to improve the plant nutrition and resistance to stress (Farrar et al. 2014; Smith et al. 2017).

The advent of metagenomic technologies like next-generation sequencing has accelerated the investigations of crucial unknown structure and function of phyto-microbiome. Deep sequencing of metagenome from diverse environments using next-generation technologies exhibited a continuous rise in the diversity estimates for microorganisms, and the number of total predicted "species" reaches well into the millions (Pedrós-Alió 2006; Rinke et al. 2013). Metagenomic tools revealed the enormous diversity of the microorganisms that are associated with plant and make up the phytomicrobiome (Berendsen et al. 2012). Recent studies based on next-generation sequencing exhibited the unexplored community composition of different plant microbiomes, including *Arabidopsis*, *Populus*, maize, rice, soybean, and wheat (Bais et al. 2006; Mendes et al. 2011; Berendsen et al. 2012; Bulgarelli et al. 2012; Edwards et al. 2015; Rascovan et al. 2016). The numbers of bacterial cells estimated by microscopy of soil samples directly and of those grown on agar plates have high level of discrepancies, but recent studies based on high-throughput sequencing confirmed the presence of uncultivable majority often referred as "microbial dark matter" (Rinke et al. 2013; Solden et al. 2016). A comparison of culture-dependent and culture-independent studies has made it clear that only a small number of microorganisms could be cultivated in vitro and the majority of microorganism could not be cultured by using standard cultivation methods (Vartoukian et al. 2010). Novel cultivation techniques are being developed to cultivate these not-yet-cultivated microorganisms present in an ecosystem, including dominant and rare species (Vartoukian et al. 2010; Dubourg et al. 2014; Lagier et al. 2016; Dickson 2017). These techniques involve multiple culture conditions, including simulation of natural environmental conditions by inclusion of essential nutrients and signaling molecules (Vartoukian et al. 2010; Lagier et al. 2016). Such high-throughput cultivation techniques generally generate microbial isolates in large numbers, which require instant and error-free characterization of these microorganisms. Matrix-assisted laser desorption/ionization time-of-flight mass spectrometry (MALDI-TOF MS)-based biotyping has offered a solution for high-throughput and rapid microbial identification (Lagier et al. 2016; Rahi et al. 2016). It is expected that in the near future, researchers will adopt high-throughput cultivation techniques to study the phytomicrobiome and will successfully cultivate several new members of phytomicrobiome. The advantage of cultivation is that the isolates can be used for further applications based on their potentials to manipulate the phytomicrobiome to achieve better growth and yield of crop plants in a sustainable way. In this chapter, we highlighted the strategies to understand and manipulate the phytomicrobiome for nutrient uptake, defense of pathogen and pest attacks, and resource-use efficiency, thus improving the quality and quantity of crop yield.

7.2 Rhizosphere Microbiome

All terrestrial plants get their energy through photosynthesis, which took place in the aboveground part of plants in the presence of sunlight. Roots typically lie below the surface and perform several important functions for plant, including absorption of nutrients. In 1904, Lorenz Hiltner, a German agronomist and plant physiologist, coined the term "rhizosphere," which he described as the area around a plant root inhabited by a microbial population influenced by the root exudates secreted by plant roots (Hiltner 1904; Hartmann et al. 2008). The rhizosphere microbiome play a crucial role in improving plant health and growth by facilitating nutrient acquisition, providing defense against biotic factors, and helping plants to tolerate abiotic stresses. Nearly 20% of plant photosynthates are released in the form of root exudates (Kuzyakov and Domanski 2000), and this makes rhizosphere an energy-rich niche, which attracts a huge diversity of microorganisms (Bulgarelli et al. 2012; Hirsch and Mauchline 2012; Lundberg et al. 2012; Lareen et al. 2016). Plant domestication and crop breeding under intense agriculture have largely decoupled the rhizosphere microbiome from plant selection (Wallenstein 2017).

7.2.1 Beneficial Rhizosphere Microbiome

Plants obtain most of their carbon and oxygen from the air, and after the availability of water, nitrogen, phosphorus, and potassium are the primary nutrients required for plant growth and productivity. Fertile soils generally have decent amount of these nutrients in the forms, which can be absorbed by plants, but the pool of available nutrients in soils is not permanent. The rhizosphere microbiome plays an important role in the availability of these primary nutrients by fixing atmospheric nitrogen, solubilizing phosphorus and potassium, and enhancing their uptake. Several rhizosphere and root-endophytic bacteria are known for nitrogen fixation. The members of group rhizobia (*Rhizobium*, *Mesorhizobium*, *Bradyrhizobium*, *Azorhizobium*, *Allorhizobium*, and *Sinorhizobium*) are the key players in nitrogen fixation in most of the legume plants (Oldroyd 2013; Laranjo et al. 2014; Sprent et al. 2017). The microbiomes of rhizocompartments (nodule endophytes, root endophytes, rhizosphere, and root zone) in soyabean and alfalfa analyzed using high-throughput sequencing reveal a clear hierarchical filtration of microbiota by plants (Xiao et al. 2017). In the same study, the microbiomes of root zone and rhizosphere soils belong to 30–31 phyla, which is similar to the number in the bulk soils (29 phyla), and root and root nodule endosphere microbiomes were composed of 28 and 27 phyla, respectively. Root nodules are the key compartment where nitrogen fixation took place. Cultivation-dependent and cultivation-independent studies have shown super-dominance of nitrogen-fixing rhizobia, which is generally specific to host plants (Rahi et al. 2012; Sprent et al. 2017; Xiao et al. 2017). In addition to legumes, several other plants like rice, maize, wheat, and sugarcane also host diverse endophytic bacteria in very high proportions and are also involved in nitrogen fixations (Charpentier and Oldroyd 2010; Brusamarello-Santos et al. 2017; Defez et al. 2017). *Gluconacetobacter diazotrophicus* is an endophytic plant growth-promoting bacteria found in many of unrelated monocot and dicot plant species, which can fix nitrogen, secret phytohormone, and solubilize mineral nutrient (Rangel de Souza et al. 2016; Yoon et al. 2016). Rhizosphere bacteria such as *Azospirillum*, *Klebsiella*, *Burkholderia*, *Bacillus*, and *Pseudomonas* have also been reported for nitrogen fixation (Montañez et al. 2009; Piromyou et al. 2011; Arruda et al. 2013; Kuan et al. 2016).

Several rhizobacterial metabolites and various mechanisms are involved in promoting plant growth. Solubilization of insoluble phosphates and production of indole-3-acetic acid (IAA) are triggered by microbial metabolites (Hariprasad and Niranjana 2009; Gulati et al. 2010; Oteino et al. 2015). Several rhizosphere microorganisms secrete organic acids, which contribute in the solubilization of insoluble complexes of phosphate and make phosphorus available for plants (Richardson et al. 2009; Gulati et al. 2010). The inoculation of *Pisum sativum* L. rhizosphere with phosphate-solubilizing and gluconic acid-producing endophytic bacteria resulted in an increase in plant growth (Oteino et al. 2015). However it is difficult to ascertain the involvement of a specific plant growth-promoting activity to the increase in plant growth and yield as there are several other mechanisms, which could also contribute to such improvement. Correlation-based analysis provides much better support to prove the impact of a specific plant growth-promoting attribute resulting in increase in plant growth and productivity (Norman et al. 2017). It is further also recommended to perform detailed analyses to prove the effect of a particular plant growth-promoting mechanisms. Microbiome data analysis revealed that plants hosted significantly different microbiomes in the rhizosphere as well as in root compartment; however, differences were more pronounced in the root compartment (Samad et al. 2017).

In addition to rhizosphere bacteria, arbuscular mycorrhiza (AM) interactions with plants are one of the most widespread and well-documented symbioses. AM fungi colonize the roots of nearly 80% of terrestrial plant species (Williams et al. 2017). The roots of host plant are colonized by fungi and form intracellular structures (i.e., arbuscules) in the cortical cells to facilitate exchange of nutrients (Parniske 2008). AM fungi promote plant growth and productivity by improving absorption and assimilation of essential elements, which are less soluble and non-available to the plants, i.e., N, P, Zn, Cu, etc., from the rhizosphere (Smith and Smith 2012; Johnson et al. 2013; Bender et al. 2016). AM fungi help their host plants by improving nutrient uptake and water absorption and inducing disease resistance; in exchange of this, the host plant supports fungal growth and reproduction. They contribute significantly by absorbing soil nutrients and making them available to the plant host. Nitrogen is an important element absorbed by AM fungi and supplied to their host plants. Transportation of nitrogen has been confirmed by the identification of genes responsible for transportation of ammonium and amino acids (Lopez-Pedrosa et al. 2006; Cappellazzo et al. 2008). A recent study has indicated that plants can regulate carbon allocation to AM fungi in response to P fertilization (Williams et al. 2017). Under N fertilization, plants allocate an increasing amount of C to AM fungi and receive relatively less P, which suggested an alteration in the terms of P-C exchange under N fertilization regardless of soil P status (Williams et al. 2017). Importance of organic matter inputs in structuring AM fungal communities was proved by removal of litter, which led to alteration of AM fungal community composition (Sheldrake et al. 2017).

7.2.2 Harmful Rhizosphere Microbiome

Plant pathogenic microorganisms cause serious diseases for many major agriculture crops and fruit trees throughout the world. There are several examples exhibiting the impact of plant diseases on human activities. Potato late blight caused by *Phytophthora infestans* triggered the nineteenth-century Irish famine, which is associated with massive North American immigration (Bourke 1964; Saville et al. 2016; Larousse and Galiana 2017). Spread of plant pathogens into new geographic ranges and expansion into new hosts are major factors in the emergence of novel virulent lineages that threaten food security (Stukenbrock and McDonald 2009). Pathogens evolve in agricultural ecosystems by several mechanisms, including tracking hosts, infecting a new host and horizontal gene transfer. It is possible that first plant pathogens were also domesticated simultaneously with their hosts during the beginning of agriculture practices (approximately 10,000 years ago). Plant diseases also have consequences for biodiversity conservation, in addition to severe agricultural and economic loss (Anderson et al. 2004). Significant losses in crop yield have been recorded due to various soilborne plant pathogens (Weller et al. 2002; Raaijmakers et al. 2009; Latz et al. 2016). Several root-related diseases like rot, wilt, stunting, and seedling damping off in plants are caused by soilborne pathogens (Haas and Défago 2005). The rhizosphere bacteria belonging to *Pseudomonas*, *Actinobacteria*, and *Bacillus* have shown their potential in the suppression of soilborne pathogens (Haas and Défago 2005; Mendes et al. 2011; Latz et al. 2016). Root-colonizing beneficial fungi like *Fusarium* and *Trichoderma* spp. are also key players in protecting plants from root pathogens (Fravel et al. 2003; Harman et al. 2004). Plants determine the structure and function of their rhizosphere microbial communities and can recruit specialized microbes, which can control the growth of pathogens (Garbeva et al. 2006; Latz et al. 2015). A recent study exhibited that plant communities can affect soil pH and the abundance of *Actinomyces*, both of which are positively related to the suppression against pathogenic fungus *R. solani* (Latz et al. 2016). This study suggests that plant communities alter the composition of microbes in the rhizosphere and can also influence the soil abiotic properties.

7.3 Phyllosphere Microbiome

The aboveground parts of plants are colonized by diverse microbial communities and are referred as phyllosphere (Newton et al. 2010; Karlsson et al. 2017). Like rhizosphere, the phyllosphere microorganisms are also important for plant health and productivity and include several plant pathogens, saprotrophs, and antagonists (Vorholt 2012; Peñuelas and Terradas 2014). The diversity of phyllospheric microorganisms is as high as that of in the roots or of the human gut (Berlec 2012). Valuable information can be generated about the composition and function of phyllosphere microbial communities, which can be utilized in a wide range of agricultural applications (Peñuelas and Terradas 2014). Substrate-driven selection underlies the establishment of phyllosphere communities (Bulgarelli et al. 2013). The phyllosphere microbes can have positive, negative, or neutral influences on plant. The activities of the phyllosphere microbiome in and on leaves influence plant health and growth significantly (Vorholt 2012; Bulgarelli et al. 2013; Rastogi et al. 2013; Saleem et al. 2017). Beneficial phyllosphere microbes can improve plant health by competing with pathogens for nutrients and by producing antimicrobial agents (Berlec 2012; Rastogi et al. 2013). The phyllosphere microbiome also provides specific ecosystem services such as phytoremediation of toxic pollutants like atmospheric hydrocarbons (Ali et al. 2012; Gandolfi et al. 2017). Several researchers have already reported the ability of phyllospheric bacteria to degrade aliphatic (Al-Awadhi et al. 2012) and aromatic hydrocarbons, namely, phenolic compounds, toluene, xylene, and phenanthrene (De Kempeneer et al. 2004; Sandhu et al. 2007; Sangthong et al. 2016; Scheublin et al. 2014; Waight et al. 2007; Yutthammo et al. 2010). Phyllosphere microbiome can exhibit the important developments between the interface of plants, microorganisms, and the atmosphere in diverse ecosystems (Bringel and Couée 2015).

The phyllosphere environment is exposed to fluctuations of different seasons, diagonal changes in day and night, and the morphological and developmental changes of the plant, and the phyllosphere microbiome adjusts to all these fluctuations very efficiently. Being an open system phyllosphere microbial community includes many transient microbes, especially fungal spores. Several recent studies both cultivation dependent and cultivation independent have exhibited insights into the composition of phyllosphere microbial communities. From these studies, it is apparent that these communities do not represent random assemblies of microorganisms but instead undergo selection that results in predictable microbial communities in different parts of plants and even in different types of plants with few dominant phyla and other subgroups as core phyllosphere microbiome. There are evidences that plant genotype can shape and determine the phyllosphere microbiome structure (Whipps et al. 2008). Highly variable composition of bacterial community was observed during the growing season of cottonwood trees (Redford and Fierer 2009). In this study , consistent patterns in the bacterial communities in early and late season were similar to the mid-season, confirming the predictable successional patterns in the phyllosphere communities (Redford and Fierer 2009). The phyllosphere supported a relatively greater abundance of beneficial rather than pathogenic bacteria under common garden conditions that included the presence of fungus gnat herbivory (Saleem et al. 2017). This observation was in consonant to previous studies, which confirmed that plant recruits beneficial microbes in relatively greater numbers under stressed and normal conditions to obtain the maximum mutualistic benefits (Vorholt 2012; Ortega et al. 2016). Plants grown with the beneficial bacteria (*Bacillus*) performed relatively poor in their growth, developmental, and reproductive traits than with the pathogenic bacteria (*Pseudomonas* and *Xanthomonas*) under herbivory, indicating that pathogens may have induced the host defense leading to higher growth (Saleem et al. 2017). High-throughput sequencing-based study on quantification of bacterial communities in the leaves of 57 tree species growing in a forest revealed that leaves from individual trees hosted more than 400 bacterial operational taxonomic units (Kembel et al. 2014). A core microbiome could be identified in the phyllosphere of these trees, which include *Actinobacteria*, *Proteobacteria*, and *Sphingobacteria* (Kembel et al. 2014). The main components of the fungal community in wheat phyllosphere were similar throughout the 350 km-long sampling area and seven operational taxonomic units (OTUs) comprising *Zymoseptoria*, *Dioszegia fristingensis*, *Cladosporium*, *Dioszegia hungarica*, *Cryptococcus*, *Ascochyta*, and *Dioszegia* representing the core fungal members of wheat phyllosphere microbiome (Karlsson et al. 2017).

Various plants produce especially fresh fruits, and leafy green vegetables are significant vectors of human pathogens, leading to several food-borne diseases (Teplitski et al. 2011; Williams and Marco 2014; Ximenes et al. 2017). Frequent outbreaks of illnesses caused by *Escherichia coli* O157:H7 and different non-typhoidal serovars of *Salmonella* have been linked to the food items, which consumed raw or after minimum processing like sprouts, salads, fruits, leafy greens, and nuts (Teplitski et al. 2011; Koukkidis et al. 2017). The latest dietary guidelines recommend eating healthy food, which include a variety of vegetables, legumes-beans and peas, and fresh fruits either raw or minimally processed. Plant produce gets expose to microbial contaminants in the field during cultivation in contaminated soils or by irrigation water. The pathogenic microbes interact with the plant, penetrate the plant through roots and translocate, and can colonize the aboveground tissues. The main sources of human pathogen contamination of plants are manures and marginal irrigation waters such as treated or untreated wastewater (Ibenyassine et al. 2007; Gallegos-Robles et al. 2008; Teplitski et al. 2011). The market for organically cultivated leafy greens is increasing day by day throughout the world, and in the similar fashion, the reports on food-borne infections are also growing, which are directly or indirectly related to the consumption of leafy greens. The persistence of enteric pathogens, such as *Escherichia coli* and *Salmonella* in soil and subsequently in the plant produce, is a serious food safety issue (Nicholson et al. 2015). Prolonged survivability of *Escherichia coli* strain O157:H7 in soil was recorded as it can persist up to 76 days following the irrigation with contaminated water (Erickson et al. 2010). Similarly, *Salmonella enterica* serovar *Typhimurium* can stay in the contaminated compost-amended soil up to 231 days, and the strain was also detected after several days on the surface of cultivated plants (Islam et al. 2004). There are many potential routes for human pathogens that can reach plants, but very little is known about how they are able to colonize and persist on plant hosts (Ximenes et al. 2017). Some studies have suggested that plants show disease symptoms or reduction in vigor response to colonization by human pathogenic bacteria, indicating that human pathogens could act as weak plant pathogens (Barak and Schroeder 2012). Many studies also exhibited that specific genotypes of plants can recognize the human pathogens and initiate their defense responses to prevent colonization and internalization (Melotto et al. 2014; Han and Micallef 2016). Many nonhuman pathogens including symbionts like rhizobia and AM fungi, endophytes, and plant pathogens colonize and enter into plant tissues. These microorganisms could provide a passage to the human pathogens to enter and colonize the plant tissues (Melotto et al. 2014; Nicholson et al. 2015).

Conclusions

The continuously increasing world population is posing a tremendous pressure on already vulnerable agriculture systems to meet world food security. The global trends in agriculture at present support the monoculture agriculture systems, which facilitate ease of mechanization in planting, fertilization, weed control, and harvest. The modernization-driven intensification of agriculture has placed a considerable burden on the fragile agroecosystems. Continuous efforts are being made toward the development of ecologically safe, efficient, and cheap biological alternatives to improve the agriculture productivity. Microbes beneficial to plants have been exploited as an alternative to the application of chemical fertilizers and pesticides, but the lack of consistency in their field performance has always been a big challenge. With the availability of next-generation sequencing techniques, it is now possible to unravel the previously unknown majority of microbial communities. Several plant microbiome studies have revealed that a huge part of microbial diversity is still not cultivated and thus not screened for their possible role. Based on these studies, the desirability of using these not-yet-cultivated microbes to aid plant growth has become apparent to shape the future of global agriculture. The fundamental knowledge generated on structure and function of microbial communities can be translated to develop strategies to alter the plant microbiome, to improve their health and productivity, and to reduce chemical inputs and emissions of greenhouse gases, resulting in sustainable agriculture. Recent developments in the microbial identification based on MALDI-TOF MS have offered an opportunity to cultivate the microorganisms in large numbers by using "culturomic" approaches. It is expected that this will allow the cultivation of several newer microorganisms, with multiple plant growth-promoting activities, which can be used to manipulate the plant microbiomes for better crop yields. In addition to this, there is an upsurge in the demand for organic crop produced throughout the world, and to achieve this, farmers are depending on organic manures and recycled sewage water, which are the potential source of several human pathogens. The persistence of human pathogenic microorganisms is becoming a very serious problem, which leads to frequent disease outbreaks. The characterization of phytomicrobiome in crop plants is becoming highly important, and application of plant beneficial microorganisms, which are nonhuman pathogens, holds the key to success to improve crop productivity for sustainable agriculture.

Acknowledgments This work was supported by the Department of Biotechnology, Government of India (BT/Coord.II/01/03/2016), and Department of Science and Technology, Government of India (YSS/2015/000149).

References

Akiyama K, Matsuzaki K, Hayashi H (2005) Plant sesquiterpenes induce hyphal branching in arbuscular mycorrhizal fungi. Nature 435:824–827. doi:10.1038/nature03608

Al-Awadhi H, Al-Mailem D, Dashti N, Hakam L, Eliyas M, Radwan S (2012) The abundant occurrence of hydrocarbon-utilizing bacteria in the phyllospheres of cultivated and wild plants in Kuwait. Int Biodeterior Biodegrad 73:73–79. doi:10.1016/j.ibiod.2012.05.016

Ali N, Sorkhoh N, Salamah S, Eliyas M, Radwan S (2012) The potential of epiphytic hydrocarbon-utilizing bacteria on legume leaves for attenuation of atmospheric hydrocarbon pollutants. J Environ Manage 93:113–120. doi:10.1016/j.jenvman.2011.08.014

Anderson PK, Cunningham AA, Patel NG, Morales FJ, Epstein PR, Daszak P (2004) Emerging infectious diseases of plants: pathogen pollution, climate change and agrotechnology drivers. Trends Ecol Evol 19:535–544. doi:10.1016/j.tree.2004.07.021

Arruda L, Beneduzi A, Martins A, Lisboa B, Lopes C, Bertolo F, Passaglia LMP, Vargas LK (2013) Screening of rhizobacteria from maize (*Zea mays* L.) in Rio Grande do Sul State (South Brazil) and analysis of their potential to improve plant growth. Appl Soil Ecol 63:15–22. doi:10.1016/j.apsoil.2012.09.001

Bais HP, Weir TL, Perry LG, Gilroy S, Vivanco JM (2006) The role of root exudates in rhizosphere interactions with plants and other organisms. Annu Rev Plant Biol 57:233–266. doi:10.1146/annurev.arplant.57.032905.105159

Barak JD, Schroeder BK (2012) Interrelationships of food safety and plant pathology: the life cycle of human pathogens on plants. Annu Rev Phytopathol 50:241–266. doi:10.1146/annurev-phyto-081211-172936

Bascom-Slack CA, Arnold AE, Strobel SA (2012) Student-directed discovery of the plant microbiome and its products. Science 338:485–486. doi:10.1126/science.1215227

Bender SF, Wagg C, van der Heijden MGA (2016) An underground revolution: biodiversity and soil ecological engineering for agricultural sustainability. Trends Ecol Evol 31:440–452. doi:10.1016/j.tree.2016.02.016

Berendsen RL, Pieterse CM, Bakker PA (2012) The rhizosphere microbiome and plant health. Trends Plant Sci 17:478–486. doi:10.1016/j.tplants.2012.04.001

Berlec A (2012) Novel techniques and findings in the study of plant microbiota: search for plant probiotics. Plant Sci 193:96–102. doi:10.1016/j.plantsci.2012.05.010

Bourke PMA (1964) Emergence of potato blight, 1843-46. Nature 203:805–808. doi:10.1038/203805a0

Bringel F, Couée I (2015) Pivotal roles of phyllosphere microorganisms at the interface between plant functioning and atmospheric trace gas dynamics. Front Microbiol 6:486. doi:10.3389/fmicb.2015.00486

Brusamarello-Santos LC, Gilard F, Brulé L, Quilleré I, Gourion B, Ratet P, de Souza EM, Lea PJ, Hirel B (2017) Metabolic profiling of two maize (*Zea mays* L.) inbred lines inoculated with the nitrogen fixing plant-interacting bacteria *Herbaspirillum seropedicae* and *Azospirillum brasilense*. PLoS One 12(3):e0174576. doi:10.1371/journal.pone.0174576

Bulgarelli D, Rott M, Schlaeppi K, van Themaat EVL, Ahmadinejad N, Assenza F, Rauf P, Huettel B, Reinhardt R, Schmelzer E, Peplies J, Gloeckner FO, Amann R, Eickhorst T, Schulze-Lefert P (2012) Revealing structure and assembly cues for *Arabidopsis* root-inhibiting bacterial microbiota. Nature 488:91–95. doi:10.1038/nature11336

Bulgarelli D, Schlaeppi K, Spaepen S, van Themaat EV, Schulze-Lefert P (2013) Structure and functions of the bacterial microbiota of plants. Annu Rev Plant Biol 64:807–838. doi:10.1146/annurev-arplant-050312-120106

Burgess TI, Scott JK, Mcdougall KL, Stukely MJC, Crane C, Dunstan WA, Brigg F, Andjic V, White D, Rudman T, Arentz F, Ota N, Hardy GESJ (2017) Current and projected global distribution of *Phytophthora cinnamomi*, one of the world's worst plant pathogens. Glob Chang Biol 23:1661–1674. doi:10.1111/gcb.13492

Cappellazzo G, Lanfranco L, Fitz M, Wipf D, Bonfante P (2008) Characterization of an amino acid permease from the endomycorrhizal fungus *Glomus mosseae*. Plant Physiol 147:429–437

Chaparro JM, Badri VD, Vivanco MJ (2014) Rhizosphere microbiome assemblage is affected by plant development. ISME J 8:790–803. doi:10.1038/ismej.2013.196

Charpentier M, Oldroyd G (2010) How close are we to nitrogen-fixing cereals? Curr Opin Plant Biol 13:556–564. doi:10.1016/j.pbi.2010.08.003

De Kempeneer L, Sercu B, Vanbrabant W, Van Langenhove H, Verstraete W (2004) Bioaugmentation of the phyllosphere for the removal of toluene from indoor air. Appl Microbiol Biotechnol 64:284–288. doi:10.1007/s00253-003-1415-3

Defez R, Andreozzi A, Bianco C (2017) The overproduction of indole-3-acetic acid (IAA) in endophytes upregulates nitrogen fixation in both bacterial cultures and inoculated rice plants. Microb Ecol (in press). doi:10.1007/s00248-017-0948-4

Dickson I (2017) Gut microbiota: culturomics: illuminating microbial dark matter. Nat Rev Gastroenterol Hepatol 4:3. doi:10.1038/nrgastro.2016.189

Dubourg G, Lagier JC, Robert C, Armougom F, Hugon P, Metidji S, Dione N, Dangui NP, Pfleiderer A, Abrahao J, Musso D, Papazian L, Brouqui P, Bibi F, Yasir M, Vialettes B, Raoult D (2014) Culturomics and pyrosequencing evidence of the reduction in gut microbiota diversity in patients with broad-spectrum antibiotics. Int J Antimicrob Agents 44:17–124. doi:10.1016/j.ijantimicag.2014.04.020

Edwards J, Johnson C, Santos-Medellín C, Lurie E, Podishetty NK, Bhatnagar S, Eisen JA, Sundaresan V (2015) Structure, variation, and assembly of the root-associated microbiomes of rice. Proc Natl Acad Sci U S A 112:E911–E920. doi:10.1073/pnas.1414592112

Erickson MC, Webb CC, Diaz-Perez JC, Phatak SC, Silvoy JJ, Davey L, Payton AS, Liao J, Ma L, Doyle MP (2010) Infrequent internalization of *Escherichia coli* O157:H7 into field-grown leafy greens. J Food Prot 73:500–506. doi:10.4315/0362-028X-73.3.500

Farrar K, Bryant D, Cope-Selby N (2014) Understanding and engineering beneficial plant–microbe interactions: plant growth promotion in energy crops. Plant Biotechnol J 12:1193–1206. doi:10.1111/pbi.12279

Fravel D, Olivain C, Alabouvette C (2003) *Fusarium oxysporum* and its biocontrol. New Phytol 157:493–502. doi:10.1046/j.1469-8137.2003.00700.x

Gallegos-Robles MA, Morales-Loredo A, Alvarez-Ojeda G, Vega PA, Chew MY, Velarde S, Fratamico P (2008) Identification of *Salmonella* serotypes isolated from cantaloupe and Chile pepper production systems in Mexico by PCR-restriction fragment length polymorphism. J Food Prot 71:2217–2222. doi:10.4315/0362-028X-71.11.2217

Gandolfi I, Canedoli C, Imperato V, Tagliaferri I, Gkorezis P, Vangronsveld J, Schioppa EP, Papacchini M, Bestetti G, Franzetti A (2017) Diversity and hydrocarbon-degrading potential of epiphytic microbial communities on *Platanus* x *acerifolia* leaves in an urban area. Environ Pollut 220:650–658. doi:10.1016/j.envpol.2016.10.022

Garbeva P, Postma J, van Veen JA, van Elsas JD (2006) Effect of above-ground plant species on soil microbial community structure and its impact on suppression of *Rhizoctonia solani* AG3. Environ Microbiol 8:233–246. doi:10.1111/j.1462-2920.2005.00888.x

Gulati A, Sharma N, Vyas P, Sood S, Rahi P, Pathania V, Prasad R (2010) Organic acid production and plant growth promotion as a function of phosphate solubilization by *Acinetobacter rhizosphaerae* strain BIHB 723 isolated from the cold deserts of the trans-Himalayas. Arch Microbiol 192:975–983. doi:10.1007/s00203-010-0615-3

Haas D, Défago G (2005) Biological control of soil-borne pathogens by fluorescent pseudomonads. Nat Rev Microbiol 3:307–319. doi:10.1038/nrmicro1129

Han S, Micallef SA (2016) Environmental metabolomics of the tomato plant surface provides insights on *Salmonella enterica* colonization. Appl Environ Microbiol 82:3131–3142. doi:10.1128/AEM.00435-16

Hariprasad P, Niranjana SR (2009) Isolation and characterization of phosphate solubilizing rhizobacteria to improve plant health of tomato. Plant Soil 316:13–24. doi:10.1007/s11104-008-9754-6

Harman GE, Howell CR, Viterbo A, Chet I, Lorito M (2004) *Trichoderma* species: opportunistic, avirulent plant symbionts. Nat Rev Microbiol 2:43–56. doi:10.1038/nrmicro797

Hartmann A, Rothballer M, Schmid M (2008) Lorenz Hiltner, a pioneer in rhizosphere microbial ecology and soil bacteriology research. Plant Soil 312:7–14. doi:10.1007/s11104-007-9514-z

Hiltner L (1904) Ueber neuere Erfahrungen und Probleme auf dem Gebiete der Bodenbakteriologie und unter besonderer BerUcksichtigung der Grundungung und Brache. Arb Deut Landw Gesell 98:59–78

Hirsch PR, Mauchline TH (2012) Who's who in the plant root microbiome? Nat Biotechnol 30:961–962. doi:10.1038/nbt.2387

Hofmann A, Fischer D, Hartmann A, Schmid M (2014) Colonization of plants by human pathogenic bacteria in the course of organic vegetable production. Front Microbiol 5:191. doi:10.3389/fmicb.2014.00191

Ibenyassine K, Mhand RA, Karamoko Y, Anajjar B, Chouibani MM, Ennaji M (2007) Bacterial pathogens recovered from vegetables irrigated by wastewater in Morocco. J Environ Health 69:47–51

Islam M, Morgan J, Doyle MP, Phatak SC, Millner P, Jiang X (2004) Persistence of *Salmonella enterica* serovar Typhimurium on lettuce and parsley and in soils on which they were grown in fields treated with contaminated manure composts or irrigation water. Foodborne Pathog Dis 1:27–35. doi:10.1089/153531404772914437

Johnson NC, Angelard C, Sanders IR, Kiers ET (2013) Predicting community and ecosystem outcomes of mycorrhizal responses to global change. Ecol Lett 16:140–153. doi:10.1111/ele.12085

Karlsson I, Friberg H, Kolseth A-K, Steinberg C, Persson P (2017) Organic farming increases richness of fungal taxa in the wheat phyllosphere. Mol Ecol (in press). doi:10.1111/mec.14132

Kembel SW, O'Connor TK, Arnold HK, Hubbell SP, Wright SJ, Green JL (2014) Relationships between phyllosphere bacterial communities and plant functional traits in a neotropical forest. Proc Natl Acad Sci 111:13715–13720. doi:10.1073/pnas.1216057111

Koukkidis G, Haigh R, Allcock N, Jordan S, Freestone P (2017) Salad leaf juices enhance Salmonella growth, colonization of fresh produce, and virulence. Appl Environ Microbiol 83:e02416–e02416. doi:10.1128/AEM.02416-16

Kuan KB, Othman R, Abdul Rahim K, Shamsuddin ZH (2016) Plant growth-promoting rhizobacteria inoculation to enhance vegetative growth, nitrogen fixation and nitrogen remobilisation of maize under greenhouse conditions. PLoS One 11(3):e0152478. doi:10.1371/journal.pone.0152478

Kuzyakov Y, Domanski G (2000) Carbon input by plants into the soil. Review. J Plant Nutr Soil Sci 163:421–431. doi:10.1002/1522-2624

Lagier JC, Khelaifia S, Alou MT, Ndongo S, Dione N, Hugon P, Caputo A, Cadoret F, Traore SI, Seck EH, Dubourg G, Durand G, Mourembou G, Guilhot E, Togo A, Bellali S, Bachar D, Cassir N, Bittar F, Delerce J, Mailhe M, Ricaboni D, Bilen M, Dangui Nieko NP, Dia Badiane NM, Valles C, Mouelhi D, Diop K, Million M, Musso D, Abrahão J, Azhar EI, Bibi F, Yasir M, Diallo A, Sokhna C, Djossou F, Vitton V, Robert C, Rolain JM, La Scola B, Fournier PE, Levasseur A, Raoult D (2016) Culture of previously uncultured members of the human gut microbiota by culturomics. Nat Microbiol 1:16203. doi:10.1038/nmicrobiol.2016.203

Laranjo M, Alexandre A, Oliveira S (2014) Legume growth-promoting rhizobia: an overview on the *Mesorhizobium* genus. Microbiol Res 169:2–17. doi:10.1016/j.micres.2013.09.012

Lareen A, Burton F, Schafer P (2016) Plant root-microbe communication in shaping root microbiomes. Plant Mol Biol 90:575–587. doi:10.1007/s11103-015-0417-8

Larousse M, Galiana E (2017) Microbial partnerships of pathogenic oomycetes. PLoS Pathog 13:e1006028. doi:10.1371/journal.ppat.1006028

Latz E, Eisenhauer N, Scheu S, Jousset A (2015) Plant identity drives the expression of biocontrol factors in a rhizosphere bacterium across a plant diversity gradient. Funct Ecol 29:1225–1234. doi:10.1111/1365-2435.12417

Latz E, Eisenhauer N, Rall BC, Scheu S, Jousset A (2016) Unravelling linkages between plant community composition and the pathogen-suppressive potential of soils. Sci Rep 6:23584. doi:10.1038/srep23584

Lecomte C, Alabouvette C, Edel-Hermann V, Robert F, Steinberg C (2016) Biological control of ornamental plant diseases caused by *Fusarium oxysporum*: a review. Biol Control 101:17–30. doi:10.1016/j.biocontrol.2016.06.004

Lopez-Pedrosa A, Gonzalez-Guerrero M, Valderas A, Azcon-Aguilar C, Ferrol N (2006) GintAMT1 encodes a functional high-affinity ammonium transporter that is expressed in the extra radical mycelium of *Glomus intraradices*. Fungal Genet Biol 43:102–110. doi:10.1016/j.fgb.2005.10.005

Lundberg DS, Lebeis SL, Paredes SH, Yourstone S, Gehring J, Malfatti S, Tremblay J, Engelbrektson A, Kunin V, del Rio TG, Edgar RC, Eickhorst T, Ley RE, Hugenholtz P, Tringe SG, Dangl JL (2012) Defining the core *Arabidopsis thaliana* root. Nature 488:86–90. doi:10.1038/nature11237

Mansfield J, Genin S, Magori S, Citovsky V, Sriariyanum M, Ronald P, Dow M, Verdier V, Beer SV, Machado MA, Toth I, Salmond G, Foster GD (2012) Top 10 plant pathogenic bacteria in molecular plant pathology. Mol Plant Pathol 13:614–629. doi:10.1111/J.1364-3703.2012.00804.X

Melotto M, Panchal S, Roy D (2014) Plant innate immunity against human bacterial pathogens. Front Microbiol 5:411. doi:10.3389/fmicb.2014.00411

Mendes R, Kruijt M, de Bruijn I, Dekkers E, van der Voort M, Schneider JHM, Piceno YM, DeSantis TZ, Andersen GL, Bakker PAHM, Raaijmakers JM (2011) Deciphering the rhizosphere microbiome for disease-suppressive bacteria. Science 332:1097–1100. doi:10.1126/science.1203980

Mendes R, Garbeva P, Raaijmakers JM (2013) The rhizosphere microbiome: significance of plant beneficial, plant pathogenic, and human pathogenic microorganisms. FEMS Microbiol Rev 37:634–663. doi:10.1111/1574-6976.12028

Montañez A, Abreu C, Gill PR, Hardarson G, Siracdi M (2009) Biological nitrogen fixation in maize (*Zea mays* L.) by 15N isotope-dilution and identification of associated culturable diazotrophs. Biol Fertil Soils 45:253–263. doi:10.1007/s00374-008-0322-2

Neal AL, Ahmad S, Gordon-Weeks R, Ton J (2012) Benzoxazinoids in root exudates of maize attract *Pseudomonas putida* to the rhizosphere. PLoS One 7:e35498. doi:10.1371/journal.pone.0035498

Newton AC, Gravouil C, Fountaine JM (2010) Managing the ecology of foliar pathogens: ecological tolerance in crops. Ann Appl Biol 157:343–359. doi:10.1111/j.1744-7348.2010.00437.x

Nicholson AM, Gurtler JB, Bailey RB, Niemira BA, Douds DD (2015) Influence of mycorrhizal fungi on fate of *E. coli* O157:H7 and *Salmonella* in soil and internalization into Romaine lettuce plants. Int J Food Microbiol 192:95–102. doi:10.1016/j.ijfoodmicro.2014.10.001

Norman JS, Hare JR, Friesen ML (2017) Comment: isolation and screening of bacteria for their diazotrophic potential and their influence on growth promotion of maize seedlings in greenhouses. Front Plant Sci 8:212. doi:10.3389/fpls.2017.00212

Oldroyd GED (2013) Speak, friend, and enter: signalling systems that promote beneficial symbiotic associations in plants. Nat Rev Microbiol 11:252–263. doi:10.1038/nrmicro2990

Ongeng D, Haberbeck LU, Mauriello G, Ryckeboer J, Springael D, Geeraerd AH (2014) Modeling the fate of *Escherichia coli* O157:H7 and *Salmonella enterica* in the agricultural environment: current perspective. J Food Sci 79:R421–R427. doi:10.1111/1750-3841.12392

Ortega RA, Mahnert A, Berg C, Müller H, Berg G (2016) The plant is crucial: specific composition and function of the phyllosphere microbiome of indoor ornamentals. FEMS Microbiol Ecol 92:fiw173. doi:10.1093/femsec/ fiw173

Oteino N, Lally RD, Kiwanuka S, Lloyd A, Ryan D, Germaine KJ, Dowling DN (2015) Plant growth promotion induced by phosphate solubilizing endophytic *Pseudomonas* isolates. Front Microbiol 6:745. doi:10.3389/fmicb.2015.00745

Parniske M (2008) Arbuscular mycorrhiza: the mother of plant root endosymbioses. Nat Rev Microbiol 6:763–775. doi:10.1038/nrmicro1987

Pedrós-Alió C (2006) Marine microbial diversity: can it be determined? Trends Microbiol 14:257–263. doi:10.1016/j.tim.2006.04.007

Peñuelas J, Terradas J (2014) The foliar microbiome. Trends Plant Sci 19:278–280. doi:10.1016/j.tplants.2013.12.007

Piromyou P, Buranabanyat B, Tantasawat P, Tittabutr P, Boonkerd N, Teaumroong N (2011) Effect of plant growth promoting rhizobacteria (PGPR) inoculation on microbial community structure in rhizosphere of forage corn cultivated in Thailand. Eur J Soil Biol 47:44–54. doi:10.1016/j.ejsobi.2010.11.004

Raaijmakers JM, Paulitz TC, Steinberg C, Alabouvette C, Moënne-Loccoz Y (2009) The rhizosphere: a playground and battlefield for soilborne pathogens and beneficial microorganisms. Plant Soil 321:341–361. doi:10.1007/s11104-008-9568-6

Rahi P, Kapoor R, Young JPW, Gulati A (2012) A genetic discontinuity in root-nodulating bacteria of cultivated pea in the Indian trans-Himalayas. Mol Ecol 21(1):145–159. doi:10.1111/j.1365-294X.2011.05368.x

Rahi P, Sharma OP, Shouche YS (2016) Matrix-Assisted Laser Desorption/Ionization Time-of-Flight Mass-Spectrometry (MALDI-TOF MS) based microbial identifications: challenges and scopes for microbial ecologists. Front Microbiol 7:1359. doi:10.3389/fmicb.2016.01359

Rangel de Souza ALS, De Souza SA, De Oliveira MVV, Ferraz TM, Figueiredo FAMMA, Da Silva ND, Rangel PL, Panisset CRS, Olivares FL, Campostrini E, De Souza FGA (2016) Endophytic colonization of *Arabidopsis thaliana* by *Gluconacetobacter diazotrophicus* and its effect on plant growth promotion, plant physiology, and activation of plant defense. Plant Soil 399:257–270. doi:10.1007/s11104-015-2672-5

Rascovan N, Carbonetto B, Perrig D, Díaz M, Canciani W, Abalo M, Alloati J, González-Anta G, Vazquez MP (2016) Integrated analysis of root microbiomes of soybean and wheat from agricultural fields. Sci Rep 6:28084. doi:10.1038/srep28084

Rastogi G, Coaker GL, Leveau JHJ (2013) New insights into the structure and function of phyllosphere microbiota through high-throughput molecular approaches. FEMS Microbiol Lett 348:1–10. doi:10.1111/1574-6968.12225

Redford AJ, Fierer N (2009) Bacterial succession on the leaf surface: a novel system for studying successional dynamics. Microb Ecol 58:189. doi:10.1007/s00248-009-9495-y

Reinhold-Hurek B, Bünger W, Burbano CS, Sabale M, Hurek T (2015) Roots shaping their microbiome: global hotspots for microbial activity. Annu Rev Phytopathol 53:403–424. doi:10.1146/annurev-phyto-082712-102342

Richardson A, Barea J-M, McNeill A, Prigent-Combaret C (2009) Acquisition of phosphorus and nitrogen in the rhizosphere and plant growth promotion by microorganisms. Plant Soil 321:305–339. doi:10.1007/s11104-009-9895-2

Rinke C, Schwientek P, Sczyrba A, Ivanova NN, Anderson IJ, Cheng J-F, Darling A, Malfatti S, Swan BK, Gies EA, Dodsworth JA, Hedlund BP, Tsiamis G, Sievert SM, Liu W-T, Eisen JA, Hallam SJ, Kyrpides NC, Stepanauskas R, Rubin EM, Hugenholtz P, Woyke T (2013) Insights into the phylogeny and coding potential of microbial dark matter. Nature 499:431–437. doi:10.1038/nature12352

Saleem M, Meckes N, Pervaiz ZH, Traw MB (2017) Microbial interactions in the phyllosphere increase plant performance under herbivore biotic stress. Front Microbiol 8:41. doi:10.3389/fmicb.2017.00041

Samad A, Trognitz F, Compant S, Antonielli L, Sessitsch A (2017) Shared and host-specific microbiome diversity and functioning of grapevine and accompanying weed plants. Environ Microbiol 19:1407–1424. doi:10.1111/1462-2920.13618

Sandhu A, Halverson LJ, Beattie GA (2007) Bacterial degradation of airborne phenol in the phyllosphere. Environ Microbiol 9:383–392. doi:10.1111/j.1462-2920.2006.01149.x

Sangthong S, Suksabye P, Thiravetyan P (2016) Air-borne xylene degradation by *Bouganvillea buttiana* and the role of epiphytic bacteria in the degradation. Ecotoxicol Environ Saf 126:273–280. doi:10.1016/j.ecoenv.2015.12.017

Saville AC, Martin MD, Ristaino JB (2016) Historic late blight outbreaks caused by a widespread dominant lineage of *Phytophthora infestans* (Mont.) de Bary. PLoS One 11(12):e0168381. doi:10.1371/journal.pone.0168381

Scheublin TR, Deusch S, Moreno-Forero SK, Müller JA, van der Meer JR, Leveau JHJ (2014) Transcriptional profiling of Gram-positive *Arthrobacter* in the phyllosphere: induction of pollutant degradation genes by natural plant phenolic compounds. Environ Microbiol 16:2212–2225. doi:10.1111/1462-2920.12375

Sheldrake M, Rosenstock NP, Revillini D, Olsson PA, Mangan S, Sayer EJ, Wallander H, Turner BL, Tanner EVJ (2017) Arbuscular mycorrhizal fungal community composition is altered by long-term litter removal but not litter addition in a lowland tropical forest. New Phytol 214:455–467. doi:10.1111/nph.14384

Singh BK, Munro S, Potts JM, Millard P (2007) Influence of grass species and soil type on rhizosphere microbial community structure in grassland soils. Appl Soil Ecol 36:147–155. doi:10.1016/j.apsoil.2007.01.004

Smith SE, Smith FA (2012) Fresh perspectives on the roles of arbuscular mycorrhizal fungi in plant nutrition and growth. Mycologia 104:1–13. doi:10.3852/11-229

Smith DL, Gravel V, Yergeau E (2017) Signalling in the phytomicrobiome: Understanding and harnessing the plant holobiont. Front Plant Sci 8:611. doi:10.3389/fpls.2017.00611

Solden L, Lloyd K, Wrighton K (2016) The bright side of microbial dark matter: lessons learned from the uncultivated majority. Cur Opin Microbiol 31(2):17–226. doi:10.1016/j.mib.2016.04.020

Sprent JI, Ardley J, James EK (2017) Biogeography of nodulated legumes and their nitrogen-fixing symbionts. New Phytol (in press). doi:10.1111/nph.14474

Stukenbrock EH, McDonald B (2009) The origins of plant pathogens in agroecosystems. Annu Rev Phytopathol 46:75–100. doi:10.1146/annurev.phyto.010708.154114

Teplitski M, Warriner K, Bartz J, Schneider KR (2011) Untangling metabolic and communication networks: interactions of enterics with phytobacteria and their implications in produce safety. Trends Microbiol 19:121–127. doi:10.1016/j.tim.2010.11.007

Vartoukian SR, Palmer RM, Wade WG (2010) Strategies for culture of 'unculturable' bacteria. FEMS Microbiol Let 309:1–7. doi:10.1111/j.1574-6968.2010.02000.x

Vorholt JA (2012) Microbial life in the phyllosphere. Nat Rev Microbiol 10:828–840. doi:10.1038/nrmicro2910

Vuong HB, Thrall PH, Barrett LG (2017) Host species and environmental variation can influence rhizobial community composition. J Ecol 105:540–548. doi:10.1111/1365-2745.12687

Waight K, Pinyakong O, Luepromchai E (2007) Degradation of phenanthrene on plant leaves by phyllosphere bacteria. J Gen Appl Microbiol 53:265–272. doi:10.2323/jgam.53.265

Wallenstein MD (2017) Managing and manipulating the rhizosphere microbiome for plant health: a systems approach. Rhizo (in press). doi:10.1016/j.rhisph.2017.04.004

Weller DM, Raaijmakers JM, Gardener BBM, Thomashow LS (2002) Microbial populations responsible for specific soil suppressiveness to plant pathogens. Annu Rev Phytopathol 40:309–348. doi:10.1146/annurev.phyto.40.030402.110010

Whipps JM, Hand P, Pink D, Bending GD (2008) Phyllosphere microbiology with special reference to diversity and plant genotype. J Appl Microbiol 105:1744–1755. doi:10.1111/j.1365-2672.2008.03906.x

Williams TR, Marco ML (2014) Phyllosphere microbiota composition and microbial community transplantation on lettuce plants grown indoors. mBio 5:e01564-14. doi:10.1128/mBio.01564-14

Williams A, Manoharan L, Rosenstock NP, Olsson PA, Hedlund K (2017) Long-term agricultural fertilization alters arbuscular mycorrhizal fungal community composition and barley (*Hordeum vulgare*) mycorrhizal carbon and phosphorus exchange. New Phytol 213:874–885. doi:10.1111/nph.14196

Xiao X, Chen W, Zong L, Yang J, Jiao S, Lin Y, Wang E, Wei G (2017) Two cultivated legume plants reveal the enrichment process of the microbiome in the rhizocompartments. Mol Ecol 26:1641–1651. doi:10.1111/mec.14027

Ximenes E, Hoagland L, Ku S, Li X, Ladisch M (2017) Human pathogens in plant biofilms: formation, physiology, and detection. Biotechnol Bioeng (in press). doi:10.1002/bit.26247

Yoon V, Tian G, Vessey JK, Macfie SM, Dangi OP, Kumer AK, Tian L (2016) Colonization efficiency of different sorghum genotypes by *Gluconacetobacter diazotrophicus*. Plant Soil 398:243–256. doi:10.1007/s11104-015-2653-8

Yutthammo C, Thongthammachat N, Pinphanichakarn P, Luepromchai E (2010) Diversity and activity of PAH-degrading bacteria in the phyllosphere of ornamental plants. Microb Ecol 59:357–368. doi:10.1007/s00248-009-9631-8

Rare Biosphere in Human Gut: A Less Explored Component of Human Gut Microbiota and Its Association with Human Health

8

Shrikant S. Bhute, Saroj S. Ghaskadbi, and Yogesh S. Shouche

Abstract

With the advent of next-generation sequencing technologies, we are able to ascertain the vast microbial diversity of various habitats. These studies have led to the discovery of presence and persistence of microbial communities that are extremely low in the abundance. These rare microbial taxa in soil and marine environments are known to act as seed-bank for the organisms that are waiting for the favorable conditions to make their move. The human body is one of the diverse ecosystems known to us supporting the growth of myriads of microbes in or on our body. Studies in the last decade or so have significantly contributed to our understanding of the involvement of these human-associated microbes with our health and diseases. So far, the studies on human gut microbiota have focused largely on dominated microbial species, and the information about the rare taxa associated with the human body is still lurking. This chapter provides evidence that microbes from the three domains of life (bacteria, archaea, and eukarya) existing in the gut have the rare components relevant to human health. Hence, we advocate the need for more studies concerning the rare taxa present in the human gut and their association with the general well-being.

Keywords

Human gut microbiome • NGS sequencing • Rare microbial communities • Health and diseases

S.S. Bhute • S.S. Ghaskadbi
Department of Zoology, Savitribai Phule Pune University, Pune, India

Y.S. Shouche (✉)
Microbial Culture Collection, National Centre for Cell Science, Pune, India
e-mail: yogesh@nccs.res.in

V.C. Kalia et al. (eds.), *Mining of Microbial Wealth and MetaGenomics*,
DOI 10.1007/978-981-10-5708-3_8

8.1 Rare Biosphere

Study dealing with the diversity analysis of unicellular and multicellular organisms and their association with the functioning of an ecosystem is one of the important aspects of ecological science. Earlier, the exploration of microbial ecosystems was largely dominated by traditional microscopic and culture-dependent methods, due to which a large number of microbial ecosystems were underexplored. Nonetheless, in the past few decades, this scenario has changed dramatically due to the establishment of 16S rRNA gene as a molecular chronometer (Woese and Fox 1977) and advent of next-generation sequencing technologies, both of which are assisting us to look more deeper into complex microbial ecosystems (Roh et al. 2010). Thus, exploration of microbial assemblages has become one of the major areas of research across the globe, and microbial diversity analysis of almost every possible habitat on the planet earth is underway. Studies such as these are contributing to the enormous amount of information about spatial and temporal variations in microbial communities of these habitats.

A consistent observation from most of the high-throughput sequencing studies dealing with the characterization of microbial communities of different habitats, such as marine, soil, and human gut, is the presence of a vast amount of sequences being affiliated to low-abundant microbial taxa. Rank abundance curves produced using such a data often indicate that indeed many habitats including the human gut are dominated by few microbial taxa and have a long tail due to the presence of less abundant taxa (Fig. 8.1). As can be seen from Fig. 8.1, few microbial taxa dominate almost all types of human-associated samples, while the majority of them are present in low abundance contributing significantly to the observed diversity and are often termed as “rare biosphere” (Sogin et al. 2006). At the same time, it can easily

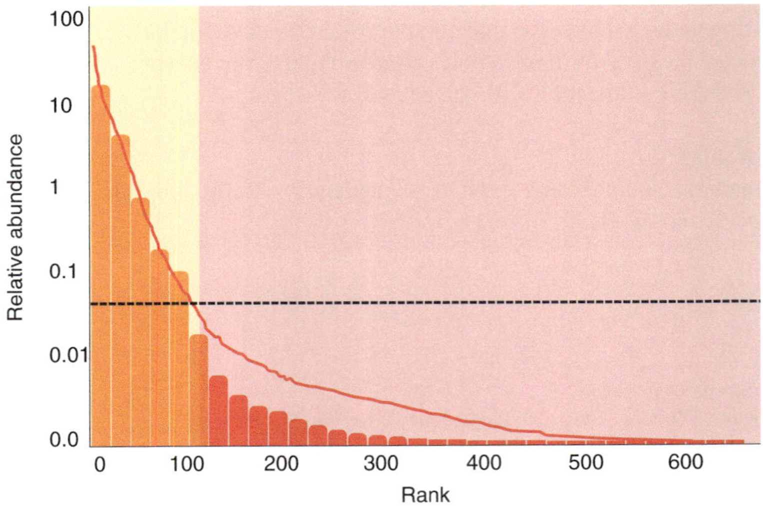

Fig. 8.1 Hypothetical rank abundance curve showing the existence of large amount of rare microbial taxa

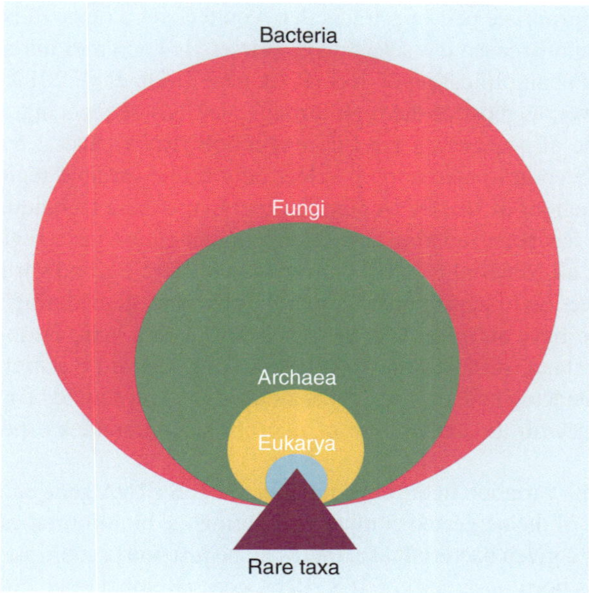

Fig. 8.2 Rare taxa belong to all the three domains of life in the human gut

be perceived that although they are less abundant, rare taxa are more diverse for any microbial community and, hence, are equally, at times, more crucial in the functioning of the particular ecosystem.

8.2 Scope of the Chapter

This chapter deals with the rare biosphere in the human gut, which is one of the richest and diverse ecosystems known to us. Therefore, finding the rare taxa and their association with human health is equivalent to finding the needle in a haystack. In Sect. 8.3, we first bring to the surface the major concerns during identification of truly rare taxa. Section 8.4 deals with the general characteristics of the human gut microbiome and relevance to human health. In Sect. 8.5, we describe few specific examples of rare bacterial taxa associated with human well-being, while in Sects. 8.6 and 8.7, we shade some light on rare taxa from largely ignored microbial components, viz., archaea and eukarya (Fig. 8.2).

8.3 Major Concerns in Detecting Rare Gut Biosphere

Although the contemporary sequencing methodologies allow us to sequence the 16S rRNA marker gene to a very high sequencing depth, most of these methods are still based on the amplification of the 16S rRNA gene prior to the sequencing. In order to detect the rare taxa successfully and confidently, it is therefore very critical to select the appropriate pair of primers that would cover a range of bacterial phyla. This is important because the selection of primers is known to introduce selective biases during the amplification of 16S rRNA gene (Mao et al. 2012; Schloss et al. 2011). Moreover, as most of the high-throughput DNA sequencing methodologies are not capable of sequencing the full-length 16S rRNA gene (1.5 kb fragment), hence it also becomes mandatory to select only a few variable regions out of the nine variable regions of 16S rRNA gene in order to use these techniques for molecular surveys. It has been reported that there is no single variable region capable of differentiating all bacterial taxa (Chakravorty et al. 2007; Klindworth et al. 2013); hence, the selection of appropriate region of 16S rRNA gene during the amplification could also have an impact on the discovery of rare taxa. Further, in order to detect the rare taxa, sufficient depth of sequencing has to be achieved which can inflate the sequencing budget, and hence, a surplus budget should be incorporated in studies dealing with the exploration of microbial communities especially of very high diversity.

Likewise, the variation in the copy numbers of 16S rRNA gene among the bacterial taxa is one of the neglected components during the bioinformatics analysis. It is known that for a given bacterial taxa, there can be just one or multiple copies (maximum of 15 has been observed) of 16S rRNA gene (Kembel et al. 2012); hence, an account of this during the analysis is advocated which will significantly enhance the possibility of the detection of rare ones. Further, it is essential to tune the algorithms used for OTU picking, such that they are trained to discriminate the sequencing artifacts or chimeras from the true rare taxon (Hugenholtz and Huber 2003).

8.4 Rare Biosphere in the Human Gut

Microbial community structure and their activities in the human gut are important in determining the healthy state of the host (Cho and Blaser 2012). Our current knowledge about these gut microbial communities leaves us with the impression that they are composed of few microbial genera dominating this ecosystem, while there are many microbial genera that are relatively low in abundance. Studies dealing with the exploration of gut microbial communities largely define the dominant members of the gut, but the distribution of rare taxa in the gut and their association with the different disease cannot be overlooked. So far, most of the empirical investigations dealing with gut microbiota largely focused on the dominant bacterial phyla: *Firmicutes* and *Bacteroidetes* (Qin et al. 2010), and we are largely unaware of the immensely large low-abundant communities in the human gut and their association with human diseases.

These rare taxa are phylogenetically diverse and could independently or collectively participate in varied metabolic functions important to human health. For instance, microbial taxa belonging to this rare biosphere in the gut may be participating in some crucial or yet unknown physiological function which can be a potential candidate for medical intervention that is currently underway. Since the pathophysiology of the different human diseases differs, it would be also interesting to see whether rare taxa in one disease condition could actually be the abundant one in another disease condition, i.e., they are conditionally rare. Or indeed, there are some microbial taxa that are truly rare and remain rare irrespective of the health status of the individuals.

8.5 Bacterial Rare Taxa Associated with Health and Disease

Oxalobacter formigenes (OF) is a Gram-negative obligate anaerobic bacterium that inhabits in the colon of humans and other animals (Allison et al. 1985). Several whole-genome short-gun and 16S rRNA sequencing studies have revealed that OF is one of the members of rare taxa present in the human gut (Barnett et al. 2016). OF has a unique ability to use oxalate as a principal carbon source for its growth and, hence, participates in oxalate homeostasis by preventing its absorption through the gut (Sidhu et al. 1999). OF possesses two key enzymes that are required for oxalate metabolism: oxalyl-CoA decarboxylase (*oxc*) and formyl-CoA transferase (*frc*). *frc* transfers CoA to oxalate from formyl-CoA and converts it to oxalyl-CoA; *oxc* catalyzes the formation of formyl-CoA and CO_2 (Stewart et al. 2004). Hyperoxaluria is a complex disorder related to oxalate metabolism and often poses significant challenges to medical practitioners in controlling the hyperoxaluria. Hence, novel therapies are aimed at controlling the recurrent stone formation in the hyperoxaluric subjects (Asplin 2016). Lower colonization of the OF has been linked with excess oxalate secretion and in the development of oxalate stones (Siva et al. 2009). In this regard, OF has attracted attention for its oxalate metabolizing property. In one interesting study, the rats with chronic hyperoxaluria were fed OF as a probiotic, and rapid decrease in urinary oxalate was noticed in a dose-dependent manner, suggesting the important role of oxalobacter in oxalate clearance (Sidhu et al. 2001). OF has also been found promising in reducing the levels of oxalate in the humans and could be an important probiotics species for controlling the hyperoxaluria and associated disorders (Jairath et al. 2015). Thus, OF, one of the rare species found in the gut, has both prophylactic and therapeutic benefits for hyperoxaluria subjects.

Global burden of cancer is increasing tremendously in all parts of the world. While cancer is a complex disease, a link between some infectious agents and cancer is unambiguous; several viruses such as hepatitis B and C and human papillomavirus are known to cause cancer in humans (Parkin 2006). Of the different forms of cancer, colorectal cancer is the third most diagnosed cancer, and the incidence rate of which is alarming (Jemal et al. 2011). Evidence potentiates the role of gut microbiota in the development of colonic adenomas and colonic adenocarcinomas. *Fusobacterium nucleatum* (FN), one of the rare members of gut microbiota, has been shown to be linked with the different forms of colorectal cancers both in humans and in animals (Castellarin et al. 2012; Kostic et al. 2013). An animal study utilizing mice that developed intestinal tumors due to the mutation in tumor suppressor gene Apc (C57BL/6 $Apc^{Min/+}$) was conducted to show whether the introduction of FN promotes the tumorogenesis. This study confirms that mice receiving FN accelerated the onset of tumors compared to the mice receiving the *Streptococcus* species (Kostic et al. 2013). The same study also reports that the immune cells of myeloid lineage significantly infiltrate in the tumors of FN-fed mice compared to *Streptococcus*-fed mice. The myeloid lineages like macrophages, dendritic cells (DCs), and granulocytes are crucial in promoting tumor progression.

Escherichia coli (*E. coli*) is one of the prototypic commensals characterized from the human gut. It is a Gram-negative, non-sporulating facultative anaerobe, and even though it is considered as an indicator of fecal contamination, it is outnumbered by the anaerobes in the human distal gut (Tenaillon et al. 2010). Although *E. coli* are normally considered as the commensal in the gut, genomic diversity of some pathogenic *E. coli* is significant, which makes them highly pathogenic and at times lethal (Buchholz et al. 2011). Various strains of *E. coli* belonging to several phylogenetic groups have been found to be associated with human disorders both inside and outside the gastrointestinal tract. Of note, pathogenesis of Crohn's disease and ulcerative colitis has been linked with adherent-invasive *E. coli* (AIEC) pathotype (Sasaki et al. 2007). The AIEC strain of *E. coli* reported in this study was found to induce significant TNF-alpha expression in macrophage supernatant and IL-8 in epithelial cells. In addition, this invasive strain was associated with the decreased epithelial barrier function. Stimulation of IL-8 and TNF-alpha expression and concomitant drop in the barrier function are commonly observed pathophysiological features of IBD. A surveillance study based on more than 500 published papers of virulence-associated genes and pathogenicity islands among the fecal *E. coli* revealed that the *E. coli* strains present in the human gut can act as a reservoir of the extraintestinal pathogenic *E. coli* (ExPEC) (Starčič Erjavec et al. 2015).

Members of the genus *Propionibacteria* belongs to the order Actinomycetales of the phylum *Actinobacteria*. This and other members of Actinomycetales have been regarded as rare species in the human gut (Rajilić-Stojanović and de Vos 2014). Species of *Propionibacteria* have been used as probiotics due to their ability to produce beneficial metabolites and the expression of key surface proteins involved in immunomodulation (Saraoui et al. 2013). Two species of *Propionibacteria*, viz., *P. acidipropionici* and *P. freudenreichii*, have been proposed as a strong probiotic candidate in digestive cancer prophylaxis (Jan et al. 2002). These two species are involved in killing colorectal carcinoma cell lines through selectively inducing apoptosis in these cell lines but not in the normal cells. The short-chain fatty acids (SCFAs) produced by these *Propionibacteria* (namely, propionate and acetate) are identified as major cytotoxic compounds responsible for the production of typical signs of apoptosis described elsewhere (Jan et al. 2002).

8.6 Archaeal Rare Taxa Associated with Health and Disease

Archaea, especially methanogenic archaea, are known to inhabit the human gut; the two most studied of them are *Methanobrevibacter smithii* and *Methanosphaera stadtmanae* (Gaci et al. 2014). Archaea under anaerobic condition use metabolic by-products of bacterial metabolism such as CO_2, H_2, and methylated compounds (methylamines and methyl sulfides) during methanogenesis. In addition to these dominant taxa, many low-abundance archaea are being reported that include *Methanomassiliicoccus luminyensis* (Dridi et al. 2012) and members of genus *Nitrososphaera*.

Methanomassiliicoccus luminyensis and its close relatives are dependent on H_2 and methanol for methanogenesis. But the unusual property of this organism is the use of trimethylamine as a substrate for methanogenesis. Trimethylamine is produced by gut microbes upon utilization of nutrients such as lecithin, choline, TMAO, and L-carnitine from the diet. The consequences of high levels of trimethylamine are the two diseases, trimethylaminuria and cardiovascular disease. The genome analysis of *Methanomassiliicoccus luminyensis* revealed that it contains the genes necessary for the reduction of tri-, di-, and monomethylamine with hydrogen for methanogenesis (Gorlas et al. 2012). This suggests that this organism has the capacity to reduce the levels of trimethylamine as it forms in the gut, and if this is true, it can naturally be used to control the levels of trimethylamine and related complications. In fact the new concept called archaebiotics has already been put forth which suggests the importance of rare taxa in controlling the human ailments (Brugère et al. 2014).

Recently, a member of ammonia-oxidizing archaea of the phylum *Thaumarchaeota* called *Nitrososphaera* has been reported in the gut microbiome (Hoffmann et al. 2013). The study reports that it was present only in those individuals for whom apparently there was no detection of the dominant archaea, i.e., *Methanobrevibacter smithii*, suggesting a possible antagonistic relationship between them. The same study has also demonstrated that this organism has a positive association with the ingestion of proteins and amino acids indicating the possible role of this organism in the digestion of these food ingredients.

8.7 Eukaryal Rare Taxa Relevant to Human Health

In addition to bacteria and archaea, the human gut is also a home for many macro- and microeukaryotes. Although the studies dealing with these components are lagging behind the eubacterial studies, few studies using high-throughput sequencing have contributed to our understanding of this generally neglected but important component in the human gut. This is especially true because the parasitic infections caused by intestinal parasites are among the most prevalent forms of infections in both developing and developed countries (Haque 2007). While the microeukaryotes in the gut are generally regarded as pathogenic, a major proportion of them do not cause any harm, and they live as a commensal in the gut. Particularly, the yeast *Saccharomyces boulardii* is used as one of the important ingredients of many probiotic preparations (Kelesidis and Pothoulakis 2011).

Host's response to the eukaryotic infections varies greatly from the asymptomatic phase to the one that leads to mortality and largely depends upon the health status, preexposure to the pathogen, and the mixed infections with other parasites (Pritt and Clark 2008). A longitudinal study conducted to understand the association of giardiasis on the growth of Brazilian children (Prado et al. 2005) suggests that the less abundant eukaryote in the gut (*Giardia duodenalis*) of school-going children leads to impede child growth through the malabsorption and hence has the direct effect on child's health.

8.8 Perspective

Gut microbiota makes an immense contribution to the proper functioning of the human body. As more and more diseases are found to be associated with the changes in the diversity of gut microbial communities, the medical science is experiencing a paradigm shift in understanding the pathophysiology of various diseases. In order to develop the microbiome-based biomarkers, it is essential to have a thorough knowledge of vastly diverse but low-abundant microbial inhabitants in the gut. In this context, it becomes essential to identify these rare taxa with confidence; therefore, the need of development of new methods to sequence full-length 16S rRNA gene to the satisfying depth of sequencing is being experienced. In addition, development and/or improvisation of associated bioinformatics tools to analyze the sequence data is also expected. Further, as many studies suggest archaea and eukarya as important components of the human gut, their assessment in the context with rare taxa and human health needs to be assessed in detail. Once the rare taxa are found to be associated with particular diseases, a targeted approach needs to be taken in order to unravel the exact role of those taxa in the disease. This is especially important because many studies dealing with the exploration of gut microbial communities often describe relative changes in gut microbial communities leaving us with no clue about whether the microbial changes are cause or consequence of the disease. In this regard, it is essential to obtain the taxa in concern in pure form and to find its specific metabolic pathways relevant to the disease development. Studies such as these could help us in developing future microbiome-based strategies in managing the disorders.

Acknowledgments The authors thank the Department of Biotechnology, Government of India, New Delhi, for necessary funding of Microbial Culture Collection project, grant no. BT/PR10054/NDB/52/94/2007. Shrikant Bhute is thankful to UGC, India, for the financial support to carry out the research work.

References

Allison MJ, Dawson KA, Mayberry WR, Foss JG (1985) Oxalobacter formigenes gen. nov., sp. nov.: oxalate-degrading anaerobes that inhabit the gastrointestinal tract. Arch Microbiol 141:1–7

Asplin JR (2016) The management of patients with enteric hyperoxaluria. Urolithiasis 44:33–43

Barnett C, Nazzal L, Goldfarb DS, Blaser MJ (2016) The presence of oxalobacter formigenes in the microbiome of healthy young adults. J Urol 195:499–506

Brugère J-F, Borrel G, Gaci N, Tottey W, O'Toole PW, Malpuech-Brugère C (2014) Archaebiotics: proposed therapeutic use of archaea to prevent trimethylaminuria and cardiovascular disease. Gut Microbes 5:5–10

Buchholz U, Bernard H, Werber D, Böhmer MM, Remschmidt C, Wilking H, Deleré Y, an der Heiden M, Adlhoch C, Dreesman J et al (2011) German outbreak of *Escherichia coli* O104:H4 associated with sprouts. N Engl J Med 365:1763–1770

Castellarin M, Warren RL, Freeman JD, Dreolini L, Krzywinski M, Strauss J, Barnes R, Watson P, Allen-Vercoe E, Moore RA et al (2012) Fusobacterium nucleatum infection is prevalent in human colorectal carcinoma. Genome Res 22:299–306

Chakravorty S, Helb D, Burday M, Connell N (2007) A detailed analysis of 16S ribosomal RNA gene segments for the diagnosis of pathogenic bacteria. J Microbiol Methods 69:330–339

Cho I, Blaser MJ (2012) The human microbiome: at the interface of health and disease. Nat Rev Genet 13:260–270

Dridi B, Fardeau ML, Ollivier B, Raoult D, Drancourt M (2012) Methanomassiliicoccus luminyensis gen. nov., sp. nov., a methanogenic archaeon isolated from human faeces. Int J Syst Evol Microbiol 62:1902–1907

Gaci N, Borrel G, Tottey W, O'Toole PW, Brugère J-F (2014) Archaea and the human gut: new beginning of an old story. World J Gastroenterol 20:16062–16078

Gorlas A, Robert C, Gimenez G, Drancourt M, Raoult D (2012) Complete genome sequence of Methanomassiliicoccus luminyensis, the largest genome of a human-associated Archaea species. J Bacteriol 194:4745

Haque R (2007) Human intestinal parasites. J Health Popul Nutr 25:387–391

Hoffmann C, Dollive S, Grunberg S, Chen J, Li H, Wu GD, Lewis JD, Bushman FD (2013) Archaea and fungi of the human gut microbiome: correlations with diet and bacterial residents. PLoS One 8:e66019

Hugenholtz P, Huber T (2003) Chimeric 16S rDNA sequences of diverse origin are accumulating in the public databases. Int J Syst Evol Microbiol 53:289–293

Jairath A, Parekh N, Otano N, Mishra S, Ganpule A, Sabnis R, Desai M (2015) Oxalobacter formigenes: opening the door to probiotic therapy for the treatment of hyperoxaluria. Scand J Urol 49:334–337

Jan G, Belzacq AS, Haouzi D, Rouault A, Metivier D, Kroemer G, Brenner C (2002) Propionibacteria induce apoptosis of colorectal carcinoma cells via short-chain fatty acids acting on mitochondria. Cell Death Differ 9:179–188

Jemal A, Bray F, Center MM, Ferlay J, Ward E, Forman D (2011) Global cancer statistics. CA Cancer J Clin 61:69–90

Kelesidis T, Pothoulakis C (2011) Efficacy and safety of the probiotic Saccharomyces boulardii for the prevention and therapy of gastrointestinal disorders. Therap Adv Gastroenterol 5:111–125

Kembel SW, Wu M, Eisen JA, Green JL (2012) Incorporating 16S gene copy number information improves estimates of microbial diversity and abundance. PLoS Comput Biol 8:16–18

Klindworth A, Pruesse E, Schweer T, Peplies J, Quast C, Horn M, Glöckner FO (2013) Evaluation of general 16S ribosomal RNA gene PCR primers for classical and next-generation sequencing-based diversity studies. Nucleic Acids Res 41:1–11

Kostic AD, Chun E, Robertson L, Glickman JN, Gallini CA, Michaud M, Clancy TE, Chung DC, Lochhead P, Hold GL et al (2013) Fusobacterium nucleatum potentiates intestinal tumorigenesis and modulates the tumor-immune microenvironment. Cell Host Microbe 14:207–215

Mao D-P, Zhou Q, Chen C-Y, Quan Z-X (2012) Coverage evaluation of universal bacterial primers using the metagenomic datasets. BMC Microbiol 12:66

Parkin DM (2006) The global health burden of infection-associated cancers in the year 2002. Int J Cancer 118:3030–3044

Prado MS, Cairncross S, Strina A, Barreto ML, Oliveira-Assis AM, Rego S (2005) Asymptomatic giardiasis and growth in young children; a longitudinal study in Salvador, Brazil. Parasitology 131:51–56

Pritt BS, Clark CG (2008) Amebiasis. Mayo Clin Proc 83:1154–1160

Qin J, Li R, Raes J, Arumugam M, Burgdorf KS, Manichanh C, Nielsen T, Pons N, Levenez F, Yamada T et al (2010) A human gut microbial gene catalogue established by metagenomic sequencing. Nature 464:59–65

Rajilić-Stojanović M, de Vos WM (2014) The first 1000 cultured species of the human gastrointestinal microbiota. FEMS Microbiol Rev 38:996–1047

Roh SW, Abell GCJ, Kim K-H, Nam Y-D, Bae J-W (2010) Comparing microarrays and next-generation sequencing technologies for microbial ecology research. Trends Biotechnol 28:291–299

Saraoui T, Parayre S, Guernec G, Loux V, Montfort J, Le Cam A, Boudry G, Jan G, Falentin H (2013) A unique in vivo experimental approach reveals metabolic adaptation of the probiotic Propionibacterium freudenreichii to the colon environment. BMC Genomics 14:911

Sasaki M, Sitaraman SV, Babbin BA, Gerner-Smidt P, Ribot EM, Garrett N, Alpern JA, Akyildiz A, Theiss AL, Nusrat A et al (2007) Invasive Escherichia coli are a feature of Crohn's disease. Lab Investig 87:1042–1054

Schloss PD, Gevers D, Westcott SL (2011) Reducing the effects of PCR amplification and sequencing artifacts on 16s rRNA-based studies. PLoS One 6(12):e27310

Sidhu H, Schmidt ME, Cornelius JG, Thamilselvan S, Khan SR, Hesse A, Peck AB (1999) Direct correlation between hyperoxaluria/oxalate stone disease and the absence of the gastrointestinal tract-dwelling bacterium Oxalobacter formigenes: possible prevention by gut recolonization or enzyme replacement therapy. J Am Soc Nephrol 10(Suppl 1):S334–S340

Sidhu H, Allison MJ, Chow JM, Clark A, Peck AB (2001) Rapid reversal of hyperoxaluria in a rat model after probiotic administration of oxalobacter formigenes. J Urol 166:1487–1491

Siva S, Barrack ER, Reddy GPV, Thamilselvan V, Thamilselvan S, Menon M, Bhandari M (2009) A critical analysis of the role of gut Oxalobacter formigenes in oxalate stone disease. BJU Int 103:18–21

Sogin ML, Morrison HG, Huber JA, Mark Welch D, Huse SM, Neal PR, Arrieta JM, Herndl GJ (2006) Microbial diversity in the deep sea and the underexplored "rare biosphere". Proc Natl Acad Sci U S A 103:12115–12120

Starčič Erjavec M, Žgur-Bertok D, Bagger-Skjøt L, Sandvang D, Frimodt-Møller N, Bélanger L, Garenaux A, Harel J, Dixit S, Gordon D et al (2015) Virulence potential for extraintestinal infections among commensal Escherichia coli isolated from healthy humans--the Trojan horse within our gut. FEMS Microbiol Lett 362:724–727

Stewart CS, Duncan SH, Cave DR (2004) Oxalobacter formigenes and its role in oxalate metabolism in the human gut. FEMS Microbiol Lett 230:1–7

Tenaillon O, Skurnik D, Picard B, Denamur E (2010) The population genetics of commensal Escherichia coli. Nat Rev Microbiol 8:207–217

Woese CR, Fox GE (1977) Phylogenetic structure of the prokaryotic domain: the primary kingdoms. Proc Natl Acad Sci 74:5088–5090

Metagenomic Insights into Microbial Diversity and Metabolic Potential of Hot Spring Ecosystems

9

Jasleen Saini, Ramanpreet Kaur,
and Pushpender Kumar Sharma

Abstract

Hot water springs or hydrothermal springs are places where warm or hot water comes out of earth surfaces regularly or for a significant period, in a year. These ecosystems present an epitome of extreme environments and are extensively distributed all over the globe. Geographically, these ecosystems encompass unique physical and chemical characteristics. Interestingly, 16S rRNA gene analysis in combination with next-generation sequencing has provided in-depth knowledge about phylogeny and the metabolic potential of a particular environment, including the hot springs. Every hot spring is unique and dynamic in its characteristics compare to the other. Investigation of metagenome from diverse ecological habitats, using high-throughput sequencing or library construction, has led to the discovery of a number of novel biocatalysts. Metagenomic studies in recent years have achieved two major goals: first it has resulted in deep understanding about structural and functional dynamics of microbial communities, and secondly, it has led to the discovery of diverse novel bioactive molecules. This book chapter will shed light into the role of metagenome gene cloning in revealing the true and comprehensive diversity and the metabolic potential of microbes in hot spring ecosystems.

Keywords

Metagenomics • Hot spring • Bacterial phylogeny • Ecosystems • 16S rRNA

J. Saini • R. Kaur • P.K. Sharma, Ph.D. (✉)
Department of Biotechnology, Sri Guru Granth Sahib World University,
Fatehgarh Sahib, Punjab 140 406, India
e-mail: pushpg_78@rediffmail.com; pushpg78@sggswu.edu.in

V.C. Kalia et al. (eds.), *Mining of Microbial Wealth and MetaGenomics*,
DOI 10.1007/978-981-10-5708-3_9

9.1 Introduction

The approximate number of prokaryotic cells in biosphere may exceed ~4–6 × 10^{30} (Whitman et al. 1998). The major fractions of prokaryotic organisms yet remain unexplored, and thus it presents an exciting challenge for scientific community to explore the genetic and metabolic diversity from various ecosystems. Investigation and analysis of 16S rRNA genes from diverse ecological habitats have demonstrated the presence of novel sequences with unique origin (Felske et al. 1999). It was established in a number of studies that 90–95% bacteria remain uncultured (Amann et al. 1995; Hugenholtz et al. 1998a; Hugenholtz 2002). The term metagenomics was coined by Jo Handelsman (Handelsman et al. 1998) and refers to genomic analysis of environmental DNA. Hot water springs or hydrothermal springs are the places where warm or hot groundwater comes out of earth surface regularly or for a significant period, in a year. These ecosystems present an epitome of extreme environment and are extensively distributed all over the world and are quite distinct with respect to their physicochemical characteristics (Hugenholtz et al. 1998b; Marteinsson et al. 2001).

These ecosystems hold variety of microflora with vast gene pool which can be explored for biotechnological applications. Metagenomic studies thus offer major facade for microbiologists, to connect phylogeny with ecological functions (Sharma et al. 2005).

Microbial community structure in hot spring is strongly dependent on the gradients of temperature, ecological interactions, chemistry of the underlying rocks, pH, oxidation-reduction potential or concentrations of various dissolved sulphides and inorganic carbons (Fouke et al. 2000; Dick and Shock 2013). Microorganisms possess propensity to append and aggregate to surfaces, when they come under the vicinity of water that results in formation of intricate networks (Gerbersdorf and Wieprecht 2015). Hot spring ecosystem holds enormous microbial diversity capable of surviving and blossom under array of environmental conditions (Wang et al. 2013; Chan et al. 2015). However, the range of mesophilic or thermophilic diversity in these ecosystems, as discussed above, is strongly dependent upon the temperature (Hobel et al. 2005). Construction and functional screening of metagenomic library from such ecosystems have already provided access to untapped wealth of active molecules (Simon et al. 2009; Xie et al. 2011; Tirawongsaroj et al. 2008; Steele et al. 2009; Jimenez et al. 2012a, b). Altogether, in recent years, culture-independent studies have achieved two major goals: firstly, it has enabled researchers in better understanding of structure and functioning of the microbes, and secondly, it has resulted in identification of novel active biomolecules (Neelakanta and Sultana 2013).

9.2 Microbial Diversity in Hot Springs

The major hot spring in the hot springs are found in Canada, New Zealand, United States, Chile, Japan, India and Malaysia (Song et al. 2010). India is referred to be one of the most tectonically active areas in the world, and according to geological surveys, it harbours ~340 hot water springs, which are classified into six geothermal provinces (Bisht et al. 2011). Here, in this book chapter, insights would be provided into the latest study being carried out to investigate microbial diversity and the metabolic potential of hot spring ecosystems. Starting with Indian subcontinents, a hot spring analysed in West Bengal predominantly demonstrated the presence of *Proteobacteria* and *Cyanobacteria* (Ghosh et al. 2003). 16S rRNA investigation of two hot springs, Tulsi Shyam and Lasundra of Gujarat state, in India, employing tag-encoded FLX amplicon pyrosequencing (bTEFAP) and shotgun sequencing approach, respectively, revealed variation in distribution of microbial diversity in these two hot springs, whereas Tulsi Shyam hot spring predominantly revealed the presence of *Firmicutes* (65.38%), *Proteobacteria* (21.21%) and an unclassified bacterial population ~10.69% (Ghelani et al. 2015). High-throughput sequencing of community DNA from Lasundra hot spring using an Ion Torrent PGM platform revealed predominantly *Bacillus* (86.7%), *Geobacillus* (2.4%) and *Paenibacillus* (1.0%) (Mangrola et al. 2015). Hot springs of Tibetan Plateau investigated employing Cluster and SIMPER divulges that temperature can greatly affect all over distribution of microbial diversity. Analysis of microbial distribution pattern using non-metric multidimensional scaling (NMDS) and principal coordinates analysis (PCoA) at species-level OTUs depicted a total of 42 bacterial phyla. Temperature range of 66–75 °C supported *Aquificae*, *Archaea* and GAL35 (a novel bacterial lineage), whose abundance exhibited a positive correlation with increasing temperature. In sharp contrast to this, *Deinococcus-Thermus*, *Cyanobacteria* and *Chloroflexi* showed its predominance in the temperature range 22–60 °C and were negatively associated with the temperature (Wang et al. 2013). Investigation of microbial diversity from Yellowstone National Park (YNP), United States, revealed varied composition of thermophilic microorganisms. Analysis of 16S rRNA genes from metagenome sample derived from 19 positions of 11 geothermal springs of YNP showed foremost presence of *Metallosphaera* and *Sulfolobus*. The bacterial genus *Hydrogenobaculum* showed its marked presence, followed by *Acidimicrobium*, *Acidovorax*, *Acidicaldus*, *Methylacidiphilum*, *Meiothermus*, *Geothermobacterium* and *Sulfobacillus*. Interestingly, four novel lineages that represented *Sulfolobus*, *Sulfolobales*, *Sulfobacillus* and *Acidicaldus* revealed maximum similarity to *Sulfolobus* sp. T1 (96.6%), *Sulfolobus islandicus* (88%), *Sulfobacillus acidophilus* (98%) and *Acidicaldus organivorans* (97.6%), respectively (Kozubal et al. 2012).

Interestingly, several hot springs characterized from YNP portrayed the dominant existence of photosynthetic microorganisms. A specific example is the mushroom spring, where four members of the bacterial community, *Cyanobacteria*, *Chloroflexi*, *Chlorobi* and *Acidobacteria*, showed their predominant occurrence in a phototrophic mat community (Liu et al. 2011). Furthermore, 16S rRNA and RFLP analysis of metagenome sample derived from three hot springs of Indonesia showed presence of *Proteobacteria*, *Bacillus* and *Flavobacterium*; interestingly, all these bacterial groups are not usually related with the thermophilic lineages (Baker et al. 2001). The site near volcanic eruption offers opportunity to explore, understand and compare the structural and functional dynamics of extremophilic archaea. Analysis of microbial communities from two such extreme sites (Mutnovsky and Uzon) of Kamchatka Peninsula revealed the presence of notable members from various communities. Interestingly, the phylum *Thaumarchaeota* comprises 57% of the total community at Mutnovsky, whereas it constituted 68% of the total community at Uzon sample, and members of phylum *Euryarchaeota* dominated the Mutnovsky by 34.7%. Among the bacterial lineages, *Thermotogae* showed its abundant presence in Mutnovsky, whereas it was negligible in Uzon. In sharp contrast to this, *Proteobacteria* followed by *Enterobacteriaceae*, *Aquificae* and *Thermodesulfobacteria* showed their marked presence at Uzon (Wemheuer et al. 2013). Yet in another study, 16S rRNA gene sequencing, in combination with the next-generation sequencing of metagenome from the hot spring of Sungai Klah, Malaysia, revealed foremost presence of *Firmicutes* (37.15%) and *Proteobacteria* (19.26%), whereas *Aquificae*, *Verrucomicrobia*, *Thermotogae* and 29 other members demonstrated less abundance. Notably, the study reports the presence of several phototrophic bacteria like *Roseiflexus*, *Porphyrobacter* and *Chloroflexus*. In addition, various pathogenic microbes like *Clostridium hiranonis*, *Brucella suis*, *Legionella pneumophila*, *Leptospira licerasiae*, *Leptospira wolffii*, *Pseudomonas fluorescens*, *Rickettsia montanensis*, *Rickettsiales*, etc. were also observed. Nevertheless, several microorganisms involved in carbon, sulphur and nitrogen metabolism were also found in this ecosystem (Chan et al. 2015). Microbial phylogeny investigation of hot springs from China revealed the presence of distinct monophyletic bacterial groups and several unidentified lineages. The archaea identified from this hot spring belong to *Euryarchaeota*, *Crenarchaeota* and *Korarchaeota* (Pagaling et al. 2012). The group *Crenarchaeota* is suggested to play a major role in the nitrification process during the nitrogen cycle (Leininger et al. 2006; Reigstad et al. 2008). Metagenome investigation of thermal spring from South Africa showed dominant presence of genera *Stenotrophomonas*, *Hydrogenophaga*, *Flectobacillus*, *Rheinheimera*, *Pseudomonas*, *Zavarzinella*, *Aquaspirillum* and *Limnobacter* (Tekere et al. 2015).

In another study, community analysis of environmental DNA from three hot springs, Tshipise, Mphephu and Sagole of South Africa, demonstrated the dominant presence of *Bacteroidetes* and *Proteobacteria* in Mphephu. In contrast, *Proteobacteria* and *Cyanobacteria* showed their prominent occurrence in Tshipise and Sagole. Several other phyla recovered revealed their presence <0.20%. (Tekere

Table 9.1 Major representative phyla having physiological roles in hot springs

Organism	Functions	Physiological type	References
Chloroflexi (e.g. *Chloronema*)	BChl-c and BChl-a biosynthesis, oxidize sulphide to polysulphides	Chlorophototrophs	Bryant et al. (2012), Klatt et al. (2011)
Aquificae (e.g. *Aquificales*)	Biological oxidation of sulphur compounds	Autotrophs	Skirnisdottir et al. (2000)
Proteobacteria (e.g. *Thiobacillus*)	Calvin-Benson cycle, reductive tricarboxylic cycle	Aerobic chemoorganotrophs	Chan et al. (2015)
Cyanobacteria (e.g. *Synechococcus*)	Oxygenic photosynthesis, nitrogen metabolism	Oxygenic phototrophs	Steunou et al. (2006), Bhaya et al. (2007)
Planctomycetes	Reductive acetyl-CoA pathway	Oligotrophic aerobic chemoorganotroph	Chan et al. (2015)
Euryarchaeota (e.g. *Methanobacteriales*)	Methanogenesis, reductive acetyl-CoA pathway	–	Ward et al. (1998), Chan et al. (2015)
Thaumarchaeota (e.g. *Nitrososphaera*)	Oxidation of ammonia to nitrite in the nitrogen cycle	Chemolithoautotrophs	Reigstad et al. (2008)
Crenarchaeota (e.g. *Desulfurococcales*)	Nitrification, carbon fixation, sulphur respiration	Organotrophs, chemolithotrophs	Huber and Stetter (2001), Leininger et al. (2006)

et al. 2012). Altogether, it was observed that various hot springs investigated for the microbial diversity all over the world demonstrated significant level of variations in distribution pattern of microbial community. Table 9.1 further reports microbial diversity from few more hot springs explored recently.

9.3 Metabolic Potential of Hot Spring Environments

With the advent of next-generation sequencing (NGS), it has become possible to determine the metabolic potential of any microbiome. NGS investigation of metagenome and metatranscriptome from hot springs provides inventory of microbial communities inhabiting in such habitats.

A classic example is the investigation of hot spring from Shi-Huang-ping from Taiwan that showed abundant presence of *Hydrogenobaculum* as a principle microorganism in this hot spring. This study also demonstrated the presence of genes related to carbon assimilation, nitrogen fixation and sulphur and hydrogen metabolism (Lin et al. 2015). Investigation of metabolic and functional potential of mushroom and octopus thermal springs from Yellowstone National Park (YNP), along with numerous other hot springs, revealed that almost all microbes principally transcribed genes for chlorophototrophy (Klatt et al. 2011). Furthermore, a comparative study of phototrophic, streamer and archaeal communities from 20 geothermal areas of YNP demonstrated variations in the numerous functional categories, like cell replication, energy metabolism, nitrogen fixation, cofactor biosynthesis, fatty acid biosynthesis, nitrogen metabolism, amino acid biosynthesis, etc. (William et al. 2013). Comparative genomics of microbes from alkaline hot springs revealed presence of 3-hydroxypropionate autotrophic pathway in bacteria. These microbial mats present archetype for studying microbial community ecology in siliceous hot springs of YNP. Interestingly, molecular and microscopic analysis of microbial mats established the dominant presence of unicellular *Synechococcus* species and filamentous anoxygenic phototrophs (FAPs). This study further signifies that there is cross feeding of metabolites among different organisms (Klatt et al. 2007). Functional analysis of Lasundra hot spring from Gujarat, India, revealed the presence of several genes that participate in the metabolism of aromatic compounds (Mangrola et al. 2015).

Understanding biological processes that involve nitrification and ammonia oxidation can enhance our understanding about the biogeochemical nitrogen cycling. Initially, these processes were thought to be restricted to few bacterial groups of *Proteobacteria* (Purkhold et al. 2000). Recent development in the molecular biology, however, has depicted that archaea are also efficient in oxidation of ammonia into nitrite (Dodsworth et al. 2011). Interestingly, archaea now have been implicated dominant component of the ammonia oxidation in terrestrial and marine environments. In this context, a study conducted from 22 hot springs showed that out of 22 hot spring, only 14 showed positive ammonia monooxygenase gene (AMO) from terrestrial hot springs, and most of these genes were observed at temperature range of 82–97 °C and pH range of 2.5–7 (Reigstad et al. 2008). KEGG analysis of enzymes involved in different metabolic pathways from acidic hot spring of Colombian Andean region elucidated several genes that encode enzymes responsible for nitrogen and sulphur cycle (Jimenez et al. 2012a, b). Several studies correlate geochemistry with the microbiological processes (Vick et al. 2010; Swingley et al. 2012).

Microbes have been recognized as major source of bioactive compounds (Bottone and Peluso 2003; Volk 2006; Volk and Furkert 2006; Williams 2009). Microbial mats from hot spring environment have received much attention, due to their vast potential towards synthesis of novel bioactive compounds. In this context, investigation of antimicrobial potential of cyanobacterial mats was evaluated using direct microscopy, from four hot springs located in the Sultanate of Oman. Active components extracted resulted in isolation and identification of 74 chemical compounds that displayed inhibitory activities against a diverse range of bacterial species and a diatom *Amphora coffeaeformis*. Determination of bacterial community composition showed that cyanobacterial species identified has shared homology mainly with *Chroococcus*, *Phormidium*, *Leptolyngbya*, *Spirulina* and *Lyngbya* (Dobretsov et al. 2011). The metabolic potential of microorganisms associated with various metabolic pathways is further enlisted in Table 9.2.

Table 9.2 List of metabolic functions performed by microbes dominated in hot springs

Metabolic function	Bacteria	References
Carbon cycle		
Reductive citrate cycle	*Hydrogenobaculum*	Lin et al. (2015)
Hydroxypropionate-hydroxybutyrate cycle	*Sulfolobus*	Alber et al. (2006)
	Metallosphaera	Alber et al. (2006)
Reductive citrate cycle and dicarboxylate-hydroxybutyrate cycle	*T. uzoniensis*, *T. tenax*	Mardanov et al. (2011), Siebers et al. (2011)
Calvin cycle	*Acidithiobacillus*	You et al. (2011)
	Thiomonas	Duquesne et al. (2008)
Nitrogen cycle		
Fixation of nitrogen	*A. ferrooxidans*	Lin et al. (2015)
Transformation of nitroalkane compounds (R-NO_2) to nitrite	*Hydrogenobaculum*, *A. ferrooxidans*, *Thiomonas*	Lin et al. (2015)
Assimilatory nitrate reduction	*Sphingomonas* sp., *Candidatus*, *Koribacter versatilis*, *Acidobacterium capsulatum*, *Pseudochlorella* sp., *Thalassiosira pseudonana*, *Chthoniobacter flavus*	Jimenez et al. (2012a, b)
Sulphur metabolism		
Transformation of trithionate into sulphite with sulphite reductase	*Vulcanisaeta archaea*, *Thermoproteus tenax*, *Caldivirga maquilingensis*	Lin et al. (2015)
Conversion of thiosulphate into sulphate	*Thiomonas*	Lin et al. (2015)
Conversion of tetrathionate or trithionate into thiosulphate	*Hydrogenobaculum*, *S. tokodaii*, *Metallosphaera*	Auernik and Kelly (2008), Lin et al. (2015)
Sulphur oxidation	*Phaeodactylum tricornutum*	Jimenez et al. (2012a, b)
Sulphate reduction	*Pyrobaculum* spp. *Caldivirga* spp.	William et al. (2010)
	Thermodesulfovibrio yellowstonii	Henry et al. (1994)
	Thermodesulfovibrio aggregans	Sekiguchi et al. (2008)
	Desulfomicrobium thermophilum	Thevenieau et al. (2007)
	Desulfotomaculum carboxydivorans	Parshina et al. (2005)
	Desulfotomaculum kuznetsovii	Visser et al. (2013)
	Thermodesulfatator indicus	Moussard et al. (2004)
	Thermodesulfobacterium commune	Zeikus et al. (1983)
	Thermodesulfobium narugense	Mori et al. (2003)
	Archaeoglobus veneficus	Huber et al. (1997)
	Caldivirga maquilingensis	Itoh et al. (1999)
Reduction of sulphur to hydrogen sulphide	*Hippea maritima*	Miroshnichenko et al. (1999)
	Thermococcus gammatolerans	Jolivet et al. (2003)
	Thermofilum pendens	Anderson et al. (2008)
	Caldivirga maquilingensis	Itoh et al. (1999)
	Vulcanisaeta distributa	Itoh et al. (2002)
	Vulcanisaeta moutnovskia	Gumerov et al. (2011)

(continued)

Table 9.2 (continued)

Metabolic function	Bacteria	References
Iron metabolism		
Oxidation of iron	*Metallosphaera yellowstonensis*	Kozubal et al. (2011)
Reduction of ferric iron under anaerobic conditions	*Sulfolobales* str. MK5, *Acidicaldus* str. MK6	Kozubal et al. (2012)

9.4 Insights into Carbon, Nitrogen and Sulphur Cycle

The investigation of the major elemental cycles can help in predicting diverse microbial functions. The following paragraphs provide insights into various studies that report the role of metagenomics in understanding such processes.

9.5 Carbon Cycle

To date, several carbon assimilation pathways have been identified employing metagenomic studies. Reductive citrate cycle in *Hydrogenobaculum* was reported from hot springs of Taiwan (Lin et al. 2015). Microbial species *Sulfolobus* and *Metallosphaera* were reported to harbour genes that participate in hydroxypropionate-hydroxybutyrate cycle (Alber et al. 2006; Teufel et al. 2009). Interestingly, both reductive citrate cycle and dicarboxylate-hydroxybutyrate cycle take place in *T. uzoniensis* and *T. tenax* (Mardanov et al. 2011; Siebers et al. 2011). The presence of genes related to Calvin cycle in *Acidithiobacillus* and *Thiomonas* indicates that these microbes are dynamically involved in carbon metabolism (Duquesne et al. 2008; You et al. 2011).

9.6 Nitrogen Cycle

Nitrogen metabolism can provide insights about the biotransformation of various nitrogenous compounds. The processes like nitrogen fixation are well studied in *A. ferrooxidans*. Biological transformation of nitroalkane compounds to nitrite has been notably found in *Hydrogenobaculum*, *A. ferrooxidans* and *Thiomonas* (Lin et al. 2015). Major genes involved in nitrogen cycle include narG, narH, narI, norB, norE, norC, nifD, nifK, nirB, nirA and nirS encoding different enzymes. Importantly, microorganisms involved in the nitrogen metabolism belong to *Proteobacteria*, *Acidobacteria*, *Firmicutes*, *Nitrospira*, *Spartobacteria*, *Trebouxiophyceae*, *Coscinodiscophyceae*, etc. in acidic hot spring of Colombian Andes (Jimenez et al. 2012a, b).

9.7 Sulphur Metabolism

In addition to carbon and nitrogen metabolism, the presence of sulphur-metabolizing enzymes has been identified and mapped in various hot springs. Dominant microorganism involved in sulphur oxidation is *Phaeodactylum tricornutum* (Jimenez et al. 2012a, b). Several microbes have revealed the presence of key enzymes involved in sulphate as well as sulphur reduction. Other sulphur-related metabolic pathways that involve transformation of trithionate into sulphite are reported in *Vulcanisaeta*, *Thermoproteus* and *Caldivirga*. Furthermore, *Thiomonas* has key enzymes for converting thiosulphate into sulphate (Lin et al. 2015).

9.8 Biocatalysts Isolated from Hot Springs

Microorganisms from thermophilic environments are the major source of thermostable enzymes. *Taq* polymerase, the first enzyme isolated from the thermophilic strain *Thermus aquaticus*, has been an innovation towards the discovery of polymerase chain reaction (Chien et al. 1976). Enzymes obtained from these microorganisms have great potential to be used as biocatalysts for biotechnology and industrial purposes. With the increasing demand of thermostable enzyme in various chemical industries, their recovery from hot springs has increased tremendously. Thermostable enzymes have been extensively used in food, pharmaceuticals, cosmetics, geochemicals and leather, dairy, pulp and paper industries and for brewing and baking purposes (Haki and Rakshit 2003).

Various hot springs have been explored to obtain novel thermostable enzymes, e.g. investigation of metagenome from Lobios hot spring revealed the presence of 11 ORFs homologous to lipolytic enzymes. The enzyme showed sequence similarity to β-lactamase irrespective of showing any β-lactamase activity (López-López et al. 2015). Several other esterases were isolated from several hot springs all over the world (Rhee et al. 2005; Tirawongsaroj et al. 2008; Leis et al. 2015).

PCR-based cloning has also been successfully used in screening the novel enzymes directly from the metagenomic samples (Lorenz et al. 2002). A novel cyclomaltodextrinase gene was cloned from environmental DNA that has the ability to hydrolyse cyclodextrins and starch (Tang et al. 2006).

A gene encoding lipase enzyme was cloned from metagenome sample of Manikaran Sahib (Himachal Pradesh). Sequence analysis of the cloned gene revealed its identity with lipase gene of *Geobacillus*. Biochemical analysis of the lipase demonstrated its maximum activity at 60 °C (Sharma et al. 2011). Interestingly, the thermostability of the enzyme was further enhanced employing directed evolution (Sharma et al. 2012). Recently, thermostable protease isolated from *Bacillus licheniformis* of Unnai hot springs has been employed in various industrial settings (Dudhagara et al. 2014).

Recent studies from hot springs of Manikaran reported isolation of a *Bacillus altitudinis* IARI-MB-9 and *Gulbenkiania mobilis* IARI-MB-18 which produces

Table 9.3 List of enzymes isolated from hot springs

Enzyme	Source	Optimum temperature (°C)	Optimum pH	References
Taq Polymerase	Yellowstone National Park, USA	80	7.8	Chien et al. (1976)
Thermoalkaliphilic lipases	Hot springs of Southern Sinai	60	10	Deyaa et al. (2016)
Lipase	Hot springs of Manikaran	50	9.0	Sharma et al. (2011)
Esterase (Est1)	Hot springs in Tangkuban Perahu	90	6.0	Rhee et al. (2005)
Patatin-like phospholipase (PLP)	Thailand hot spring	70	9.0	Tirawongsaroj et al. (2008)
Esterase (Est1)	Thailand hot spring	70	9.0	Tirawongsaroj et al. (2008)
α-Amylase	Omer hot spring, Afyonkarahisar in Turkey	80	5.0	Ozdemir et al. (2015)
α-Amylase	Hot spring of Larijan, Iran	80	5.0–7.0	Mollania et al. (2009)
Amylase	Hot spring at Purwokerto, Central Java Province, Indonesia	60	8.0	Amin and Zusfahair (2012)
α-Amylase	Hot spring sources in Yangmingshan National Park, Northern Taiwan	70	5.5–6.5	Shaw et al. (1995)
Amylase	Wondo Genet hot spring	75–80	5.5	Mamo and Gessese (1999)
Lipase	Hot springs in Indonesia	50	7.5	Lee et al. (1999)
GH5 cellulase	Hot spring in Grensdalur, Iceland	70	5	Zarafeta et al. (2016)
Cellulase (EBI-244)	Great Boiling Spring, Gerlach, Nevada	109	–	Graham et al. (2011)
Cyclomaltodextrinase	Bor Khleung hot spring in Ratchaburi province, Thailand	50–55	6–7	Tang et al. (2006)
Xylanase	Hot spring in Yongtai (Fuzhou, China)	75	8.2	Liu et al. (2012)
Neopullulanase-like enzyme (Env Npu193A)	Bor Khleung hot spring in Thailand	75	7.0	Tang et al. (2008)
Alkaline Protease	Unnai hot spring	50	9.0	Dudhagara et al. (2015)
β-D-galactosidase (MbgI)	Geothermal springs in Northern Himalayan Region of India	65	8.0	Gupta et al. (2012)

Table 9.3 (continued)

Enzyme	Source	Optimum temperature (°C)	Optimum pH	References
Esterase (EstA2)	Hot springs in the town of Furnas, Azores, Portugal	80	8.0	Leis et al. (2015)
Esterase (EstB1)	Hot springs in the town of Furnas, Azores, Portugal	75	8.0	Leis et al. (2015)
Esterase (LOB4Est)	Lobios hot spring, in Ourense (Galician region, Spain)	40	7.5	López-López et al. (2015)
Lipolytic enzyme (PlpBW1)	Hot springs located in Kamchatka Peninsula	85	10	Wemheuer et al. (2013)
Lipolytic enzyme (EstBW1)	Hot springs located in Kamchatka Peninsula	90	7.0	Wemheuer et al. (2013)
Lipolytic enzyme (EstBW2)	Hot springs located in Kamchatka Peninsula	65	7.0	Wemheuer et al. (2013)

thermostable hydrolytic enzymes like CMCase, Xylanase, FPase and Cellobiose that display activities at high temperatures (Verma et al. 2015). In addition, various metagenomic studies also report isolation of two lipolytic enzymes—phospholipase and esterase from hot spring of Thailand (Thevenieau et al. 2007). Metagenome investigation of two hot springs from Kamchatka Peninsula resulted in isolation of novel genes encoding lipolytic and proteolytic enzymes that displayed maximum activities at 85, 90 and 65 °C, respectively (Wemheuer et al. 2013). Table 9.3 further enlists various enzymes reported from various hot spring ecosystems.

Conclusion

The biological diversity of bacteria can help us recognize the way of shaping and survival of these microbes in hot springs through various physicochemical conditions and biological interactions. The existence of biotechnological significant species in the metagenome suggests the impending applications of the hot spring bacteria that evoke the continuing research in this field.

Acknowledgement Financial assistance to Dr. P.K.S. by the Science and Engineering Research Board (SERB) New Delhi, with project file number: SB/YS/LS-63/2013, under fast track scheme for the young scientists, is highly acknowledged.

References

Alber B, Olinger M, Rieder A, Kockelkorn D, Jobst B, Hugler M, Fuchs G (2006) Malonylcoenzyme A reductase in the modified 3-hydroxypropionate cycle for autotrophic carbon fixation in archaeal Metallosphaera and Sulfolobus spp. J Bacteriol 188:8551–8559. doi:10.1128/JB.00987-06

Amann RI, Ludwig W, Schleifer KH (1995) Phylogenetic identification and in situ detection of individual microbial cells without cultivation. Microbiol Rev 59:143–169

Amin F, Zusfahair (2012) Thermophilic amylase from Thermus sp. isolation and its potential application for bioethanol production. J Sci Technol 34:525–531

Anderson I, Rodriguez J, Susanti D, Porat I, Reich C, Ulrich LE, Elkins JG, Mavromatis K, Lykidis A, Kim E, Thompson LS, Nolan M, Land M, Copeland A, Lapidus A, Lucas S, Detter C, Zhulin IB, Olsen GJ, Whitman W, Mukhopadhyay B, Bristow J, Kyrpides N (2008) Genome sequence of Thermofilum pendens reveals an exceptional loss of biosynthetic pathways without genome reduction. J Bacteriol 190:2957–2965. doi:10.1128/JB.01949-07

Auernik KS, Kelly RM (2008) Identification of components of electron transport chains in the extremely thermoacidophilic crenarchaeon Metallosphaera sedula through iron and sulfur compound oxidation transcriptomes. Appl Environ Microbiol 74:7723–7732. doi:10.1128/AEM.01545-08

Baker GC, Gaffar S, Cowan DA, Suharto AR (2001) Bacterial community analysis of Indonesian hot springs. FEMS Microbiol Lett 200:103–109. doi:10.1111/j.1574-6968.2001.tb10700.x

Bhaya D, Grossman AR, Steunou AS, Khuri N, Cohan FM, Hamamura N, Melendrez MC, Bateson MM, Ward DM, Heidelberg JE (2007) Population level functional diversity in a microbial community revealed by comparative genomic and metagenomic analyses. ISME J 1:703–713. doi:10.1038/ismej.2007.46

Bisht SS, Das NN, Tripathy NK (2011) Indian hot-water springs: a bird's eye view. J Energy Environ Carbon Credits 1:1–15

Bottone EJ, Peluso RW (2003) Production by Bacillus pumilus (MSH) of an antifungal compound that is active against Mucoraceae and Aspergillus species: preliminary report. J Med Microbiol 52:69–74. doi:10.1099/jmm.0.04935-0

Bryant DA, Klatt CG, Frigaard NU, Liu Z, Li T, Zhao F, Garcia Costas AM, Overmann J, Ward DM (2012) Comparative and functional genomics of anoxygenic green bacteria from the taxa Chlorobi, Chloroflexi, and Acidobacteria. Adv Photosynth Respir 33:47–102

Chan CS, GanChan K, LingTay Y, HengChua Y, MauGoh K (2015) Diversity of thermophiles in a Malaysian hot spring determined using 16S rRNA and shotgun metagenome sequencing. Front Microbiol 6:177. doi:10.3389/fmicb.2015.00177

Chien A, Edgar DB, Trela JM (1976) Deoxyribonucleic acid polymerase from the extreme thermophile Thermus aquaticus. J Bacteriol 127:1550–1557

Deyaa M, Fotouh A, Bayoumi RA, Hassan MA (2016) Production of thermoalkaliphilic lipase from Geobacillus thermoleovorans DA2 and application in Leather Industry. Enzyme Res 2016:9034364. doi:10.1155/2016/9034364

Dick JM, Shock EL (2013) A metastable equilibrium model for the relative abundances of microbial phyla in a hot spring. PLoS One 8:e72395. doi:10.1371/journal.pone.007239

Dobretsov S, Abed RMM, Al Maskari SMS, Al Sabahi JN, Victor R (2011) Cyanobacterial mats from hot springs produce antimicrobial compounds and quorum-sensing inhibitors under natural conditions. J Appl Phycol 23:983–993. doi:10.1007/s10811-010-9627-2

Dodsworth JA, Hungate BA, Hedlund BP (2011) Ammonia oxidation, denitrification and dissimilatory nitrate reduction to ammonium in two US Great Basin hot springs with abundant ammonia-oxidizing archaea. Environ Microbiol 13:2371–2386. doi:10.1111/j.1462-2920.2011.02508.x

Dudhagara PR, Bhavasar S, Ghelani A, Bhatt S (2014) Isolation, characterization and investing the industrial applications of thermostable and solvent tolerant serine protease from hot spring isolated thermophilic Bacillus licheniformis U1. Int J Appl Sci Biotechnol 2:75–82. doi:10.3126/ijasbt.v2i1.9519

Dudhagara PR, Bhavasar S, Ghelani A (2015) Hide dehairing and laundry detergent compatibility testing of thermostable and solvents tolerant alkaline protease from hot spring isolate bacillus cohniiU3. OnLine J Biol Sci 15:152–161. doi:10.3844/ojbsci.2015.152.161

Duquesne K, Lieutaud A, Ratouchniak J, Muller D, Lett MC, Bonnefoy V (2008) Arsenite oxidation by a chemoautotrophic moderately acidophilic Thiomonas sp.: from the strain isolation to the gene study. Environ Microbiol 10:228–237. doi:10.1111/j.1462-2920.2007.01447.x

Felske A, Wolterink A, Lis RV, de Vos WM, Akkermans ADL (1999) Searching for the predominant soil bacteria: 16S rDNA cloning versus strain cultivation. FEMS Microbiol Ecol 30:137–145. doi:10.1111/j.1574-6941.1999.tb00642.x

Fouke BW, Farmer JD, Des Marais DJ, Pratt L, Sturchio NC, Burns PC, Discipulo MK (2000) Depositional facies and aqueous–solid geochemistry of travertine-depositing hot springs (Angel Terrace, Mammoth Hot Springs, Yellowstone National Park, U.S.A.) J Sediment Res 70:565–585. doi:10.1306/2DC40929-0E47-11D7-8643000102C1865D

Gerbersdorf SU, Wieprecht S (2015) Biostabilization of cohesive sediments: revisiting the role of abiotic conditions, physiology and diversity of microbes, polymeric secretion, and biofilm architecture. Geobiology 13:68–97. doi:10.1111/gbi.12115

Ghelani A, Patel R, Mangrola A, Dudhagara P (2015) Cultivation-independent comprehensive survey of bacterial diversity in Tulsi Shyam Hot Springs, India. Genomics Data 4:54–56. doi:10.1016/j.gdata.2015.03.003

Ghosh D, Bal B, Kashyap VK, Pal S (2003) Molecular phylogenetic exploration of bacterial diversity in a Bakreshwar (India) Hot Spring and culture of Shewanella-related thermophiles. Appl Environ Microbiol 69:4332–4336. doi:10.1128/AEM.69.7.4332-4336.2003

Graham JE, Clark ME, Nadler DC, Huffer S, Chokhawala HA, Rowland SE, Blanch HW, Clark DS, Robb FT (2011) Identification and characterization of a multidomain hyperthermophilic cellulase from an archaeal enrichment. Nat Commun 2:375. doi:10.1038/ncomms1373

Gumerov VM, Mardanov AV, Beletsky AV, Prokofeva MI, Bonch-Osmolovskaya EA, Ravin NV, Skryabin KG (2011) Complete genome sequence of "Vulcanisaeta moutnovskia" strain 768-28, a novel member of the hyperthermophilic crenarchaeal genus Vulcanisaeta. J Bacteriol 193:2355–2356. doi:10.1128/JB.00237-11

Gupta R, Govil T, Capalash N, Sharma P (2012) Characterization of a glycoside hydrolase family 1 β-galactosidase from hot spring metagenome with transglycosylation activity. Appl Biochem Biotechnol 168:1681–1693. doi:10.1007/s12010-012-9889-z

Haki GD, Rakshit SK (2003) Developments in industrially important thermostable enzymes: a review. Bioresour Technol 89:17–34. doi:10.1016/S0960-8524(03)00033-6

Handelsman J, Rondon MR, Brady SF, Clardy J, Goodman RM (1998) Molecular biological access to the chemistry of unknown soil microbes: a new frontier for natural products. Chem Biol 5:R245–R249. http://biomednet.com/elecref/10745521005R0245

Henry EA, Devereux R, Maki JS, Gilmour CC, Woese CR, Mandelco L, Schauder R, Remsen CC, Mitchell R (1994) Characterization of a new thermophilic sulfate-reducing bacterium. Arch Microbial 161:62–69. doi:10.1007/BF00248894

Hobel CFV, Marteinsson VT, Hreggvidsson GO, Kristjansson JK (2005) Investigation of the microbial ecology of intertidal hot springs by using diversity analysis of 16S rRNA and chitinase genes. Appl Environ Microbiol 71:2771–2776. doi:10.1128/AEM.71.5.2771-2776.2005

Huber H, Stetter KO (2001) Desulfurococcales. In: Garrity GM, Boone DR, Castenholz RW (eds) Bergey's manual of systematic bacteriology. Springer, New York, pp 179–210

Huber H, Jannasch H, Rachel R, Fuchs T, Stetter KO (1997) Archaeoglobus veneficus sp. Nov., a novel facultative chemolithoautotrophic hyperthrmophilic sulphite reducer, isolated from abyssal black smokers. Syst Appl Microbiol 20:374–380. doi:10.1016/S0723-2020(97)80005-7

Hugenholtz P (2002) Exploring prokaryotic diversity in the genomic era. Genome Biol 3:Reviews0003. doi:10.1186/gb-2002-3-2-reviews0003

Hugenholtz P, Goebel BM, Pace NR (1998a) Impact of culture independent studies on the emerging phylogenetic view of bacterial diversity. J Bacteriol 180:4765–4774

Hugenholtz P, Pitulle C, Hershberger KL, Pace NR (1998b) Novel division level bacterial diversity in a Yellowstone hot spring. J Bacteriol 180:366–376

Itoh T, Suzuki KI, Sanchez PC, Nakase T (1999) Caldivirga maquiligenesis gen. nov., sp. nov., a new genus of rod-shaped crenarchaeote isolated from a hot spring in the Philippines. Int J Syst Bacteriol 49:1157–1163. doi:10.1099/00207713-49-3-1157

Itoh T, Suzuki KI, Nakase T (2002) Vulcanisaeta distribute gen. nov., sp. Nov., and Vulcanisaeta souniana sp. nov., novel hyperthermophilic, rod shaped crenarchaeota isolated from hot springs in Japan. Int J Syst Evol Microbiol 52:1097–1104. doi:10.1099/ijs.0.02152-0

Jimenez DJ, Andreote FD, Chaves D, Montana JS, Forero CS, Junca H, Zambrano MM, Baena S (2012a) Structural and functional insights from the metagenome of an acidic hot spring microbial planktonic community in the Colombian Andes. PLoS One 7:e52069. doi:10.1371/journal.pone.0052069

Jimenez DJ, Montaña JS, Alvarez D, Baena S (2012b) A novel cold active esterase derived from high Andean forest soil metagenome. World J Microbiol Biotechnol 28:361–370. doi:10.1007/s11274-011-0828-x

Jolivet EL, Haridon S, Corre E, Forterre P, Prieur D (2003) Thermococcus gammatolerans sp. nov., a hyperthermophilic archaeon from a deep-sea hydrothermal vent that resist ionizing radiation. Int J Syst Evol Microbiol 53:847–851. doi:10.1099/ijs.0.02503-0

Klatt CG, Bryant DA, Ward DM (2007) Comparative genomics provides evidence for the 3-hydroxypropionate autotrophic pathway in filamentous anoxygenic phototrophic bacteria and in hot spring microbial mats. Environ Microbiol 9:2067–2078. doi:10.1111/j.1462-2920.2007.01323.x

Klatt CG, Wood JM, Rusch DB, Bateson MM, Hamamura N, Heidelberg JF, Grossman AR, Bhaya D, Cohan FM, Kuhl M, Byrant DA, Ward DM (2011) Community ecology of hot spring cyanobacterial mats: predominant populations and their functional potential. ISME J 5:1262–1278. doi:10.1038/ismej.2011.73

Kozubal MA, Dlakic M, Macur RE, Inskeep WP (2011) Terminal oxidase diversity and function in "Metallosphaera yellowstonensis": gene expression and protein modeling suggest mechanisms of Fe (II) oxidation in the sulfolobales. Appl Environ Microbiol 77:1844–1853. doi:10.1128/AEM.01646-10

Kozubal MA, Macur RE, Jay ZJ, Beam JP, Malfatti SA, Tringe SG, Kocar BD, Borch T, Inskeep WP (2012) Microbial iron cycling in acidic geothermal springs of Yellowstone National Park: integration molecular surveys, geochemical processes and isolation of novel Fe-active microorganisms. Front Microbiol 3:109. doi:10.3389/fmicb.2012.00109

Lee D-W, Koh Y-S, Kim K-J, Kim B-C, Choj H-J, Kim D-S, Suhartono MT, Pyun Y-R (1999) Isolation and characterisation of thermophilic lipase from Bacillus thermoleovorans ID-1. FEMS Microbiol Lett 179:393–400. doi:10.1111/j.1574-6968.1999.tb08754.x

Leininger S, Urich T, Schloter M, Schwark L, Qi J, Nicol GW, Prosser JI, Schuster SC, Schleper C (2006) Archaea predominate among ammonia-oxidizing prokaryotes in soils. Nature 442:806–809. doi:10.1038/nature04983

Leis B, Angelov A, Mientus M, Li H, Pham VTT, Lauinger B, Bongen P, Pietruszka J, Gonçalves LG, Santos H, Liebl W (2015) Identification of novel esterase-active enzymes from hot environments by use of the host bacterium Thermus thermophilus. Front Microbiol 6:275. doi:10.3389/fmicb.2015.00275

Lin KH, Liao BY, Chang HW, Huang SW, Chang TY, Yang CY, Wang YB, Lin YTK, Wu YW, Tang SL, Yu H-T (2015) Metabolic characteristics of dominant microbes and key rare species from an acidic hot spring in Taiwan revealed by metagenomics. BMC Genomics 16:1029. doi:10.1186/s12864-015-2230-9

Liu Z, Klatt CG, Wood JM, Rusch DB, Ludwig M, Wittekindt N, Tomsho LP, Schuster SC, Ward DM, Bryant DA (2011) Metatranscriptomic analyses of chlorophototrophs of a hot-spring microbial mat. ISME J 5:1279–1290. doi:10.1038/ismej.2011.37

Liu B, Zhang N, Zhao C, Lin B, Xie L, Huang Y (2012) Characterization of a recombinant thermostable xylanase from hot spring thermophilic Geobacillus sp. TC-W7. J Microbiol Biotechnol 22:1388–1394. doi:10.4014/jmb.1203.03045

López-López O, Knapik K, Cerdán ME, González-Siso MI (2015) Metagenomics of an Alkaline Hot Spring in Galicia (Spain): microbial diversity analysis and screening for novel lipolytic enzymes. Front Microbiol 6:1291. doi:10.3389/fmicb.2015.01291

Lorenz P, Liebeton K, Niehaus F, Eck J (2002) Screening for novel enzymes for biocatalytic processed: accessing the metagenome as a resource of novel functional sequence space. Curr Opin Biotechnol 13:572–577

Mamo G, Gessese A (1999) A highly thermostable amylase from a newly isolated thermophilic Bacillus sp. WN11. J Appl Microbiol 86:557–560. doi:10.1046/j.1365-2672.1999.00685.x

Mangrola AV, Dudhagara P, Koringa P, Joshi CG, Patel RK (2015) Shotgun metagenomic sequencing based microbial diversity assessment of Lasundra hot spring, India. Genomics Data 4:73–75. doi:10.1016/j.gdata.2015.03.005

Mardanov AV, Gumerov VM, Beletsky AV, Prokofeva MI, Bonch-Osmolovskaya EA, Ravin NV, Skryabin KG (2011) Complete genome sequence of the thermoacidophilic crenarchaeon Thermoproteus uzoniensis 768-20. J Bacteriol 193:3156–3157. doi:10.1128/JB.00409-11

Marteinsson VT, Hauksdottir S, Hobel CF, Kristmannsdottir H, Hreggvidsson GO, Kristjansson JK (2001) Phylogenetic diversity analysis of subterranean hot springs in Iceland. Appl Environ Microbiol 67:4242–4248. doi:10.1128/AEM.67.9.4242-4248.2001

Miroshnichenko ML, Rainey FA, Rhode M, Bonch-Osmolovskaya EA (1999) Hippeamaritima gen. nov., sp. nov., a new genus of thermophilic, sulfur-reducing bacterium from submarine hotvents. Int J Syst Bacteriol 49:1033–1038. doi:10.1099/00207713-49-3-1033

Mollania N, Khajeh K, Hosseinkhani S, Dabirmanesh B (2009) Purification and characterization of a thermostable phytate resistant alpha-amylase from Geobacillus sp. LH8. Int J Biol Macromol 46:27–36. doi:10.1016/j.ijbiomac.2009.10.010

Mori K, Kim H, Kakegawa T, Hanada S (2003) A novel lineage of sulfate-reducing microorganisms: Thermodesulfobiaceaefam. Nov., Thermodesulfobium narugense, gen. nov., sp. nov., a new thermophilic isolate from a hot spring. Extremophiles 7:283–290. doi:10.1007/s00792-003-0320-0

Moussard HL, Haridon S, Tindall BJ, Banta A, Schumann P, Stackebrandt E, Reysenbach AL, Jeanthon C (2004) Thermodesulfatator indicus gen. nov., sp. nov., a novel thermophilic chemolithoautotrophic sulfate-reducing bacterium isolated from the Central Indian Ridge. Int J Syst Evol Microbiol 54:227–233. doi:10.1099/ijs.0.02669-0

Neelakanta G, Sultana H (2013) The use of metagenomic approaches to analyze changes in microbial communities. Microbiol Insights 6:37–48. doi:10.4137/MBI.S10819

Ozdemir S, Okumus V, Ulutas MS, Dundar A, Akarsubası AT, Dumonted S (2015) Isolation of a novel thermophilic anoxybacillus flavithermus so-13, production, characterization and industrial applications of its thermostable α-amylase. Bioprocess Biotech 5:7. doi:10.4172/2155-9821.1000237

Pagaling E, Grant WD, Cowan DA, Jones BE, Ma Y, Ventosa A, Heaphy S (2012) Bacterial and archaeal diversity in two hot spring microbial mats from the geothermal region of Tengchong, China. Extremophiles 16:607–618. doi:10.1007/s00792-012-0460-1

Parshina SN, Sipma J, Nakashimada Y, Henstra AM, Smidt H, Lysenko AM, Lens PN, Lettinga G, Stams AJ (2005) Desulfotomaculum carboxydivorans sp. nov., a novel sulfate-reducing bacterium capable of growth at 100% CO. Int J Syst Evol Microbiol 55:2159–2165. doi:10.1099/ijs.0.63780-0

Purkhold U, Ser AP, Juretschko S, Schmid MC, Koops HP, Wagner M (2000) Phylogeny of all recognized species of ammonia oxidizers based on comparative 16S rRNA and amoA sequence analysis: implications for molecular diversity surveys. Appl Environ Microbiol 66:5368–5382. doi:10.1128/AEM.66.12.5368-5382.2000

Reigstad LJ, Richter A, Daims H, Urich T, Schwark L, Schleper C (2008) Nitrification in terrestrial hot springs of Iceland and Kamchatka. FEMS Microbiol Ecol 64:167–174. doi:10.1111/j.1574-6941.2008.00466.x

Rhee JK, Ahn DG, Kim YG, Oh JW (2005) New thermophilic and thermostable esterase with sequence similarity to the hormone sensitive lipase family, cloned from a metagenomic library. Appl Environ Microbiol 71:817–825. doi:10.1128/AEM.71.2.817-825.2005

Sekiguchi Y, Muramatsu M, Imachu H, Narihiro T, Ohashi A, Harada H, Hanada S, Kamagata Y (2008) Thermodesulfovibrio aggregans sp. nov. and Thermodesulfovibrio thiophilus sp. nov., anaerobic, thermophilic, sulfate-reducing bacteria isolated from thermophilic methanogenic sludge, and emended description of the genus Thermodesulfovibrio. Int J Syst Evol Microbiol 58:2541–2548. doi:10.1099/ijs.0.2008/000893-0

Sharma R, Ranjan R, Kapardar RK, Grover A (2005) Unculturable bacterial diversity: an untapped resource. Curr Sci 89:72–77

Sharma PK, Singh K, Singh R, Capalash N, Ali A, Mohammad O, Kaur J (2011) Characterization of a thermostable lipase showing loss of secondary structure at ambient temperature. Mol Biol Rep 39:2795–2804. doi:10.1007/s11033-011-1038-1

Sharma PK, Kumar R, Kumar R, Mohammad O, Singh R, Kaur J (2012) Engineering of a metagenome derived lipase toward thermal tolerance: effect of asparagine to lysine mutation on the protein surface. Gene 491:264–271. doi:10.1016/j.gene.2011.09.028

Shaw J, Pang L, Chen S, Chen I (1995) Purification and properties of an extracellular a-amylase from Thermus sp. Bot Bull Acad Sci 36:195–200

Siebers B, Zaparty M, Raddatz G, Tjaden B, Albers SV, Bell SD, Blombach F, Kletzin A, Kyrpides N, Lanz C, Plagens A, Rampp M, Rosinus A, Jan MV, Makaroya KS, Klenk HP, Schuster SC, Hensel R (2011) The complete genome sequence of Thermoproteus tenax: a physiologically versatile member of the Crenarchaeota. PLoS One 6:e24222. doi:10.1371/journal.pone.0024222

Simon C, Wiezer A, Strittmatter AW, Daniel R (2009) Phylogenetic diversity and metabolic potential revealed in a glacier ice metagenome. Appl Environ Microbiol 75:7519–7526. doi:10.1128/AEM.00946-09

Skirnisdottir S, Hreggvidsson GO, Hjörleifsdottir S, Marteinsson VT, Petursdottir SK, Holst O, Kristjansson JK (2000) Influence of sulfide and temperature on species composition and community structure of hot spring microbial mats. Appl Environ Microbiol 66:2835–2841

Song ZQ, Chen JQ, Jiang HC, Zhou EM, Tang SK, Zhi XY, Zhang LX, Zhang CL, Li WJ (2010) Diversity of Crenarchaeota in terrestrial hot springs in Tengchong, China. Extremophiles 14:287–296. doi:10.1007/s00792-010-0307-6

Steele HE, Jaeger JE, Daniel R, Streit WR (2009) Advances in recovery of novel biocatalysts from metagenomes. J Mol Microbiol Biotechnol 16:25–37. http://dx.doi.org/10.1159/000142892

Steunou AS, Bhaya D, Bateson MM, Melendrez MC, Ward DM, Brecht E et al (2006) In situ analysis of nitrogen fixation and metabolic switching in unicellular thermophilic cyanobacteria inhabiting hot spring microbial mats. Proc Natl Acad Sci U S A 103:2398–2403

Swingley WD, Meyer-Dombard DR, Shock EL, Alsop EB, Falenski HD, Havig JR, Raymond J (2012) Coordinating environmental genomics and geochemistry reveals metabolic transitions in a Hot Spring Ecosystem. PLoS One 7:E38108. doi:10.1371/journal.pone.0038108

Tang K, Utairungsee T, Kanokratana P, Sriprang R, Champreda V, Eurwilaichitr L, Tanapongpipat S (2006) Characterization of a novel cyclomaltodextrinase expressed from environmental DNAisolated from Bor Khleung hot spring in Thailand. FEMS Microbiol Lett 260:91–99. doi:10.1111/j.1574-6968.2006.00308.x

Tang K, Kobayashi RS, Champreda V, Eurwilaichitr L, Tanapongpipat S (2008) Isolation and characterization of a novel thermostable neopullulanase-like enzyme from a Hot Spring in Thailand. Biosci Biotechnol Biochem 72:1448–1456. doi:10.1271/bbb.70754

Tekere M, Prinsloo A, Olivier J, Jonker N, Venter S (2012) An evaluation of the bacterial diversity at Tshipise, Mphephu and Sagole hot water springs, Limpopo Province, South Africa. Afr J Microbiol Res 6:4993–5004. doi:10.5897/AJMR12.250

Tekere M, Lötter A, Olivier J, Venter S (2015) Bacterial diversity in some South African Thermal Springs: a metagenomic analysis. In: Proceedings world geothermal congress 2015, Melbourne, Australia, pp 19–25

Teufel R, Kung JW, Kockelkorn D, Alber BE, Fuchs G (2009) 3-Hydroxypropionylcoenzyme A dehydratase and acryloyl-coenzyme A reductase, enzymes of the autotrophic 3-hydroxypropionate/4-hydroxybutyrate cycle in the Sulfolobales. J Bacteriol 191:4572–4581

Thevenieau F, Fardeau ML, Ollivier B, Joulian C, Baena S (2007) Desulfomicrobium thermophilum sp. nov., a novel thermophilic sulphate-reducing bacterium isolated from a terrestrial hot spring in Colombia. Extremophiles 11:295–303. doi:10.1007/s00792-006-0039-9

Tirawongsaroj PR, Sriprang P, Harnpicharnchai T, Thongaram V, Champreda S, Tanapongpipat K, Pootanakit L, Eurwilaichitr L (2008) Novel thermophilic and thermostable lipolytic enzymes from a Thailand hot spring metagenomic library. J Biotechnol 133:42–49. doi:10.1016/j.jbiotec.2007.08.046

Verma P, Yadav AN, Shukla L, Saxena AK, Suman A (2015) Hydrolytic enzymes production by thermotolerant Bacillus altitudinis IARI-MB-9 and Gulbenkianiamobilis IARI-MB-18 isolated from Manikaran hot springs. Int J Adv Res 3:1241–1250

Vick TJ, Dodsworth JA, Costa KC, Shock EL, Hedlund BP (2010) Microbiology and geochemistry of Little Hot Creek, a hot spring environment in the Long Valley Caldera. Geobiology 8:140–154. doi:10.1111/j.1472-4669.2009.00228.x

Visser M, Worm P, Muyzer G, Pereira IA, Schaap PJ, Plugge CM, Kuever J, Parshina SN, Nazina TN, Ivanova AE, Bernier-Latmani R, Goodwin LA, Kyrpides NC, Woyke T, Chain P, Davenport KW, Spring S, Klenk HP, Stams AJ (2013) Genome analysis of Desulfotomaculum kuznetsovii strain 17T reveals a physiological similarity with Pelotomaculum thermopropionicum SI(T). Stan Genomic Sci 8:69–87. doi:10.4056/sigs.3627141

Volk R (2006) Antialgal activity of several cyanobacterial exometabolites. J Appl Phycol 18:145–151. doi:10.1007/s10811-006-9085-z

Volk R, Furkert FH (2006) Antialgal, antibacterial and antifungal activity of two metabolites produced and excreted by cyanobacteria during growth. Microbiol Res 161:180–186. doi:10.1016/j.micres.2005.08.005

Wang S, Hou W, Dong H, Jiang H, Huang L, Wu G, Zhang C, Song Z, Zhang Y, Ren H, Zhang J, Zhang L (2013) Control of temperature on microbial community structure in Hot Springs of the Tibetan. PLoS One 8:e62901. doi:10.1371/journal.pone.0062901

Ward DM, Ferris MJ, Nold SC, Bateson MM (1998) A natural view of microbial biodiversity within hot spring cyanobacterial mat communities. Microbiol Mol Biol Rev 62:1353–1370

Wemheuer B, Taube R, Akyol P, Wemheuer F, Daniel R (2013) Microbial diversity and biochemical potential encoded by thermal spring metagenomes derived from the Kamchatka Peninsula. Hindawi Publishing Corporation. Archaea Article ID 136714, 13p. doi:10.1155/2013/136714

Whitman WB, Coleman DC, Wiebe WJ (1998) Prokaryotes: the unseen majority. Proc Natl Acad Sci U S A 95:6578–6583

William PI, Rusch DB, Jay ZJ, Herrgard MJ, Kozubal MA, Richardson TH, Macur RE, Hamamura N, Jennings RD, Fouke BW, Reysenbach AL, Roberto F, Young M, Schwartz A, Boyd ES, Badger JH, Mathur EJ, Ortmann AC, Bateson M, Geesey G, Frazier M (2010) Metagenomes from high-temperature chemotrophic systems reveal geochemical controls on microbial community structure and function. PLoS One 5:39773. doi:10.1371/journal.pone.0009773

William PI, Jay ZJ, Tringe SG, Herrgård MJ, Rusch DB, YNP Metagenome Project Steering Committee, Working Group Members (2013) The YNP metagenome project: environmental parameters responsible for microbial distribution in the Yellowstone geothermal ecosystem. Front Microbiol 4:67. doi:10.3389/fmicb.2013.00067

Williams PG (2009) Panning for chemical gold: marine bacteria as a source of new therapeutics. Trends Biotechnol 27:45–52. doi:10.1016/j.tibtech.2008.10.005

Xie W, Wang F, Guo L, Chen Z, Sievert SM, Meng J, Huang G, Li Y, Yan Q, Wu S, Wang X, Chen S, He G, Xiao X, Xu A (2011) Comparative metagenomics of microbial communities inhabiting deep-sea hydrothermal vent chimneys with contrasting chemistries. ISME J 5:414–426. doi:10.1038/ismej.2010.144

You XY, Guo X, Zheng HJ, Zhang MJ, Liu LJ, Zhu YQ, Zhu B, Wang SY, Zhao GP, Poetsch A, Jiang CY, Liu SJ (2011) Unraveling the Acidithiobacillus caldus complete genome and its central metabolisms for carbon assimilation. J Genet Genomics 38:243–252. doi:10.1016/J.JGG.2011.04.006

Zarafeta D, Kissas D, Sayer C, Gudbergsdottir SR, Ladoukakis E, Isupov MN, Chatziioannou A, Peng X, Littlechild JA, Skretas G, Kolisis FN (2016) Discovery and characterization of a thermostable and highly halotolerant GH5 cellulase from an Icelandic Hot Spring isolate. PLoS One 11:e0146454. doi:10.1371/journal.pone.0146454

Zeikus JG, Dawson MA, Thompson TE, Ingvorsen K, Hatchikian EC (1983) Microbial ecology of volcanic sulphidogenesis: ISOLATION AND CHARACTERization of Thermodesulfobacterium commune gen. nov. and sp. nov. J Gen Microbiol 129:1159–1169. doi:10.1099/00221287-129-4-1159

10 Bioprospecting Gastrointestinal Microflora of Common Fishes for Disease Control in Aquaculture

Jiun Yan Loh and Adeline Su Yien Ting

Abstract

The aquaculture industry faces many challenges related to diseases, which leads to serious economic losses worldwide. Massive use of antimicrobial agents encourages the natural emergence of antibiotic-resistant bacteria. As such, there is an urgent demand on the use of beneficial bacteria as a better strategy for disease control. Commercially available probiotics are mostly sourced from terrestrial origins. However, their extensive usage in fish farming could cause unexpected adverse effects to the aquaculture species. Therefore, there is an immediate need to select probiotics originating from fishes. This chapter discusses the importance of beneficial bacteria in disease prevention and management. The role of probiotic in aquaculture is explicitly elucidated in this section. The complex microbial profile and dynamics of the microflora in the fish microbiota is also discussed here, particularly with the aid of metagenomics. Bioprospecting of beneficial microflora as probiotics is described based on the various rigorous tests (e.g. antagonistic tests, antibiotic susceptibility and haemolytic assays) performed. Other aspects examined to select for beneficial bacteria include colonization and proliferation capability, host immunomodulation, stimulation of enzymatic production and amenability to biotransformation for efficient delivery. All these propel the interest in understanding gastrointestinal microflora for probiotic development to control diseases in aquaculture.

J.Y. Loh
Faculty of Applied Sciences, UCSI University,
Jalan Menara Gading, UCSI Heights, 56000 Kuala Lumpur, Malaysia

School of Science, Monash University Malaysia,
Jalan Lagoon Selatan, 47500 Bandar Sunway, Selangor, Malaysia

A.S.Y. Ting (✉)
School of Science, Monash University Malaysia,
Jalan Lagoon Selatan, 47500 Bandar Sunway, Selangor, Malaysia
e-mail: adeline.ting@monash.edu, adelsuyien@yahoo.com

V.C. Kalia et al. (eds.), *Mining of Microbial Wealth and MetaGenomics*,
DOI 10.1007/978-981-10-5708-3_10

Keywords
Aquaculture • Bioformulation • Intestinal microflora • Metagenomics • Probiotics

Abbreviations

ARISA	Automated ribosomal intergenic spacer analysis
BLIS	Bacteriocin-like inhibitory substances
cBKD	Bacterial kidney disease
DGGE	Denaturing gradient gel electrophoresis
FISH	Fluorescence in situ hybridization
GALT	Gut-associated lymphoid tissues
gDNA	Genomic DNA
GI	Gastrointestinal
GI	Green fluorescence protein gene
HPLC	High-performance liquid chromatography
Ig	Immunoglobulins
MHS	Milky haemolymph disease of spiny lobster/milky haemolymph syndrome
mt	Millions of tons
NGS	Next-generation sequencing
NHP	Necrotizing hepatopancreatitis
PCoA	Principal coordinate analysis
T-RFLP	Terminal restriction fragment length polymorphism
WSSV	White spot syndrome virus

10.1 Introduction

Aquaculture is the fastest developing food production sector in the world, contributing 90% (in the form of seafood) to the global animal production for human consumption (Sahu et al. 2008). The aquaculture industry faces many challenges related to disease infection, which leads to serious environmental consequences and economic losses. Disease in aquaculture is commonly caused by bacteria. The extent of infection ranges from topical injuries (e.g. necrosis, lesions) to severe organ malfunction (e.g. bacterial kidney disease). Table 10.1 summarizes the important common fish diseases in aquaculture.

The increase in fish diseases catalysed the use of antibiotics as chemotherapeutic agents. In recent years, the application of antibiotics has been extended to preventive measures. Antibiotics are derived from natural sources or synthetic origins and are mainly categorized as therapeutic, prophylactic or metaphylactic to either treat established infections, prevent development of infections or medicate infected groups, respectively (Cabello 2006; Romero et al. 2012). One most significant drawback of the rampant use of antibiotics is the emergence of antibiotic-resistant bacteria through horizontal and promiscuous flow of resistance genes (Mohanty

Table 10.1 Common bacterial disease in aquaculture

Fish disease	Causative agent	Common affected fish groups	Symptoms
Bacterial kidney disease (BKD)	*Renibacterium salmoninarum*	Trout and salmon	Affected fish exhibit protruding eyes (exophthalmia), darker skin pigmentation and haemorrhage at the base of the fins. Gills are pale and anaemic. Fluid accumulation may occur internally in the abdominal cavity, causing the enlargement of the kidney. Grey nodules appear on the kidney, liver, spleen and heart
Enteric red mouth disease	*Yersinia ruckeri*	All fishes	Affected fish shows dark body colour and haemorrhage at base of the fins, gill, mouth corner, gums, palate and tongue. Fishes exhibit exophthalmos and orbital haemorrhage, swollen abdomen, and loss of appetite. Pinpoint haemorrhage is also observed on the pancreas, liver, swim bladder and lateral musculature surfaces. Haematopoietic tissue in the kidney and spleen is also affected by necrosis
Enteric septicaemia of Catfish	*Edwardsiella ictaluri*	Catfish, sea bass, salmon and trout	Swelling on top of the head, occasionally progressing to erosion of tissue, resulting in the exposure of the brain ("hole-in-the-head" disease). Fish develop pale gills, but skin darkens with multiple small white spots detected. Ulcers can be spotted on the flanks and head, and haemorrhaging occurs at the base of the fins, mouth, throat, operculum and abdomen. Affected fish shows exophthalmos and swollen abdomen
Edwardsiellosis	*Edwardsiella tarda*	Freshwater and marine fishes, reptiles, amphibians and mammals	Affected fish shows haemorrhage in the body cavities, muscles and organs (liver and kidneys). Necrotic white/grey lesions can be seen on the kidneys and spleen. In adult fish, organomegaly, pale inflamed gills, exophthalmia and cataract are observed. In some infected fishes, haemorrhagic lesions on the skin and fins, erosion of the skin, systemic oedema and ascites can occur. Swelling and hyperaemic at the anal region of certain species can also be detected

(continued)

Table 10.1 (continued)

Fish disease	Causative agent	Common affected fish groups	Symptoms
Furunculosis	*Aeromonas salmonicida*	Freshwater and marine salmonid species	Crater lesions on skin and/or muscle and haemorrhage on the skin, mouth and fin bases. Body pigmentation darkens but gills turn pale. Nares and vent exude bloody fluid. Haemorrhages can be detected in the muscle and internal organs. Internally, enlarged spleen develops, the liver and stomach can be full of mucus, blood and sloughed epithelial cells and the intestines are congested. Symptoms on gills include fusion of gill lamellae and inflamed gills
Milky haemolymph disease of spiny lobster/milky haemolymph syndrome (MHS)	*Rickettsia*-like bacterium	Black tiger prawn, European shore crab and tropical spiny lobsters	Milky haemolymph exuding from wounds and may be visible in affected fish under swollen abdominal pleura of the exoskeleton. White hypertrophied connective tissues observed in major organs and tissues
Necrotizing hepatopancreatitis (NHP)	β-Proteobacterium	Pacific white shrimp, Western blue shrimp, some American species, for example, *Farfantepenaeus aztecus*, *Farfantepenaeus californiensis* and *Litopenaeus setiferus*	Affected shrimps have flaccid body and soft shells. The gills are black, and the intestinal tract is empty. Often, hepatopancreas degeneration occurs with black streaks on the pancreas. Hepatopancreatic tubules are severely affected where multifocal granulomatous lesions and atrophy of adjacent hepatopancreatic tubule epithelial develop. The cytoplasm is typically filled with basophilic organisms and tubule epithelial cells. The intratubular spaces experience severe haemocytic inflammation
Piscirickettsiosis	*Piscirickettsia salmonis*	Salmon, trout and sea bass	Affected fish exhibits skin lesions on raised scales. Flanks and heads appear with shallow ulcers, skin darkens but gills are pale. Swollen abdomen, spleen and kidney are observed. Mottled liver or ring-shaped white to pale-yellow lesions, ascites and signs of peritonitis are detected. Pinpoint haemorrhages occur in the gastrointestinal tract, swim bladder and visceral fat. Vasculitis and necrosis occur on the liver and kidney

Table 10.1 (continued)

Fish disease	Causative agent	Common affected fish groups	Symptoms
Xenohaliotis californiensis	*Xenohaliotis californiensis*	Wild and farmed abalones, *Haliotis* spp.	Diseased abalones develop intracytoplasmic bacterial inclusions found internally (oesophagus, intestine, and the epithelia of digestive gland). Degeneration (metaplastic changes) within the digestive gland may occur in moderate to severe cases. Pedal muscle atrophy is also commonly observed in susceptible species
Columnaris	*Flavobacterium columnare*	All fishes	Affected fish exhibits frayed and ragged fins. Ulceration occurs on the oral mucosa, resulting in mouth rot and epidermal loss. This disease is also known as "cotton wool disease" or "mouth fungus disease". Other symptoms include the accumulation of mucus on the external structures (gills, head and dorsal regions) with gills becoming light or dark brown in colour

Information is compiled and modified from http://www.thefishsite.com and related sources (Melba et al. 2001; Declercq et al. 2013)

and Sahoo 2007; Romero et al. 2012). The resistance genes carried over subsequently lead to the inevitable emergence of antibiotic-resistant bacteria (Mohanty and Sahoo 2007; Sahu et al. 2008). The emergence of antibiotic resistance in pathogens has devastating effects to the hosts and the entire aquatic ecosystem. In fact, gastrointestinal (GI) microbiome of the fishes from natural habitats are vastly different from cultured farmed fishes due to the use of antibiotics. Loh (2015) revealed similar findings using metagenomics approach where microflora diversity in farmed fish was inferior to those in the natural habitats. This presumably caused by antimicrobial agent used in the farms and the availability of different food sources.

In the past few decades, the use of beneficial bacteria, predominantly probiotic bacteria, has been explored as a more sustainable and environmental-friendly approach to control fish diseases (Planas et al. 2004; Burr and Gatlin 2005; Pirarat et al. 2006; Taoka et al. 2006; Pintado et al. 2010). Probiotics are defined as live bacterial cells which, when applied, render beneficial health effect to the host, through either altering the microflora community, improving the feed and/or enhancing nutritional value, or stimulating host's immune response towards disease or through improvements to the environment. In simpler words, a probiotic is a microorganism or a microorganism producing components/derivatives that

Fig. 10.1 The mechanisms of action of probiotics in aquaculture

render health benefits to the host (Devereux 2002). Probiotics suppress pathogen via competitive exclusion, improve host well-being by modulating immunoresponse, enhance digestion for better nutrient absorption and combat against viruses (Balcázar et al. 2006, 2008; Ziaei-Nejad et al. 2006; Wang and Xu 2006; Klaenhammer 2007; Lauzon et al. 2008). Interestingly, some strains are of particular use in water quality management (Harikrishnan et al. 2010; Andani et al. 2012) (Fig. 10.1).

10.2 Mechanisms of Actions of Probiotics

The mechanisms of action of probiotics in aquaculture encompass the following: competitive exclusion, nutrients and enzymatic assimilation, modulation of immune response, antiviral effects and water quality improvement (Balcázar et al. 2006; Nayak 2010; Cruz et al. 2012). Competitive exclusion is the prevention of opportunistic pathogens from colonizing the GI system, achieved through the antimicrobial compounds, e.g. bacteriocins, hydrogen peroxide, proteases and lysozymes, or by competitively excluding similar nutrient resources or mucosal space for colonization (Bandyopadhyay and Mohapatra 2009). In aquaculture, the use of probiotics improved the survival of many aquatic species, such as crab larvae (*Portunus trituberculatus*), through the application of probiotic *Thalassobacter utilis* (Nogami et al. 1997). The survival of turbot, *Scophthalmus maximus*, was also improved through the use of *Lactobacillus* sp., *Bifidobacterium* sp. and *Streptococcus* sp. (Gatesoupe 1994; Olsson et al. 1998) and the vibriosis cases in tiger shrimp (*Penaeus monodon*) with the application of *Bacillus subtilis* BT23 (Vaseeharan and Ramasamy 2003). The application of the *Bacillus* strain in larviculture also proved to be effective for larval survival, promoted growth and reduced the occurrence of pathogenic bacteria in common snook (*Centropomus undecimalis*) (Kennedy et al. 1998). Douillet and Langdon (1994) found that *Alteromonas* sp. could reduce the infection of *Vibrio* sp. in Pacific oyster (*Crassostrea gigas*) larviculture. Other studies, such as using *Pseudomonas fluorescens* strains, have been found to reduce the mortality of the rainbow trout fingerlings (*Oncorhynchus mykiss*) infected by *V. anguillarum* (Gram et al. 1999).

Clearly, the host organisms benefit greatly from the presence of probiotics and beneficial microbes in the gut system. Many studies showed probiotic can modulate host's nutritional and digestion processes through supplying fatty acids and vitamins to the host cells and/or participating in digestive enzyme production. These beneficial bacteria include *Agrobacterium* sp., *Brevibacterium* sp., *Clostridium* sp., *Microbacterium* sp., *Pseudomonas* sp., *Staphylococcus* sp. and *Lactococcus lactis* subsp. *lactis*, which are known to contribute to the host's nutritional processes (Ringø et al. 1995; Loh and Ting 2016). Some enzyme-producing microbiota such as *Bacillus* and *Enterobacteriaceae* (*Aeromonas* sp., *Acinetobacter* sp., *Flavobacterium* sp., *Microbacterium* sp., *Micrococcus* sp., *Photobacterium* sp., *Pseudomonas* sp., *Staphylococcus* sp., *Vibrio* sp. and some unidentified anaerobes and yeasts) are also possible contributors to stimulate the production of enzymes (Rawls et al. 2004, 2006; Bates et al. 2006; Ray et al. 2012).

The immune response of fish could be up-regulated through the supplementation of various probiotics either in the form of monospecies or multispecies (Das et al. 2006; Nayak 2010; Butprom et al. 2013). It is an established fact that probiotics can stimulate various cytokines in the fishes as well as to express phagocytosis, or to produce lysozyme, as means of defence mechanism against pathogens. This was clearly demonstrated in the earlier study by Sakai et al. (1995). The authors found that the phagocytic activity of leucocytes increased through administration of the bacterium *Clostridium butyricum* in rainbow trout cultivation, which subsequently improved their resistance to vibriosis. Similar work by Balcázar (2003) and Balcázar et al. (2007) reported administration of bacterial strains (*Bacillus* and *Vibrio* sp.) in a mix consortium positively influenced the protection against the white spot syndrome virus (WSSV), *V. harveyi* and early mortality syndrome (EMS) caused by *V. parahaemolyticus* in white shrimp. Nikoskelainen et al. (2003) also demonstrated that with the application of *Lactobacillus rhamnosus* (strain ATCC 53103), the occurrence of respiratory burst specifically in rainbow trout (*Oncorhynchus mykiss*) is significantly stimulated. This possibly could be attributed to some bacteriocin-like inhibitory substances (BLIS) such as antimicrobial peptides, proteins or protein complexes excreted/synthesized by probiotics, which can effectively control several fish diseases caused by *Vibrio parahaemolyticus* (Carraturo et al. 2006), *Flavobacterium* sp. (Balakrishna and Keerthi 2012) and *Aeromonas hydrophila* (Selvendran and Michael Babu 2013).

In recent years, the indigenous microflora communities in the GI systems of fishes are explored as potential candidates of beneficial microbes (probiotics) for disease control. Sanchez et al. (2012) showed that fish intestinal bacteria might present an alternative for the discovery of natural products. Loh et al. (2014) successfully isolated probiotic strains from fish GI system, particularly the probiotic *L. lactis* subsp. *lactis*, which has shown potential to control *Edwardsiella tarda* infection in fishes. Probiotics from fish origins are theoretically more acceptable to the target hosts and to the culture environment (Schulze et al. 2006). Hence, probiotics originating from the GI of fishes are highly favoured. The following sections discuss the approaches in bioprospecting beneficial microbes from the indigenous microflora of the gut system for potential application to control of fish diseases and, more specifically, on the technologies and functions of understanding the diversity of microflora in GI of fishes and the screening exercises and application of beneficial microbes isolated from gut system for control of fish diseases.

10.3 Diversity of Gastrointestinal Communities

The gastrointestinal microflora in a fish is made up of a complex community of microbiota forming commensalistic or mutualistic relationships with their hosts (Spor et al. 2011). Both allochthonous and autochthonous microbiota can be found in the GI of fishes, as bacteria can transiently pass through the passage of digestive tract without permanently residing in the GI, or they could reside and proliferate in the gut system. These bacteria influence a broad range of host biological processes, such as food digestion and assimilation (Nelson et al. 2012). Nevertheless, our understanding of the microbiota and their microbial functions in the fish gut is very limited compared to that of reports from terrestrial vertebrates (Gatesoupe 2005; Ley et al. 2008; Nayak 2010; Qin et al. 2010; Ray et al. 2012). As such, there is potential in exploring and harnessing useful microflora of the GI as bioactive agents.

Early investigations on gastrointestinal microbiota of fish species mainly relied on culture-dependent methods. These methods are, however, time consuming and, in some instances, lack the accuracy to identify species correctly. In addition, most of the gastrointestinal microbes are non-culturable. Thus, identification of complex microbial communities is best determined using culture-independent techniques such as molecular identification approaches. These methods require only genomic DNA (gDNA), and the analysis provides more accurate and relatively unbiased information on their species and phylogenetic affiliation (Nayak 2010). High-throughput DNA (e.g. denaturing gradient gel electrophoresis (DGGE) (Giatsis et al. 2014; Forberg et al. 2016; Ringø et al. 2016), ribosomal intergenic spacer analysis (ARISA) (Arias et al. 2006; Jami et al. 2014; Larsen et al. 2014) and terminal restriction fragment length polymorphism (T-RFLP) (Formosa et al. 2010; Liu et al. 2016)) and the sequencing technologies (e.g. next-generation sequencing (NGS) (Star et al. 2013; Xia et al. 2014; Tarnecki et al. 2016), Ion Torrent sequencing (Standen et al. 2015; Gajardo et al. 2016; Lokesh and Kiron 2016) and Sanger sequencing (Lan and Love 2012; Givens et al. 2015; Alikunhi et al. 2016)) have enabled analysis of millions of nucleotides from both culturable and non-culturable microorganisms. In recent years, microbial diversity is also studied via metagenomics analyses using NGS Illumina® technology (Degnan and Ochman 2012; Sinclair et al. 2015; Wu et al. 2015). Loh (2015) successfully unveiled the taxonomic composition of intestinal bacteria communities in both wild- and farmed-types of adult climbing perch, *Anabas testudineus*. With NGS, three predominant bacterial phyla were determined as *Proteobacteria*, *Firmicutes* and *Fusobacteria*, and this microbial distribution was consistent in climbing perch from different geographical locations (Fig. 10.2a). These findings closely resembled microbial communities reported in other fishes as well (Clements et al. 2007; Sullam et al. 2012; Ye et al. 2013). Metagenomics analysis also revealed the patterns of bacterial distribution and allows the comparison between bacterial groups via principal coordinate analysis (PCoA) (Yao et al. 2012). Similarity of bacterial community can then be analysed and grouped based on the similarity of microbial genotypes (Fig. 10.2b). Bacterial variation in terms of bacterial composition, richness and diversity can be summarized through such analytical pipeline as well (Smriga et al. 2010; Newton et al. 2011; Wang et al. 2012; Rungrassamee et al. 2014).

10.4 Screening for Beneficial Bacteria as Probiotics for Disease Control

The first probiotics applied in aquaculture was *Bacillus toyoi*; it was used as a feed additive to enhance the growth rate of yellow tail fish, *Seriola quinqueradiata*. In the subsequent years, *Bacillus* spp. adopted to improve productivity of Tiger prawn, *Penaeus monodon*, and water quality (Porubcan 1991). Other important genera include *Bifidobacterium*, *Lactobacillus*, *Streptococcus*, *Alteromonas* and *Carnobacterium*, which have potential to be used as probiotic in aquaculture. The primary interest of probiotic research is largely focused on their use as growth promoters and to improve the health of animals. The selection of beneficial bacteria for development into probiotics is typically initiated by the in vitro screening process (Fig. 10.3). A vast number of isolates are first derived from samples and subjected to rigorous screening assays, beginning with antimicrobial (antagonistic) assays. This is often accompanied by haemolytic and antibiotic-resistance assays, which further helps in the selection of desired probiotics.

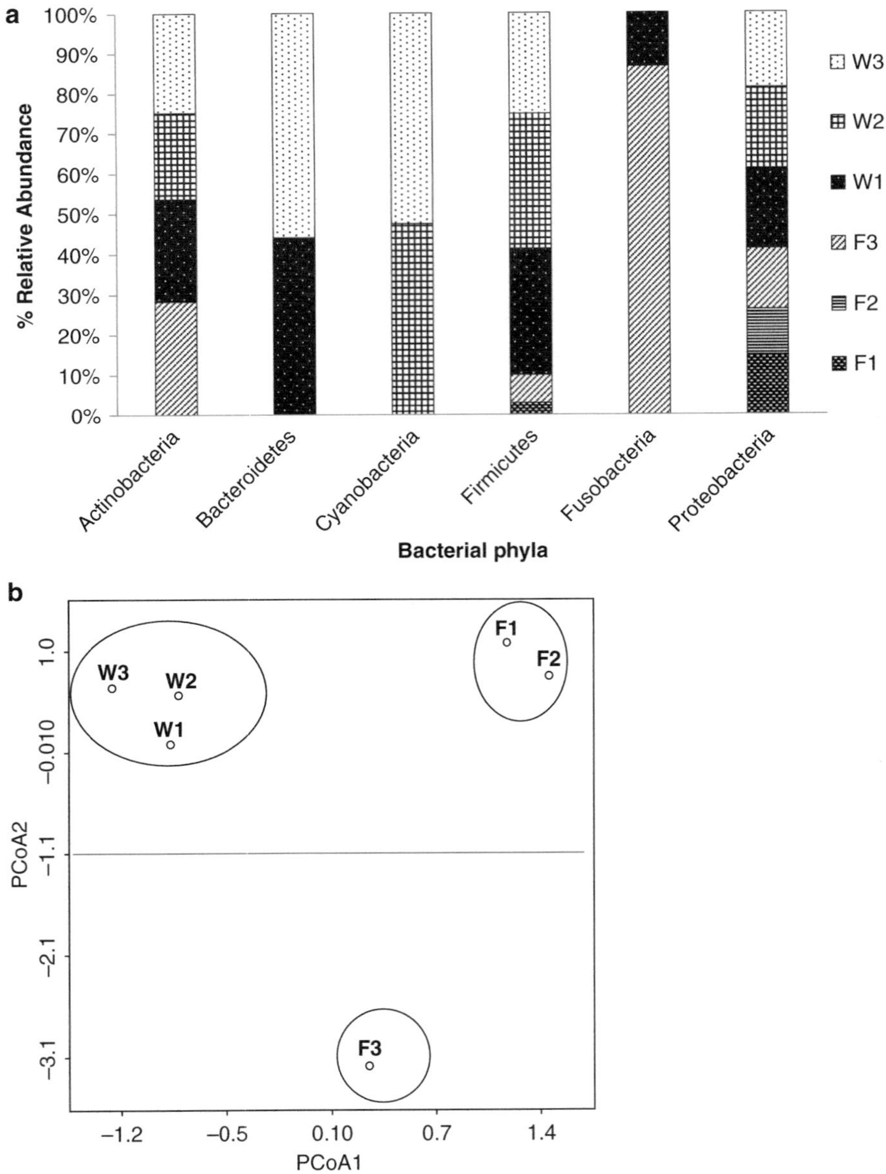

Fig. 10.2 (**a**) Relative abundance (%) of bacterial phyla in climbing perch intestines (farmed-type sample 1 (F1), sample 2 (F2) and sample 3 (F3); wild-type sample 1 (W1), sample 2 (W2) and sample 3 (W3)). (**b**) Principal component analysis (PCoA) of microbial population diversity based on the relative abundance of phyla associated with the wild- and farmed-type climbing perch intestines (compiled and modified from Loh 2015)

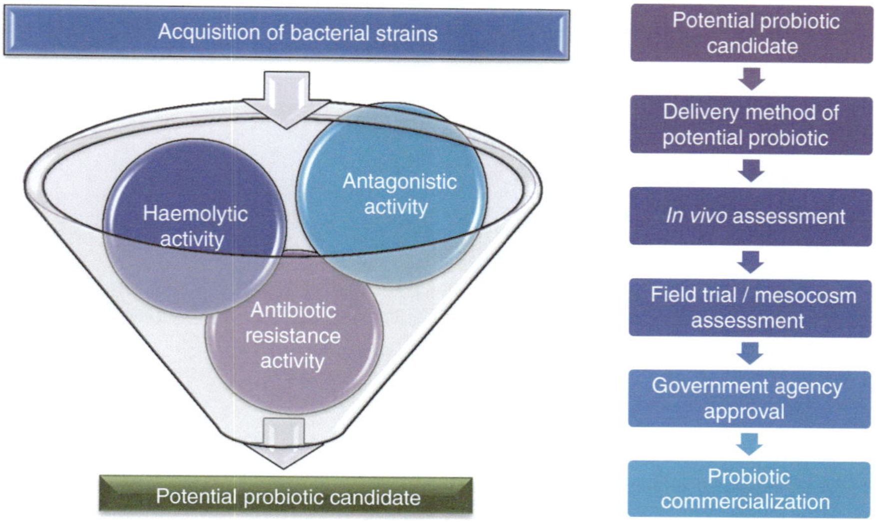

Fig. 10.3 Development of probiotics for use in aquaculture

The antimicrobial assay is crucial as it identifies the isolates and compounds that are responsible for antagonistic activities. The streak test (Fig. 10.4a, b), disc diffusion (Fig. 10.4c, d) and microdilution methods are examples of antimicrobial assays frequently performed to detect the bioactivity of the tested isolates. Once the antimicrobial activity is established, the antimicrobial compounds can be further characterized. This often includes lactic acid, hydrogen peroxide (H_2O_2), diacetyl (2,3-butanedione), acetaldehyde, D-isomers of various amino acids, bacteriocins and reuterin, which are known to inhibit the growth or kill a broad range of pathogenic bacteria (Cintas et al. 2001; Loh 2015). These bioactive compounds can be characterized through chromatographic techniques such as high-performance liquid chromatography (HPLC) to profile and fingerprint compounds of interest (Brusotti et al. 2013).

In addition to antimicrobial assay, other in vitro assessments could also be conducted to further validate the potential of the isolates as probiotic candidates. Common tests conducted include haemolytic assay and antibiotic resistance test. Haemolytic assay is performed by inoculating isolate on blood agar, in which gradual haemolytic zones formed were measured (either in mm or cm) and classified as α-, β-, or γ-haemolysis (Fig. 10.5a–c). This test is crucial to determine the virulence of the isolates as blood-haemolytic bacteria would destroy the epithelial layer of its host cells and is less likely a candidate for probiotic development (Maslow et al. 1999; Loh et al. 2014). The antibiotic resistance test is another important test in the screening process. This assay involves common antibiotics in aquaculture such as tetracycline, chloramphenicol, kanamycin, streptomycin, gentamicin and ampicillin

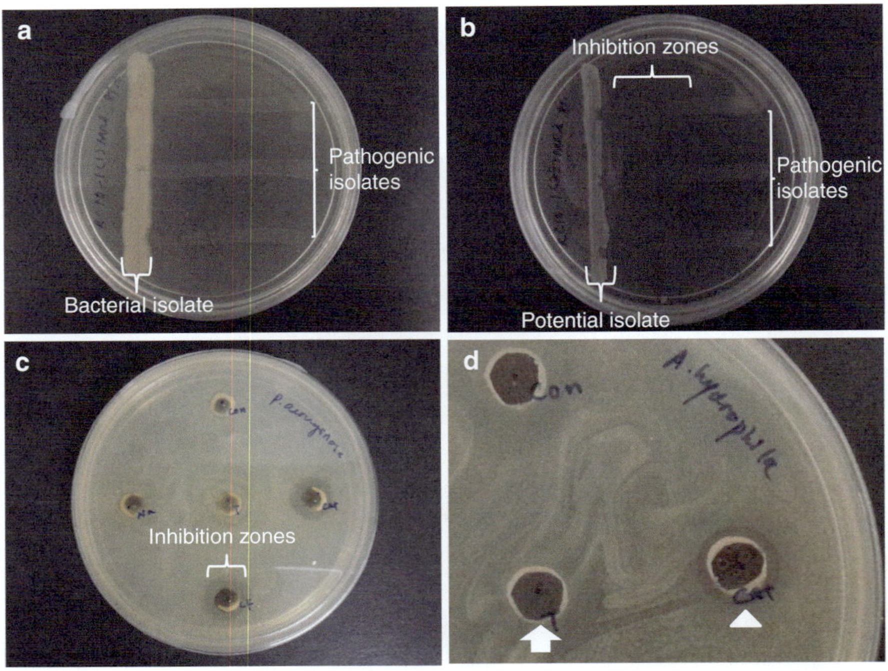

Fig. 10.4 Common laboratory techniques used in the in vitro screenings to detect antagonistic activities. (**a**) Streaking method on a control plate (no inhibition against target pathogen). (**b**) Bacterial inhibitory activity is shown by the potential probiotic candidate—no pathogen growth adjacent to the vertical line of inoculum. (**c**) Disc diffusion method; one to several wells are made on the plate to test for antibacterial activity. (**d**) Example of antimicrobial activities of bacterial supernatant against *Aeromonas hydrophila*—no inhibition was found on wells labelled T (allow) and Con (allow head); an inhibitory effect was noticed on the right well (Cat) indicated with a clearing zone (Complied and modified from Loh 2015)

(Fig. 10.5d) (Loh et al. 2014). This screening procedure eliminates the bacteria carrying antibiotic-resistance genes, so that antibiotic-resistant bacteria are not used as probiotics (Romero et al. 2012; Loh et al. 2014).

In sourcing for probiotics for aquaculture, specific assessments such as on their colonization ability in the host's gut system, improvements to digestibility and feed utilization and immunomodulation or immunostimulatory function are performed occasionally. Modern molecular approaches such as fluorescence in situ hybridization (FISH) and genetically transformed bacteria (e.g. *gfp* gene—green fluorescence protein gene) have been developed to understand and visualize the colonization and adhesion of bacteria in the gut system. These techniques provide a real-time observation to understand the dynamics as well as the adhesive sites for the bacteria. Figure 10.6 shows a recent study on the colonization pattern and adhesion ability of *L. lactis* in *Artemia franciscana* via application of *gfp* gene insertion. This molecular approach elucidated the efficacy of colonization of *L. lactis* in the gastrointestinal tract and on the surfaces of *Artemia* (Loh 2015).

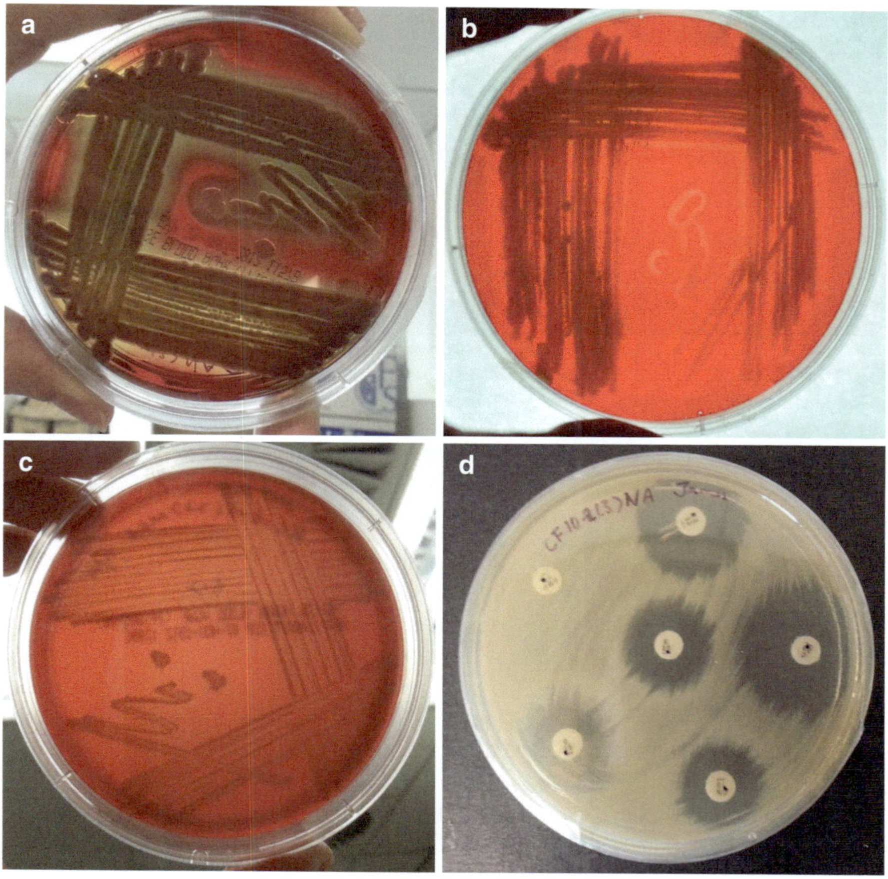

Fig. 10.5 (**a**) Haemolytic screening on blood agar. Blood cells were lysed, and agar under the colony appears dark and greenish, indicating α-haemolysis. (**b**) β-Haemolysis; a complete lysis of red cells in the media around and under the colonies, resulting in agar showing clearing zone or yellowish colour. (**c**) γ-Haemolysis; no lysis activity, the agar under and around the colony remained unchanged. (**d**) Antibiotic susceptibility assay; different antibiotic discs were then placed on the seeded plates. Only bacteria susceptible to particular antibiotics form clearing zones (Compiled and modified from Loh et al. (2013). Isolation and evaluation of intestinal microflora from farmed fishes against edwardsiellosis. *In*: International Congress of the Malaysian Society for Microbiology, ICMSM. 12–15 December 2013, Langkawi, Malaysia)

For the improvement of digestion and feed utilization, it can be determined by performing enzymatic assays on the major enzymatic production such as protease, amylase, lipase and cellulase. *Lactococcus lactis* subsp. *lactis*, for example, a potential probiotic to control *V. anguillarum* (Touraki et al. 2013), *Vibrio alginolyticus* (Villamil et al. 2003) and *Edwardsiella tarda* (Loh et al. 2014), could produce extracellular digestive enzymes such as protease (Sun et al. 2011; Loh and Ting 2016), amylase (Cho et al. 2007; Loh and Tiny 2016), lipase (Mahmoud et al. 2011; Loh and Tiny 2016) and cellulase (Loh and Tiny 2016). This is of particular importance

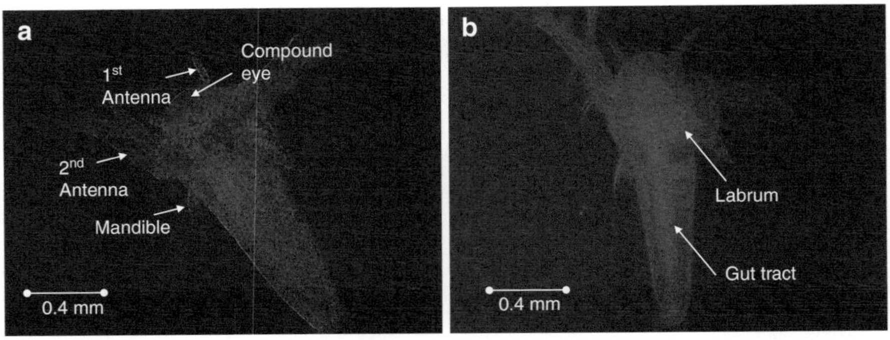

Fig. 10.6 (**a**) Wild-type *A. franciscana* without bioencapsulation with *gfp-L. lactis* (lactococci inserted with green fluorescent protein gene) acts as the control under *dark-field-view* microscopy. (**b**) *A. franciscana* bioencapsulated with *gfp-L. lactis*. Putative adhesion of bacteria in the midgut and hindgut areas under exposure of UV ranged from 450 to 490 nm (Compiled and modified from Loh 2015)

to aquaculture, as microbial enzymatic production is now gaining attention especially in promoting fish nutritional uptake. Furthermore, it is also an added value if these probiotics can produce useful enzymes, as this will produce protective effect against fish diseases.

10.5 Application of Probiotics for Disease Control

The gastrointestinal microflora (developed as probiotics) with potential for use in disease control is often applied in combination with the fish or shrimp feed. In larviculture, probiotic is almost always delivered through bioencapsulation with live feed such as *Artemia* and rotifers. Direct ingestion of the probiotic-supplemented live feed would help to establish colonization and proliferation of probiotic, which can further nourish larval gut system (Picot and Lacroix 2004). Probiotics are generally sensitive to the various environmental stresses and to the aquatic microbiota (Cruz et al. 2012; Ibrahem 2013). Therefore, with bioencapsulation, a physical barrier is present to resist the harsh gastrointestinal and aquatic environment (Loh et al. 2012; Loh 2015). In aquaculture hatchery, bacterial strains such as *Lactobacillus* sp. and *Bacillus* sp. are commonly bioencapsulated with live feed to improve the survival, growth and the balance of gut microflora of fishes and shrimps (Iranshahi et al. 2011; Dagá et al. 2013). In older stage of finfish cultivation practices, probiotics can be encapsulated via spray-dried method, microencapsulation, emulsification or micronization (Tsen et al. 2007; Picot and Lacroix 2004; Doleyres et al. 2004) as a supplementation or an additive in aquafeed.

To achieve effective bioencapsulation, the encapsulation media should be harmless to the live feed. The encapsulation media should also effectively entrap bacterial cells to allow the recovery of bacteria during the digestion. Lastly, the encapsulation media should also be free from chemical residues in fish and crustacean larvae tissues (Loh and Ting 2015). Commercial food grade materials such as polysaccharides (alginate, *k*-carrageenan), and materials from various sources such as bacteria (xanthan, gellan), animal proteins (milk, gelatin), and plants (starch, gum Arabic), are known to be useful materials for bacteria encapsulation (Rokka and Rantamäki 2010). Loh and Ting (2015) tested the efficacy of sodium alginate, palm oil, starch, gum Arabic and gelatin for bioencapsulation (Fig. 10.7). Results showed that with bioencapsulation, a higher recovery of probiotics was obtained compared to without encapsulation.

Other advantages of bioencapsulation include the role of encapsulation material (e.g. alginate) as immunostimulant to enhance innate immunity of certain aquatic species (Ringø et al. 2012; Loh 2015). Feeding fish larvae with probiotic-alginate-enriched *Artemia* showed better performance in terms of survival, growth and lysozyme production (Fig. 10.8a–c). The level of lysozyme, a key enzyme in immune defence of both invertebrates and vertebrates (Magnadóttir 2006), detected in the body fluid corresponds to the survival of larvae. This suggested that synergistic effects may occur between the immunostimulant and the probiotic. Several other studies also reported similar enhancement of nonspecific defence effect against fish pathogens such as common carp, *Cyprinus carpio* (Fujiki and Yano 1997), and orange-spotted grouper, *Epinephelus coioides* (Yeh et al. 2008), when fed with sodium alginate. Other immunostimulants such as

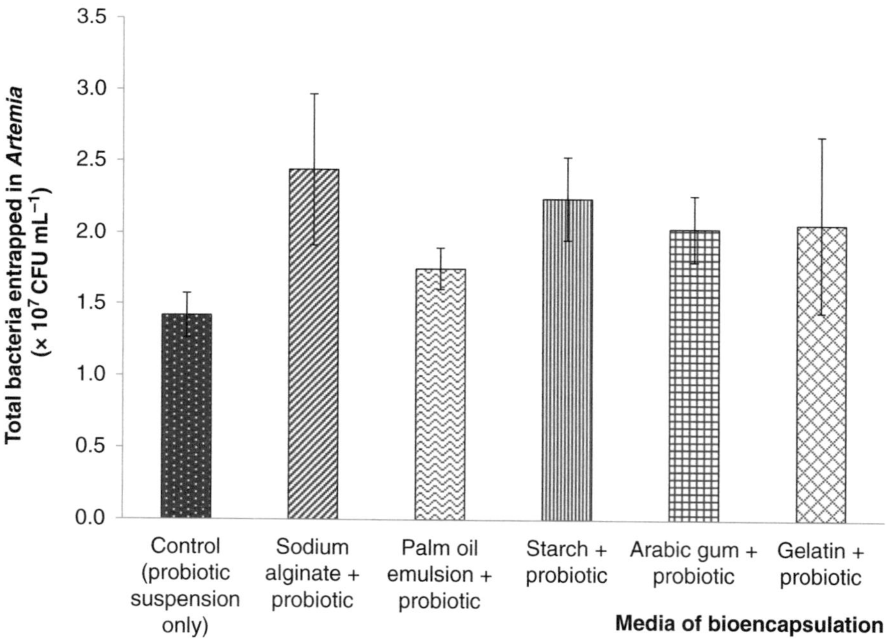

Fig. 10.7 Recovery of probiotic (*L. lactis*) from *A. franciscana* after 8 h of bioencapsulation. *Vertical bars* indicate standard deviation of means (Compiled and modified from Loh and Ting 2015)

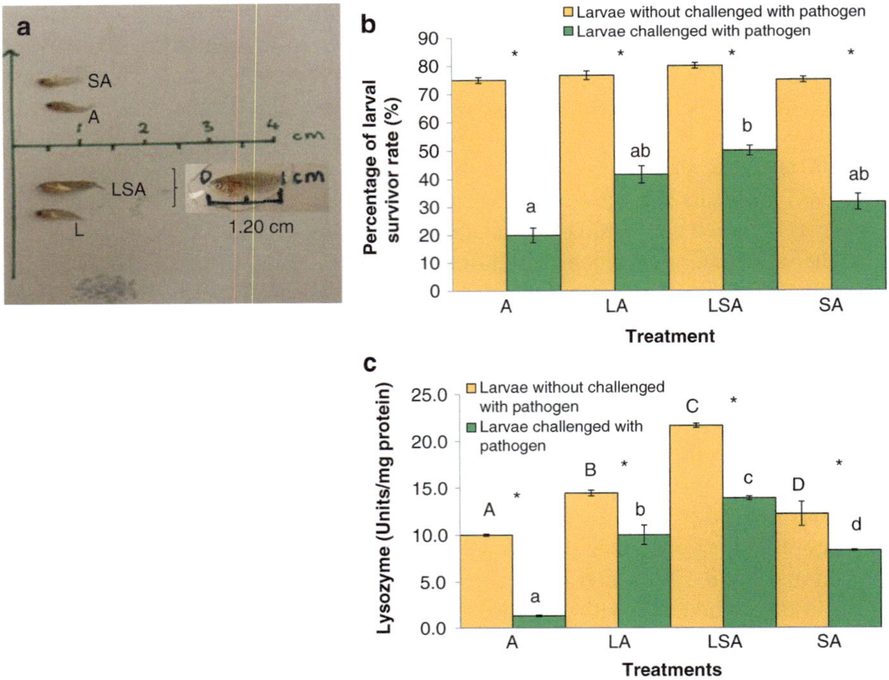

Fig. 10.8 (**a**) Growth of climbing perch, *Anabas testudineus*, under the influence of probiotic and immunostimulant (i.e. alginate). (**b**) Survival of *A. testudineus* larvae. (**c**) Lysozyme-like activity of *A. testudineus* larvae fed with four treatments (*A Artemia* only, *LA Artemia* enriched with *Lactococcus lactis* subsp. *lactis* as probiotic, *LSA Artemia* enriched with sodium alginate containing *L. lactis*, *SA Artemia* enriched with sodium alginate). Mean values with asterisks and letters show significant difference to others ($HSD_{0.05}$). *Vertical bars* indicate standard deviation of means (Compiled and modified from Loh 2015)

seaweed mixtures and β-D-glucose polysaccharides are also found to elicit similar response in Atlantic salmon, *Salmo salar* L. (Gabrielsen and Austreng 1998), carp *Labeo rohita* (Misra et al. 2006), grass carp *Ctenopharyngodon idella* and tilapia *Tilapia aureus* P. (Wang and Wang 1997).

Conclusion

Exploring potential beneficial intestinal bacteria for the use in maintaining healthy intestinal microflora or managing fish diseases requires a comprehensive understanding of the bacterial diversity in fish intestines. With the advancement of technology, the microbial taxonomic profile and GI microbiome can be revealed to a greater extent. With effective screening assays/measures (antagonistic, enzymatic, colonization assays as discussed here), the selection of desirable probiotics is more pronounced and may lead to greater success of controlling fish diseases.

Acknowledgements The authors would like to express their gratitude to the Monash University for the Higher Degree Research (HDR) grant.

References

Alikunhi, NM, Batang ZB, Aljahdali HA, Aziz MAM, Al-Suwailem AM (2016) Culture-dependent bacteria in commercial fishes: qualitative assessment and molecular identification using 16S rRNA gene sequencing. Saudi J Biol Sci. doi:10.1016/j.sjbs.2016.05.017

Andani HRR, Tukmechi A, Meshkini S, Sheikhzadeh N (2012) Antagonistic activity of two potential probiotic bacteria from fish intestines and investigation of their effects on growth performance and immune response in rainbow trout (*Oncorhynchus mykiss*). J Appl Ichthyol 2012:1–7. doi:10.1111/j.1439-0426.2012.01974.x

Arias CR, Abernathy JW, Liu Z (2006) Combined use of 16S ribosomal DNA and automated ribosomal intergenic spacer analysis to study the bacterial community in catfish ponds. Lett Appl Microbiol 43:287–292. doi:10.1111/j.1472-765X.2006.01955.x

Balakrishna A, Keerthi TR (2012) Screening of potential aquatic probiotics from the major microflora of guppies (*Poecilia reticulata*). Front Chem Sci Eng 6(2):163–173. doi:10.1007/s11705

Balcázar JL (2003) Evaluation of probiotic bacterial strains in *Litopenaeus vannamei*. Final Report, National Center for Marine and Aquaculture Research, Guayaquil, Ecuador

Balcázar JL, de Blas I, Ruiz-Zarzuela I, Cunningham D, Vendrell D, Múzquiz JL (2006) The role of probiotics in aquaculture. Vet Microbiol 114:173–186. doi:10.1016/j.vetmic.2006.01.009

Balcázar JL, Rojas-Luna T, Cunningham DP (2007) Effect of the addition of four potential probiotic strains on the survival of pacific white shrimp (*Litopenaeus vannamei*) following immersion challenge with *Vibrio parahaemolyticus*. J Invertebr Pathol 96:147–150. doi:10.1016/j.jip.2007.04.008

Balcázar JL, Vendrell D, de Blas I, Ruiz-Zarzuela I, Muzquiz JL, Girones O (2008) Characterization of probiotic properties of lactic acid bacteria isolated from intestinal microbiota of fish. Aquaculture 278:188–191. doi:10.1016/j.aquaculture.2008.03.014

Bandyopadhyay P, Mohapatra PKD (2009) Effect of a probiotic bacterium *Bacillus circulans* PB7 in the formulated diets: on growth, nutritional quality and immunity of *Catla catla* (Ham.) Fish Physiol Biochem 35:467–478. doi:10.1007/s10695-008-9272-8

Bates JM, Mittge E, Kuhlman J, Baden KN, Cheesman SE, Guillemin K (2006) Distinct signals from the microbiota promote different aspects of zebrafish gut differentiation. Dev Biol 297:374–386. doi:10.1016/j.ydbio.2006.05.006

Brusotti G, Cesari I, Dentamaro A, Caccialanza G, Massolini G (2013) Isolation and characterization of bioactive compounds from plant resources: the role of analysis in the ethnopharmacological approach. J Pharm Biomed Anal 87:218–228. doi:10.1016/j.jpba.2013.03.007

Burr G, Gatlin D (2005) Microbial ecology of the gastrointestinal tract of fish and the potential application of prebiotics and probiotics in finfish aquaculture. J World Aquacult Soc 36:425–436. doi:10.1111/j.1749-7345.2005.tb00390.x

Butprom S, Phumkhachorn P, Rattanachaikunsopon P (2013) Effect of *Lactobacillus plantarum* C014 on innate immune response and disease resistance against *Aeromonas hydrophila* in hybrid catfish. Sci World J 2013:392523. doi:10.1155/2013/392523

Cabello FC (2006) Heavy use of prophylactic antibiotics in aquaculture: a growing problem for human and animal health and for the environment. Environ Microbiol 8:1137–1144. doi:10.1111/j.1462-2920.2006.01054.x

Carraturo A, Raieta K, Ottaviani D, Russo GL (2006) Inhibition of *Vibrio parahaemolyticus* by a bacteriocin-like inhibitory substance (BLIS) produced by *Vibrio mediterranei 1*. J Appl Microbiol 101(1):234–241. doi:10.1111/j.1365-2672.2006.02909.x

Cho MH, Park SE, Lee MH, Ha SJ, Kim HY, Kim MJ, Lee SJ, Madsen SM, Park CS (2007) Extracellular secretion of a maltogenic amylase from *Lactobacillus gasseri* ATCC33323 in

Lactococcus lactis MG1363 and its application on the production of branched maltooligosaccharides. J Microbiol Biotechnol 17(9):1521–1526
Cintas LM, Casaus MP, Herranz C, Nes IF, Hernández PE (2001) Review: bacteriocins of lactic acid bacteria. Food Sci Technol Int 7:281–305. doi:10.1106/R8DE-P6HU-CLXP-5RYT
Clements KD, Pasch IBY, Moran D, Turner SJ (2007) Clostridia dominate 16S rRNA gene libraries prepared from the hindgut of temperate marine herbivorous fishes. Mar Biol 150:1431–1440. doi:10.1007/s00227-006-0443-9
Cruz PM, Ibáñez AL, Hermosillo OAM, Saad HCR (2012) Use of probiotics in aquaculture. ISRN Microbiol 2012:916845. doi:10.5402/2012/916845
Dagá P, Feijoo G, Moreira MT, Costas D, Villanueva AG, Lema JM (2013) Bioencapsulated probiotics increased survival, growth and improved gut flora of turbot (*Psetta maxima*) larvae. Aquacult Int 21:337–345. doi:10.1007/s10499-012-9556-y
Das BK, Samal SK, Samantaray BR, Sethi S, Pattnaik P, Mishra BK (2006) Antagonistic activity of cellular components of *Pseudomonas* species against *Aeromonas hydrophila*. Aquaculture 253:17–24. doi:10.1016/j.aquaculture.2005.01.028
Declercq AM, Haesebrouck F, Van den Broeck W, Bossier P, Decostere A (2013) Columnaris disease in fish: a review with emphasis on bacterium-host interactions. Vet Res 44(1):27. doi:10.1186/1297-9716-44-27
Degnan PH, Ochman H (2012) Illumina-based analysis of microbial community diversity. ISME J 6(1):183–194. doi:10.1038/ismej.2011.74
Devereux G (2002) The immune system an overview. In: Calder P, Field C, Gill H (eds) Nutrition and immune function. CABI Publishing, Wallingford, pp 1–20
Doleyres Y, Fliss I, Lacroix C (2004) Continuous production of mixed lactic starters containing probiotics using immobilized cell technology. Biotechnol Prog 20:145–150. doi:10.1021/bp020096w
Douillet PA, Langdon CJ (1994) Use of a probiotic for the culture of larvae of the pacific oyster (*Crassostrea gigas* Thunberg). Aquaculture 119:25–40. doi:10.1016/0044-8486(94)90441-3
Forberg T, Sjulstad EB, Bakke I, Olsen Y, Hagiwara A, Sakakura Y, Vadsteina O (2016) Correlation between microbiota and growth in Mangrove Killifish (*Kryptolebias marmoratus*) and Atlantic cod (*Gadus morhua*). Sci Rep 6:21192. doi:10.1038/srep21192
Formosa R, Ravi H, Happe S, Huffman D, Novoradovskaya N, Kincaid R, Garrett S (2010) DNA-based fish species identification protocol. J Vis Exp 38:e1871. doi:10.3791/1871
Fujiki K, Yano T (1997) Effects of sodium alginate on the non-specific defense system of the common carp, (*Cyprinus carpio* L.) Fish Shellfish Immunol 17:417–427. doi:10.1006/fsim.1997.0095
Gabrielsen BO, Austreng E (1998) Growth, product quality and immune status of Atlantic salmon, *Salmo salar* L., fed wet feed with alginate. Aquac Res 29:397–401. doi:10.1046/j.1365-2109.1998.00215.x
Gajardo K, Rodiles A, Kortner TM, Krogdahl Å, Bakke AM, Merrifield DL, Sørum H (2016) A high-resolution map of the gut microbiota in Atlantic Salmon (*Salmo sonar*): a basis for comparative gut microbial research. Sci Rep 6:30893. doi:10.1038/srep30893
Gatesoupe FJ (1994) Lactic acid bacteria increased the resistance of turbot larvae, *Scophthalmus maximus,* against pathogenic vibrio. Aquat Living Resour 7:277–282. doi:10.1051/alr:1994030
Gatesoupe FJ (2005) Probiotics and prebiotics for fish culture, at the parting of the ways. Aqua Feeds Formul Beyond 2(3):3–5. http://www.feedware.com/index.html
Giatsis C, Sipkema D, Smidt H, Verreth J, Verdegem M (2014) The colonization dynamics of the gut microbiota in Tilapis larvae. PLoS One 9(7):e103641. doi:10.1371/journal.pone.0103641
Givens CE, Ransom B, Bano N, Hollibaugh JT (2015) Comparison of the gut microbiomes of 12 bony fish and 3 shark species. Mar Ecol Prog Ser 518:209–223. doi:10.3354/meps11034
Gram L, Melchiorsen J, Spanggaard B, Huber I, Nielsen TF (1999) Inhibition of *Vibrio anguillarum* by *Pseudomonas fluorescens* AH2, a possible probiotic treatment of fish. Appl Environ Microbiol 65:969–973
Harikrishnan R, Balasundaram C, Heo MS (2010) *Lactobacillus sakei* BK19 enriched diet enhances the immunity status and disease resistance to streptococcosis infection in kelp

grouper, *Epinephelus bruneus*. Fish Shellfish Immun 29:1037–1043. doi:10.1016/j.fsi.2010.08.017
Ibrahem MD (2013) Evolution of probiotics in aquatic world: potential effects, the current status in Egypt and recent prospective. J Adv Res 6(6):765–791. doi:10.1016/j.jare.2013.12.004
Iranshahi F, Faramarzi M, Kiaalvandi S, Boloki ML (2011) The enhancement of growth and feeding performance of Persian sturgeon (*Acipenser persicus*) larvae by *Artemia urmiana* nauplii boencapsulated via baker's yeast (*Saccharomyces cerevisiae*). J Anim Vet Adv 10(20):2730–2735. doi:10.3923/javaa.2011.2730.2735
Jami E, Shterzer N, Mizrahi I (2014) Evaluation of automated ribosomal intergenic spacer analysis for bacterial fingerprinting of rumen microbiome compared to pyrosequencing technology. Pathogens 3:109–120. doi:10.3390/pathogens3010109
Kennedy SB, Tucker JW, Neidig CL, Vermeer GK, Cooper VR, Jarrell JL, Sennett DG (1998) Bacterial management strategies for stock management of warm water marine fish: a case study with common snook (*Centropomus undecimalis*). Bull Mar Sci 62:573–588
Klaenhammer TR (2007) Probiotics and prebiotics. In: Doyle DP, Beuchat LR (eds) Food microbiology—fundamental and frontiers, 3rd edn. ASM Press, Washington, DC, pp 891–910
Lan C-C, Love DR (2012) Molecular characterisation of bacterial community structure along the intestinal tract of Zebrafish (*Danio rerio*): a pilot study. ISRN Microbiol 2012:590385. doi:10.5402/2012/590385
Larsen AM, Mohammed HH, Arias CR (2014) Comparison of DNA extraction protocols for the analysis of gut microbiota in fishes. FEMS Microbiol Lett 362(5). doi: 10.1093/femsle/fnu031
Lauzon HL, Gudmundsdottir S, Pedersen MH, Budde BB, Gudmundsdottir BK (2008) Isolation of putative probionts from cod rearing environment. Vet Microbiol 132:328–339. doi:10.1016/j.vetmic.2008.05.014
Ley RE, Lozupone CA, Hamady M, Knight R, Gordon JI (2008) Worlds within worlds: evolution of the vertebrate gut microbiota. Nat Rev Microbiol 6(10):776–788. doi:10.1038/nrmicro1978
Liu H, Guo X, Gooneratne R, Lai R, Zeng C, Zhan F, Wang W (2016) The gut microbiome and degradation enzyme activity of wild freshwater fishes influenced by their trophic levels. Sci Rep 6:24340. doi:10.1038/srep24340
Loh JY (2015) Bioencapsulation and disease suppression studies of beneficial bacteria against *Edwardsiella tarda* infection in climbing perch larvae (*Anabas testudineus*, Bloch). Doctorate Thesis, Monash University, 293 pp
Loh JY, Ting ASY (2015) Bioencapsulation of probiotic *Lactococcus lactis* subsp. *lactis* on *Artemia franciscana* nauplii: effects of encapsulation media on nauplii survival and probiotic recovery. Malays J Microbiol 11(2):121–127. doi:10.21161/mjm.12314
Loh JY, Ting ASY (2016) Effects of potential probiotic *Lactococcus lactis* subsp. *lactis* on digestive enzymatic activities of live feed *Artemia franciscana*. Aquacult Int 24(5):1341–1351. doi:10.1007/s10499-016-9991-2
Loh JY, Ong HKA, Hii YS, Smith TJ, Lock MW, Khoo G (2012) Highly unsaturated fatty acid (HUFA) retention in the freshwater cladoceran, *Moina macrocopa*, enriched with lipid emulsions. Isr J Aquacult-Bamid IJA 64(2012):637. 9 pages
Loh JY, Lim YY, Harmin SA, Ting ASY (2013) Isolation and evaluation of intestinal microflora from farmed fishes against edwardsiellosis. In: International Congress of the Malaysian Society for Microbiology, ICMSM, 12–15 December 2013, Langkawi, Malaysia
Loh JY, Lim YY, Harmin SA, Ting ASY (2014) In vitro assessment on intestinal microflora from commonly farmed fishes for control of the fish pathogen *Edwardsiella tarda*. Turk J Vet Anim Sci 38:257–263. doi:10.3906/vet-1312-53
Lokesh J, Kiron V (2016) Transition from freshwater to seawater reshapes the skin associated microbiota of Atlantic salmon. Sci Rep 6:19707. doi:10.1038/srep19707
Magnadóttir B (2006) Innate immunity of fish (overview). Fish Shellfish Immunol 20:137–151. doi:10.1016/j.fsi.2004.09.006
Mahmoud E-H, El-Tanboly ES, Abd-Rabou NS (2011) Industrial application of lipases in cheese making: a review. Int J Food Safety 13:293–302

Maslow JN, Dawson D, Carlin EA, Holland SM (1999) Haemolysin as a virulence factor for systemic infection with isolates of *Mycobacterium avium* complex. J Clin Microbiol 37(2):445–446
Melba G, Bondad-Reantaso SE, McGladdery I East, Subasinghe RP (2001) "Chapter 4". Asia diagnostic guide to aquatic animal diseases. FAO Fisheries Technical Paper 402/2, NACA/FAO 2001. ISBN 92-5-104620-4
Misra CK, Das BK, Mukherjee SC, Pattnaik P (2006) Effect of long term administration of dietary β-glucan on immunity, growth and survival of *Labeo rohita* fingerlings. Aquaculture 255:82–94. doi:10.1016/j.aquaculture.2005.12.009
Mohanty BR, Sahoo PK (2007) Edwardsiellosis in fish: a brief review. J Biosci 32(7):1331–1344. doi:10.1007/s12038-007-0143-8
Nayak SK (2010) Probiotics and immunity: a fish perspective. Fish Shellfish Immunol 29(1):2–14. doi:10.1016/j.fsi.2010.02.017
Nelson TM, Rogers TL, Carlini AR, Brown MV (2012) Diet and phylogeny shape the microbiota of Antarctic seals: a comparison of wild and captive animals. Environ Microbiol 15:1132–1145. doi:10.1111/1462-2920.12022
Newton RJ, Jones SE, Eiler A, McMahon KD, Bertilsson S (2011) A guide to the natural history of freshwater lake bacteria. Microbiol Mol Biol Rev 75:14–49. doi:10.1128/MMBR.00028-10
Nikoskelainen S, Ouwehand A, Bylund G, Salminen S, Lilius EM (2003) Immune enhancement in rainbow trout (*Oncorhynchus mykiss*) by potential probiotic bacteria (*Lactobacillus rhamnosus*). Fish Shellfish Immunol 15:443–452. doi:10.1016/S1050-4648(03)00023-8
Nogami K, Hamasaki K, Maeda M, Hirayama K (1997) Biocontrol method in aquaculture for rearing the swimming crab larvae *Portunus trituberculatus*. Hydrobiologia 358:291–295. doi: 10.1023/A:1003149306511
Olsson JC, Jöborn K, Westerdahl A, Blomberg L, Kjelleberg S, Conway PL (1998) Survival, persistence and proliferation of *Vibrio anguillarum* in juvenile turbot, *Scophthalmus maximus* (L.), intestine and faeces. J Fish Dis 21:1–9. doi:10.1046/j.1365-2761.1998.00327.x
Picot A, Lacroix C (2004) Encapsulation of *Bifidobacteria* in whey protein-based microcapsules and survival in stimulated gastrointestinal conditions and in yoghurt. Int Dairy J 14(6):505–515. doi:10.1016/j.idairyj.2003.10.008
Pintado J, Pérez-Lorenzo M, Luna-González A, Sotelo CG (2010) Monitoring of the bioencpsulation of a probiotic *Phaeobacter* strain in the rotifer *Brachionus plicatilis* using denaturing gradient gel electrophoresis. Aquaculture 302:182–194. doi:10.1016/j.aquaculture.2010.02.014
Pirarat N, Kobayashi T, Katagiri T, Manita M, Endo M (2006) Protective effects and mechanisms of a probiotic bacterium *Lactobacillus rhamnosus* against experimental *Edwardsiella tarda* infection in tilapia (*Oreochromis niloticus*). Vet Immunol Immunopathol 113:339–347. doi:10.1016/j.vetimm.2006.06.003
Planas M, Vazqueuz JA, Marques J, Perez-Lomba R, Gonzalez MP, Murado M (2004) Enhancement of rotifer (*Brachionus plicatilis*) growth by using terrestrial lactic acid bacteria. Aquaculture 240:313–329. doi:10.1016/j.aquaculture.2004.07.016
Porubcan RS (1991) Reduction of ammonia nitrogen and nitrite in tanks of *Penaeus monodon* using floating biofilters containing processed diatomaceous earth media pre-inoculated with nitrifying bacteria. In: Proceedings of the program and abstracts of the 22nd annual conference and exposition; June 1991; Puerto Rico, Spain. World Aquaculture Society
Qin J, Li R, Raes J, Arumugam M, Burgdorf KS, Manichanh C, Nielsen T, Pons N, Levenez F, Yamada T, Mende DR, Li J, Xu J, Li S, Li D, Cao J, Wang B, Liang H, Zheng H, Xie Y, Tap J, Lepage P, Bertalan M, Batto JM, Hansen T, Le Paslier D, Linneberg A, Nielsen HB, Pelletier E, Renault P (2010) A human gut microbial gene catalogue established by metagenomic sequencing. Nature 464(7285):59–65. doi:10.1038/nature08821
Rawls JF, Samuel BS, Gordon JI (2004) Gnotobiotic zebrafish reveal evolutionarily conserved responses to the gut microbiota. Proc Natl Acad Sci U S A 101:4596–4601. doi:10.1073/pnas.0400706101

Rawls JF, Mahowald MA, Ley RE, Gordon JI (2006) Reciprocal gut microbiota transplants from zebrafish and mice to germ-free recipients reveal host habitat selection. Cell 127:423–433. doi:10.1016/j.cell.2006.08.043

Ray AK, Ghosh K, Ringø E (2012) Enzyme-producing bacteria isolated from fish gut: a review. Aquac Nutr 18(5):465–492. doi:10.1111/j.1365-2095.2012.00943.x

Ringø E, Strøm E, Tabacheck J (1995) Intestinal microflora of salmonids: a review. Aquac Res 26:773–789. doi:10.1111/j.1365-2109.1995.tb00870.x

Ringø E, Olsen RE, Gonzalez Vecino JL, Wadsworth S, Song SK (2012) Use of immunostimulants and nucleotides in aquaculture: a review. J Mar Sci Res Dev 1:104. doi:10.4172/2155-9910.1000104

Ringø E, Zhou Z, Vecino JLG, Wadsworth S, Romero J, Krogdahl Å, Olsen RE, Dimitroglou A, Foey A, Davies S, Owen M, Lauzon HL, Martinsen LL, De Schryver P, Bossier P, Sperstad S, Merrifield DL (2016) Effect of dietary components on the gut microbiota of aquatic animals. A never-ending story? Aquac Nutr 22:219–282. doi:10.1111/anu.12346

Rokka S, Rantamäki P (2010) Protecting probiotic bacteria by microencapsulation: challenges for industrial applications. Eur Food Res Technol 231:1–2. doi:10.1007/s00217-010-1246-2

Romero J, Feijoó CG, Navarrete P (2012) Antibiotics in aquaculture—use, abuse and alternatives. In: Carvalho ED, Silva David G, da Silva RJ (eds) Health and environment in aquaculture. InTech, Croatia, 414 pages

Rungrassamee W, Klanchui A, Maibunkaew S, Chaiyapechara S, Jiravanichpaisal P, Karoonuthaisiri N (2014) Characterization of intestinal bacteria in wild and domesticated adult black tiger shrimp (*Penaeus monodon*). PLoS One 9(3):e91853. doi:10.1371/journal.pone.0091853

Sahu MK, Swarnakumar NS, Sivakumar K, Thangaradjou T, Kannan L (2008) Probiotics in aquaculture: importance and future perspectives. Indian J Microbiol 48:299–308. doi:10.1007/s12088-008-0024-3

Sakai M, Yoshida T, Astuta S, Kobayashi M (1995) Enhancement of resistance to vibriosis in rainbow trout, *Oncorhynchus mykiss* (Walbaum) by oral administration of *Clostridium butyricum* bacteria. J Fish Diseases 18:187–190. doi:10.1111/j.1365-2761.1995.tb00276.x

Sanchez LM, Wong WR, Riener RM, Schulze CJ, Linington RG (2012) Examining the fish microbiome: vertebrate-derived bacteria as an environmental niche for the discovery of unique marine natural products. PLoS One 7:e35398. doi:10.1371/journal.pone.0035398

Schulze AD, Alabi AO, Sheldrake AR, Miller KM (2006) Bacterial diversity in a marine hatchery: balance between pathogenic and potentially probiotic bacteria strains. Aquaculture 256:50–73. doi:10.1016/j.aquaculture.2006.02.008

Selvendran M, Michael Babu M (2013) Studies on novel bacteriocin like inhibitory substance (BLIS) from microalgal symbiotic *Vibrio* spp MMB2 and its activity against aquatic bacterial pathogens. J Appl Pharm Sci 3(2):169–175. doi:10.7324/JAPS.2013.30230

Sinclair L, Osman OA, Bertilsson S, Eiler A (2015) Microbial community composition and diversity via 16S rRNA gene amplicons: evaluating the illumina platform. PLoS One 10(2):e0116955. doi:10.1371/journal.pone.0116955

Smriga S, Sandin SA, Azam F (2010) Abundance, diversity, and activity of microbial assemblages associated with coral reef fish guts and feces. FEMS Microbiol Ecol 73:31–42. doi:10.1111/j.1574-6941.2010.00879.x

Spor A, Koren O, Ley R (2011) Unraveling the effects of the environment and host genotype on the gut microbiome. Nat Rev Microbiol 9(4):279–290. doi:10.1038/nrmicro2540

Standen BT, Rodiles A, Peggs DL, Davies SJ, Santos GA, Merrifield DL (2015) Modulation of the intestinal microbiota and morphology of tilapia, *Oreochromis niloticus*, following the application of a multi-species probiotic. Appl Microbiol Biotechnol 99:8403. doi:10.1007/s00253-015-6702-2

Star B, Haverkamp THA, Jentoft S, Jakobsen KS (2013) Next generation sequencing shows high variation of the intestinal microbial species composition in Atlantic cod caught at a single location. BMC Microbiol 13:248. doi:10.1186/1471-2180-13-248

Sullam KE, Essinger SD, Lozupone CA, O'Connor MP, Rosen GL, Knight R, Kilham SS, Russell JA (2012) Environmental and ecological factors that shape the gut bacterial communities of fish: a meta-analysis. Mol Ecol 21:3363–3378. doi:10.1111/j.1365-294X.2012.05552.x

Sun YZ, Yang HL, Ma RL, Song K, Li JS (2011) Effect of *Lactococcus lactis* and *Enterococcus faecium* on growth performance, digestive enzymes and immune response of grouper *Epinephelus coioides*. Aquac Nutr 18(3):281–289. doi:10.1111/j.1365-2095.2011.00894.x

Taoka Y, Maeda H, Jo JY, Kim SM, Park SI, Yoshikawa T, Sakata T (2006) Use of live and dead probiotic cells in tilapia *Oreochromis niloticus*. Fisheries Sci 72:755–766. doi:10.1111/j.1444-2906.2006.01215.x

Tarnecki AM, Patterson WF, Arias CR (2016) Microbiota of wild-caught Red Snapper *Lutjanus campechanus*. BMC Microbiol 16:245. doi:10.1186/s12866-016-0864-7

Touraki M, Karamanlidou G, Koziotis M, Christidis I (2013) Antibacterial effect of *Lactococcus lactis* subsp. *lactis* on *Artemia fransciscana* and *Dicentrarchus labrax* larvae against the fish pathogen *Vibrio anguillarum*. Aquacult Int 21:481–495. doi:10.1007/s10499-012-9579-4

Tsen JH, Huang HY, Lin YP, King VA (2007) Freezing resistance improvement of *Lactobacillus reuteri* by using cell immobilization. J Microbiol Methods 70:561–564. doi:10.1016/j.mimet.2007.06.004

Vaseeharan B, Ramasamy P (2003) Control of pathogenic *Vibrio* spp. by *Bacillus subtilis* BT23, a possible probiotic treatment for black tiger shrimp *Penaeus monodon*. Lett Appl Microbiol 36:83–87. doi:10.1046/j.1472-765X.2003.01255.x

Villamil L, Figueras A, Planas M, Novoa B (2003) Control of *Vibrio alginolyticus* in *Artemia* culture by treatment with bacterial probiotics. Aquaculture 219:43–56. doi:10.1016/S0044-8486(02)00515-X

Wang W, Wang D (1997) Enhancement of the resistance of tilapia and grass carp to experimental *Aeromonas hydrophila* and *Edwardsiella tarda* infections by several polysaccharides. Comp Immunol Microbiol Infect Dis 20:261–270. doi:10.1016/S0147-9571(96)00035-5

Wang YB, Xu Z (2006) Effect of probiotics for common carp (*Cyprinus carpio*) based on growth performance and digestive enzyme activities. Anim Feed Sci Technol 127:283–292. doi:10.1016/j.anifeedsci.2005.09.003

Wang Y, Sheng HF, He Y, Wu JY, Jiang YX, Tam NFY, Zhou HW (2012) Comparison of the levels of bacterial diversity in freshwater, intertidal wetland, and marine sediments by using millions of illumina tags. Appl Environ Microbiol 78:23. doi:10.1128/AEM.01821-12

Wu LY, Wen CQ, Qin YJ, Yin HQ, Tu QC, Van Nostrand JD, Yuan T, Yuan MT, Deng Y, Zhou JZ (2015) Phasing amplicon sequencing on Illumina Miseq for robust environmental microbial community analysis. BMC Microbiol 15:125. doi:10.1186/s12866-015-0450-4

Xia JH, Lin G, Fu GH, Wan ZY, Lee M, Wang L, Liu XJ, Yue GH (2014) The intestinal microbiome of fish under starvation. BMC Genomics 15:266. doi:10.1186/1471-2164-15-266

Yao FZ, Coquery J, Lê Cao KA (2012) Independent principal component analysis for biologically meaningful dimension reduction of large biological data sets. BMC Bioinform 13:24. doi:10.1186/1471-2105-13-24

Ye L, Amberg J, Chapman D, Gaikowski M, Liu W-T (2013) Fish gut microbiota analysis differentiates physiology and behavior of invasive Asian carp and indigenous American fish. ISME J 2013:1–11. doi:10.1038/ismej.2013.181

Yeh SP, Chang CA, Chang CY, Liu CH, Cheng W (2008) Dietary sodium alginate administration affects fingerling growth and resistance to *Streptococcus* sp. and iridovirus, and juvenile non-specific immune responses of the orange-spotted grouper, *Epinephelus coioides*. Fish Shellfish Immunol 25:19–27. doi:10.1016/j.fsi.2007.11.011

Ziaei-Nejad S, Rezaei MH, Takami GA, Lovett DL, Mirvaghefi AR, Shakouri M (2006) The effect of *Bacillus* spp. bacteria used as probiotics on digestive enzyme activitiy, survival and growth in the Indian white shrimp (*Fenneropenaeus indicus*). Aquaculture 252:516–524. doi:10.1016/j.aquaculture.2005.07.021

Marine Sponge-Associated Microbiome: Reservoir of Novel Bioactive Compounds

11

Uttara Lele-Rahalkar and Shrikant Pawar

Abstract

Oceans consist of around half a million to ten million species out of which sponges, being the oldest and almost omnipresent of the marine metazoans, make an integral part of marine benthic biodiversity and play an important role in benthic–pelagic coupling. Sponges are the most popular invertebrate microbial hosts of today. A symbiotic relationship between the two is predicted. Sponge-associated microbes in league with the sponges themselves are believed to produce a large array of bioactive compounds. However, the true origin of most of these compounds is still ambiguous and remains a key issue in further developmental stages of biotechnological applications. Also most of the microorganisms associated with marine sponges remain so far uncultivable. Today, with advances in the molecular biology techniques, nearly complete microbial diversity can be accessed for understanding diversity and abundance within the microbiome, and functional screening approaches based on homology-based screening give further insights into the specific traits which may be useful in the discovery of novel natural products.

Keywords

Marine sponges • Sponge associated microorganisms • Sponge microbiome • Marine microorganisms • Natural products

---

U. Lele-Rahalkar
Indian Institute of Science Education and Research, Pune, India

S. Pawar (✉)
National Centre for Microbial Resource, National Centre for Cell Science, Central Tower, Sai Trinity Building Garware Circle, Sutarwadi, Pashan Pune, Maharashtra 411 021, India
e-mail: shrikantpawar@nccs.res.in

V.C. Kalia et al. (eds.), *Mining of Microbial Wealth and MetaGenomics*,
DOI 10.1007/978-981-10-5708-3_11

Oceans cover around 70% of the planet surface and more than 95% of the biosphere. Thirty per cent of the existing phyla are exclusively marine as against only one phylum being exclusively terrestrial; thus, marine system has much higher phylogenetic diversity. Oceans include many types of ecological habitats such as coral reefs, hydrothermal vents, sponge reefs, sea grass beds, mangroves and soft sediments on the ocean floor deep below its surface. It makes one of the largest natural carbon reservoirs as it stocks around 15 times more CO_2 than terrestrial habitats. Deep sea life which consists of around half a million to ten million species is the most important player in nutrient recycling, biogeochemical cycles and climate moderation ('Facts and Figures on Marine Biodiversity | UNESCO' 2016).

Amongst the ocean ecosystems, the tropical oceans are the most diverse and dynamic. These ecosystems harbour peculiar coral reefs and high benthic biodiversity regions that are characterized by high competition for space on the ocean floor and intense predation. Sponges make an important class of this community and may cover up to 80% of the ocean bed in some places (Taylor et al. 2007). Poriferans or sponges are predicted to have existed since 700–800 million years making them the most ancient metazoans with longest evolutionary history (Hentschel et al. 2002). Sponges occupy most of the aquatic habitats right from the tropical oceans to the temperate waters and freshwater bodies. Within marine environments, sponges are found on mangrove shores, on intertidal pools, and in the deep seas to the hadal depths of thousands of meters (Fig. 11.1a–f). Being the oldest and almost omnipresent of the marine metazoans, sponges make an integral part of marine benthic biodiversity and play an important role in benthic–pelagic coupling (Bell 2008). Sponges are classified into three taxonomic classes: Demospongiae, Calcarea and Hexactinellida. There are more than 15,000 described sponge species out of which only 1% have been described for the freshwater habitats (Hentschel et al. 2002; Chambers 2003). Therefore, we would focus mainly on the marine sponges and associated microbes in this chapter.

11.1 Sponges as Hosts for Microorganisms

It is a well-known fact that sponges host diverse and abundant microorganisms within their body cavity. Some sponges are known to have microorganisms cover up to 40% of their body volume (Vacelet and Donadey 1977; Bayer et al. 2014). This particular ecological association is mainly due to the body plan and aquiferous system of sponges (Fig. 11.2). A typical sponge body is made up of spicules and several different cell layers. As the name suggests, sponges have many pores and extending canals arranged in a specific manner covering the outer surface. These pores and canals are carved so as to channelize filtering of the flowing water. Pinacoderm, the outer surface, comprises of epithelial cells known as pinacocytes. These pinacocytes also extend through the smaller pores on the surface through the extending canals. Many chambers collectively called as choanoderms are found in the interior and are connected via these canals. These choanoderms are covered with the

Fig. 11.1 (**a**) Intertidal marine encrusting sponge *Hyrtios cavernosus* found in intertidal pool; (**b**) typical intertidal rocky pools exposed during the lowest low tide; (**c**) subtidal marine barrel sponge *Xestospongia* sp.; (**d**) intertidal marine encrusting sponge *Haliclona* sp. growing in association with the barnacle–gastropod encrustation, exposed during the lowest low tide; (**e**) intertidal marine encrusting sponge *Haliclona* sp. growing in association with marine algae, exposed during the lowest low tide; (**f**) typical subtidal habitat of coral and sponge reef from Andaman Islands

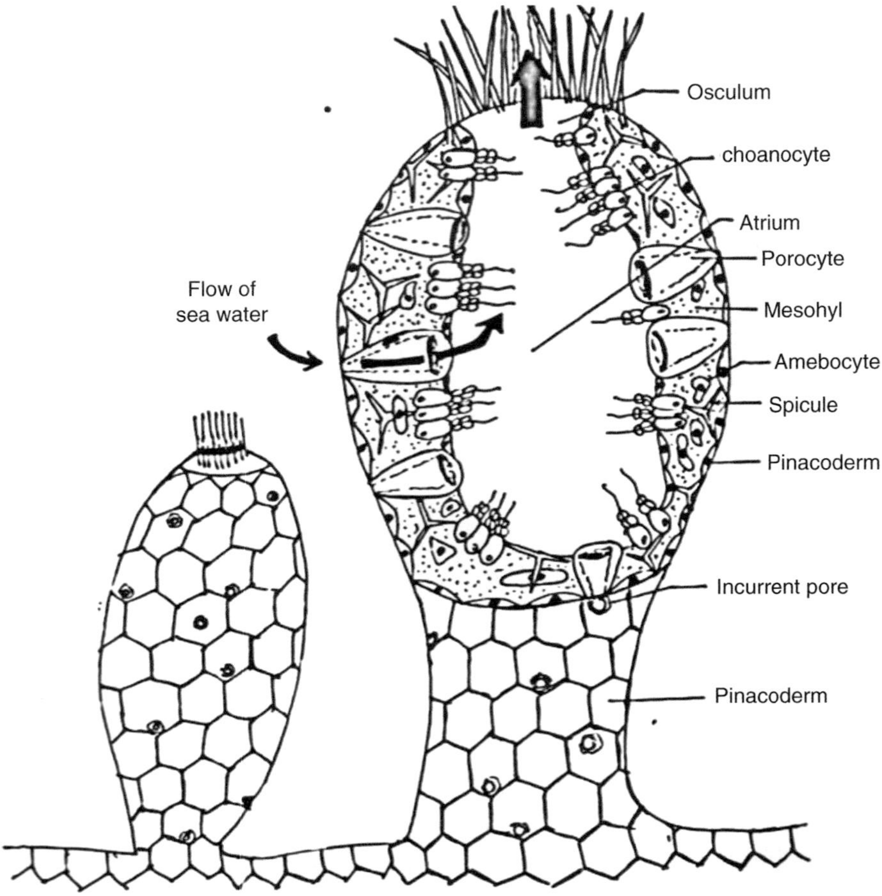

Fig. 11.2 Diagram showing sponge morphology and aquiferous system (reproduced as per R D Barnes)

flagellated epithelial cells called as choanocytes. With the help of the flagella, choanocytes pump the water inside through the ostia. Choanocytes also filter the incoming water for small particles which may include organic matter as well as microorganisms. The connective tissue between the choanoderms is called mesohyl. Mesohyl is mainly made up of gelatinous matrix and phagocytic cells called archeocytes. The filtered food particles and microorganisms are pushed by choanocytes into the mesohyl where most of the food particles are digested by archeocytes. Mesohyl contains the maximum load of microorganisms. The morphology of sponge body is carved and maintained by skeletal structures called as spicules. These spicules are made up of calcium carbonate or silica. A large variety of spicules are secreted by a sponge most of which are taxon specific. These spicules come in numerous shapes and sizes and make the basis for sponge biology and taxonomy. In addition to the spicules, collagenous tissue called as spongin also helps in forming structural support especially for some large individuals that are found in tropical regions (Taylor et al. 2007).

11.2 Relationship between Sponges and Microorganisms

As mentioned earlier, sponges are filter feeders and they continuously filter the surrounding seawater, and along with particulate matter they take up microorganisms from seawater. The general estimate of the water being filtered by 1 kg of sponge per day is estimated to be up to 24,000 L, with many sponges pumping orders of magnitude more volume of water than its own body volume (Weisz et al. 2007). Both sponges and microorganisms being ancient taxa, the relationship or association between these two is said to be the longest of the microbe–metazoan relationships (Webster and Blackall 2009). As compared to other host–microbe associations, sponge–microbe association is more widespread and more diverse. Same sponge species generally hosts many different microorganisms, and the same microbial species may be found associated with many sponge species (Hentschel et al. 2002). Many times sponge-associated microorganisms are called as 'sponge symbionts'; however, the term here does not indicate the classical ecological symbiotic relationship. This symbiosis does not necessarily mean 'give and take' or benefit to each other, but it simply means that two different organisms are found in a stable relationship, living together for a long period. Benefits to each other may be found but are not necessary in this case.

However, the predicted symbiotic relationship is such that the sponge provides microbes with shelter and colonization substratum. Some nutrient benefits between the two are also predicted, e.g. sponges may acquire food from residing primary producers (Taylor et al. 2007). The biomolecules produced by microorganisms may render protection to the sponges from harm by other microbes (Wang 2006).

11.3 Sponge-Associated Microorganisms: Composition, Abundance and Diversity

The study of marine sponges and associated microorganisms presents an interesting study area mainly because (a) ecological relationship between sponges and associated microorganisms is not yet completely understood; (b) sponges harbour highly diverse, specific and also some exclusive microbial groups, and this diversity itself presents a great potential to study the biotechnological applications derived from those (Webster and Thomas 2016). Although any actual role of microorganisms in the symbiosis has not been clearly demonstrated, several functions have been suggested so far (Hentschel et al. 2006). These suggested roles mainly include provision of nutrition, production of photosynthates, provision of structural rigidity and protection from sunlight (Reiswig 1975; Wilkinson 1978a, b; Hinde et al. 1994). The role carried out by sponge-associated microbes in cycling of sulphur and nitrogen has been explored to large extent so far (Nicol and Schleper 2006; Hoffmann et al. 2009).

11.3.1 Cultured and Uncultured Sponge-Associated Microbial Diversity

Frequent efforts have been made so far to cultivate and identify microorganisms associated with different species of sponges across habitats. These methods include conventional culturing techniques as well as modified culturing techniques like the use of specific media components or the use of various incubation temperatures and durations. In spite of the use of these novel culturing methods, a large proportion of microbes associated with poriferans is anticipated to be unexplored (Amann et al. 1995). This statement is not restricted only to the sponge-associated microorganisms but is true for all the microbiomes.

The very first thorough study of sponge–microbe relationship was done by Wilkinson and group in the 1980s (Wilkinson 1978a, b). This work predicted association of complex microorganisms with sponges and predicted the role of symbiotic microorganisms in metabolic processes like sulphur reduction, photosynthesis, nitrogen fixation and nitrification. Some sponge-associated bacterial species from the following phyla have been successfully cultured using pure culture techniques—*Cyanobacteria*, *Actinobacteria*, *Planctomycetes*, *Bacteroidetes*, *Firmicutes*, *Proteobacteria* and *Verrucomicrobia* (Hentschel et al. 2002; Taylor et al. 2007; Schmitt et al. 2012).

Along with cultivation using conventional media, some mechanical separation techniques have also been used for observation and cultivation of sponge-associated microorganisms. These mainly include differential centrifugation and gradient centrifugation to separate sponge cells from microbial cells. So far various species of heterotrophic bacteria like *pseudomonads*, *bacilli*, some previously unidentified bacteria and *micrococci* have been isolated and cultivated from sponge species *Theonella swinhoei*, *Halichondria panicea*, *Tedania ignis*, *Hyatella* sp. and more. Other microorganisms like red algae, diatoms, and cyanobacteria have also been isolated, and some of them have been successfully cultivated in the laboratory (Kennedy et al. 2007). Examples of these include red macroalga *Ceratodictyon spongiosum* and cyanobacterium *Oscillatoria spongeliae* isolated from sponges *Sigmadocia symbiotica* and *Dysidea herbacea*, respectively (Price et al. 1984; Hinde et al. 1994).

There are many such studies that have successfully cultivated some microorganisms; however, these studies show considerably small part of the true microbial diversity. If these studies are taken into account while analysing the sponge-associated microbial diversity, we are most likely to get a biased picture because of the limitations of the culturing techniques. This issue has been tackled successfully with the emergence of molecular techniques which allow researchers to bypass limitations of culturing methods. The use of fluorescence in situ hybridization (FISH), denaturing gradient gel electrophoresis (DGGE), group-specific 16S rDNA gene-targeted oligonucleotide probes and amplicon sequencing (Webster et al. 2001, 2004; Taylor et al. 2004) has unearthed many more microbial phyla and some of them almost exclusive to marine sponges. Using these techniques, the distribution of microbial symbionts across sponge species and across habitats has also been studied.

11.3.2 High and Low Microbial Abundance Sponges

In the beginning, sponge-associated microorganisms were studied by observing sponge tissues under the electron microscope. Microbial symbionts within various sponge parts were observed in order to study associated microbial community composition. This type of study gave rise to a system of classifying sponges according to the diversity and abundance of the microorganisms they host. It was found that sponges having high microbial abundance (HMA) and low microbial abundance (LMA) have different morphology and physiology. It was observed that the HMA sponges host microbial densities more than that of the seawater by a magnitude of 3 or 4. On the other hand, LMA sponges have microbial loads very close to that of the surrounding waters (Vacelet and Donadey 1977; Gloeckner et al. 2014; Bayer et al. 2014). HMA sponges were found to have relatively thick walls and small choanocyte chambers. On the other hand, LMA sponges were found to have advanced aquiferous system and mesohyl with low microbial density. Hence, the hypothesis is that the type of microbe–sponge relationship determines sponge morphology and development of aquiferous system (Vacelet and Donadey 1977). The HMA sponges typically have low pumping rates, and LMA sponges have higher pumping rates and high rate of heterotrophic feeding on particulate organic matter (Webster et al. 2010).

11.3.3 Metagenomic Approaches to Explore Sponge-Associated Microbial Diversity

The use of metagenomic approaches has successfully discovered a large information on microbiomes from soils, plants, animal guts and marine habitats too (Dinsdale et al. 2008; Hugenholtz and Tyson 2008; Woyke et al. 2009). In metagenomic studies, one analyses specific genomic fragments (amplified) from complex microbial communities from the source bypassing the cultivation of organisms. Since the past decade, huge data is generated on the sponge-associated microbial diversity and has enabled in-depth exploration. Hentschel et al. (2002) found evidence for a core of shared prokaryotic species amongst poriferans that was not correlated with the host biogeography. This work followed many others demonstrating different microbial communities in sponges when compared to the surrounding waters. In 2002, Hentschel et al. reanalysed 190 16S rDNA sequences derived from sponges. Seventy per cent of these sequences belonged to the sponge-specific clusters, with phyla including *Acidobacteria*, *Actinobacteria*, *Chloroflexi*, *Nitrospira*,

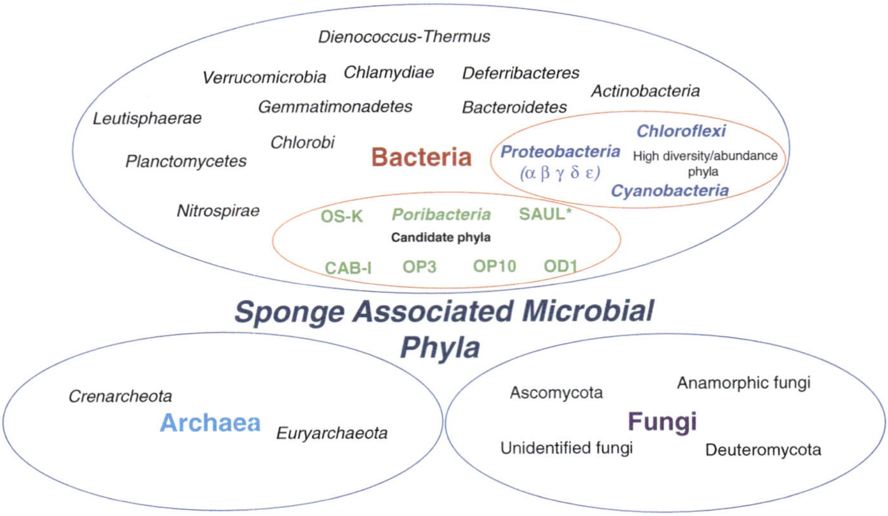

Fig. 11.3 Some of the representative microbial phyla found associated with sponges (*SAUL stands for sponge-associated unidentified lineage)

and *Proteobacteria*. This trend was found consistent irrespective of the sponge species or geographical location (Taylor et al. 2007).

In 2012, Schmitt et al. analysed 16S rDNA amplicon pyrosequencing data taken from 32 sponge species found at eight sites worldwide. The pyrosequencing generated 2567 OTUs (operational taxonomic units) at 97% similarity. In this data, up to 364 OTUs were found to be associated per sponge species, and 25 bacterial phyla based on the 97% similarity OTUs were detected after the analysis. Amongst these *Proteobacteria* (α, β, γ, δ and ε) and *Chloroflexi* were found to be the most diverse phyla. *Poribacteria*, *Acidobacteria*, *Actinobacteria*, *Cyanobacteria*, *Gemmatimonadetes* and *Bacteroidetes* were found to be next diverse phyla, respectively. They also found nine candidate phyla out of which six were found to be associated with sponges for the first time. Out of which TM 7 was found to be the most diverse phylum (Schmitt et al. 2012). Figure 11.3 shows most of the so far detected microbial phyla associated with marine sponges.

11.4 Sponge-Microbe Association and Derived Bioactive Molecules

Phylum Porifera remains the most important and promising phylum when marine natural products are discussed. Since 1990, poriferans made up for 48.8% of all the discovered marine natural products (Leal et al. 2012). Within these products, the number of sponge-derived compounds that made to clinical and preclinical trials is higher than other marine phyla (Blunt et al. 2006). Sponges being very primitive and sessile organisms lack mechanical defence systems and have developed chemical ways (secondary metabolites) to tackle competition and predation (Kennedy et al. 2007; Jensen and Fenical 1994; Bernan et al. 1997; Haygood et al. 1999; Osinga et al. 2001). The biomedical and pharmaceutical compounds isolated from this duo so far have been shown to have antiviral, antitumor, antimicrobial and general cytotoxic properties (Kennedy et al. 2008; Thomas et al. 2010; Fuerst 2014). A landmark discovery in natural product chemistry from sponges is that of two nucleosides spongothymidine and spongouridine in a Caribbean sponge *Tectitethya crypta*. These structures provided the basis for the development of drugs like ara-A (Vidarabine®, Vidarabin Thilo®), a drug active against Herpes simplex encephalitis virus, and ara-C (cytarabine, Alexan®, Udicil®), an anticancer drug (Bergmann and Burke 1955). One of the significant sponge-derived metabolites in clinical trials is the polyketide discodermolide from the Caribbean sponge *Discodermia dissoluta* (Gunasekera et al. 1990). Two more sponge-derived anticancer compounds—glycolipid KRN7000 from the *Agelas mauritianus* and halichondrin B, isolated from *Halichondria okadai*—have also been under serious investigation (Hart et al. 2000; Hayakawa et al. 2003).

11.4.1 Bioactive Compounds from the Microbes Hosted by Marine Sponges

Since many microbial taxa that are also found in association with sponges are known producers of bioactive compounds, microbial origin of such compounds cannot be denied. However, identifying and confirming the actual origin just from the circumstantial evidence is difficult. Discovery of similar chemical compounds from unrelated sponges with variable chemical yield that were previously reported exclusively from microorganisms strengthened the speculation of the microbial origin (Bewley and Faulkner 1998; Haygood et al. 1999; Piel et al. 2004). However, the true origin of most of these compounds is still ambiguous and remains a key issue in further developmental stages of drug discovery (Taylor et al. 2007). Table 11.1 summarizes some of the natural products derived from marine sponges and associated microbes.

Manzamine alkaloids were isolated from geographically distant and unrelated sponge species. This suggested a microbial origin for this class of compounds (Hill et al. 2005). Hill et al. discovered that manzamine A was produced by a *Micromonospora* sp. strain M42 from an Indonesian sponge *Acanthostrongylophora ingens* (Hu et al. 2003). An antifungal compound theopalauamide derived from a Philippine sponge *Theonella swinhoei* was detected in a fraction obtained by differential centrifugation. This fraction mostly contained filamentous *delta-proteobacterium* that has been assigned to '*Candidatus Entotheonella palauensis*' suggesting bacterial origin (Schmidt et al. 1998). However, mere co-occurrence of compound in cellular fractionation cannot prove its microbial origin. It is crucial to demonstrate production by the bacterial culture for further applications to be explored.

Table 11.1 Compilation of some of the important products derived from marine sponges or associated microorganisms and synthetic compounds based on structures derived from the marine sponges

Sr. no.	Name of the compound	Source sponge	Associated microorganism	Action	Reference
1	2,4,4′-Trichloro-2′-hydroxydiphenylether (triclosan) and acyl-1-(acyl-6′- mannobiosyl)-3-glycerol (lutoside)	*Xestospongia* sp.	*Micrococcus luteus* R-1588-10 (bacterium)	Antibacterial	Bultel-Poncé et al. (1997)
2	3,3′-Bicoumarin (bicoumanigrin A)	*Axinella damicornis*	*Aspergillus niger* (fungus)	Cytotoxic	Jadulco et al. (2004), Hiort et al. (2004)
3	Anthraquinone, evariquinone	*Haliclona valliculata*	*Emericella variecolor* (fungus)	Anticancer	Bringmann et al. (2003a, b)
4	Asperazine	*Hyrtios proteus*	*Aspergillus niger* (fungus)	Antileukemic	Varoglu and Crews (2000)
5	Chitinase (high pH and salinity tolerance)	*Craniella australiensis*	*Streptomyces* sp. DA11 (bacterium)	Antifungal	Han et al. (2009)
8	Dictyodendrins (A-E)	*Dictyodendrilla verongiformis*	–	Telomerase inhibitor	Warabi et al. (2003)
9	Discodermolide	*Discodermia dissoluta*	–	Anticancer	Gunasekera et al. (1990)
10	KRN7000 (agelasphin derivative)	*Agelas mauritiana*	–	Anticancer	Natori et al. (1993)
11	Laulimalide	*Cacospongia mycofijiensis*	–	Microtubule stabilizer	Gollner et al. (2009)
12	Leucamide A	*Leucetta microraphis*	–	Antitumor	König et al. (2006)
13	Lunatin and cytoskyrin A (anthraquinones)	*Niphates olemda*	*Curvularia lunata* (fungus)	Broad spectrum of antibacterial activity	Jadulco et al. (2002), Bhadury et al. (2006)
14	Manzamine A, B	*Several sponges*	*Micromonospora* sp. (bacterium)	Antimalarial, anticancer, respectively	Sakai et al. (1986), Ang et al. (2000)
15	NVP-LAQ824, psammaplin derivative	*Psammaplysilla* sp.	–	Histone deacetylase (HDAC) inhibition	Remiszewski (2003)
16	Peloruside A	*Mycale hentscheli*	–	Antimitotic	West et al. (2000)
18	Polybrominated diphenyl ether	*Lamellodysidea herbacea*	*Oscillatoria spongeliae* (cyanobacterium)	Broad spectrum of antibacterial activity	Arillo et al. (1993)

19	Metacycloprodigiosin and undecylprodigiosin (prodigiosin analogues)	*Mycale plumose*	*Saccharopolyspora* sp. nov. (bacterium)	Cytotoxic	Liu et al. (2005)
20	Roridin A and D (macrocyclic trichothecenes)	*Axinella* sp.	*Myrothecium* sp. JS9 (fungus)	Antifungal	Xie et al. (2008)
21	Salicylihalamides A, B	*Haliclona* sp.	–	Anticancer	Erickson et al. (2001)
22	Sorbicillactone A	*Ircinia fasciculata*	*Penicillium chrysogenum* (fungus)	Antileukemic	Bringmann et al. (2003a, b)
23	Swinholide A, theopalauamide	*Theonella swinhoei*	*Candidatus Entotheonella palauensis* (bacterium)	Cytotoxic, antifungal, respectively	Bewley et al. (1996), Bewley and Faulkner (1998)
24	Trichodenone A, B and C	*Halichondria okadai*	*Trichoderma harzianum* OUPSN115 (fungus)	Cytotoxic	Amagata et al. (1998), Usami et al. (2000), Thakur et al. (2003)
25	Variolins	*Kirkpatrickia variolosa*	–	Cdk inhibitor	Jayatilake et al. (1995)
26	Xestodecalactone B	*Xestospongia exigua*	*Penicillium cf. montanense* (fungus)	Antifungal	Edrada et al. (2002)
Synthetic compounds constructed from structures derived from marine sponges					
27	Cytarabine (Ara-C)	*Tectitethya crypta*	–	Anticancer	Bergmann and Burke (1955), Tsimberidou et al. (2002)
28	Eribulin mesylate (E7389)	*Halichondria okadai*	–	Anticancer	Kuznetsov et al. (2007)
29	Hemiasterlin (E7974)	*Cymbastela* sp.	–	Anticancer	Kuznetsov et al. (2009)
30	Vidarabine (Ara-A)	*Tectitethya crypta*	–	Antiviral	Bergmann and Burke (1955), Fiume et al. (1988)

11.4.2 Bioactive Compounds and Metagenomic Approaches

Some of the prime hindrances in the path of discovery of marine natural products are accessibility, efficient screening methods and the supply problem, i.e. sustainable production of the active compounds. Lack of thorough taxonomic knowledge of marine species is also a major problem. For pharmacological purposes, correct assignment of lineage to organisms is crucial. An incorrect classification of species might make it impossible to reproduce the isolation and analytical processes done on the species and associated compound, compromising the entire drug discovery process (Martins et al. 2014).

The use of metagenomic techniques has been one of the most successful approaches to tackle some of these problems. Metagenome analysis methods can detect a large proportion of the so far uncultivated organisms along with cultured organisms. This gives us access to the information on previously unreachable bioresource. A bulk of information on the bacterial diversity has been extracted using metagenome libraries by analysing and screening: (a) the sequences themselves to give information on the diversity and phylogeny and (b) the specific functional traits from the microbiome (Taylor et al. 2007).

One of the most popular functional trait analyses has been focused around sequence analysis of polyketide synthase (PKS) genes. The PKS genes have been known to be present exclusively in bacteria. These genes are involved in the synthesis of some well-known antibiotics, antitumor agents, immunosuppressants, etc. (Kumar et al. 2004). In 2004, Piel et al. found genes closely resembling pederin, a defensive polyketide in metagenome sequences of *Theonella swinhoei* (Piel et al. 2004). They also found compounds omnamides and theopederins in the sponge *T. swinhoei* and successfully cloned PKS genes for biosynthesis of antitumor polyketides of omnamide series from *T. swinhoei* and traced them back to a prokaryote. This provided support to the theory that sponge-associated bacteria may be the true producers of the majority of the sponge-derived compounds.

Nearly complete microbial diversity can be accessed using standard metagenomic approaches like sequencing and analysing16S rDNA amplicon libraries, whereas functional screening approaches constructed on homology-based screening give further insights into the specific traits which may be useful in the discovery of novel natural products. With improved screening and DNA isolation techniques and availability of good cloning vectors like bacterial artificial chromosomes (BACs) and cosmids, it is now possible to express large fragments of DNA and subsequently screen large clone libraries for functional activities from the microbiome (Lorenz and Eck 2005).

11.4.2.1 Way ahead

In the last two decades, a huge amount of data on sponges and associated microbes has been generated using classical as well as next-generation sequencing approaches. However, many of the ecological and evolutionary questions regarding the nature of the relationship between these two remain unanswered. It is clear from the wealth of literature available that the bioprospecting potential of this duo has been very well appreciated; however, large-scale production of any of the sponge–microbe-derived compounds remains challenging. Along with these molecular approaches, one also needs to study the ecology of the sponge–microbe interactions as well as intraspecies interactions in order to be able identify relevant functional genes for exploration. The study of these ecological interactions would also help in fine-tuning many steps of the bioprospecting process.

Acknowledgements The authors are thankful to the RGSTC—Maharashtra Gene Bank Programme and the entire team; NCMR-NCCS, Pune; and IISER, Pune, for the funding and support provided. The authors also wish to thank Mr. Chinmay S. Kulkarni for the underwater photographs and Ms. Neha Shintre for the photographs shared.

References

Amagata T, Usami Y, Minoura K, Ito T, Numata A (1998) Cytotoxic substances produced by a fungal strain from a sponge: physico-chemical properties and structures. J Antibiot 51(1):33–40

Amann RI, Ludwig W, Schleifer KH (1995) Phylogenetic identification and in situ detection of individual microbial cells without cultivation. Microbiol Rev 59(1):143–169

Ang KKH, Holmes MJ, Higa T, Hamann MT, Kara UAK (2000) In vivo antimalarial activity of the beta-carboline alkaloid manzamine A. Antimicrob Agents Chemother 44(6):1645–1649. doi:10.1128/AAC.44.6.1645-1649.2000

Arillo A, Bavestrello G, Burlando B, Sarà M (1993) Metabolic integration between symbiotic cyanobacteria and sponges: a possible mechanism. Mar Biol 117(1):159–162. doi:10.1007/BF00346438

Bayer K, Kamke J, Hentschel U (2014) Quantification of bacterial and archaeal symbionts in high and low microbial abundance sponges using real-time PCR. FEMS Microbiol Ecol 89(3):679–690. doi:10.1111/1574-6941.12369

Bell JJ (2008) The functional roles of marine sponges. Estuar Coast Shelf Sci 79(3):341–353. doi:10.1016/j.ecss.2008.05.002

Bergmann W, Burke DC (1955) Contributions to the study of marine products. XXXIX. The nucleosides of sponges. III. Spongothymidine and spongouridine. J Org Chem 20(11):1501–1507. doi:10.1021/jo01128a007

Bernan VS, Greenstein M., Maiese WM (1997) Marine microorganisms as a source of new natural products. In: Neidleman SL, Laskin AI (eds) Advances in applied microbiology, vol 43. Academic Press, New York, pp 57–90. http://www.sciencedirect.com/science/article/pii/S0065216408702235

Bewley CA, Holland ND, Faulkner DJ (1996) Two classes of metabolites from Theonella swinhoei are localized in distinct populations of bacterial symbionts. Experientia 52(7):716–722

Bewley CA, John Faulkner D (1998) Lithistid sponges: star performers or hosts to the stars. Angew Chem Int Ed 37(16):2162–2178

Bhadury P, Mohammad BT, Wright PC (2006) The current status of natural products from marine fungi and their potential as anti-infective agents. J Ind Microbiol Biotechnol 33(5):325–337. doi:10.1007/s10295-005-0070-3

Biabani M, Farooq A, Laatsch H (1998) Advances in chemical studies on low-molecular weight metabolites of marine fungi. J Prakt Chem 340(7):589–607. doi:10.1002/prac.19983400702

Blunt JW, Copp BR, Munro MHG, Northcote PT, Prinsep MR (2006) Mar Nat Prod. http://dx.doi.org/10.1039/b502792f

Bringmann G, Lang G, Steffens S, Günther E, Schaumann K (2003b) Evariquinone, isoemericellin, and stromemycin from a sponge derived strain of the fungus Emericella variecolor. Phytochemistry 63(4):437–443

Bringmann G, Lang G, Mühlbacher J, Schaumann K, Steffens S, Rytik PG, Hentschel U, Morschhäuser J, Müller WEG (2003a) Sorbicillactone A: a structurally unprecedented bioactive novel-type alkaloid from a sponge-derived fungus. Prog Mol Subcell Biol 37:231–253

Bultel-Poncé V, Debitus C, Blond A, Cerceau C, Guyot M (1997) Lutoside: an acyl-1-(acyl-6′-mannobiosyl)-3-glycerol isolated from the sponge-associated bacterium Micrococcus luteus. Tetrahedron Lett 38(33):5805–5808. doi:10.1016/S0040-4039(97)01283-5

Chambers S (2003) In: Hooper JNA, van Soest RWM (eds) Systema porifera. A guide to the classification of sponges. Kluwer Academic, Dordrecht, 1756 pp. ISBN 0-306-47260-0. Aquat Conserv Mar Freshwat Ecosyst 13(5):461–462. doi:10.1002/aqc.593

Cheng X-C, Varoglu M, Abrell L, Crews P, Lobkovsky E, Clardy J (1994) Chloriolins A-C, chlorinated sesquiterpenes produced by fungal cultures separated from a Jaspis marine sponge. J Org Chem 59(21):6344–6348. doi:10.1021/jo00100a041

Dinsdale EA, Edwards RA, Hall D, Angly F, Breitbart M, Brulc JM, Furlan M et al (2008) Functional metagenomic profiling of nine biomes. Nature 452(7187):629–632. doi:10.1038/nature06810

Edrada RA, Heubes M, Brauers G, Wray V, Berg A, Grafe U, Wohlfarth M et al (2002) Online analysis of xestodecalactones A-C, novel bioactive metabolites from the fungus Penicillium cf. montanense and their subsequent isolation from the sponge Xestospongia exigua. J Nat Prod 65(11):1598–1604

Erickson KL, Beutler JA, Cardellina JH, Boyd MR (2001) Salicylihalamides A and B, novel cytotoxic macrolides from the marine sponge Haliclona sp. J Org Chem 66(4):1532. doi:10.1021/jo004040x

Facts and Figures on Marine Biodiversity | United Nations Educational, Scientific and Cultural Organization (2016) http://www.unesco.org/new/en/natural-sciences/ioc-oceans/priority-areas/rio-20-ocean/blueprint-for-the-future-we-want/marine-biodiversity/facts-and-figures-on-marine-biodiversity/. Accessed June 24

Fiume L, Bonino F, Mattioli A, Chiaberge E, Cerenzia MRT, Busi C, Brunetto MR, Verme G (1988) Originally Published as Volume 2, Issue 8601 Inhibition of hepatitis b virus replication by vidarabine monophosphate conjugated with lactosaminated serum albumin. Lancet 332(8601):13–15. doi:10.1016/S0140-6736(88)92946-7

Fuerst JA (2014) Diversity and biotechnological potential of microorganisms associated with marine sponges. Appl Microbiol Biotechnol 98(17):7331–7347. doi:10.1007/s00253-014-5861-x

Gloeckner V, Wehrl M, Moitinho-Silva L, Gernert C, Schupp P, Pawlik JR, Lindquist NL, Erpenbeck D, Wörheide G, Hentschel U (2014) The HMA-LMA dichotomy revisited: an electron microscopical survey of 56 sponge species. Biol Bull 227(1):78–88

Gollner A, Altmann K-H, Gertsch J, Mulzer J (2009) The laulimalide family: total synthesis and biological evaluation of neolaulimalide, isolaulimalide, laulimalide and a nonnatural analogue. Chem Eur J 15(24):5979–5997. doi:10.1002/chem.200802605

Gunasekera SP, Gunasekera M, Longley RE, Schulte GK (1990) Discodermolide: a new bioactive polyhydroxylated lactone from the marine sponge discodermia dissoluta. J Org Chem 55(16):4912–4915. doi:10.1021/jo00303a029

Han Y, Yang B, Zhang F, Miao X, Li Z (2009) Characterization of antifungal chitinase from marine streptomyces sp. DA11 associated with South China Sea sponge Craniella australiensis. Mar Biotechnol (NY) 11(1):132–140. doi:10.1007/s10126-008-9126-5

Hart JB, Lill RE, Hickford SJH, Blunt JW, Munro MHG (2000) The halichondrins: chemistry, biology, supply and delivery. In: Fusetani N (ed) Drugs from the sea. Karger, Basel, pp 134–153. http://www.karger.com/doi/10.1159/000062488

Hayakawa Y, Rovero S, Forni G, Smyth MJ (2003) α-Galactosylceramide (KRN7000) suppression of chemical- and oncogene-dependent carcinogenesis. Proc Natl Acad Sci 100(16):9464–9469. doi:10.1073/pnas.1630663100

Haygood MG, Schmidt EW, Davidson SK, John Faulkner D (1999) Microbial symbionts of marine invertebrates. J Mol Microbiol Biotechnol 1(1):33–43. https://utah.pure.elsevier.com/en/publications/microbial-symbionts-of-marine-invertebrates-opportunities-for-mic

Hentschel U, Hopke J, Horn M, Friedrich AB, Wagner M, Hacker J, Moore BS (2002) Molecular evidence for a uniform microbial community in sponges from different oceans. Appl Environ Microbiol 68(9):4431–4440

Hentschel U, Usher KM, Taylor MW (2006) Marine sponges as microbial fermenters. FEMS Microbiol Ecol 55(2):167–177. doi:10.1111/j.1574-6941.2005.00046.x

Hill R, Peraud O, Hamann M, Kasanah N (2005) Manzamine-producing actinomycetes. US20050244938 A1, filed August 1, 2003, and issued November 3, 2005

Hinde R, Pironet F, Borowitzka MA (1994) Isolation of Oscillatoria spongeliae, the filamentous cyanobacterial symbiont of the marine sponge Dysidea herbacea. Mar Biol 119(1):99–104. doi:10.1007/BF00350111

Hiort J, Maksimenka K, Reichert M, Perović-Ottstadt S, Lin WH, Wray V, Steube K et al (2004) New natural products from the sponge-derived fungus Aspergillus niger. J Nat Prod 67(9):1532–1543. doi:10.1021/np030551d

Hoffmann F, Larsen O, Thiel V, Rapp HT, Pape T, Michaelis W, Reitner J (2005) An anaerobic world in sponges. Geomicrobiol J 22(1–2):1–10. doi:10.1080/01490450590922505

Hoffmann F, Radax R, Woebken D, Holtappels M, Lavik G, Rapp HT, Schläppy M-L, Schleper C, Kuypers MMM (2009) Complex nitrogen cycling in the sponge Geodia barretti. Environ Microbiol 11(9):2228–2243. doi:10.1111/j.1462-2920.2009.01944.x

Hugenholtz P, Tyson GW (2008) Microbiology: metagenomics. Nature 455(7212):481–483. doi:10.1038/455481a

Hu J-F, Hamann MT, Hill R, Kelly M (2003) The manzamine alkaloids. In: Cordell GA (eds) The alkaloids: chemistry and biology, vol 60. Academic, New York, pp 207–285

Jadulco R, Brauers G, Edrada RA, Ebel R, Wray V, Sudarsono S, Proksch P (2002) New metabolites from sponge-derived fungi Curvularia lunata and Cladosporium herbarum. J Nat Prod 65(5):730–733

Jadulco R, Edrada RA, Ebel R, Berg A, Schaumann K, Wray V, Steube K, Proksch P (2004) New communesin derivatives from the fungus Penicillium sp. derived from the Mediterranean sponge Axinella verrucosa. J Nat Prod 67(1):78–81. doi:10.1021/np030271y

Jayatilake GS, Baker BJ, McClintock JB (1995) Isolation and identification of a stilbene derivative from the antarctic sponge Kirkpatrickia variolosa. J Nat Prod 58(12):1958–1960. doi:10.1021/np50126a028

Jensen PR, Fenical W (1994) Strategies for the discovery of secondary metabolites from marine bacteria: ecological perspectives. Annu Rev Microbiol 48(1):559–584. doi:10.1146/annurev.mi.48.100194.003015

Kennedy J, Marchesi JR, Dobson ADW (2007) Metagenomic approaches to exploit the biotechnological potential of the microbial consortia of marine sponges. Appl Microbiol Biotechnol 75(1):11–20. doi:10.1007/s00253-007-0875-2

Kennedy J, Marchesi JR, Dobson ADW (2008) Marine metagenomics: strategies for the discovery of novel enzymes with biotechnological applications from marine environments. Microb Cell Factories 7:27. doi:10.1186/1475-2859-7-27

König GM, Kehraus S, Seibert SF, Abdel-Lateff A, Müller D (2006) Natural products from marine organisms and their associated microbes. Chembiochem 7(2):229–238. doi:10.1002/cbic.200500087

Kumar P, Khosla C, Tang Y (2004) Manipulation and analysis of polyketide synthases. Methods Enzymol Protein Eng 388:269–293

Kuznetsov G, TenDyke K, Towle MJ, Cheng H, Liu J, Marsh JP, Schiller SER et al (2009) Tubulin-based antimitotic mechanism of E7974, a novel analogue of the marine sponge natural product hemiasterlin. Am Assoc Cancer Res 8(10):2852–2860. doi:10.1158/1535-7163.MCT-09-0301

Kuznetsov G, TenDyke K, Yu M, Littlefield B (2007) Antiproliferative effects of halichondrin B analog eribulin mesylate (E7389) against paclitaxel-resist ant human cancer cells in vitro. Am Assoc Cancer Res 6(11 Suppl):C58

Leal MC, Puga J, Serôdio J, Gomes NCM, Calado R (2012) Trends in the discovery of new marine natural products from invertebrates over the last two decades—where and what are we bioprospecting? PLoS One 7(1):e30580. doi:10.1371/journal.pone.0030580

Liu R, Cui C-B, Lin D, Qian-Qun G, Zhu W-M (2005) Potent in vitro anticancer activity of metacycloprodigiosin and undecylprodigiosin from a sponge-derived actinomycete Saccharopolyspora sp. nov. Arch Pharm Res 28(12):1341–1344

Li Z (2009) Advances in marine microbial symbionts in the China Sea and related pharmaceutical metabolites. Mar Drugs 7(2):113–129. doi:10.3390/md7020113

Lorenz P, Eck J (2005) Metagenomics and industrial applications. Nat Rev Microbiol 3(6):510–516. doi:10.1038/nrmicro1161

Martins A, Vieira H, Gaspar H, Santos S (2014) Marketed marine natural products in the pharmaceutical and cosmeceutical industries: tips for success. Mar Drugs 12(2):1066–1101. doi:10.3390/md12021066

Natori T, Koezuka Y, Higa T (1993) Agelasphins, novel α-galactosylceramides from the marine sponge Agelas mauritianus. Tetrahedron Lett 34(35):5591–5592. doi:10.1016/S0040-4039(00)73889-5

Nicol GW, Schleper C (2006) Ammonia-oxidising crenarchaeota: important players in the nitrogen cycle? Trends Microbiol 14(5):207–212. doi:10.1016/j.tim.2006.03.004

Osinga R, Evelyn A, Grant Burgess J, Hoffmann F, Reitner J, Schumann-Kindel G (2001) Sponge–microbe associations and their importance for sponge bioprocess engineering. Hydrobiologia 461(1–3):55–62. doi:10.1023/A:1012717200362

Piel J, Hui D, Wen G, Butzke D, Platzer M, Fusetani N, Matsunaga S (2004) Antitumor polyketide biosynthesis by an uncultivated bacterial symbiont of the marine sponge Theonella swinhoei. Proc Natl Acad Sci U S A 101(46):16222–16227. doi:10.1073/pnas.0405976101

Price IR, Friker RL, Wilkinson CR (1984) Ceratodictyon spongiosum (rhodofhyta), the macroalgal partner in an alga-sponge symbiosis, grown in unialgal culture12. J Phycol 20(1):156–158. doi:10.1111/j.0022-3646.1984.00156.x

Reiswig HM (1975) Bacteria as food for temperate-water marine sponges. Can J Zool 53(5):582–589. doi:10.1139/z75-072

Remiszewski SW (2003) The discovery of NVP-LAQ824: from concept to clinic. Curr Med Chem 10(22):2393–2402. doi:10.2174/0929867033456675

Sakai R, Higa T, Jefford CW, Bernardinelli G (1986) Manzamine A, a novel antitumor alkaloid from a sponge. J Am Chem Soc 108(20):6404–6405. doi:10.1021/ja00280a055

Schmidt EW, Bewley CA, John Faulkner D (1998) Theopalauamide, a bicyclic glycopeptide from filamentous bacterial symbionts of the lithistid sponge Theonella swinhoei from Palau and Mozambique. J Org Chem 63(4):1254–1258. doi:10.1021/jo9718455

Schmitt S, Tsai P, Bell J, Fromont J, Ilan M, Lindquist N, Perez T et al (2012) Assessing the complex sponge microbiota: core, variable and species-specific bacterial communities in marine sponges. ISME J 6(3):564–576. doi:10.1038/ismej.2011.116

Selvin J (2009) Exploring the antagonistic producer streptomyces MSI051: implications of polyketide synthase gene type II and a ubiquitous defense enzyme phospholipase A2 in the host sponge Dendrilla nigra. Curr Microbiol 58(5):459–463. doi:10.1007/s00284-008-9343-1

Selvin J, Shanmughapriya S, Gandhimathi R, Seghal Kiran G, Rajeetha Ravji T, Natarajaseenivasan K, Hema TA (2009) Optimization and production of novel antimicrobial agents from sponge associated marine actinomycetes Nocardiopsis dassonvillei MAD08. Appl Microbiol Biotechnol 83(3):435–445. doi:10.1007/s00253-009-1878-y

Taylor MW, Radax R, Steger D, Wagner M (2007) Sponge-associated microorganisms: evolution, ecology, and biotechnological potential. Microbiol Mol Biol Rev 71(2):295–347. doi:10.1128/MMBR.00040-06

Taylor MW, Schupp PJ, Dahllöf I, Kjelleberg S, Steinberg PD (2004) Host specificity in marine sponge-associated bacteria, and potential implications for marine microbial diversity. Environ Microbiol 6(2):121–130. doi:10.1046/j.1462-2920.2003.00545.x

Thakur NL, Hentschel U, Krasko A, Pabel CT, Anil AC, Mueller WEG (2003) Antibacterial activity of the sponge Suberites domuncula and its primmorphs: potential basis for epibacterial chemical defense. Aquat Microb Ecol 31:83–87. http://drs.nio.org/drs/handle/2264/1203

Thomas TRA, Kavlekar DP, LokaBharathi PA (2010) Marine drugs from sponge-microbe association—a review. Mar Drugs 8(4):1417–1468. doi:10.3390/md8041417

Tsimberidou AM, O'Brien SM, Cortes JE, Faderl S, Andreeff M, Kantarjian HM, Keating MJ, Giles FJ (2002) Phase II study of fludarabine, cytarabine (Ara-C), cyclophosphamide, cisplatin and GM-CSF (FACPGM) in patients with Richter's syndrome or refractory lymphoproliferative disorders. Leuk Lymphoma 43(4):767–772. doi:10.1080/10428190290016872

Usami Y, Ikura T, Amagata T, Numata A (2000) First total syntheses and configurational assignments of cytotoxic trichodenones A–C. Tetrahedron Asymmetry 11(18):3711–3725. doi:10.1016/S0957-4166(00)00329-3

Vacelet J, Donadey C (1977) Electron microscope study of the association between some sponges and bacteria. J Exp Mar Biol Ecol 30(3):301–314. doi:10.1016/0022-0981(77)90038-7

Varoglu M, Crews P (2000) Biosynthetically diverse compounds from a saltwater culture of sponge-derived Aspergillus niger. J Nat Prod 63(1):41–43. doi:10.1021/np9902892

Wang G (2006) Diversity and biotechnological potential of the sponge-associated microbial consortia. J Ind Microbiol Biotechnol 33(7):545–551. doi:10.1007/s10295-006-0123-2

Warabi K, Matsunaga S, van Soest RWM, Fusetani N (2003) Dictyodendrins A–E, the first telomerase-inhibitory marine natural products from the sponge Dictyodendrilla verongiformis. J Org Chem 68(7):2765–2770. doi:10.1021/jo0267910

Webster NS, Blackall LL (2009) What do we really know about sponge-microbial symbioses? ISME J 3(1):1–3. doi:10.1038/ismej.2008.102

Webster NS, Negri AP, Munro MMHG, Battershill CN (2004) Diverse microbial communities inhabit antarctic sponges. Environ Microbiol 6(3):288–300. doi:10.1111/j.1462-2920.2004.00570.x

Webster NS, Taylor MW, Behnam F, Lücker S, Rattei T, Whalan S, Horn M, Wagner M (2010) Deep sequencing reveals exceptional diversity and modes of transmission for bacterial sponge symbionts. Environ Microbiol 12(8):2070–2082. doi:10.1111/j.1462-2920.2009.02065.x

Webster NS, Thomas T (2016) The sponge hologenome. mBio 7(2):e00135-16. doi:10.1128/mBio.00135-16

Webster NS, Wilson KJ, Blackall LL, Hill RT (2001) Phylogenetic diversity of bacteria associated with the marine sponge Rhopaloeides odorabile. Appl Environ Microbiol 67(1):434–444. doi:10.1128/AEM.67.1.434-444.2001

Weisz JB, Lindquist N, Martens CS (2007) Do associated microbial abundances impact marine demosponge pumping rates and tissue densities? Oecologia 155(2):367–376. doi:10.1007/s00442-007-0910-0

West LM, Northcote PT, Battershill CN (2000) Peloruside A: a potent cytotoxic macrolide isolated from the New Zealand marine sponge Mycale sp. J Org Chem 65(2):445–449. doi:10.1021/jo991296y

Wilkinson CR (1978a) Microbial associations in sponges. I. Ecology, physiology and microbial populations of coral reef sponges. Mar Biol 49(2):161–167. doi:10.1007/BF00387115

Wilkinson CR (1978b) Microbial associations in sponges. II. Numerical analysis of sponge and water bacterial populations. Mar Biol 49(2):169–176. doi:10.1007/BF00387116

Woyke T, Xie G, Copeland A, González JM, Han C, Kiss H, Saw JH et al (2009) Assembling the marine metagenome, one cell at a time. PLoS One 4(4):e5299. doi:10.1371/journal.pone.0005299

Zhang Y, Jun M, Feng Y, Kang Y, Zhang J, Peng-Juan G, Yu W, Ma L-F, Zhu Y-H (2009) Broad-spectrum antimicrobial epiphytic and endophytic fungi from marine organisms: isolation, bioassay and taxonomy. Mar Drugs 7(2):97–112. doi:10.3390/md7020097

Xie LW, Shu MJ, Hong HZ, Wei S, Yong CO, Shi KD, and Xiang L (2008) Potential inhibitors against sclerotinia sclerotiorum, produced by the fungus myrothecium Sp. Associated with the marine sponge axinella Sp. Eur J Plant Pathol 122(4):571–78. doi:10.1007/s10658-008-9326-x

Metagenomic Insights into Herbivore Gut: An Application-Based Perspective

12

Birbal Singh, Gorakh Mal, Dixit Sharma, Rinku Sharma, Philem S. Kapoor, Gagandeep Singh, and Sarbjit S. Kanwar

Abstract

'Herbivores possess a fascinating gut ecosystem explicitly evolved to convert plant biomass into high quality proteins and energy constituents. In post-genomic era, the herbivore gut microbiota is viewed as capable sources of precious enzymes, genes and therapeutic biomolecules. Culture-independent metagenomic approaches have proven to be useful to unravel and utilize the gut microbiota for nutrition, health and industrial applications'.

Keywords

Herbivores • Gut microbiome • Metagenomics • Plant biomass

12.1 Introduction

The enormous microbial diversity, including extremophiles, occupying a range of niches is largely an underutilized genetic and biological pool for obtaining novel genes, biocatalysts and therapeutics. The resident microbes inhabiting the

B. Singh (✉) • G. Mal • R. Sharma • P.S. Kapoor
ICAR-Indian Veterinary Research Institute, Regional Station, Palampur 176 061, India
e-mail: bsbpalampur@yahoo.co.in

D. Sharma • G. Singh
Centre for Computational Biology and Bioinformatics, School of Life Sciences, Central University of Himachal Pradesh, Shahpur 176 206, India

S.S. Kanwar
CSK HPKVV, Palampur 176 061, India

V.C. Kalia et al. (eds.), *Mining of Microbial Wealth and MetaGenomics*,
DOI 10.1007/978-981-10-5708-3_12

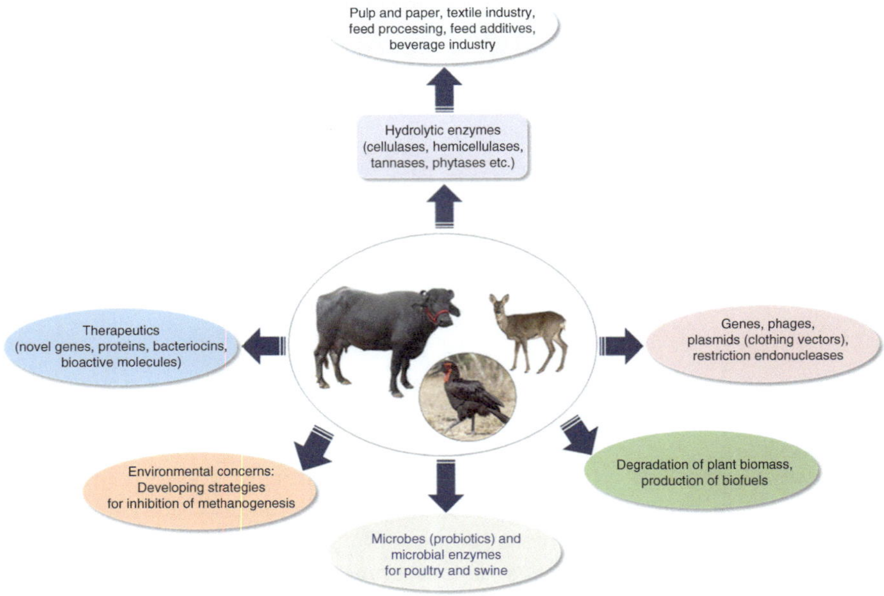

Fig. 12.1 A diagrammatic depiction of herbivore gut ecosystem as a source of regimented microbiome for digesting lignocellulosic plant biomass containing various anti-nutritional phytometabolites. Buffaloes and some feral ungulates efficiently digest fibrous forage and hence could be a promising source of fibrolytic enzymes and microorganisms for food, feed, textile and pulp industry. The herbivorous birds and goats due to their habit to consume multiple plant ingredients may harbour microbes for degrading phytometabolites

digestive tract of herbivores benefit the host in profound and diverse ways. The system-wide investigations into microbial communities specialized in lignocellulose degradation and detoxification of anti-nutritional phytometabolites have both basic and applied prospects (Fig. 12.1). Therefore, a thorough understanding of complex gut ecosystem is of interest to microbial ecologists, molecular biologists and industrialists.

The term metagenomics ('meta' Greek, for transcending; more comprehensive), which constitutes a challenging domain to discover enzymes, genes and metabolic pathways from diverse niches, was suggested by Handelsman et al. (1998) for investigating the genomes of whole microbial consortia. Metagenomics has been applied to analyse various microbial ecosystems, such as soil, compost, sewerage and marine microorganisms. This chapter discusses metagenomic mining of gut microbiota for novel enzymes and microorganisms with potential biotechnological applications. As methodology of metagenomics is described by other authors in this book, this chapter is restricted to insights and promising outcomes from metagenomic mining of herbivore gut microbiota.

12.2 Why Study the Herbivore Gut Ecosystem?

The fact that food and feed processing, textile and surfactants, biofuels and pharmaceutical industries need a sustained supply of enzymes has urged the researchers to discover microbial communities for their beneficial attributes. The gut microbiota of herbivores virtually serve as a 'metabolic organ' that provides the host with a wide range of metabolic capabilities such as degradation of complex plant biomass (White et al. 2014), production of short-chain volatile fatty acids, amino acids and microbial proteins from nonprotein nitrogen sources such as urea and uric acid. As culture-dependent microbiological studies are unable to explore the complex microbial species, genomic methodologies are developed to gain a complete blueprint of gut and other complex microbiomes and understanding the microbe-environment interactions.

The very basic purpose of metagenomics is to decipher the genome and traits of noncultivable microorganisms in an ecosystem and eventually exploring their use for industrial applications. The technique (Fig. 12.2), comprising of construction, screening and analysis of metagenomic DNA/RNA libraries, has grown at faster rate and now emerged as a powerful tool to discover biocatalysts, microorganisms and therapeutic biomolecules. At present, the metagenomics is applied to study extremophiles, marine microbial resources, soil and compost, gut, skin and genitourinary microbiota (Shanks et al. 2006; Singh et al. 2008; Yatsunenko et al. 2012; Kohl et al. 2016). Technical advances in basic and applied sciences, technologies construction of high-efficiency cloning vectors (cosmids, phosmids, bacterial artificial chromosomes, yeast artificial chromosomes (Box 12.1)) (Babcock et al. 2007; Xu 2006), allowing cloning and expression of larger eukaryotic genes, and development of powerful bioinformatics methods have entirely transformed the concept of metagenomics to a practical approach. Figure 12.2 summarizes the major technical aspects of a representative metagenomic analysis.

Box 12.1. The innovations in basic science and allied technologies that have advanced the field of metagenomic studies

1. Development of kits for *in situ* extraction of high purity intact gene frames from cells and microbes
2. Removal of contaminants from environmental samples
3. The use of pre-cultivation step to improve the representation of microbial communities in complex microbial ecosystem
4. Development of high-capacity gene cloning vectors such as bacterial artificial chromosomes (BACs) and yeast artificial chromosomes (YACs)
5. Availability of high-throughput, NGS technology to sequence whole genome of microbiota and development of high accuracy computer programs for analysis of data
6. Availability of data in the gene sequence data banks and their accessibility to users
7. Development of comprehensive functional gene arrays (GeoChips) that can be used for analysing functional diversity, community composition, structure and metabolic potential of microbial ecosystems

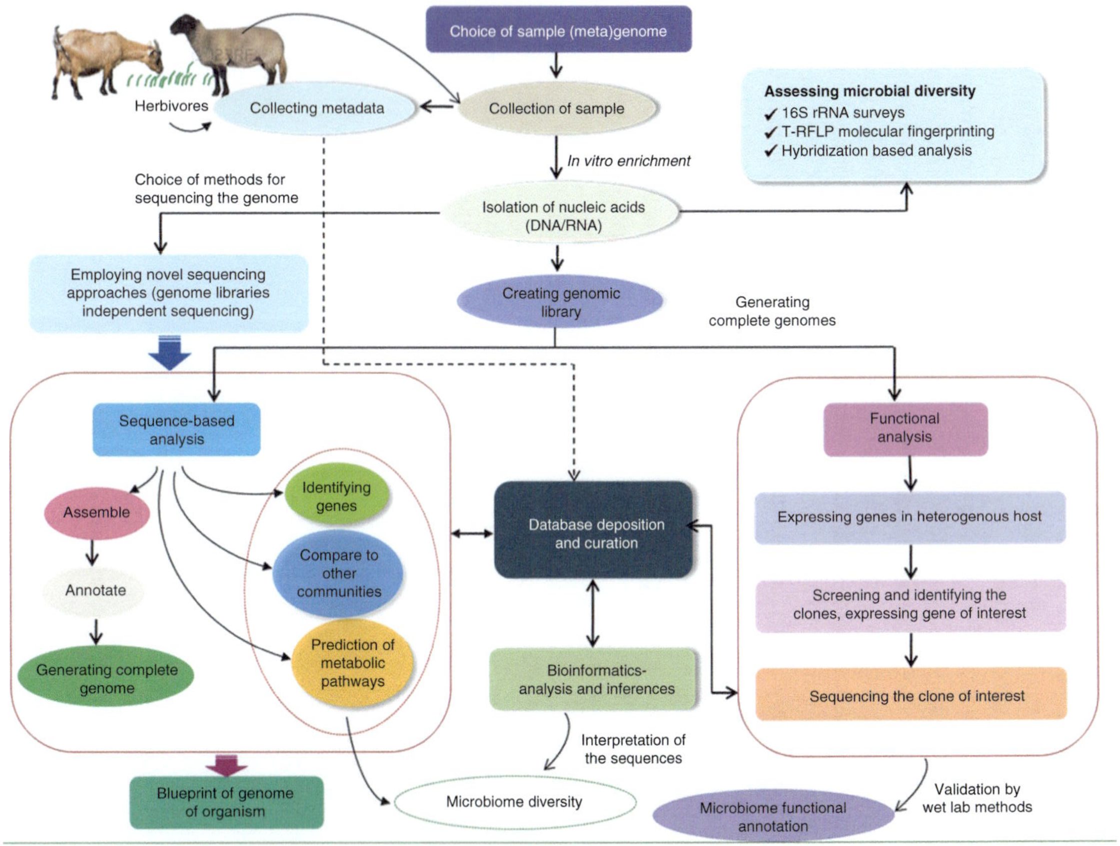

Fig. 12.2 A diagrammatic presentation of metagenomic profiling and utilization of herbivore GI microbiome. Metagenomics is a powerful tool to view structural and functional microbial diversity, and obtaining useful enzymes. The standard protocols may vary slightly depending on the requirements and methods used

12.3 Metagenome Sequence Data Analysis

Metagenomics involves isolation of genetic material directly from complex microbial communities, sequencing it and analysing the sequence database by bioinformatics and computer-assisted programs. A typical metagenomic analysis is based on two types of analyses, namely, sequence-based analysis (screening the metagenomic libraries for nucleotide sequences) and function-based analysis (screening the metagenome clones or libraries for a particular trait) or a combination of both depending on requirements.

12.3.1 Sequence-Based Analysis

Identifying potential enzymes or a biomolecule of interest in a metagenome based on sequence similarity is a rewarding strategy. This approach entails sequencing and analysis of genome or specific phylogenetic clusters (such as 16S rRNA) from a taxonomic group or particular microbial community (Margulies et al. 2005; Hall 2007). The whole genome sequencing (WGS) produces enormous data which needs reliable bioinformatic tools for analysis and interpretation. The classic projects may involve multiple samples and billions of sequence reads.

The metagenomes are analysed based on sequences corresponding to a particular activity or sequences already available in database. Various software packages have been developed for primary and secondary analysis of metagenomic sequence data. For example, a computer-aided program, DOTUR, was developed for studying operational taxonomic units and estimating the microbial species abundance in a microbial ecosystem (Schloss and Handelsman 2005). A series of MEGAN (MEtaGenome ANalyzers) programs, originally developed to provide as tool for investigating the taxonomic content of single dataset, are available (Huson et al. 2007, 2016). Several datasets, including Sargasso Sea data, data obtained from woolly mammoth (*Mammuthus primigenius*) bones, were analysed by using this program (Huson et al. 2007). The 'transposon-aided capture' (TRACA), a culture-independent tool, was developed for studying the mobile genetic elements, such as plasmids and plasmid-determined traits in human gut microbial communities (Jones and Marchesi 2007).

The web-based MG-RAST (Metagenomic Rapid Annotation using Subsystem Technology) facilitates processing, sharing and analysis of metagenomic data (Meyer et al. 2008).

Currently, a number of software packages are available, and many more are likely to come in the future (https://www.biostars.org/p/58279/, accessed August 29, 2016). The readers should refer to updates online for developments in software packages and bioinformatic tools for microbial diversity analysis, sequence similarity, functional annotation, mapping to reference genomes and quality analysis of metagenomic data.

12.3.2 Function-Driven Analysis

Enzymes are the backbone of industrial development. The industrial enzyme market valued at 4.2 billion USD in 2014 is envisioned to grow at annual growth rate of 7.0% from 2015 to 2020. Food and beverage, cleaning agents, consumer goods and biofuel and animal feed processing industry will need a sustained supply of potential biocatalysts. Bioengineering of microbial strains and computational enzyme engineering are in increasing demand to fill the gap in demand and supply of potential biocatalysts and therapeutics. However, the scientists should continue investigating microbial niches for enzymes for industries.

Identification of genes and the enzymes in a metagenomic data is of paramount interest. Visual screening methods are most widely used for selecting a particular gene of interest. The functional active screening is based on identifying the clones and characterization of positive clones by molecular or biochemical techniques. This facilitates identification of genes for functional applications such as production of pharmaceuticals, agriculture and industrially important products. However, this method has certain limitations including low expression of the cloned genes and, hence, requires additional strategies to improve the gene expression and detection of the products. Under certain circumstances, it becomes necessary to cluster the genes encoding a single protein.

Metagenomics has potential for bioprospecting the microbial diversity of gut ecosystem (Table 12.1). Potentially useful hydrolytic enzymes are from the metagenomic analysis of digestive tracts of domestic and feral herbivores (Table 12.2).

Table 12.1 Summarized presentation of prospects of metagenomics in herbivore gut ecosystem

Targets	Future prospects
Hydrolytic enzymes	Identification and characterization of enzymes for feed processing, textile, paper and pulp industry
	Enhancing the utility of plant biomass for animals such as poultry and swine
	Using hydrolytic enzymes for improving the quality of silages
Novel microbial species	Establishing new culture conditions for isolation of novel microflora for use as microbial feed supplements or direct-fed microbials (DFM)
	Unravelling the gut ecosystem of species (migratory goats and feral animals) exhibiting evolutionary adaptation to dietary resources containing anti-nutritional phytometabolites
Novel genes, enzymes and antimicrobials	Identifying genes, restriction endonucleases, DNA modifying enzymes and genetic elements as gene cloning vectors
	Identification of antimicrobial proteins, peptides (AMPs) and bacteriocins for use as inhibitors of gut pathogens in monogastric animals and as rumen modulators
Developing alternative feed systems	Utilizing fermentation and enzymes and 'omics' data for investigating the nutritive value of unconventional plants and weeds as alternative feeds
Methanogenesis	Studying the diversity of rumen methanogens and methane emissions in herbivores and evolving the strategies for mitigating enteric greenhouse gas (GHG) emissions

Table 12.2 A summary of enzymes and microbes identified from the digestive tract of various animals

Enzymes/microorganisms studied	Source	Remarks (references)
Rumen microbial ecosystem	Bovine rumen	Molecular analysis of complexity of rumen archaea is revealed (Lammle et al. 2007)
Acetylxylan esterase (R.4) family carbohydrate esterase (CE6) enzymes	Rumen	Description of novel hydrolytic enzymes (Beloqui et al. 2006)
Hybrid glycosylases	Cattle rumen	Identification of industrially relevant hydrolytic enzymes (Lopez-Cortes et al. 2007)
Alpha-amylase family (RA.04) enzyme	Cattle rumen	Detection and description of enzymes (Lan et al. 2006; Palackal et al. 2007)
Polyphenol oxidase encoding enzymes	Cattle rumen	Description of polyphenol oxidases and their use in industrial applications (Feng et al. 2007)
umcel3G, a gene encoding β-glucosidase	Bubaline rumen	Fermentative production of biofuel from plant biomass (Guo et al. 2008)
Low G + C bacteria and *Cytophaga-Flexibacter-Bacteroides* phyla	Guangxi buffalo rumen	Cellulose hydrolysis, abundance of cellulases in rumen microbes (Liu et al. 2009a, b)
β-glucosidases from fibre-adherent bacteria	Cattle rumen	Saccharification of lignocellulose (Del Pozo et al. 2012)
Novel cellulases	Buffalo rumen	Characterization and purification of enzyme expressed in *E. coli* for industrial applications (Duan et al. 2009)
Umbgl3B (β-glucosidase)	Rabbit cecum	Characterization of some hydrolytic enzymes (Feng et al. 2009)
Cel A, *xyl A* genes and their products	Cow rumen	Purification and characterization of enzymes Cel5 A and xyl A from the cattle rumen (Shedova et al. 2009)
Novel glycosyl hydrolases	Cow rumen	Identification of genes producing novel glycosyl hydrolases (Zhao et al. 2010)
Identification of genes responsible for producing glycoside hydrolases (*GH*)	Yak rumen	Detection of genes *GH5, GH9 and GH10* and their enzymes in yak rumen (Dai et al. 2012)
		The study envisions the role of rumen enzymes in degradation of lignocellulose
Multifunctional glycosyl hydrolase family (GHF)	Cattle rumen	The study explores the role of GHF in digestion of plant biomass in rumen (Ferrer et al. 2012)
Cellulase-encoding genes *cel5A* and *cel5B*, belonging to the glycosyl hydrolase family 5	Cow rumen	Identification and characterization (pH optima, substrate specificity, molecular mass) of enzymes encoded by these genes
		Possibility of using the bifunctional properties of enzymes in degrading β-1,4 bonds of cellulose and hemicellulose (Rashamuse et al. 2013)

(continued)

Table 12.2 (continued)

Enzymes/microorganisms studied	Source	Remarks (references)
Identification of a novel feruloyl esterase (FAE) gene	China Holstein cow rumen	Identification, cloning and expression of *FAE* gene encoding enzyme FAE-SH1 in *Escherichia coli* and its possible applications in plant biomass degradation (Cheng et al. 2012)
β-glucosidase/xylosidase (*RuBX1*) gene belonging to GHF 3 β-glucosidase/xylosidase family	Yak rumen	Cloning of gene in *Escherichia coli*, demonstration of its hydrolytic activity on various substrates (Zhou et al. 2012)
Glycosyl hydrolase family 5 (GHF5)	Buffalo rumen	Elucidation of GHF5 as predominant fibrolytic enzymes, possibility of utilizing the enzymes as feed supplements (Nguyen et al. 2012)
RlipE1 and *RlipE2* genes and their products	Cow rumen	Purification and characterization of recombinant lipases and their applications in rumen lipid metabolism (Liu et al. 2009a, b)
Lipolytic esterases	Sheep rumen	Identification and characterization of esterase genes using activity-based cluster screening approach (Bayer et al. 2010)
Lipases-encoding genes	Bovine rumen	Biochemical and genetic characterization of lipase/esterase/phospholipases with biotechnological importance (Privé et al. 2015)
Cellulase gene (*cel28a*)	Goat rumen	Molecular characterization of the gene and the enzymes produced, experimental validation of the enzyme activity (Cheng et al. 2016)
Antibiotic resistance and virulence factors	Buffalo rumen	Identification of repertoire of microbial genes associated with antibiotic resistance and bacterial virulence in rumen microbes (Singh et al. 2012)
Multiple recombinant family 26 glycohydrolase	Buffalo rumen	Characterization of multifunctional activities of the enzyme (Patel et al. 2016)
Novel methanogens	Cattle, sheep rumen	Identification of methanogens in rumen (Ferrer et al. 2007)
Fungal taxa	Murine GI tract	Elucidation of diverse fungal taxa and their role in the GI tract (Toyoda et al. 2009)
Protozoal glycoside hydrolases	Bovine rumen protozoa	Activity and characterization of the enzymes revealed (Findley et al. 2011)
Glycosyl hydrolases	*Rhinopithecus bieti* faecal microbiota	Abundance of a broad diversity of bacterial glycoside hydrolases with the ability to degrade lignocellulosic plant biomass (Xu et al. 2015)

Using functional metagenomic screening, some antibiotics and drug-resistance genes were identified (Gillespie et al. 2002; Diaz-Torres et al. 2003). In addition, gut and faecal microorganisms are also reported to be the promising sources of key enzymes.

Certain parts of GI tract (rumen and large intestine) are densely populated by anaerobic bacteria, archaea, fungi and some unicellular eukaryotic microbes (Flint et al. 2008). These regions are sites of the microbial metabolic activities and metabolites (short-chain fatty acids, amino acids, microorganisms) that are important for host. Hence, gut ecosystem of herbivores offers an inexhaustible source of microorganisms, genes and enzymes.

12.4 Insights from Gut Metagenomics

The herbivore gut, especially the rumen, is a naturally evolved ecosystem for microorganisms that are specialized in rapid hydrolysis of plant biomass for providing energy, proteins and other nutrients to the host. The rumen microbiota (bacteria, fungi, protozoa) mediate the digestion of fibrous plant cell wall and detoxification of some phytometabolites for producing short-chain fatty acids and microorganisms that are digested and assimilated in the intestine.

Rumen is a promising source of fibrolytic enzymes, restriction enzymes, gens and therapeutic microbiota and biomolecules (Singh et al. 2008; Hess et al. 2011; Prajapati et al. 2016). The next-generation sequencing (NGS), reference databases and bioinformatic tools have contributed to metagenomic analysis of rumen microbiome. The following areas are among the current hot topics of metagenomic investigations.

12.4.1 Deriving Commercially Important Enzymes

As stated above, the metagenomics have made remarkable advances in studying the complex microbial niches that may have important industrial prospects. Sequence analysis of the metagenomic expression library from cattle rumen revealed that 36% (8/22) gene sequences belonged to novel phylogenetic lineages (Ferrer et al. 2007). Polyfunctional glycosyl hydrolases of cattle and buffalo rumen may have applications in textile and pulp industry (Ferrer et al. 2007; Duan et al. 2010). The β-glucosidase from bubaline rumen metagenome may have applications in fermentative production of biofuel from lignocellulose (Guo et al. 2008).

The genomes of rumen origin fibrolytic bacteria, *viz.*, *Fibrobacter succinogenes*, *Ruminococcus albus* and *Prevotella ruminicola*, are already sequenced. These sequences can be used in the future for comparing the sequences from new rumen bacteria with similar metabolic characteristics (Nelson et al. 2003).

12.4.2 Processing the Feed for Improved Utilization

Metagenomics could be a valuable tool for identifying superior microbial strains, genes and microbial metabolic pathways for augmenting nutrient utilization. Feral herbivores or ungulates can utilize diverse forages without apparent adverse effects and may therefore be the sources of elite gut microbes to enhance utilization of fibrous grasses and unconventional forage.

High milk-yielding females during early lactation due to being in negative energy balance need energy-rich diets. DFM can be most suitable supplements to provide energy from roughage-based diets. Identification and use of lactate-utilizing bacteria as microbial supplements may have important implications when they are offered grain-enriched diets. At present, *Megasphaera elsdenii* is the major species known to utilize lactate in the rumen. Lactate-utilizing microbial feed supplements may help animals overcome acidosis.

12.4.3 Demand for Alternative Therapeutics

In the era emerging multiple antibiotic resistances among pathogens, it is important to evolve alternative therapeutic interventions that are effective and safer (Kumar et al. 2016; Singh et al. 2016). In probiotics and probiotic metabolites, viz., bacteriocins, bioactive peptides serve as potential antimicrobial and anticancer candidates. The bacteria with abilities to produce antimicrobial compounds (organic acids, hydrogen peroxide, diacetyl and antibiotics or antibiotic-like compounds) are ubiquitously distributed in all habitats ranging from plants to gut ecosystem. Some rumen bacteria produce antimicrobial peptides and bacteriocins that are in high demand for livestock health and food industrial applications.

At present, a large number of bacteriocinogenic microorganisms are identified from the gut. When used as alternatives to ionophores in the feedlot, the bacteriocins can improve the environmental sustainability of milk and meat production. As the bacteriocins are readily digestible by gastric enzymes, no residues are released into milk or meat. Rumen microbiota could be a promising source of AMPs and peptides. Metagenomics can identify new bacterial species for production of bacteriocins for use as dietary supplements to reduce faecal pathogens and as feed additive to promote animal growth. Genome mining of the sequences of 224 rumen bacteria and 5 rumen archaea showed a total of 46 bacteriocin gene clusters in 33 bacterial strains including *Streptococcus* sp. and *Ruminococcus albus* (Azevedo et al. 2015).

12.4.4 Environmental Concerns

Ruminants have environmental impact. Because of enteric, anaerobic microbial fermentation of fibrous forage, volatile fatty acids, H_2 and CO_2 are produced which after reduction produce CH_4. According to Intergovernmental Panel on Climate

Change Synthesis Report (2014), the CH_4 has a global warming potential 28-fold that of CO_2. Hence, lowering methane emissions is a major concern while planning for augmenting livestock rearing.

The metagenomic investigations may lead to identification of rumen methanogens (Nicholson et al. 2007; Firkins et al. 2007). Understanding the adaptation of the methanogenic archaea to dietary ingredients and cellular and molecular mechanisms of interaction between gut archaea and protozoa is a topic of interest to minimize methane emissions by rumen fermentation. Once ecology of the methanogens and the methane production pathways are identified, novel strategies to mitigate emissions, including dietary interventions, chemical and biological feed additives, chemogenomics and antimethane vaccines, may be developed (Wallace et al. 2015).

12.5 Outlook and Challenges

Metagenomics has contributed immensely to find diversity and novelty within microbial ecosystems and genomes of representative species therein. Metagenomics can be used to analyse microbial communities regardless of their ability to grow in vitro. As wild herbivores are evolutionarily more efficient in lignocellulose dietary ingredients, it is envisaged that compared to domestic species, the former possess strong hydrolytic enzymes.

The metagenomic analysis has certain bottlenecks that limit its wider applications. A major problem of metagenomic analysis is that it requires advanced instrumentation for genome sequencing, and the data generated is enormous. Cost and time of analysis of data generated by NGS are more compared to sequencing cost and time. Although some novel enzymes and catalytic functions are already reported, the majority of them are validated by standard culture-dependent microbiological methods.

In conclusion, the microbial communities in the GI tract of herbivores are of economic concern because of their important metabolic attributes. As culture-dependent microbiological techniques are incapable of describing the intact microbial consortia, metagenomics has increasing role in exploring gut ecosystem. The metagenomic studies should focus on investigating the potential microbiota and microbial mechanisms for producing biofuels and enhancing the utilization of fibrous plant biomass. Another important area for metagenomic interventions is evolving strategies to improve fermentation of fibrous roughage in vivo and minimize enteric methane emissions.

References

Azevedo AC, Bento CB, Ruiz JC, Queiroz MV, Mantovani HC (2015) Distribution and genetic diversity of bacteriocin gene clusters in rumen microbial genomes. Appl Environ Microbiol 81:7290–7304. doi:10.1128/AEM.01223-15

Babcock DA, Wawrik B, Paul JH, McGuinness L, Kerkhof LJ (2007) Rapid screening of a large insert BAC library for specific 16S rRNA genes using TRFLP. J Microbiol Methods 71:156–161

Bayer S, Kunert A, Ballschmiter M, Greiner-Stoeffele T (2010) Indication for a new lipolytic enzyme family: isolation and characterization of two esterases from a metagenomic library. J Mol Microbiol Biotechnol 18:181–187. doi:10.1159/000315459

Beloqui A, Pita M, Polaina J, Martinez-Arias A, Golyshina OV, Zumarraga M (2006) Novel polyphenol oxidase mined from a metagenome expression library of bovine rumen: biochemical properties, structural analysis, and phylogenetic relationship. J Biol Chem 281:22933–22942

Cheng F, Sheng J, Cai T, Jin J, Liu W, Lin Y, Du Y, Zhang M, Shen L (2012) A protease-insensitive feruloyl esterase from China Holstein cow rumen metagenomic library: expression, characterization, and utilization in ferulic acid release from wheat straw. J Agric Food Chem 60:2546–2553. doi:10.1021/jf204556u

Cheng J, Huang S, Jiang H, Zhang Y, Li L, Wang J, Fan C (2016) Isolation and characterization of a non-specific endoglucanase from a metagenomic library of goat rumen. World J Microbiol Biotechnol 32(1):12. doi:10.1007/s11274-015-1957-4

Dai X, Zhu Y, Luo Y, Song L, Liu D, Liu L, Chen F, Wang M, Li J, Zeng X, Dong Z, Hu S, Li L, Xu J, Huang L, Dong X (2012) Metagenomic insights into the fibrolytic microbiome in yak rumen. PLoS One 7:e40430. doi:10.1371/journal.pone.0040430

Del Pozo MV, Fernández-Arrojo L, Gil-Martínez J, Montesinos A, Chernikova TN, Nechitaylo TY, Waliszek A, Tortajada M, Rojas A, Huws SA, Golyshina OV, Newbold CJ, Polaina J, Ferrer M, Golyshin PN (2012) Microbial beta-glucosidases from cow rumen metagenome enhance the saccharification of lignocellulose in combination with commercial cellulase cocktail. Biotechnol Biofuels 5:73. doi:10.1186/1754-6834-5-73

Diaz-Torres ML, McNab R, Spratt DA, Villedieu A, Hunt N, Wilson M, Mullany P (2003) Novel tetracycline resistance determined from the oral metagenome. Antimicrob Agents Chemother 47:1430–1432

Duan CJ, Liu JL, Wu X, Tang JL, Feng JX (2010) Novel carbohydrate-binding module identified in a ruminal metagenomic endoglucanase. Appl Environ Microbiol 76:4867–4870. doi:10.1128/AEM.00011-10

Duan CJ, Xian L, Zhao GC, Feng Y, Pang H, Bai XL, Tang JL, Ma QS, Feng JX (2009) Isolation and partial characterization of novel genes encoding acidic cellulases from metagenome of buffalo rumens. J Appl Microbiol 107:245–256. doi:10.1111/j.1365-2672.2009.04202.x

Feng Y, Duan CJ, Liu L, Tang JL, Feng JX (2009) Properties of a metagenome-derived beta-glucosidase from the contents of rabbit cecum. Biosci Biotechnol Biochem 73:1470–1473

Feng Y, Duan CJ, Pang H, Mo XC, Wu CF, Yu Y, Hu YL, Wei J, Tang JL, Feng JX (2007) Cloning and identification of novel cellulase genes from uncultured microorganisms in rabbit caecum and characterization of the expressed cellulases. Appl Microbiol Biotechnol 75:319–328

Ferrer M, Ghazi A, Beloqui A, Vieites JM, López-Cortés N, Marín-Navarro J, Nechitaylo TY, Guazzaroni ME, Polaina J, Waliczek A, Chernikova TN, Reva ON, Golyshina OV, Golyshin PN (2012) Functional metagenomics unveils a multifunctional glycosyl hydrolase from the family 43 catalysing the breakdown of plant polymers in the calf rumen. PLoS One 7:e38134. doi:10.1371/journal.pone.0038134

Ferrer M, Beloqui A, Golyshina OV, Plou FJ, Neef A, Chernikova TN (2007) Biochemical structure features of a novel cyclodextrinase from cow rumen metagenome. Biotechnol J 2:207–213

Findley SD, Mormile MR, Sommer-Hurley A, Zhang XC, Tipton P, Arnett K, Porter JH, Kerley M, Stacey G (2011) Activity-based metagenomic screening and biochemical characterization of bovine ruminal protozoan glycoside hydrolases. Appl Environ Microbiol 77:8106–8113. doi:10.1128/AEM.05925-11

Firkins JL, Yu Z, Morrison M (2007) Ruminal nitrogen metabolism: perspectives for integration of microbiology and nutrition for dairy. J Dairy Sci 90(Suppl 1):E1–E16

Flint HJ, Bayer EA, Rincon MT, Lamed R, White BA (2008) Polysaccharide utilization by gut bacteria: potential for new insights from genomic analysis. Nat Rev Microbiol 6:121–131. doi:10.1038/nrmicro1817

Gillespie DE, Brady SF, Bettermann AD, Cianciotto NP, Liles MR, Rondon MR (2002) Isolation of antibiotics turbomycin A and B from a metagenome library of soil microbial DNA. Appl Environ Microbiol 68:4301–4306

Guo H, Feng Y, Mo X, Duan C, Tang J, Feng J (2008) Cloning and expression of beta-glucosidase gene umcel3G from metagenome of buffalo rumen and characterization of the translated product. Sheng Wu Gong Cheng Xue Bao 24:232–8 (Article in Chinese, abstract in English)

Hall N (2007) Advanced sequencing technologies and their wider impact in microbiology. J Exp Biol 210:1518–1525

Handelsman J, Rondon MR, Brady SF, Clardy J, Goodman RM (1998) Molecular biological access to the chemistry of unknown soil microbes: a new frontier for natural products. Chem Biol 5:R245–R249

Hess M, Sczyrba A, Egan R, Kim TW, Chokhawala H, Schroth G, Luo S, Clark DS, Chen F, Zhang T, Mackie RI, Pennacchio LA, Tringe SG, Visel A, Woyke T, Wang Z, Rubin EM (2011) Metagenomic discovery of biomass-degrading genes and genomes from cow rumen. Science 331:463–467. doi:10.1126/science.1200387

Huson DH, Auch AF, Qi J, Schuster SC (2007) MEGAN analysis of metagenomic data. Genome Res 17:377–376

Huson DH, Beier S, Flade I, Górska A, El-Hadidi M, Mitra S, Ruscheweyh HJ, Tappu R (2016) MEGAN community edition - interactive exploration and analysis of large-scale microbiome sequencing data. PLoS Comput Biol 12(6):e1004957. doi:10.1371/journal.pcbi.1004957

Jones BV, Marchesi JR (2007) Transposon-aided capture (TRACA) of plasmids resident in the human gut mobile metagenome. Nat Methods 4:55–61

Kohl KD, Connelly JW, Dearing MD, Forbey JS (2016) Microbial detoxification in the gut of a specialist avian herbivore, the Greater Sage-Grouse. FEMS Microbiol Lett 363(14):fnw144. doi:10.1093/femsle/fnw144

Kumar M, Yadav A, Verma V, Singh B, Mal G, Nagpal R, Hemalatha R (2016) Bioengineered probiotics as a new hope for health and diseases: potential and prospects: an overview. Future Microbiol 11:585–600. doi:10.2217/fmb.16.4

Lammle K, Zipper H, Breuer M, Hauer B, Buta C, Brunner H, Rupp S (2007) Identification of novel enzymes with different hydrolytic activities by metagenome expression cloning. J Biotechnol 127:575–592

Lan PT, Sakamoto M, Sakata S, Benno Y (2006) *Bacteroides barnesiae* sp. nov., *Bacteroides salanitronis* sp. nov., and *Bacteroides gallinarum* sp. nov., isolated from chicken caecum. Int J Syst Evol Microbiol 56:2853–2859

Liu K, Wang J, Bu D, Zhao S, McSweeny CS, Yu P, Li D (2009a) Isolation and biochemical characterization of two lipases from a metagenomic library of China Holstein cow rumen. Biochem Biophys Res Commun 385:605–611. doi:10.1016/j.bbrc.2009.05.110

Liu L, Tang J, Feng J (2009b). Bacterial diversity in Guangxi buffalo rumen. Wei Sheng Wu Xue Bao 49:251–256 (Article in Chinese, abstract in English)

Lopez-Cortes N, Reyes-Duarte D, Beloqui A, Polaina J, Ghazi I, Golyshina OV (2007) Catalytic role of conserved HQGE motif in the CE6 carbohydrate esterase family. FEBS Lett 581:4657–4662

Margulies M, Egholm M, Altman WE, Attiya S, Bader JS, Bemben LA (2005) Genome sequencing in microfabricated high-density picolitre reactions. Nature 437:376–380

Meyer F, Paarmann D, D'Souza M, Olson R, Glass EM, Kubal M, Paczian T, Rodriguez A, Stevens R, Wilke A, Wilkening J, Edwards RA (2008) The metagenomics RAST server - a public resource for the automatic phylogenetic and functional analysis of metagenomes. BMC Bioinformatics 9:386. doi:10.1186/1471-2105-9-386

Nelson KE, Zinder SH, Hance I, Burr P, Odongo D, Wasawo D, Odenyo A, Bishop R (2003) Phylogenetic analysis of the microbial populations in the wild herbivore gastrointestinal tract: insights into an unexplored niche. Environ Microbiol 5:1212–1220

Nguyen NH, Maruset L, Uengwetwanit T, Mhuantong W, Harnpicharnchai P, Champreda V, Tanapongpipat S, Jirajaroenrat K, Rakshit SK, Eurwilaichitr L, Pongpattanakitshote S (2012) Identification and characterization of a cellulase-encoding gene from the buffalo rumen metagenomic library. Biosci Biotechnol Biochem 76:1075–1084

Nicholson MJ, Evans PN, Joblin KN (2007) Analysis of methanogens diversity in the rumen using temporal temperature gradient gel electrophoresis: identification of uncultured methanogens. Microb Ecol 54:141–150

Palackal N, Lyon CS, Zaidi S, Luginbuhl P, Dupree P, Goubet F (2007) A multifunctional hybrid glycosyl hydrolase discovered in an uncultured microbial consortium from ruminant gut. Appl Microbiol Biotechnol 74:113–124

Patel AB, Patel AK, Shah MP, Parikh IK, Joshi CG (2016) Isolation and characterization of novel multifunctional recombinant family 26 glycoside hydrolase from Mehsani buffalo rumen metagenome. Biotechnol Appl Biochem 63:257–265. doi:10.1002/bab.1358

Prajapati VS, Purohit HJ, Raje DV, Parmar N, Patel AB, Jones OA, Joshi CG (2016) The effect of a high-roughage diet on the metabolism of aromatic compounds by rumen microbes: a metagenomic study using Mehsani buffalo (*Bubalus bubalis*). Appl Microbiol Biotechnol 100:1319–1331. doi:10.1007/s00253-015-7239-0

Privé F, Newbold CJ, Kaderbhai NN, Girdwood SG, Golyshina OV, Golyshin PN, Scollan ND, Huws SA (2015) Isolation and characterization of novel lipases/esterases from a bovine rumen metagenome. Appl Microbiol Biotechnol 99:5475–5485. doi:10.1007/s00253-014-6355-6

Rashamuse KJ, Visser DF, Hennessy F, Kemp J, Roux-van der Merwe MP, Badenhorst J, Ronneburg T, Francis-Pope R, Brady D (2013) Characterisation of two bifunctional cellulase-xylanase enzymes isolated from a bovine rumen metagenome library. Curr Microbiol 66:145–151. doi:10.1007/s00284-012-0251-z

Schloss PD, Handelsman J (2005) Introducing DOTUR, a computer program for defining operational taxonomic units and estimating species richness. Appl Environ Microbiol 71:1501–1506

Shanks OC, Santo Domingo JW, Lamendella R, Kelty CA, Graham JE (2006) Competitive metagenomic DNA hybridization identifies host-specific microbial genetic markers in cow faecal samples. Appl Environ Microbiol 72:4054–4060

Shedova EN, Lunina NA, Berezina OV, Zverlov VV, Schwarz V, Velikodvorskaia GA (2009) Expression of the genes *celA* and *XylA* isolated from a fragment of metagenomic DNA in *Escherichia coli*. Mol Gen Mikrobiol Virusol 2:28–32 (Article in Russian, abstract in English)

Singh B, Gautam SK, Verma V, Kumar M, Singh B (2008) Metagenomics in animals gastrointestinal tract-potential biotechnological prospects. Anaerobe 14:138–144. doi:10.1016/j.anaerobe.2008.03.002

Singh B, Mal G, Bissi L, Marotta F (2016) The holy grail of designer probiotics: the probiotics with multiple health benefits. J Gastrointest Digestive Syst 6:2. doi:10.4172/2161-069X.1000415

Singh KM, Jakhesara SJ, Koringa PG, Rank DN, Joshi CG (2012) Metagenomic analysis of virulence-associated and antibiotic resistance genes of microbes in rumen of Indian buffalo (*Bubalus bubalis*). Gene 507:146–151. doi:10.1016/j.gene.2012.07.037

Toyoda A, Iio W, Mitsumori M, Minato H (2009) Isolation and identification of cellulose binding proteins from sheep rumen contents. Appl Environ Microbiol 75:1667–1673. doi:10.1128/AEM.01838-08

White BA, Lamed R, Bayer EA, Flint HJ (2014) Biomass utilization by gut microbiomes. Annu Rev Microbiol 68:279–296. doi:10.1146/annurev-micro-092412-155618

Xu B, Xu W, Li J, Dai L, Xiong C, Tang X, Yang Y, Mu Y, Zhou J, Ding J, Wu Q, Huang Z (2015) Metagenomic analysis of the *Rhinopithecus bieti* fecal microbiome reveals a broad diversity of bacterial and glycoside hydrolase profiles related to lignocellulose degradation. BMC Genomics 16:174. doi:10.1186/s12864-015-1378-7

Xu J (2006) Microbial ecology in the age of genomics and metagenomics: concepts, tools, and recent advances. Mol Ecol 15:1713–1731

Yatsunenko T, Rey FE, Manary MJ, Trehan I, Dominguez-Bello MG, Contreras M, Magris M, Hidalgo G, Baldassano RN, Anokhin AP, Heath AC, Warner B, Reeder J, Kuczynski J, Caporaso JG, Lozupone CA, Lauber C, Clemente JC, Knights D, Knight R, Gordon JI (2012) Human gut microbiome viewed across age and geography. Nature 486:222–227. doi:10.1038/nature11053

Wallace RJ, Rooke JA, McKain N, Duthie CA, Hyslop JJ, Ross DW, Waterhouse A, Watson M, Roehe R (2015) The rumen microbial metagenome associated with high methane production in cattle. BMC Genomics 16:839. doi:10.1186/s12864-015-2032-0

Zhao S, Wang J, Bu D, Liu K, Zhu Y, Dong Z, Yu Z (2010) Novel glycoside hydrolases identified by screening a Chinese Holstein dairy cow rumen-derived metagenome library. Appl Environ Microbiol 76:6701–6705. doi:10.1128/AEM.00361-10

Zhou J, Bao L, Chang L, Liu Z, You C, Lu H (2012) Beta-xylosidase activity of a GH3 glucosidase/xylosidase from yak rumen metagenome promotes the enzymatic degradation of hemicellulosic xylans. Lett Appl Microbiol 54:79–87. doi:10.1111/j.1472-765X.2011.03175.x

Soil Metagenomics: A Tool for Sustainable Agriculture

13

Reeta Goel, Deep Chandra Suyal, Narayan, Biplab Dash, and Ravindra Soni

Abstract

Microorganisms are considered as an unlimited, untapped, and intriguing source for development of novel genes, antibiotics, bioinoculants, biocatalysts, etc. Therefore, the identification and characterization of the microbial wealth becomes inevitable. Besides other features of bacteria that are a challenge to health, including increase in pathogenicity, antibiotic resistance, mounting host spectra, and possibility of usage for bioterrorism, it becomes a compulsion. Furthermore, a sustainable agriculture development program that frequently exploits the plant growth-promoting rhizobacteria (PGPR) also requires a correct identification and characterization of agriculturally important microorganisms. Metagenomics is defined as the study of all available genomes in an environment that contains considerably more genetic information than provided by the cultured subset. It also helps in bridging the gap between genetics and ecology, indicating that the genes of a single microorganism are connected to the genes of other members of the community.

Keywords

Metagenomics • Agriculture • Himalayan Soil

R. Goel • D.C. Suyal
Department of Microbiology, CBSH, G. B. Pant University of Agriculture and Technology, Pantnagar 263 145, Uttaranchal, India

Narayan • B. Dash • R. Soni (✉)
Department of Agricultural Microbiology, College of Agriculture, Indira Gandhi Krishi Vishwavidyalaya, Raipur, Chhattisgarh, India
e-mail: rs31693@gmail.com

V.C. Kalia et al. (eds.), *Mining of Microbial Wealth and MetaGenomics*,
DOI 10.1007/978-981-10-5708-3_13

13.1 Introduction

Soil biodiversity analysis is a very important aspect in the environmental sciences due to its significant interlinkages with other areas, like agriculture. As evident from available literature, soil biodiversity characterizes a huge underground world that contains a wide range of organisms, from prokaryotes to eukaryotes (e.g., archaea, bacteria, fungi, nematodes, insects, and earthworms). One gram of soil has been reported to contain up to 10^{10} prokaryotic cells and thousands of different species (Raynaud and Nunan 2014). This diverse microbial ecosystem plays a central role in the nutrient cycling, soil structure formation, decomposition of organic matter, soil health indicator, soilborne diseases, and plant growth promotion and thus is responsible for maintaining the biosphere integrity. Microbial diversity present in soil can be explored either through culture-based methods or recent indirect biotechnological approaches. Many indirect methods for overcoming the limitations of cultivation techniques are developed which are mainly based on nucleic acid isolation from soil and their characterization without the culture of microbes.

13.2 Metagenomics

Culture-dependent methods limit analysis of those microorganisms that can grow under laboratory conditions. It is widely accepted that only 0.1–1% (depending upon the environmental sample) of bacteria can be cultured by laboratory cultivation methods which leaves 99% or more microbial diversity unexplored. Furthermore, under environmental stress bacteria can enter a state called "viable but unculturable" which again limits the accessibility of these bacteria to traditional cultivation method. Thus, cultivation-dependent microbial identification can underestimate the microbial diversity. As per the available reports, in vitro culturability of the total microbes in freshwater and in sediments is 0.25% (Jones 1977), in sea water 0.001–0.1% (Surmann and Efferth 2014), and only less than 1% in soil (Pham and Kim 2012; Ferrari et al. 2005). Therefore, "metagenomics," i.e., direct extraction of genetic material from environment, was conceptualized (Handelsman et al. 1998) for analyzing similar but not identical genomes, in the environment.

Metagenomics may target the structure of the metagenome by cloning and sequencing strategies (structural metagenomics) and/or characterize the functions of environmental DNA by direct cloning for heterologous expression in a surrogate host organism (functional metagenomics). As the functional metagenomics method relies on the ability of the cloned environmental DNA to confer a phenotypic function to the host, no sequence homology to previously characterized genes or other a priori sequence information is required. Functional metagenomics can therefore be considered as a true discovery tool for identifying and characterizing novel gene families (Nacke et al. 2011), metabolic traits (McGarvey et al. 2012), bioactive compounds (Craig et al. 2010), or pathways (Illeghems et al. 2015) from uncultured soil microbes. With suitably long genomic fragments, functional metagenomics may also be used to define the genomic context of the functions of interest and enable their taxonomic assignment (Treusch et al. 2005).

13.3 Essential Steps of Metagenomics

13.3.1 Environmental Sampling

Sample collection is the first step of a soil metagenomic library construction for which details of physicochemical properties of the soils are required. Sometimes, enzyme activity assays or metagenomic sequencing can also be used to determine the functional diversity of the respective environment.

13.3.2 Metagenomic Library Construction

13.3.2.1 High-Quality DNA Isolation from Soil Samples

Metagenomic library construction requires a high-quality DNA from the environment. Soil heterogeneity, microbial diversity, and other soil properties make DNA extraction challenging. Moreover, soil DNA extraction often contains humic substances, which interfere and reduce the efficiency of downstream processes (Premalatha et al. 2009; Soni and Goel 2010).

To ensure efficient cloning, the isolated DNA should be purified of contaminants such as humic acids or phenolic compounds that can inhibit enzymatic reactions. The capture of full-length genes using small-insert libraries requires DNA fragments of at least 2 kb in length, whereas over 25 kb fragments are required for identification of operons using cosmid and fosmid libraries. While cell lysis directly in the soil matrix enables rapid recovery of greater amounts of DNA, indirect extraction methods typically recover DNA with larger fragment sizes, higher purity, and higher representation of many bacterial and archaeal taxa but lower representation of filamentous organisms such as fungi and *Actinobacteria* and microbes attached to soil matrix (Delmont et al. 2011).

13.3.2.2 Cloning Vector

Metagenomic libraries can be roughly categorized into small- and large-insert libraries as per the cloned DNA fragment size. Using plasmid vectors, construction of small-insert libraries is made which is able to contain up to 10 kb DNA fragments, thus making them suitable for identification of functional traits encoded by a single gene or small operon (Surmann and Efferth 2014). These plasmid-based libraries provide high transformation efficiency (>10^5 clones) and efficient expression systems using vectors that can be induced to high copy numbers and that contain promoters. Higher copy number is especially advantageous for characterization of genes, cloned without promoters or with low activity. Large-insert libraries use cosmid or fosmid vectors or bacterial artificial chromosomes (BAC) that can accommodate 25–35, 25–40, or 100–200 kb DNA fragments, respectively. Due to the large size of the cloned DNA fragments, these libraries are well suited for identification of multi-domain traits or pathways and provide linkage information for the identification of functions encoded by multiple genes and potentially allow taxonomic linkages to be determined. The vectors used to construct large-insert libraries are generally present in cells in low copy number, which allows their stable replication in the screening host and reduces the risk of overexpression of toxic gene products.

13.3.2.3 Screening Host

In the majority of applications to date, the functional screening of metagenomic libraries relies on expression in *Escherichia coli*. This well-characterized laboratory model organism has stable replication of vectors and low rates of restriction and recombination, making an attractive host microorganism for cloning and expression of foreign DNA. A variety of expression systems for *E. coli* are available as are a large number of genetically modified *E. coli* strains for highly controlled and optimized cloning and expression (Sørensen and Mortensen 2005). Alternative screening hosts for functional mining of metagenomic libraries include *Streptomyces lividans* (Wang et al. 2000), *Bacillus subtilis* (Troeschel et al. 2010), *Sulfolobus solfataricus* (Albers et al. 2006), *Thermus thermophilus* (Angelov et al. 2009), *Saccharomyces cerevisiae* (Bailly et al. 2007), etc. When using these hosts, the library is generally constructed and maintained in *E. coli* and transformed to the alternative expression hosts for screening (Craig et al. 2010).

13.3.3 Library Screening

Metagenomics libraries can be screened by several techniques based either on functional activity or on nucleotide sequence. Soil-metagenomic libraries are screened by using target-specific probes. This approach is being used extensively to identify phylogenetic markers as well as other functional genes with highly conserved domains (Premalatha et al. 2009; Soni and Goel 2010, 2011). Moreover, microarray technology is also useful for soil metagenome analysis.

Enzymatic function of clones can be monitored by adding chemical dyes or chromophore-bearing enzyme substrate derivatives into culture medium. Thus, this type of sensitive nature of screening helps in the detection of rare clones.

13.4 Metagenomics of Himalayan Soils

The entire Himalayan mountain range is well known for its biodiversity due to its unique environment. Scattered habitations, inaccessibility, and uneconomic holdings keep these regions free from any anthropogenic contamination. Nevertheless, preference for traditional farming system over chemical-based farming is responsible for emergence of hilly agriculture lands as a gold mine for potential soil microorganisms.

In Uttarakhand Himalaya perspective, a triphasic approach, viz., real-time PCR (qPCR), denaturation gradient gel electrophoresis (DGGE), and temporal gradient gel electrophoresis (TGGE), has been used for evaluation of bacterial population in different rhizospheric soil systems (Soni and Goel 2010). Moreover, several *nifH* homologs have been identified from Himalayan rhizospheric soil metagenome (Soni and Goel 2011). Recently, Goel and co-workers made two 16S rDNA clone libraries, i.e., SB1 and SB2, using rhizospheric soil samples from two different locations of Western Indian Himalaya, namely, Chhiplakot (30.70°N/80.30°E) and Munsyari (30.60°N/80.20°E), selected on the basis of qPCR analysis to characterize the total bacterial population and their community structure (Suyal et al. 2015a). The phylum *Proteobacteria* was the dominant phylum in the Himalayan soils along with *Bacteroidetes*, *Nitrospira*, *Acidobactria*, *Chloroflexi*, *Firmicutes*, *Cyanobacteria*, *Gemmatimonadetes*, *Planctomycetes*, *BRC1*, *Actinobacteria*, and *Chlorobi.* Comparative study on the bacterial diversity observed in this region with that of other Himalayan cold habitats like Tibetan plateau glacier (Liu et al. 2009), Drass, cold desert of the Western Himalaya (Shivaji et al. 2011), Puruogangri ice (Zhang et al. 2008), and Roopkund glacier (Pradhan et al. 2010) indicated that the bacterial diversity in both soils was comparable with each other.

13.5 Metagenomics as a Tool for Sustainable Agriculture

An integral constituent of integrated nutrient management (INM) and soil biodiversity system (SBS) is the soil-inhabiting microflora and microfauna, thereby playing an important role in plant growth and all-round development. Hazardous effects of chemical fertilizers and pesticides on soil and plant health having deleterious environmental impact are being frequently observed in recent years. Beneficial microbial wealth of agricultural importance can serve as a crucial alternative for achieving sustainable agriculture production. Metagenomics help in the prediction of microbial community structure and, therefore, can tackle and address fundamental scientific questions related to agriculturally important microorganisms. This approach has been successfully explored for the assessment of the diazotrophs belonging to the rhizosphere of native red kidney beans (RKB) of the Western Indian Himalaya by targeting *nif*H (Suyal et al. 2015b). This metagenomic effort has examined the community structure and diversity of N_2-fixing microorganisms in a Himalayan RKB rhizosphere, which can be explored to provide the backbone for further studies. Moreover, previous metagenomic efforts indicated the pervasiveness of *csp* and *nif* from the Himalayan soils of (Premalatha et al. 2009; Soni and Goel 2010).

The emergence of metagenome information for a rhizosphere is beginning to expose detailed information about associated community structure, dynamics, and functional activities, thereby allowing an improved perceptive of community development, interspecies coordination and competition for essential nutrients, and distribution of metabolic activities across the community members. Moreover, functional metagenomics can also be explored for reshaping the composition of rhizospheric microbial population and to readdress microbial activity, which can be referred to as "rhizosphere engineering."

13.6 Current Scenario of Metagenomics

Metagenomics has significant potential as a discovery and annotation tool for linking genes with functions and processes and providing valuable phylogenetic context that may enable the role and ecological niches of soil microorganisms to be determined. However, in order to keep up with advances in sequencing technologies and to enable discovery from the broad diversity of soil microorganisms, there are challenges to be faced. An important development would be to increase the taxonomic coverage of genes that can be expressed and screened. Most of the currently used expression vectors replicate only in *E. coli*. Given the limited ability of *E. coli* to express genes from distant taxonomic groups of organisms, shuttle vectors with extended host range are needed (Aakvik et al. 2009; Craig et al. 2010). Maintenance of these vectors is conveniently done inside *E. coli*, which can then be transferred by conjugation into another expression host for repeated screening. It will also be necessary to modify the structure and expression mechanisms of the vectors in order to accommodate larger DNA fragments, to enable expression from different orientations depending on the orientation of cloned DNA insert, and to control the gene expression level via copy number adjustment and induction (Lämmle et al. 2007). Genetic modification of existing expression hosts by ribosome engineering, co-expression of molecular chaperones, or engineering of transcription and translation factors as well as secretion systems represent a means to increase the rate of expression of foreign genes in *E. coli* (Bernstein et al. 2007). However, screening in taxonomically distant expression hosts or the use of multiple different hosts has been shown to significantly increase detection frequency for novel functional traits. Screening in a physiologically suitable host is especially useful when mining extreme environments for novel biochemical properties because proteins that function at extreme temperatures or under high salinity often require additional modifications to ensure protein stability (Angelov et al. 2009). Ideally, cell-free expression systems for universal expression of DNA originating from taxonomically different hosts could be used.

Eukaryotic genes are still in minority among the discovered genes derived from environmental DNA, although construction of functional metagenomic libraries from soil transcriptomes, i.e., total RNA isolated from a soil sample, may be used to overcome limitations in mining eukaryotic genes containing nonbacterial genetic elements and introns. Messenger RNA is captured with polyadenylated primers and reverse-transcribed to double-stranded cDNA. This cDNA is then cloned to an expression vector for expression and functional screening in either bacterial or eukaryotic hosts. Although cDNA libraries have been successfully used for identification of fungal genes from soil metatranscriptomes, RNA instability and challenges in RNA isolation often result in low recovery of full-length transcripts (Bailly et al. 2007). The separate cloning of single transcripts also limits the recovery of entire biosynthetic pathways. Regardless of the type of methodology used for library preparation or expression screening, it is obvious that the scale of a screening effort that is required to capture less abundant microbial groups and functions, and to maintain a rate of functional annotation consistent with that of metagenome sequence data acquisition, greatly exceeds the capacity of current functional metagenomic implementations. This capacity could be increased by development of more sensitive screening substrates and by increasing the throughput of screening assays. Increased assay sensitivity and throughput could be achieved through the combination of novel substrates and multiplexed assays for analysis using FACS, high-throughput liquid chromatography, or mass spectrometry, with miniaturized systems using microfluidic devices to enable nanoliter reaction volumes to improve throughput and speed and substantially reduce cost.

Conclusions

The importance of the microorganisms in the growth and development of plant ecosystems has been well known; however, the major portion of rhizosphere population is still uncharacterized and unexplored. Coupling traditional with advanced metagenomic methodologies to evaluate community structure and function will bring new insights to explore microbial life in the soil. Further, identification of the plant signals, exudates, and key factors in the rhizosphere microbial ecosystem will provide chemical and microbial markers to explain how plants recruit and stimulate beneficial microorganisms. Moreover, soil metagenomics also holds prospective to improve crop production and to uncover several yet unexploited soil microorganisms, their functions, and genes for diverse applications.

Acknowledgment The work mentioned in this chapter from author group was supported by the National Bureau of Agriculturally Important Microorganisms, India (NBAIM/ICAR), grant to R.G.

References

Aakvik T, Degnes KF, Dahlsrud R, Schmidt F, Dam R, Yu L, Völker U, Ellingsen TE, Valla S (2009) A plasmid RK2–based broad–host–range cloning vector useful for transfer of metagenomic libraries to a variety of bacterial species. FEMS Microbiol Lett 296:149–158. doi:10.1111/j.1574-6968.2009.01639.x

Albers SV, Jonuscheit M, Dinkelaker S, Urich T, Kletzin A, Tampe R, Driessen AJM, Schleper C (2006) Production of recombinant and tagged proteins in the hyperthermophilic archaeon *Sulfolobus solfataricus*. Appl Environ Microbiol 72:102–111. doi:10.1128/AEM.72.1.102-111.2006

Angelov A, Mientus M, Liebl S, Liebl W (2009) A two–host fosmid system for functional screening of (meta) genomic libraries from extreme thermophiles. Syst Appl Microbiol 3:177–185. doi:10.1016/j.syapm.2008.01.003

Bailly J, Fraissinet-Tachet L, Verner MC, Debaud JC, Lemaire M, Wésolowski-Louvel M, Marmeisse R (2007) Soil eukaryotic functional diversity, a metatranscriptomic approach. ISME J 1:632–642. doi:10.1038/ismej.2007.68

Bernstein JR, Bulter T, Shen CR, Liao JC (2007) Directed evolution of ribosomal protein S1 for enhanced translational efficiency of high GC *Rhodopseudomonas palustris* DNA in *Escherichia coli*. J Biol Chem 282:18929–18936. doi:10.1074/jbc.M701395200

Craig JW, Chang FY, Kim JH, Obiajulu SC, Brady SF (2010) Expanding small–molecule functional metagenomics through parallel screening of broad–host–range cosmid environmental DNA libraries in diverse proteobacteria. App Environ Microbiol 76:1633–1641. doi:10.1128/AEM.02169-09

Delmont TO, Robe P, Clark I, Simonet P, Vogel TM (2011) Metagenomic comparison of direct and indirect soil DNA extraction approaches. J Microbiol Methods 86:397–400. doi:10.1016/j.mimet.2011.06.013

Ferrari BC, Binnerup SJ, Gillings M (2005) Microcolony cultivation on a soil substrate membrane system selects for previously uncultured soil bacteria. Appl Environ Microbiol 71(12):8714–8720. doi:10.1128/AEM.71.12.8714-8720.2005

Handelsman J, Rondon MR, Brady SF, Clardy J, Goodman RM (1998) Molecular biological access to the chemistry of unknown soil microbes: a new frontier for natural products. Chem Biol 5:245–249. doi:10.1016/S1074-5521(98)90108-9

Illeghems K, Weckx S, Vuyst LD (2015) Applying meta-pathway analyses through metagenomics to identify the functional properties of the major bacterial communities of a single spontaneous cocoa bean fermentation process sample. Int J Food Microbiol 50:54–63. doi:10.1016/j.fm.2015.03.005

Jones JG (1977) The effect of environmental factors on estimated viable and total populations of planktonic bacteria in lakes and experimental enclosures. Freshwater Biol 7:67–91. doi:10.1111/j.1365-2427.1977.tb01659.x

Lämmle K, Zipper H, Breuer M, Hauer B, Buta C, Brunner H, Rupp S (2007) Identification of novel enzymes with different hydrolytic activities by metagenome expression cloning. J Biotechnol 127:575–592. doi:10.1016/j.jbiotec.2006.07.036

Liu Y, Yao T, Jiao N, Kang S, Xu B, Zeng Y, Huang S, Liu X (2009) Bacterial diversity in the snow over Tibetan Plateau Glaciers. Extremophiles 13:411–423. doi:10.1007/s00792-009-0227-5

McGarvey KM, Queitsch K, Fields S (2012) Wide variation in antibiotic resistance proteins identified by functional metagenomic screening of a soil DNA library. Appl Environ Microbiol 78:1708–1714. doi:10.1128/AEM.06759-11

Nacke H, Will C, Herzog S, Nowka B, Engelhaupt M, Daniel R (2011) Identification of novel lipolytic genes and gene families by screening of Metagenomic libraries derived from soil samples of the German Biodiversity Exploratories. FEMS Microbiol Ecol 78:188–201. doi:10.1111/j.1574-6941.2011.01088.x

Pham VH, Kim J (2012) Cultivation of unculturable soil bacteria. Trends Biotechnol 30(9):475–484. doi:10.1016/j.tibtech.2012.05.007

Pradhan S, Srinivas TNR, Pindi PK, Kishore KH, Begum Z, Singh PK, Singh AK, Pratibha MS, Yasala AK, Reddy GS, Shivaji S (2010) Bacterial biodiversity from Roopkund glacier, Himalayan mountain ranges, India. Extremophiles 14:377–395. doi:10.1007/s00792-010-0318-3

PremaLatha K, Soni R, Khan M, Marla SS, Goel R (2009) Exploration of csp gene(s) from temperate and glacier soils of Indian Himalaya and *in silico* analysis of encoding proteins. Curr Microbiol 58:343–348. doi:10.1007/s00284-008-9344-0

Raynaud X, Nunan N (2014) Spatial ecology of bacteria at the microscale in soil. PLoS One 9(1):e87217. doi:10.1371/journal.pone.0087217

Shivaji S, Pratibha MS, Sailaja B, Kishore KH, Singh AK, Begum Z, Anarasi U, Prabagaran SR, Reddy GSN, Srinivas TNR (2011) Bacterial diversity of soil in the vicinity of Pindari glacier, Himalayan mountain ranges, India, using culturable bacteria and soil 16S rRNA gene clones. Extremophiles 15:1–22. doi:10.1007/s00792-010-0333-4

Soni R, Goel R (2010) Triphasic approach for assessment of bacterial population in different soil systems. Ekologija 56:94–98. doi:10.2478/v10055-010-0014-8

Soni R, Goel R (2011) *nif*H homologous from soil metagenome. Ekologija 57:87–95. doi:10.6001/ekologija.v57i3.1914

Sørensen HP, Mortensen KK (2005) Advanced genetic strategies for recombinant protein expression in *Escherichia coli*. J Biotechnol 115:113–128. doi:10.1016/S0167-7799(97)01155-4
Surmann E-M, Efferth T (2014) Biodiversity and metagenomics. In: Kuete V, Efferth T (eds) Biodiversity natural, products and cancer treatment. World Scientific Publishing Co., Singapore, pp 35–69. doi:10.1142/9789814583510_0002
Suyal DC, Yadav A, Shouche Y, Goel R (2015a) Bacterial diversity and community structure of Western Indian Himalayan red kidney bean (*Phaseolus vulgaris* L.) rhizosphere as revealed by 16S rRNA gene sequences. Biologia 70:305–313. doi:10.1515/biolog-2015-0048
Suyal DC, Yadav A, Shouche Y, Goel R (2015b) Diversified diazotrophs associated with the rhizosphere of Western Indian Himalayan native red kidney beans (*Phaseolus vulgaris* L.) 3Biotech 5:433–441. doi:10.1007/s13205-014-0238-5
Treusch AH, Leininger S, Kletzin A, Schuster SC, Klenk HP, Schleper C (2005) Novel genes for nitrite reductase and Amo–related proteins indicate a role of uncultivated mesophilic crenarchaeota in nitrogen cycling. Environ Microbiol 7:1985–1995. doi:10.1111/j.1462-2920.2005.00906.x
Troeschel SC, Drepper T, Leggewie C, Streit WR, Jaeger KE (2010) Novel tools for the functional expression of metagenomic DNA. Methods Mol Biol 668:117–139. doi:10.1007/978-1-60761-823-2_8
Wang GY, Graziani E, Waters B, Pan W, Li X, McDermott J, Meurer G, Saxena G, Andersen RJ, Davies J (2000) Novel natural products from soil DNA libraries in a streptomycete host. Org Lett 10:2401–2404. doi:10.1021/ol005860z
Zhang X, Yao T, Tian L, Xu S, An L (2008) Phylogenetic and physiological diversity of bacteria isolated from Puruogangri ice core. Microbiol Ecol 55:476–488. doi:10.1007/s00248-007-9293-3

Soil Microbiome for Enhanced Crop Productivity

14

Sankalp Misra, Swapnil Pandey, Vijaykant Dixit, Shashank Kumar Mishra, Mohammad Haneef Khan, Lalit Agarwal, and Puneet Singh Chauhan

Abstract

Current agricultural practices demand for low-input technologies with an objective to scale down the synthetic fertilizers and pesticides usage in order to enhance the sustainability in food production and restore ecosystem functioning. Regardless of much understanding of the essential role played by the soil microbiome in agriculture, we still have a limited knowledge of the multifarious response of microbial heterogeneity. To explore this covert attribute of soil microbial diversity, there is a need to focus upon the infinite ways by virtue of which soil microbiome helps in sustainable agriculture. There is limited access to highly diverse and dynamic communities of microbiome in soil due to inability of culture techniques in laboratory. With the advent of next-generation sequencing (NGS) techniques and high-throughput analysis, researchers gained new opportunities to investigate undetermined composition of soil microorganisms. Among rapidly growing field of research, the role of metagenomics is crucial in studying uncultured microbes to comprehend the actual microbial diversity and pertinent cooperation, evolution, and functions in diverse environment. Soil microbiologists are putting efforts in analyzing the phylogenetic diversity of soil niches and subsequently attempting to describe the functions of these soil inhabitants at trophic levels for improvement of soil fertility and productivity for the future generation.

Keywords

16S rRNA gene • Agriculture • Diversity • Metagenomics • Microbiome Rhizosphere

S. Misra • S. Pandey • V. Dixit • S.K. Mishra • M.H. Khan • L. Agarwal • P.S. Chauhan (✉)
CSIR-National Botanical Research Institute, Rana Pratap Marg, Lucknow 226 001, Uttar Pradesh, India
e-mail: puneetnbri@gmail.com; puneet@nbri.res.in

V.C. Kalia et al. (eds.), *Mining of Microbial Wealth and MetaGenomics*,
DOI 10.1007/978-981-10-5708-3_14

14.1 Introduction

Soil microorganisms play a crucial role in agriculture primarily by improving plant nutrition and health as well as soil quality (Barea et al. 2013; Lugtenberg 2015). Exploiting the beneficial role of soil microbial communities appears as a promising and effective approach (Bardi and Malusà 2012; Owen et al. 2015), and efforts are being made to improve the agricultural production through appropriate management of soil microorganisms as a low-input biotechnology while preserving the biodiversity (Zolla et al. 2013). The present agricultural research is getting attention of several researchers to analyze the role and management of the root-associated microbiome which is essential to meet issues pertaining to both economical and ecological sustainability. Studies related to soil microbial ecology and its impact on plant ecosystem functioning, associated dynamics, and thus productivity have staged a significant landmark to the current area of research (Nautiyal et al. 2010; Chaudhry et al. 2012). In current outlook, microbial cultures and their products have been extensively used in agriculture for soil fertility (Singh et al. 2011; Bhattacharyya and Jha 2012; Vassilev et al. 2013). The comprehensive researches in the field of phyto-beneficial microorganisms have led to the development of various biofertilizers that can fulfill the crop nutritional requirement (Bashan et al. 2014; Malusà and Vassilev 2014; Owen et al. 2015). The positive response of different crops to microbial inoculation has been accessed in many experiments conducted in greenhouses and field conditions (Calvo et al. 2014).

Soil constitutes large and diverse communities which are responsible for essential plant growth functions related to global environmental well-being. These communities altogether are referred as "soil microbial biomass" which are involved in essential functions such as soil respiration and nutrient turnover (Buée et al. 2009; Raaijmakers et al. 2009). The enormous complexity of microbial life in the soil on both micro- and macroscales coupled with improper estimation of its components has challenged our current knowledge of causality between low-input farming and microbial diversity. Thus, in-depth knowledge is required for characterization of complex soil microbial diversity for better understanding of specific functions and their consequences pertaining to the soil management or the climate. Research on soil microorganisms has estimated that only 1% of the total soil microbes can be cultured in vitro while the rest (99%) is still not cultivable (Nichols 2007; Ritz 2007). Therefore, over the recent years, culture-independent techniques have emerged as an alternative to explore the soil microbial community structure by comparing genomes isolated from the soil environment. Novel sequencing technologies have expanded our understanding to explore the soil microbiome with higher resolution and coverage. It has also revolutionized our ability to classify and characterize microbial population taxonomically and functionally for better agricultural management practices (Taberlet et al. 2012).

This chapter accentuates the perception of the soil microbiome under current perspectives. Further, explicit outlooks on the advancements made in the field of latest technologies to explore the unrevealed side of soil microbiome will also be discussed. In this context, various mechanisms associated with beneficial microbes-mediated plant growth promotion have been attempted to be described. Finally, the latest paradigms of application of beneficial microorganisms in different agroecosystems under stress conditions have been presented with an aim to develop future insights.

14.2 Soil Microbiome: Benefactors and Their Beneficiaries

The soil is an extremely diverse system as assessed for its biological, physicochemical properties. Development of soil has taken place under diverse climatic regimes. In relation to microbial ecology, there exist a large numbers of niches where individual organisms and/or their group of community can be found. Soil is a self-sustaining system that harbors fundamental processes including nutrient cycling, water regulation, microbial remediation as well as biocontrol (Doran and Zeiss 2000). Among ecosystem services of soil, the microbiome plays an important role in diverse soil processes that shall determine the productivity of agricultural land in a sustainable manner (Barrios 2007; van der Heijden et al. 2008). In this context, effective management of soil microbiome may allow us to improve sustainable agricultural production. The diversity of soil microbiome affects the productivity and community dynamics via plant growth promotion, production and modification of phytohormones, and nutrient acquisition up to alleviation of abiotic and biotic stresses (Vassilev et al. 2006; Sergeeva et al. 2007; Klein et al. 2013) (Fig. 14.1).

Soil constitutes inorganic and organic matters which vary on physicochemically. Maximum portion of the soil is considered as bulk soil which is inhabited by few microbes, while the area surrounded the plant roots, also known as rhizosphere, contains myriad of microorganisms. The reason pertaining to the high density of microbial population in the rhizosphere is due to high nutrient content (Phelan et al. 2012). The rhizospheric research presents one of the most fascinating platforms for the study of microbial habitats involving the soil and plant (Chauhan et al. 2011). Microbes commonly found in the rhizosphere include fungi, bacteria, protozoa, algae, nematodes, and microarthropods (Raaijmakers et al. 2009). Rhizosphere microorganisms exhibit variety of characteristics due to which they exert beneficial effects that influence plant productivity either directly or indirectly. Additionally these plant-beneficial rhizospheric microorganisms may lower the global dependency on hazardous agricultural chemicals.

14.2.1 Colonization: A Unique Phenomenon of Soil Microbiome

Root colonization by microbes impacts many facets of plant's life including the latter's survival in ecosystem. These phenomenons are important traits for effective establishment by beneficial microbes for plants (Fig. 14.1). PGPR is one such essential characteristic for their hosts in the realm of plant ecology. However, to accomplish this task, rhizospheric competence is a necessary prerequisite. For instance, the ability to suppress disease by introduced *Pseudomonas* strains relies mainly on their

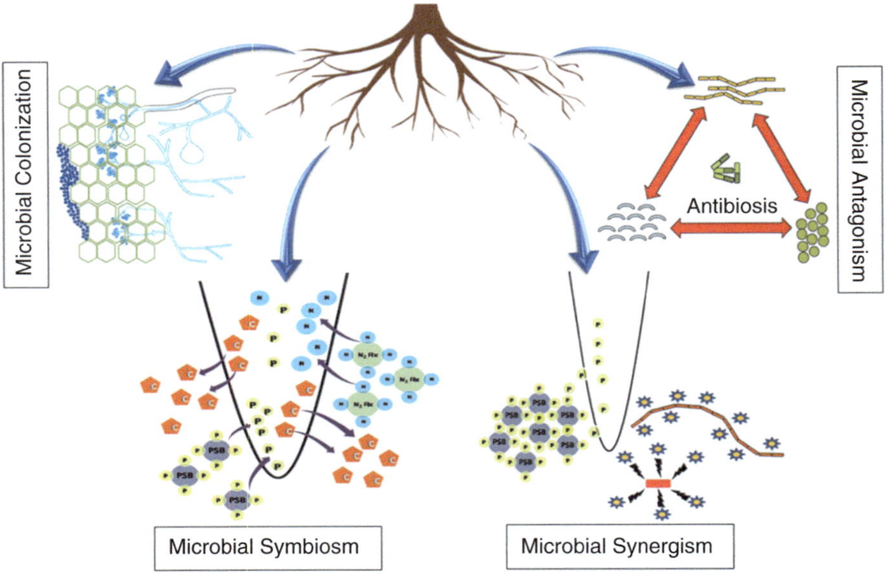

Fig. 14.1 Diagrammatic representation of different phenomenons occurring in rhizospheric soil among microbial communities and between plant-microbe. **Microbial colonization:** showing mycorrhizal association by hyphal network with plant by forming arbuscules and vesicles in the cortical cells and bacterial biofilm formation on cortical cells; **microbial symbiosis:** demonstrating exchange of nutrients between plant and microbe. Microbe represents phosphate solubilizing bacteria (PSB) and N_2-fixing bacteria responsible for transport of phosphorus (P) and nitrogen (N), respectively, to the plant. In exchange of P and N, plant provides its carbohydrate reserve to the respective microbes; **Microbial synergism:** representing PSB supplying nutrient in the form of P to plant while getting benefitted with the biocontrol activity of another microorganism. Biocontrol activity of the respective microbe comprises of antibiotic production inhibiting the growth of other competitive microbes in the rhizosphere; **microbial antagonism:** a phenomenon of particular group of microbes restricting the growth of other microbes by secreting antibiotic compounds for the survival and prosperity of the former

ability to colonize the roots (Woeng et al. 2000) and their rhizosphere population density (Raaijmakers and Weller 1998). Although some *Pseudomonas* strains produce the antibiotic phenazine, yet they fail to suppress soilborne pathogens due to their lack of motility and, consequently, rhizosphere colonization (Woeng et al. 2003). Various approaches that improve application of PGPR have narrowed down to its effect on abiotic (Howie et al. 1987) and biotic factors (Notz et al. 2001), host (Smith and Goodman 1999), and microbial genotypes (Landa et al. 2002). It was reported that PGP observed in tomato had greater influence of rhizocompetent streptomycetes secreting ACC deaminase and/or IAA than non-rhizocompetent isolates (El-Tarabily 2008). Rhizodeposition (the organic matter release) is another approach used by microbes to enhance their growth and, therefore, drives the microbial community within the rhizosphere leading to the selection of certain microbial population such as rhizobacteria. The structured rhizospheric microbial community can be grouped together as biofilm community. Biofilm community in the rhizosphere works by, first, attaching themselves to the roots with the involvement of cell components, like wall polysaccharides (capsules), membrane proteins, lipopolysaccharide (LPS), and surface agglutinin (Michiels et al. 1991). Following this, exopolysaccharide (EPS) produced by the respective community of the rhizosphere leads to enhancement in soil aggregation and water stability (Amellal et al. 1998). Biofilm community involves interaction among microbes attached with the plant. This type of interaction includes biofilm formation by *Pseudomonas fluorescens* on mycorrhizal fungi colonized in plants (Perotto and Bonfante 1997). Although the rationale for bacterial attachment to the fungi is not known, yet this might be a positioning phenomenon that allows bacteria to readily obtain nutrients and propagate (Fig. 14.1). Endophytic colonization by bacterial communities is an outcome of colonization process. However, several speculations were made for their origin other than from root zone such as anthosphere, phyllosphere, or spermosphere (Sturz et al. 2000). With regard to this process, bacteria present in rhizosphere might have a chance to enter plant root and establish their populaces. Following this establishment, they reach to the vascular system via root cortex. This endophytic colonization reflects the bacterial adeptness to selectively adapt itself for specific ecological niches with no negative effect to the host (Compant et al. 2005).

14.2.2 Microbial Antagonism: Eco-friendly Retaliation Program

The use of antibacterial and antifungal chemicals is belittled in view of sustainable agricultural productivity. Hence, biocontrol as an alternative is considered as a more environmentally friendly process. Microbial antagonists follow argumentative or inundative approach to interact with plant pathogen and achieve disease control (Johnson 2010). Argumentative approach follows introduction of considerable amount of inoculum to maintain an effective and reproducible antagonist population to suppress the pathogen population. On the other hand, inundative approach achieves control over disease by introduction in substantial quantity without sufficient propagation of the microbial antagonists.

Besides these, rhizobacteria can antagonize pathogens by different mechanisms including competition, antibiotic production, or lytic enzyme secretion (Van Loon and Bakker 2003) that make them a potent tool for reducing damages through preventing deleterious effects of phytopathogens (Fig. 14.1). Lytic enzymes produced by bacteria involve glucanases, proteases (Dunne et al. 1997), cellulases, and chitinases. These enzymes affect growth and development of phytopathogenic fungi (Frankowski et al. 2001; El-Tarabily 2006). A direct correlation was found between the antifungal activity and production of chitinase enzyme by *P. fluorescens* (Velazhahan et al. 1999). Production of cell wall-degrading enzymes has been linked to biocontrol properties of the producers (El-Tarabily 2006). In this regard, Chernin et al. (1995) demonstrated that Tn5 mutants lacking chitinolytic activity fail to provide protection to the plants against the disease. In another experiment, bacteria having various mentioned biocontrol properties were tested for synergistic impact to pathogen inhibition (Someya et al. 2007). This is due to the fact that these bacteria produce antibiotics which can be characterized as volatile and nonvolatile. Volatile ones include HCN, sulfides, alcohols, ketones, and aldehydes, while nonvolatile comprises polyketides and heterocyclic nitrogenous compounds (de Souza et al. 2003) and phenylpyrrole antibiotic (pyrrolnitrin) (Ahmad et al. 2008). For instance, *Bacillus* strains secrete various lipopeptide antibiotics such as surfactin, iturins, Zwittermicin A, and bacillomycin that can suppress the population of phytopathogens such as *F. oxysporum* and *Rhizoctonia solani* (Asaka and Shoda 1996; Kumar 1999). In addition, Weller and Cook (1983) reported that phenazine derivatives secreted by fluorescent pseudomonads can act as biocontrol of take-all disease. *P. fluorescens* 2–79, an in vitro producer of phenazine-l-carboxylate, was reported to suppress *Gaeumannomyces* (Thomashow and Weller 1988). Moreover, *P. fluorescens* protects plants from fungal phytopathogen by HCN production having a suppressive effect on black root (Ahmad et al. 2008). The antagonistic properties of *Streptomyces* against numerous phytopathogens including *Colletotrichum gloeosporioides*, *Alternaria brassicicola*, *Sclerotium rolfsii*, *Penicillium digitatum*, and *F. oxysporum* (Khamna et al. 2009) are well established. *Streptomyces* spp. are responsible for secreting well-known antifungal compounds such as Cycloheximide, Kasugamycine, Blasticidin-S, and Rhizovit, respectively.

Iron is important for plant health and metabolism. PGPR uptake iron from soil and transfer it to plant. Fluorescent pseudomonads are widely studied PGPR for their feature of siderophore production. These siderophores are low molecular mass by-products of microorganisms having high Fe affinity. They possess an iron uptake system (iron-binding ligand) able to chelate Fe^{3+} molecules. They are often induced under limiting Fe^{3+} concentrations to allow bacteria to partially fulfill their iron requirement. Siderophores comprise of biochemically diverse group produced by either plants or plant-associated microorganisms (Loper and Buyer 1991). They include hydroxamates from *Erwinia carotovora*, *Enterobacter cloacae*, and fungi; catechols from *Agrobacterium tumefaciens*, *Er. chrysanthemi*, and Enterobacteriaceae; pyoverdines from *Pseudomonas*; and rhizobactin produced by *Rhizobium meliloti*. When various plants growing on soil or nutrient solution were supplemented by pyoverdine or ferriopyoveridne, they showed, with few exceptions, enhanced chlorophyll content and enhanced iron content in the roots and ferric reductase activity (Duijff et al. 1994). Siderophores, produced by *Pseudomonas*, create competition for resource, in particular, restrict the development of rhizospheric phytopathogens, and thus are considered as indirect mechanism to pathogen management (Duijff et al. 1994). Soil suppressive to fusarium wilts exhibits low solubility toward Fe^{3+} (Alabouvette et al. 1996) leading to the strong competition of Fe. Apart from these mechanisms, known siderophores, namely, pyocyanin and pyoverdine, induce plant defense (Meziane et al. 2005; Audenaert et al. 2002).

14.2.3 Microbial Symbiosis: Thriving Together

The other most important group of mutualistic microbial symbionts comprise arbuscular mycorrhizal (AM) fungi which is known to establish relations with the roots of most land plant species and belongs to phylum *Glomeromycota* (Schüßler et al. 2001; Smith and Read 2008; van der Heijden et al. 2015). In sustainable agriculture, the AM symbiosis plays an essential role in plant productivity under adversity. This is because mycorrhizal formation may be considered as a strategy that provides the plant to adapt with an increased ability for nutrient acquisition and cycling in soil having resource limitation (Jeffries and Barea 2012). Therefore, they are known to induce an increased tolerance toward both the environmental stresses, namely, biotic or abiotic, that eventually results in improved soil structure through aggregate formation necessary to support plant growth (Jeffries et al. 2003). Similarly, AM fungi play crucial roles in forest ecosystems (Borie et al. 2010). Current researches have introduced the idea of mycorrhizosphere, which implies the site for specific microbial interactions (Barea et al. 2013). The changes contributed by AM colonization include alteration in chemical composition of root exudates, while the extracellular mycelium of AM modifies the rhizospheric environment thus affecting microbial structure and diversity. Managing the interactions between PGPR and AM fungi is recognized as mycorrhizosphere tailoring, an appropriate biotechnological tool in sustainable agriculture. Among the co-inoculation experiments that have been reported using selected mycorrhizae and other rhizosphere microorganisms include (a) symbiotic N_2 fixation, (b) phosphate mobilization, (c) phytoremediation of heavy metal polluted soils, (d) biocontrol of root pathogens, and (e) enhancement in soil quality (Barea et al. 2013) (Fig. 14.1). Therefore, AM symbiosis protects plants against pathogens, herbivorous insects, and parasitic plants. Additionally upon AM colonization, low production of strigolactones corresponds to alleviation in parasitic plant infection, which thus reduces the deleterious effect of the weeds (Jung et al. 2012; López-Ráez et al. 2012; Pozo et al. 2013). AM fungi and other rhizosphere microorganisms are gaining particular consideration to improve plant water status based on the improvement of root hydraulic conductance, which intimately depends on functioning of aquaporins (Aroca et al. 2012; Groppa et al. 2012; Ruiz-Lozano et al. 2012; Calvo-Polanco et al. 2013; Barzana et al. 2014).

The pivotal role played by rhizospheric and root endophytic microbes, including PGPR, *Trichoderma* sp., and AM fungi in sustainable agricultural development, includes protection of plants against pathogens by competition in terms of space and nutrients, antibiosis (for PGPR and *Trichoderma*), mycoparasitism (for *Trichoderma*), and by inducing plant defense mechanisms (Barea et al. 2013). The concept of defense priming is fundamental for an effective protection against pathogens as it is the preconditioning of plant immunity induced by microbial colonization. It is accomplished systemically on different parts of the root and shoots, thereby inducing systemic resistance (ISR) to protect plants efficiently against both roots and foliar pathogens (Selosse et al. 2014). Eventually, jasmonic acid (JA) plays a key role in priming plant's immunity by boosting its ability to respond to pathogen attack upon AM colonization. JA is the main hormone involved in the plant hormone signaling cross talk, which regulates plant defense and microbe-plant-insect interactions (Pangesti et al. 2013). ISR by AM symbiosis is particularly termed as mycorrhiza-induced resistance (MIR). The detection of defense regulatory elements which coordinate with AM development and MIR poses a greater challenge for an extensive research on the development of better biotechnological strategies to exploit AM fungi in the integrated management of pests and diseases (Pozo and Azcón Aguilar 2007; Jung et al. 2012; Pozo et al. 2013). Additionally, PGPR, *Trichoderma* sp., and nonpathogenic *Fusarium* strains also contribute in priming local resistance and ISR through the production of microbe-associated molecular patterns (MAMPs), which trigger immune responses by activating JA signaling pathway (Pozo et al. 2015). Interestingly, individual and combined treatment of AM fungi and vermiwash have also been exploited with a particular emphasis on suppression of plant diseases (Khan et al. 2015).

14.2.4 Synergism: Another Reinforcement Strategy

Studying soil microorganisms in a consortia manner is essential as microbes in soil do not exist and work in isolation but in communities. This involves natural processes of defense and/or resource competition in that space (Fig. 14.1). To exemplify this vibrant area of research, synergistic evaluation of responses involving improved plant growth has been observed in legumes (Babana and Antoun 2006) and cereals (Osorio and Habte 2001) with combined application of *Penicillium* sp. and AM fungi and combined application of *Penicillium* and *Rhizobium* sp. in legumes (Downey and van Kessel 1990; Rice et al. 2000). Similarly, other studies include legumes having combined inoculation of AMF and rhizobia (Wang et al. 2011), *Rhizobium* and P-solubilizing bacteria (Alagawadi and Gour 1988), and AM fungi in synergism with *Rhizobium* and phosphate solubilizing fungi (Vassilev et al. 1996). The co-inoculation of *Bradyrhizobium japonicum* and *Stenotrophomonas rhizophila* resulted to increased root growth, N uptake, and nodulation under saline condition when compared to non-saline ones in soyabean (Egamberdieva et al. 2016). Additionally, synergistic combination of *Pseudomonas* and *Bacillus* has shown to alleviate drought stress in chickpea (Kumar et al. 2016). In case of non-legumes, notable examples encompass the use of dual inoculations of AM fungi and free-living N-fixing bacteria, or other PGPR were reported to assure better nutrient uptake than chemically fertilized plants (Lisette and Germida 2003; Wu et al. 2005; Malusa et al. 2007; Adesemoye et al. 2008) (Fig. 14.1). The fascinating area in this field has revealed that microorganisms having multiple plant growth-promoting attributes employed via such consortiums are distinguished by various functions thus contributing toward a dynamically changing microbial community (De Roy et al. 2013).

14.3 Microbial Community Analysis: Insight into Rhizosphere

The diversity of soil microorganisms is critical in the sense to maintain good soil health, as these microbial communities are involved in several essential activities such as soil formation, elemental cycles of N and P, and toxin removal (Nautiyal et al. 2010). Despite much of the understanding, many challenges are yet to be faced while exploring microbial diversity of soil. The need to overcome these obstacles has resulted in the advancement of molecular and analytical techniques which in turn enhance the resolving power of soil microbial community composition (Hirsch et al. 2010). Recent progresses in next-generation sequencing (NGS) unearth the detailed information on identification and distribution of microbial taxa under various environments (Nunes da Rocha et al. 2010). Regardless of challenges either related to enormous diversity within microbial population or taxonomic ambiguity of microbes, a revolution is developing to better understand microbiology through high-throughput DNA sequencing. Latest sequencing technologies are increasing the breadth of analysis at species level for rhizospheric microbes.

In the run for developing molecular approaches to explore the soil microbial communities involves RISA, DNA cross hybridization and G + C% profiling, and community fingerprinting techniques such as DDGE and T-RFLP (Micallef et al. 2009). Phospholipid fatty acid analysis (PLFA) has been employed for estimating microbial communities and for tracing primary consumers of ^{13}C-labeled compounds (Paterson et al. 2007). Recently, for analyzing the changes related to rhizospheric soil microbial communities, application of high-throughput methods such as microarrays and pyrosequencing is being used considerably (Eilers et al. 2010). Influential works on 16S rRNA gene made it a reference for characterization of bacteria in phylogenetic context (Pace et al. 2012). Recently, assessment of prokaryotic community profiles by SSU rRNA has been validated by the Illumina sequencing method (Gloor et al. 2010; Caporaso et al. 2012). In another study, microbial communities of *Arabidopsis* have been investigated using pyrosequencing approach (Bulgarelli et al. 2012; Lundberg et al. 2012).

DNA extracted from microbial communities is usually believed as an entity, referred as metagenome, and the related study is known as metagenomics. Among recent advancements, metagenomics has revolutionized the rhizosphere research by filling the gaps of pre-NGS-based genomics, where the sequence data can be directly obtained from niches. It is a rapidly evolving scientific field that aims to unravel microbial dynamics including that of uncultured microorganisms for better understanding of the microbial diversity, their ecology, and evolution under diverse environments such as soil, water, and digestive system of animals and humans (de Bruijn 2011). Therefore, metagenomic data from rhizospheric soil may reveal more information of potential activities about the resident and uncovered microbial communities. Restoring microbial diversity can simultaneously improve crop production and soil restoration while enhancing crop resistance to environmental change. Recently, advances in DNA sequencing to exemplify the uncultured environmental samples have amplified awareness that complex microbial communities interact with plants to promote growth. A contemporary effort for refining this research area includes PhyloChip-based metagenomics unearthing the targeted microbial community composition for its phylogenetic studies (Mendes et al. 2011).

Lists of the rhizospheric microbiome reported for different host plants are summarized (Table 14.1). Briefly, PhyloChip-based method has been utilized by Brodie (Brodie et al. 2007) for extensive detection and comparison of microbial diversity in

Table 14.1 List of different host crops/plants evaluated for microbiome analysis

S.no.	Host crop/plant	Phylum of soil microbiome associated with host crop/plant	References
1.	Arabidopsis	*Proteobacteria, Actinobacteria, Acidobacteria, Cyanobacteria, Bacteroidetes, Gemmatimonadetes, Firmicutes*	Lundberg et al. (2012)
2.	Barley	*Chloroflexi, Firmicutes, Bacteroidetes, Actinobacteria, Proteobacteria*	Bulgarelli et al. (2015)
3.	Maize	*Proteobacteria, Actinobacteria, Firmicutes, Cyanobacteria, Chloroflexi*	Li et al. (2014)
4.	Rice	*Proteobacteria, Actinobacteria, Acidobacteria, Cyanobacteria, Bacteroidetes, Gemmatimonadetes, Firmicutes*	Spence et al. (2014)
5.	Wheat	*Acidobacteria, Actinobacteria, Bacteroidetes, Firmicutes, Proteobacteria, Chloroflexi*, TM7, unclassified, and others	Donn et al. (2014)
6.	Sugar beet	*Bacteroidetes, Firmicutes, Proteobacteria, Actinobacteria*, and others	Zachow et al. (2014)
7.	Soybean	*Actinobacteria, Bacteroidetes, Proteobacteria, Firmicutes, Acidobacteria, Gemmatimonadetes*, WS3, *Chloroflexi*, and unclassified bacteria	Sugiyama et al. (2014)
8.	Sweet potato	*Proteobacteria, Actinobacteria, Bacteroidetes, Firmicutes, Acidobacteria, Cyanobacteria, Gemmatimonadetes*	Marques et al. (2014)
9.	Grapevine	*Proteobacteria, Acidobacteria, Bacteroidetes, Verrucomicrobia, Planctomycetes, Actinobacteria, Gemmatimonadetes, Chloroflexi, Nitrospirae*, and others	Zarraonaindia et al. (2015)
10.	Oat	*Proteobacteria, Firmicutes, Bacteroidetes, Actinobacteria, Acidobacterium*, and others	De Angelis et al. (2009)
11.	Potato	*Proteobacteria, Firmicutes, Actinobacteria, Acidobacteria, Bacteroidetes, Spirochaetes*, and others	Weinert et al. (2011)
12.	Oak	*Proteobacteria, Actinobacteria, Acidobacteria, Bacteroidetes, Gemmatimonadetes, Verrucomicrobia, Firmicutes, Planctomycetes, Chlamydiae, Nitrospira*, unclassified bacteria, and others	Uroz et al. (2010)
13.	Cactus	*Proteobacteria, Actinobacteria, Firmicutes, Cyanobacteria*	Aguirre-Garrido et al. (2012)

rhizospheric soil of oat and potato (De Angelis et al. 2009; Weinert et al. 2011). Using pyrosequencing, the bacterial diversity in the rhizosphere associated with oak, sweet potato, soybean, sugar beet, and *Arabidopsis* was deeply investigated (Uroz et al. 2010; Lundberg et al. 2012; Marques et al. 2014; Sugiyama et al. 2014; Zachow et al. 2014). In addition, GeoChip-based functional gene arrays have been used to select functional genes responsible for characterization of the rhizospheric microbial communities in maize (Li et al. 2014). Illumina shotgun sequencing method has been used to gain insight into the soil microbiome of barley and grapevine (Bulgarelli et al. 2015; Zarraonaindia et al. 2015).

The challenge of microbial activity profiling is accomplished by metatranscriptomic approach of rhizosphere which led to the assessment of functional genes that accounts for rhizospheric interactions (Carvalhais et al. 2013). Previously, many researchers have contributed to investigate the rhizosphere microbiome by the means of metagenomics.

Wang and collaborators had made it possible by constructing and studying metagenomic library of bacterial components from tissue of *Mallotus nudiflorus* (Wang et al. 2008). The genomic library constitutes diverse bacterial phyla, represented predominantly by *Actinobacteria* and *Proteobacteria*, while *Firmicutes* and *Deinococcus-Thermus* are present in small proportions (1–3%), whereas *Cyanobacteria*, *Planctomycetes*, *Acedobacteria*, *Chloroflexi*, *Gemmatimonadetes*, and *Bacteroidetes* are found as minor phyla. Further, the metagenome study has analyzed the diversity and functions of rice root endophytes (Sessitsch et al. 2012). The metagenomic study is not only confined to rhizosphere, but it has also gained much attention in analyzing epiphytic microbial population. In this respect, the study related to characterization of different rhodopsins of epiphytic microbes pertaining to tamarisk *Arabidopsis* (*A. thaliana*), soybean (*Glycine max*), rice (*Oryza sativa*) (*Tamarix nilotica*), and clover (*Trifolium repens*) has been reported (Atamna et al. 2012). Particularly, other achievements in this area include investigation of epiphytic microbiome of *A. thaliana* by using pyrosequencing for *rrs* gene (Delmotte et al. 2009). However, combined approach of genomics and proteomics has also gained success in exploring complex microbial communities belonging to biofilms of acid mine drainage systems (Ram et al. 2005).

The above stated instances are few paradigms of progress achieved using metagenomic approach. Certainly, other omics technologies are providing new insights into a wide range of microbial functions day by day in the form of novel enzymatic activities and bioactive molecules. Clearly, these approaches will enhance our understanding of microbial community dynamics and its causality to soil functions and rhizosphere interaction in future.

14.4 Application of Rhizosphere Research in Sustainable Agriculture

Existing global reluctance to embrace food production by chemical fertilizers has provided a way to develop fertilizer of biological origin as an important tool. Furthermore, this technology is accessible to farmer worldwide (Gamalero et al. 2009). Contemporary framework for evaluating plant-microbial interactions at multiple scales has provided ecosystem services which will focus on reducing production costs for agricultural producers. Apart from this, manifested research in modeling outcomes to broadly explore potential impacts and interacting with extension and training networks to transfer microbial-based agricultural technologies across socioeconomic scales has established an integrated strategy for advancing agroecosystem sustainability.

The current advancements in applications of PGPR in diverse agroecosystems have presented overtly to gain larger perspectives concerning their functioning. The commercial exploitation of rhizospheric microorganisms for horticulture along with agriculture is still in its infancy, but it has been stated that there is worldwide market with a potential growth rate of 10% per annum (Berg 2009; Pérez-García et al. 2011). Many microorganisms including bacteria and fungi have already been commercialized. Therefore, scientific research has shown that microbial associations in rhizosphere can be managed, as low-input technologies, to improve sustainable agricultural practices (Azcón and Barea 2010).

Soil is an unpredictable environment in which interactions among members of microbial communities result in the enhancement of key processes of PGP. It has been established that all possible interactions occurring in rhizosphere are either directly or indirectly plant mediated. New understanding of this interdependence between plants and microbiomes is crucial for mitigating impacts of climate change to achieve agricultural sustainability and food security. Successful utilization of soil microorganisms requires a good understanding of their functional ecological mechanisms in rhizosphere. Furthermore, the development of crop varieties with the aim of enhancing select phyto-beneficial functions in soil microbial communities shall provide great progress in terms of yield with less chemical inputs. A more complete knowledge regarding sensing, signaling, and secretion in plant-associated soil microbial communities will fulfill the gaps in our understanding for enhanced nutrient acquisition and improved plant health.

14.5 Future Challenges

Future progress in functional genomics (metabolomics, transcriptomics, and proteomics) will be expedient to investigate functional profiling of rhizosphere that plays a crucial role in processes, namely, nutrient mobilization or plant disease management. The specific management of soil microbial associations by the design of suitable mycorrhizospheres should be a pivotal objective for applied studies in the future. In this regard, the efficiency of soil microbiome may be further improved along with its optimization and acclimatization. In the near future, they are presumed to supplant chemical fertilizers, pesticides, and artificial growth regulators. Further, research on microbe-mediated phytostimulation would pave the way for better competent isolates efficient under diverse agroecological conditions. Developments in metagenomics and metaproteomics may provide some insight into ecological engineering which will explore the functional role of microbial ecology.

Currently, diverse research approaches are addressing to declare whether the rhizosphere can be engineered to encourage beneficial organisms while preventing the presence of pathogens. The related research topics offer many challenges related to research strategies. Undoubtedly, getting biased rhizosphere opens new opportunities for future agricultural developments based on exploitation of beneficial microbial services to reduce the inputs of agrochemicals, thereby reaching sustainable environmental and economical goals. Moreover, significant benefits may be obtained by devising strategies to explore indigenous communities that hold critical clues to develop sustainable technologies with new biotechnological interventions that restore microbial diversity to local and regional agroecosystems. Communities that have retained these geographically specific knowledge bases will be well-positioned to initiate new commercial practices that involve training local farmers to produce food with low agrochemical input. Such products are desired by new generation of consumers that is increasingly interested in organic and locally grown foods. Finally, other than the scientific methodologies, administrative role in framing policies and practices that support integration of microbial ecology with crop development, agroecology, environmental sciences, nutrition, socioeconomics, and extension will promote robust, adaptable, and sustainable agroecosystems with the improved potential to mitigate impacts of globalization and climate change.

Conclusion

An increasing human population, climate change, growing per capita food and energy demands, and reduced ecosystem potential due to much dependence on chemical fertilizers have created an urgent need for improved crop production practices. To accomplish this need in a sustainable manner requires interdisciplinary approaches that integrate plant and microbial ecology with attempts to make advancements in crop production while mitigating issues of climate change and retaining microbial diversity in the rhizosphere.

The nature has bestowed us with microbial wealth which has been responsible for taking part in the formation of soil. Soil biodiversity is a multitude of diverse microorganisms living together in a realistic and effective manner, conferring multifarious characteristics and functions that may chiefly be responsible for sustainable biosphere and cost-effective agriculture. In global scenario, exploiting the interactions between soil microbial communities and plants is essential to meet the challenge of increasing food production for the growing population at low environmental costs. Essentially, this requires comprehensive knowledge of interdisciplinary strategies for managing the soil microbiome in respect to developing microbial inoculants and manipulating naturally existing microbial populations. The soil microbiome constitutes plethora of microorganisms, both rhizosphere bacteria (PGPR) and fungi, either saprophytic or endophytic symbionts (with special reference to N_2-fixing rhizobia and AM fungi) are mainstay of applied microbial biotechnology in agriculture. They also help to encounter diverse types of stress factors, including salinity, drought, nutrient deficits, contamination, diseases, and pests, which cause detrimental impacts on the functionality/productivity of agricultural systems. Despite the acknowledged value of these soil microorganisms, our knowledge of their diversity and many of their essential roles in sustaining global life support systems is still in inception. Furthermore, various underlying mechanisms of plant-microbe interactions need in-depth comprehension. These challenges are chiefly dependent on profiling the vast array of processes, predominantly contributed by unculturable soil microbial communities. Exploration, evaluation, and exploitation of soil microbial diversity as well as major interactions existing in these microenvironment are essential for scientific, industrial, and social development which is more relevant to agricultural-based countries as it abounds by enormous wealth of available biodiversity. A major breakthrough has been provided by myriad of metagenomic techniques which are currently being applied either to decipher, the unexplored diversity of soil-inhabiting microorganisms, or to characterize the molecular aspect of the plant-microbiome relationship in these microenvironments. Biotechnological strategies are being formulated on the basis of signaling network that orchestrated plant-microbiome interactions, to unravel various mechanisms for plant's adaptation and to mend the stress-mitigating adeptness of soil microbes in crops. These interactions can improve food security at the community level by promoting diversified food systems that are more responsive to culturally and environmentally diverse demands.

Conclusively, we can say that many achievements have been reached with the application of microbial biotechnology in agriculture, but many challenges as well as opportunities need to be explored for the future sustainable agricultural developments. Particularly much emphasis is being paid to formulation, quality control, and modes of application of microbial inoculants. Therefore, continued endeavor is required to describe and explore the hidden resources for the preservation of natural ecosystems and the future benefit of mankind. A necessity for framing a knowledge bank with information systems and databases on unexplored soil biodiversity necessitates the persuasion of soil biological management according to climatic conditions, socioeconomic context, and spatial and temporal scales. Construction of multi-scaled ecological models, in combination with new microbial technologies, could accelerate global progress toward food security and sustainable agriculture. Synergism of science and administration endeavors the existing efforts to address food security. This is particularly true if both are made accessible through networked, interdisciplinary, and participatory attempts that promote our understanding of biotic interactions across biophysical, temporal, and socioeconomic scales. Certainly, a well-informed public is crucial to success of any such effort. Public participation can wipe out the concept for widespread negative ecological and socioeconomic feedbacks that could stem from newly developed agricultural technologies.

Acknowledgments The study was supported by New Initiative (as a Cross Flow Technology project) "Root Biology and Its Correlation to Sustainable Plant Development and Soil Fertility" (RootSF; BSC0204) from the Council of Scientific and Industrial Research (CSIR), New Delhi, India.

References

Adesemoye AO, Torbert HA, Kloepper JW (2008) Enhanced plant nutrient use efficiency with PGPR and AMF in an integrated nutrient management system. Can J Microbiol 54:876–886. doi:10.1139/w08-081

Aguirre-Garrido JF, Montiel-Lugo D, Rodríguez CH, Torres-Cortes G, Millán V, Toro N, Abarca FM, Ramírez-Saad HC (2012) Bacterial community structure in the rhizosphere of three cactus species from semi-arid highlands in central Mexico. Antonie Van Leeuwenhoek 101:891–904. doi:10.1007/s10482-012-9705-3

Ahmad F, Ahmad I, Khan MS (2008) Screening of free-living rhizospheric bacteria for their multiple plant growth promoting activities. Microbiol Res 163:173–181. doi:10.1016/j.micres.2006.04.001

Alabouvette C, Hoper H, Lemanceau P, Steinberg C (1996) Soil suppressiveness to diseases induced by soil-borne plant pathogens. In: Stotzky G, Bollag JM (eds) Soil biochemistry. Marcel Dekker, New York, pp 371–413

Alagawadi AR, Gour AC (1988) Associative effect of *Rhizobium* and phosphate solubilizing bacteria on the yield and nutrient uptake of chickpea. Plant Soil 105:241–246. doi:10.1007/bf02376788

Amellal N, Burtin G, Bartoli F, Heulin T (1998) Colonization of wheat roots by an exopolysaccharide producing *Pantoea agglomerans* strain and its effect on rhizosphere soil aggregation. Appl Environ Microbiol 64:3740–3747

Aroca R, Porcel R, Ruiz Lozano JM (2012) Regulation of root water uptake under abiotic stress conditions. J Exp Bot 63:43–57. doi:10.1093/jxb/err266

Asaka O, Shoda M (1996) Biocontrol of *Rhizoctonia solani* damping-off of tomato with *Bacillus subtilis* RB14. Appl Environ Microbiol 62:4081–4085

Atamna IN, Finkel OM, Glaser F, Sharon I, Schneider R, Post AF (2012) Microbial rhodopsins on leaf surfaces of terrestrial plants. Environ Microbiol 14:140–146. doi:10.1111/j.1462-2920.2011.02554

Audenaert K, Pattery T, Cornelis P, Höfte M (2002) Induction of systemic resistance to *Botrytis cinerea* in tomato by *Pseudomonas aeruginosa* 7NSK2: role of salicylic acid, pyochelin and pyocyanin. Mol Plant-Microbe Interact 15:1147–1156. doi:10.1094/MPMI.2002.15.11.1147

Azcón R, Barea JM (2010) Mycorrhizosphere interactions for legume improvement. In: Khan MS, Zaidi A, Musarrat J (eds) Microbes for legume improvement. Springer, Vienna, pp 237–271

Babana AH, Antoun H (2006) Effect of Tilemsi phosphate rock solubilizing microorganisms on phosphorus uptake and yield of field grown wheat (*Triticum aestivum* L.) in Mali. Plant Soil 287:51–58. doi:10.1007/s11104-006-9060-0

Bardi L, Malusà E (2012) Drought and nutritional stresses in plant: alleviating role of rhizospheric microorganisms. In: Haryana N, Punj S (eds) Abiotic stress: new research. Nova Science, Hauppauge, pp 1–57

Barea JM, Pozo MJ, Azcón R, Azcón Aguilar C (2013) Microbial interactions in the rhizosphere. In: de Bruijn FJ (ed) Molecular microbial ecology of the rhizosphere. Wiley, Hoboken, pp 29–44

Barrios E (2007) Soil biota, ecosystem services and land productivity. Ecol Econ 64:269–285. doi:10.1016/j.ecolecon.2007.03.004

Barzana G, Aroca R, Bienert GP, Chaumont F, Ruiz-Lozano JM (2014) New insights into the regulation of aquaporins by the arbuscular mycorrhizal symbiosis in maize plants under drought stress and possible implications for plant performance. Mol Plant-Microbe Interact 27:349–363. doi:10.1094/MPMI-09-13-0268-R

Bashan Y, de-Bashan LE, Prabhu SR, Hernandez JP (2014) Advances in plant growth-promoting bacterial inoculant technology: formulations and practical perspectives (1998–2013). Plant Soil 378:1–33. doi:10.1007/s11104-013-1956-x

Berg G (2009) Plant-microbe interactions promoting plant growth and health: perspectives for controlled use of microorganisms in agriculture. Appl Microbiol Biotechnol 84:11–18. doi:10.1007/s00253-009-2092-7

Bhattacharyya PN, Jha DK (2012) Plant growth-promoting rhizobacteria (PGPR): emergence in agriculture. World J Microbiol Biotechnol 28:1327–1350. doi:10.1007/s11274-011-0979-9

Borie F, Rubio R, Morales A, Curaqueo G, Cornejo P (2010) Arbuscular mycorrhizae in agricultural and forest ecosystems in Chile. J Soil Sci Plant Nutr 10:185–206. doi:10.4067/S0718-95162010000100001

Brodie EL, DeSantis TZ, Parker JP, Zubietta IX, Piceno YM, Andersen GL (2007) Urban aerosols harbor diverse and dynamic bacterial populations. Proc Natl Acad Sci U S A 104:299. doi:10.1073/pnas.0608255104

de Bruijn FJ (2011) Handbook of molecular microbial ecology I: metagenomics and complementary approaches. Wiley, Hoboken

Buée M, De Boer W, Martin F, van Overbeek L, Jurkevitch E (2009) The rhizosphere zoo: an overview of plant-associated communities of microorganisms, including phages, bacteria, archaea, and fungi, and of some of their structuring factors. Plant Soil 321:189–212. doi:10.1007/s11104-009-9991-3

Bulgarelli D, Rott M, Schlaeppi K, Ver Loren van Themaat E, Ahmadinejad N, Assenza F, Rauf P, Huettel B, Reinhardt R, Schmelzer E (2012) Revealing structure and assembly cues for *Arabidopsis* root-inhabiting bacterial microbiota. Nature 488:91–95. doi:10.1038/nature11336

Bulgarelli D, Oter RG, McHardy AC, Lefert PS (2015) Structure and function of the bacterial root microbiota in wild and domesticated barley. Cell Host Microbe 17:392–403. doi:10.1016/j.chom.2015.01.011

Calvo P, Nelson L, Kloepper JW (2014) Agricultural uses of plant biostimulants. Plant Soil 383:3–41. doi:10.1007/s11104-014-2131-8

Calvo-Polanco M, Sánchez-Romera B, Aroca R (2013) Arbuscular mycorrhizal fungi and the tolerance of plants to drought and salinity. In: Aroca R (ed) Symbiotic endophytes. Springer, Berlin, pp 271–288

Caporaso JG, Lauber CL, Walters WA, Berg-Lyons D, Huntley J, Fierer N, Owens SM, Betley J, Fraser L, Bauer M, Gormley N, Gilbert JA, Smith G, Knight R (2012) Ultra-high-throughput microbial community analysis on the Illumina HiSeq and MiSeq platforms. ISME J 6:1621–1624. doi:10.1038/ismej.2012.8

Carvalhais LC, Dennis PG, Tyson GW, Schenk PM (2013) Rhizosphere metatranscriptomics: challenges and opportunities. In: de Bruijn FJ (ed) Molecular microbial ecology of the rhizosphere. Wiley, Hoboken, pp 1137–1144. ISBN: 978-0-470-64479-9

Chaudhry V, Rehman A, Mishra A, Chauhan PS, Nautiyal CS (2012) Changes in bacterial community structure of agricultural land due to long term organic and chemical amendments. Microb Ecol 64:450–460. doi:10.1007/s00248-012-0025-y

Chauhan PS, Chaudhry V, Mishra S, Nautiyal CS (2011) Uncultured bacterial diversity in tropical maize (*Zea mays* L.) rhizosphere. J Basic Microbiol 51:15–32. doi:10.1002/jobm.201000171

Chernin L, Ismailov Z, Haran S, Chet I (1995) Chitinolytic *Enterobacter agglomerans* antagonistic to fungal plant pathogens. Appl Environ Microbiol 61:1720–1726

Compant S, Reiter B, Sessitsch A, Nowak J, Clément C, Barka EA (2005) Endophytic colonization of *Vitis vinifera* L. by plant growth-promoting bacterium *Burkholderia* sp. strain PsJN. Appl Environ Microbiol 71:1685–1693. doi:10.1128/AEM.71.4.1685-1693.2005

De Angelis KM, Brodie EL, DeSantis TZ, Andersen GL, Lindow SE, Firestone MK (2009) Selective progressive response of soil microbial community to wild oat roots. ISME J 3:168–178. doi:10.1038/ismej.2008.103

De Roy K, Marzorati M, Van den Abbeele P, Van de Wiele T, Boon N (2013) Synthetic microbial ecosystems: an exciting tool to understand and apply microbial communities. Environ Microbiol 16:1472–1481. doi:10.1111/1462-2920.12343

Delmotte N, Knief C, Chaffron S, Innerebner G, Roschitzki B, Schlapbach R, von Mering C, Vorholt JA (2009) Community proteogenomics reveals insights into the physiology of phyllosphere bacteria. Proc Natl Acad Sci U S A 106:16428–16433. doi:10.1073/pnas.0905240106

Donn S, Kirkegaard JA, Perera G, Richardson AE, Watt M (2014) Evolution of bacterial communities in the wheat crop rhizosphere. Environ Microbiol 17:610–621. doi:10.1111/1462-2920.12452

Doran JW, Zeiss MR (2000) Soil health and sustainability: managing the biotic component of soil quality. Appl Soil Ecol 15:3–11. doi:10.1016/S0929-1393(00)00067-6

Downey J, van Kessel C (1990) Dual inoculation of *Pisum sativum* with *Rhizobium leguminosarum* and *Penicillium bilaji*. Biol Fertil Soils 10:194–196. doi:10.1007/bf00336135

Duijff BJ, Bakker PAHM, Schippers B (1994) Suppression of *Fusarium* wilt of carnation by *Pseudomonas putida* WCS358 at different levels of disease incidence and iron availability. Biocontrol Sci Tech 4:279–288. doi:10.1080/09583159409355336

Dunne C, Crowley JJ, Moënne-Loccoz Y, Dowling DN, de Bruijn FJ, O'Gara F (1997) Biological control of *Pythium ultimum* by *Stenotrophomonas maltophilia* W81 is mediated by an extracellular proteolytic activity. Microbiology 143:3921–3391. doi:10.1099/00221287-143-12-3921

Egamberdieva D, Jabborova D, Berg G (2016) Synergistic interactions between *Bradyrhizobium japonicum* and the endophyte *Stenotrophomonas rhizophila* and their effects on growth, nodulation and nutrition of soybean under salt stress. Plant Soil 405:35. doi:10.1007/s11104-015-2661-8

Eilers KG, Lauber CL, Knight R, Fierer N (2010) Shifts in bacterial community structure associated with inputs of low molecular weight carbon compounds to soil. Soil Biol Biochem 42:896–903. doi:10.1016/j.soilbio.2010.02.003

El-Tarabily KA (2006) Rhizosphere competent isolates of Streptomycete and non-streptomycete *Actinomycetes* capable of producing cell-wall degrading enzymes to control *Pythium aphanidermatum* damping off disease of cucumber. Can J Bot 84:211–222. doi:10.1139/b05-153

El-Tarabily KA (2008) Promotion of tomato (*Lycopersicon esculentum* Mill.) plant growth by rhizosphere competent 1-aminocyclopropane-1-carboxylic acid deaminase producing streptomycete *Actinomycetes*. Plant Soil 308:161–174. doi:10.1007/s11104-008-9616-2

Frankowski J, Lorito M, Scala F, Schmidt R, Berg G, Bahl H (2001) Purification and properties of two chitinolytic enzymes of *Serratia plymuthica* HRO-C48. Arch Microbiol 176:421–426. doi:10.1007/s002030100347

Gamalero E, Lingua G, Berta G, Lemanceau P (2009) Methods for studying root colonization by introduced beneficial bacteria. In: Lichtfouse E, Navarrete M, Debaeke P, Souchere V, Alberola C (eds) Sustainable agriculture. Springer, New York, pp 601–615. doi:10.1007/978-90-481-2666-8_37

Gloor GB, Hummelen R, Macklaim JM, Dickson RJ, Fernandes AD, MacPhee R, Reid G (2010) Microbiome profiling by illumina sequencing of combinatorial sequence-tagged PCR products. PLoS One 5:e15406. doi:10.1371/journal.pone.0015406

Groppa MD, Benavides MP, Zawoznik MS (2012) Root hydraulic conductance, aquaporins and plant growth promoting microorganisms: a revision. Appl Soil Ecol 61:247–254. doi:10.1016/j.apsoil.2011.11.013

van der Heijden MGA, Bardgett RD, van Straalen NM (2008) The unseen majority: soil microbes as drivers of plant diversity and productivity in terrestrial ecosystems. Ecol Lett 11:296–310. doi:10.1111/j.1461-0248.2007.01139.x

van der Heijden MGA, Martin FM, Selosse MA, Sanders IR (2015) Mycorrhizal ecology and evolution: the past, the present, and the future. New Phytol 205:1406–1423. doi:10.1111/nph.13288

Hirsch PR, Mauchline TH, Clark IM (2010) Culture independent molecular techniques for soil microbial ecology. Soil Biol Biochem 42:878–887. doi:10.1016/j.soilbio.2010.02.019

Howie WJ, Cook RJ, Weller DM (1987) Effects of soil matrix potential and cell motility on wheat root colonization by fluorescent pseudomonads suppressive to take-all. Phytopathology 77:286–292

Jeffries P, Barea JM (2012) Arbuscular mycorrhiza—a key component of sustainable plant-soil ecosystems. In: Hock B (ed) The Mycota, a comprehensive treatise on fungi as experimental systems for basic and applied research. Springer, Berlin, pp 51–75

Jeffries P, Gianinazzi S, Perotto S, Turnau K, Barea JM (2003) The contribution of arbuscular mycorrhizal fungi in sustainable maintenance of plant health and soil fertility. Biol Fertil Soils 37:1–16. doi:10.1007/s00374-002-0546-5

Johnson KB (2010) Pathogen refuge: a key to understanding biological control. Annu Rev Phytopathol 48:141–160. doi:10.1146/annurev.phyto.112408.132643

Jung SC, Martinez-Medina A, Lopez Raez JA, Pozo MJ (2012) Mycorrhiza induced resistance and priming of plant defences. J Chem Ecol 38:651–664. doi:10.1007/s10886-012-0134-6

Khamna S, Yokota A, Lumyong S (2009) *Actinomycetes* isolated from medicinal plant rhizosphere soils: diversity and screening of antifungal compounds, indole-3-acetic acid and siderophore production. World J Microbiol Biotechnol 25:649–655. doi:10.1007/s11274-008-9933-x

Khan MH, Meghvansi MK, Gupta R, Chaudhary KK, Prasad K, Siddiqui S, Veer V, Varma A (2015) Combining application of vermiwash and Arbuscular Mycorrhizal fungi for effective plant disease suppression. In: Varma A, Meghvansi MK (eds) Organic amendments and soil suppressiveness in plant disease management. Springer, Basel, pp 479–493. doi:10.1007/978-3-319-23075-7_23

Klein E, Ofek M, Katan J, Minz D, Gamliel A (2013) Soil suppressiveness to *Fusarium* disease: shifts in root microbiome associated with reduction of pathogen root colonization. Phytopathology 103:23–33. doi:10.1094/PHYTO-12-11-0349

Kumar BSD (1999) Fusarial wilt suppression and crop improvement through two rhizobacterial strains in chick pea growing in soils infested with *Fusarium oxysporum* f. sp. *ciceris*. Biol Fertil Soils 29:87–91. doi:10.1007/s003740050529

Kumar M, Mishra S, Dixit V, Kumar M, Agarwal L, Chauhan PS, Nautiyal CS (2016) Synergistic effect of *Pseudomonas putida* and *Bacillus amyloliquefaciens* ameliorates drought stress in chickpea (*Cicer arietinum* L.) Plant Signal Behav 11:e1071004. doi:10.1080/15592324.2015.1071004

Landa BB, Mavrodi OV, Raaijmakers JM, McSpadden-Gardener BB, Thomashow LS, Weller DM (2002) Differential ability of genotypes of 2,4-diacetylphloroglucinol producing *Pseudomonas fluorescens* to colonize the roots of pea. Appl Environ Microbiol 68:3226–3237. doi:10.1128/AEM.68.7.3226-3237.2002

Li X, Rui J, Xiong J, Li J, He Z, Zhou J, Anthony C, Yannarell MR (2014) Functional potential of soil microbial communities in the maize rhizosphere. PLoS One 9:e112609. doi:10.1371/journal.pone.0112609

Lisette JXC, Germida JJ (2003) Selective interactions between arbuscular mycorrhizal fungi and *Rhizobium leguminosarum* bv. *viceae* enhance pea yield and nutrition. Biol Fertil Soils 37:261–267. doi:10.1007/s00374-003-0605-6

Loper JE, Buyer JS (1991) Siderophores in microbial interactions on plant surfaces. Mol Plant-Microbe Interact 4:5–13

López-Ráez JA, Bouwmeester H, Pozo MJ (2012) Communication in the rhizosphere, a target for pest management. In: Lichtfouse E (ed) Agroecology and strategies for climate change. Springer, Dordrecht, pp 109–133. doi:10.1007/978-94-007-1905-7_5

Lugtenberg B (2015) Life of microbes in the rhizosphere. In: Lugtenberg B (ed) Principles of plant-microbe interactions. Springer, Heidelberg, pp 7–15. doi:10.1007/978-3-319-08575-3_3

Lundberg DS, Lebeis SL, Paredes SH, Yourstone S, Gehring J, Malfatti S, Tremblay J, Engelbrektson A, Kunin V, del Rio TG, Edgar RC, Eickhorst T, Ley RE, Hugenholtz P, Tringe SG, Dangl JL (2012) Defining the core *Arabidopsis thaliana* root microbiome. Nature 488:86–90. doi:10.1038/nature11237

Malusà E, Vassilev N (2014) A contribution to set a legal framework for biofertilizers. Appl Microbiol Biotechnol 98:6599–6607. doi:10.1007/s00253-014-5828-y

Malusa E, Sas-Paszt L, Popinska W, Zurawicz E (2007) The effect of a mycorrhiza-bacteria substrate and foliar fertilization on growth response and rhizosphere pH of three strawberry cultivars. Int J Fruit Sci 6:25–41. doi:10.1126/science.1203980

Marques JM, da Silva TF, Vollu RE, Blank AF, Ding GC, Seldin L, Smalla K (2014) Plant age and genotype affect the bacterial community composition in the tuber rhizosphere of field-grown sweet potato plants. FEMS Microbiol Ecol 88:424–435. doi:10.1111/1574-6941.12313

Mendes R, Kruijt M, de Bruijn I, Dekkers E, van der Voort M, Schneider JHM, Piceno YM, DeSantis TZ, Anderson GL, Bakker PAHM, Raaijmakers JM (2011) Deciphering the rhizosphere microbiome for disease-suppressive bacteria. Science 332:1097–1100. doi:10.1126/science.1203980

Meziane H, Van der Sluis I, Van Loon LC, Höfte M, Bakker PAHM (2005) Determinants of *Pseudomonas putida* WCS358 involved in inducing systemic resistance in plants. Mol Plant Pathol 6:177–185. doi:10.1111/j.1364-3703.2005.00276.x

Micallef SA, Channer S, Shiaris MP, Colón-Carmona A (2009) Plant age and genotype impact the progression of bacterial community succession in the *Arabidopsis* rhizosphere. Plant Signal Behav 4:777–780. doi:10.4161/psb.4.8.9229

Michiels KW, Croes CL, Vanderleyden J (1991) Two different modes of attachment of *Azospirillum brasilense* Sp7 to wheat roots. J Gen Microbiol 137:2241–2246. doi:10.1099/00221287-137-9-2241

Nautiyal CS, Chauhan PS, Bhatia CR (2010) Changes in soil physicochemical properties and microbial functional diversity due to 14 years of conversion of grassland to organic agriculture in semi-arid agro-ecosystem. Soil Tillage Res 109:55–60. doi:10.1016/j.still.2010.04.008

Nichols D (2007) Cultivation gives context to the microbial ecologist. FEMS Microbiol Ecol 60:351–357. doi:10.1111/j.1574-6941.2007.00332.x

Notz R, Maurhofer M, Schnider-Keel U, Duffy B, Haas D, Défago G (2001) Biotic factors affecting expression of the 2, 4-diacetylphloroglucinol biosynthesis gene phl A in *Pseudomonas fluorescens* biocontrol strain CHA0 in the rhizosphere. Phytopathology 91:873–881. doi:10.1094/PHYTO.2001.91.9.873

Nunes da Rocha U, Andreote FD, De Azevedo JL, Van Elsas JD, Van Overbeek LS (2010) Cultivation of hitherto-uncultured bacteria belonging to the Verrucomicrobia subdivision 1 from the potato (*Solanum tuberosum* L.) rhizosphere. J Soils Sediments 10:326–339. doi:10.1007/s11368-009-0160-3

Osorio NW, Habte M (2001) Synergistic influence of an arbuscular mycorrhizal fungus and P solubilizing fungus on growth and plant P uptake of *Leucaena leucocephala* in an Oxisol. Arid Land Res Manag 15:263–274. doi:10.1080/15324980152119810

Owen D, Williams AP, Griffith GW, Withers PJA (2015) Use of commercial bio-inoculants to increase agricultural production through improved phosphorus acquisition. Appl Soil Ecol 86:41–54. doi:10.1016/j.apsoil.2014.09.012

Pace NR, Sapp J, Goldenfeld N (2012) Phylogeny and beyond: scientific, historical, and conceptual significance of the first tree of life. Proc Natl Acad Sci U S A 109:1011–1018. doi:10.1073/pnas.1109716109

Pangesti N, Pineda A, Pieterse CMJ, Dicke M, van Loon JJA (2013) Two-way plant-mediated interactions between root-associated microbes and insects: from ecology to mechanisms. Front Plant Sci 4:414. doi:10.3389/fpls.2013.00414

Paterson E, Gebbing T, Abel C, Sim A, Telfer G (2007) Rhizodeposition shapes rhizosphere microbial community structure in organic soil. New Phytol 173:600–610. doi:10.1111/j.1469-8137.2006.01931.x

Pérez-García A, Romero D, de Vicente A (2011) Plant protection and growth stimulation by microorganisms: biotechnological applications of Bacilli in agriculture. Curr Opin Biotechnol 22:187–193. doi:10.1016/j.copbio.2010.12.003

Perotto S, Bonfante P (1997) Bacterial associations with mycorrhizal fungi: close and distant friends in the rhizosphere. Trends Microbiol 5:496–501. doi:10.1016/s0966-842x(97)01154-2

Phelan VV, Liu WT, Pogliano K, Dorrestein PC (2012) Microbial metabolic exchange—the chemotype phenotype link. Nat Chem Biol 8:26–35. doi:10.1038/nchembio.739

Pozo MJ, Azcón Aguilar C (2007) Unraveling mycorrhiza-induced resistance. Curr Opin Plant Biol 10:393–398. doi:10.1016/j.pbi.2007.05.004

Pozo MJ, Jung SC, Martínez-Medina A, López-Ráez JA, Azcón-Aguilar C, Barea JM (2013) Root allies: arbuscular mycorrhizal fungi help plants to cope with biotic stresses. In: Aroca R (ed) Symbiotic endophytes. Springer, Berlin, pp 289–307. doi:10.1007/978-3-642-39317-4_15

Pozo MJ, López-Ráez JA, Azcón-Aguilar C, García-Garrido JM (2015) Phytohormones as integrators of environmental signals in the regulation of mycorrhizal symbioses. New Phytol 205:1431–1436. doi:10.1111/nph.13252

Raaijmakers JM, Weller DM (1998) Natural plant protection by 2,4-diacetylphloroglucinol producing *Pseudomonas* spp. in take-all decline soils. Mol Plant-Microbe Interact 11:144–152

Raaijmakers JM, Paulitz TC, Steinberg C, Alabouvette C, Moënne-Loccoz Y (2009) The rhizosphere: a playground and battlefield for soil borne pathogens and beneficial microorganisms. Plant Soil 321:341–361. doi:10.1007/s11104-008-9568-6

Ram RJ, VerBerkmoes NC, Thelen MP, Tyson GW, Baker BJ, Blake RC II, Shah M, Hettich RL, Banfield JF (2005) Community proteomics of a natural microbial biofilm. Science 308:1915–1920. doi:10.1126/science. 1109070
Rice WA, Lupwayi NZ, Olsen PE, Schlechte D, Gleddie SC (2000) Field evaluation of dual inoculation of alfalfa with *Sinorhizobium meliloti* and *Penicillium bilaii*. Can J Plant Sci 80:303–308. doi:10.4141/p99-055
Ritz K (2007) The plate debate: cultivable communities have no utility in contemporary environmental microbial ecology. FEMS Microbiol Ecol 60:358–362. doi:10.1111/j.1574-6941.2007.00331.x
Ruiz-Lozano JM, Porcel R, Azcón C, Aroca R (2012) Regulation by arbuscular mycorrhizae of the integrated physiological response to salinity in plants: new challenges in physiological and molecular studies. J Exp Bot 63:4033–4044. doi:10.1093/jxb/ers126
Schüßler A, Schwarzott D, Walker C (2001) A new fungal phylum, the Glomeromycota, phylogeny and evolution. Mycol Res 105:1413–1421. doi:10.1017/S0953756201005196
Selosse MA, Bessis A, Pozo MJ (2014) Microbial priming of plant and animal immunity: symbionts as developmental signals. Trends Microbiol 22:607–613. doi:10.1016/j.tim.2014.07.003
Sergeeva E, Hirkala DLM, Nelson LM (2007) Production of indole-3-acetic acid, aromatic amino acid aminotransferase activities and plant growth promotion by *Pantoea agglomerans* rhizosphere isolates. Plant Soil 297:1–13. doi:10.1007/s11104-007-9314-5
Sessitsch A, Hardoim P, Döring J, Weilharter A, Krause A, Woyke T, Mitter B, Hauberg-Lotte L, Friedrich F, Rahalkar M, Hurek T, Sarkar A, Bodrossy L, van Overbeek L, Brar D, van Elsas JD, Reinhold-Hurek B (2012) Functional characteristics of an endophyte community colonizing rice roots as revealed by metagenomic analysis. Mol Plant-Microbe Interact 25:28–36. doi:10.1094/MPMI-08-11-0204
Singh JS, Pandey VC, Singh DP (2011) Efficient soil microorganisms: a new dimension for sustainable agriculture and environmental development. Agric Ecosyst Environ 140:339–353. doi:10.1016/j.agee.2011.01.017
Smith KP, Goodman RM (1999) Host variation for interactions with beneficial plant-associated microbes. Annu Rev Phytopathol 37:473–491
Smith SE, Read DJ (2008) Mycorrhizal symbiosis, 3rd edn. Elsevier, New York
Someya N, Tsuchiya K, Yoshida T, Noguchi MT, Akutsu K, Sawada H (2007) Co-inoculation of an antibiotic-producing bacterium and a lytic enzyme-producing bacterium for the biocontrol of tomato wilt caused by *Fusarium oxysporum* f. sp. *lycopersici*. Biocontrol Sci 12:1–6. doi:10.4265/bio.12.1
de Souza JTA, Arnould C, Deulvot C, Lemanceau P, Gianinazzi-Pearson V, Raaijmakers JM (2003) Effect of 2,4-diacetylphloroglucinol on *Pythium*: cellular responses and variation in sensitivity among propagules and species. Phytopathology 93:966–975. doi:10.1094/PHYTO.2003.93.8.966
Spence C, Alff E, Johnson C, Ramos C, Donofrio N, Sundaresan V, Bais H (2014) Natural rice rhizospheric microbes suppress rice blast infections. BMC Plant Biol 14:130. doi:10.1186/1471-2229-14-130
Sturz AV, Christie BR, Nowak J (2000) Bacterial endophytes: potential role in developing sustainable systems of crop production. CRC Crit Rev Plant Sci 19:1–30. doi:10.1080/07352680091139169
Sugiyama A, Ueda Y, Zushi T, Takase H, Yazaki K (2014) Changes in the bacterial community of soybean rhizospheres during growth in the field. PLoS One 9:e100709. doi:10.1371/journal.pone.0100709
Taberlet P, Prud'homme SM, Campione E, Roy J, Miquel C, Shehzad W, Gielly L, Rioux D, Choler P, Clément JC, Melodelima C, Pompanon F, Coissac E (2012) Soil sampling and isolation of extracellular DNA from large amount of starting material suitable for metabarcoding studies. Mol Ecol 21:1816–1820. doi:10.1111/j.1365-294X.2011.05317.x
Thomashow LS, Weller DM (1988) Role of a phenazine antibiotic from *Pseudomonas fluorescens* in biological control of *Gaeumannomyces graminis* var. *tritici*. J Bacteriol 170:3499–3508
Uroz S, Buée M, Murat C, Frey-Klett P, Martin F (2010) Pyrosequencing reveals a contrasted bacterial diversity between oak rhizosphere and surrounding soil. Environ Microbiol Rep 2:281–288. doi:10.1111/j.1758-2229.2009.00117.x

Van Loon LC, Bakker PAHM (2003) Signalling in rhizobacteria-plant interactions. In: De Kroon H, Visser EJW (eds) Root ecology (Ecological studies). Springer, Berlin, pp 297–330. doi:10.1007/978-3-662-09784-7_12

Vassilev N, Franco I, Vassileva M, Azcon R (1996) Improved plant growth with rock phosphate solubilized by *Aspergillus niger* grown on sugar beet waste. Bioresour Technol 55:237–241. doi:10.1016/0960-8524(96)00008-9

Vassilev N, Medina A, Azcon R, Vassileva M (2006) Microbial solubilization of rock phosphate media containing agro-industrial wastes and effect of the resulting products on plant growth and P uptake. Plant Soil 287:77–84. doi:10.1007/s11104-006-9054-y

Vassilev N, Martos E, Mendes G, Martos V, Vassileva M (2013) Biochar of animal origin: a sustainable solution of the high grade rock phosphate scarcity? J Sci Food Agric 93:1799–1804. doi:10.1002/jsfa.6130

Velazhahan R, Samiyappan R, Vidhyasekaran P (1999) Relationship between antagonistic activities of *Pseudomonas fluorescens* isolates against *Rhizoctonia solani* and their production of lytic enzymes. Z Pflanz Pflanzen 106:244–250

Wang HX, Geng ZL, Zeng Y, Shen YM (2008) Enriching plant microbiota for a metagenomic library construction. Environ Microbiol 10:2684–2691. doi:10.1111/j.1462-2920.2008.01689.x

Wang X, Pan Q, Chen F, Yan X, Liao H (2011) Effects of co-inoculation with arbuscular mycorrhizal fungi and rhizobia on soybean growth as related to root architecture and availability of N and P. Mycorrhiza 21:173–181. doi:10.1007/s00572-010-0319-1

Weinert N, Piceno Y, Ding GC, Meincke R, Heuer H, Berg G, Schloter M, Andersen G, Smalla K (2011) PhyloChip hybridization uncovered an enormous bacterial diversity in the rhizosphere of different potato cultivars: many common and few cultivar-dependent taxa. FEMS Microbiol Ecol 75:497–506. doi:10.1111/j.1574-6941.2010.01025.x

Weller DM, Cook RJ (1983) Suppression of take-all of wheat by seed treatments with fluorescent pseudomonads. Phytopathology 73:463–469

Woeng TFCCA, Bloemberg GV, Mulders IHM, Dekkers LC, Lugtenberg BJJ (2000) Root colonization by phenazine-1-carboxamide-producing bacterium *Pseudomonas chlororaphis* PCL1391 is essential for biocontrol of tomato root rot. Mol Plant-Microbe Interact 12:1340–1345. doi:10.1094/MPMI.2000.13.12.1340

Woeng TFCCA, Bloemberg GV, Lugtenberg BJJ (2003) Phenazines and their role in biocontrol by *Pseudomonas* bacteria. New Phytol 157:503–523. doi:10.1046/j.1469-8137.2003.00686.x

Wu SC, Cao ZH, Li ZG, Cheung KC, Wong MH (2005) Effects of biofertilizer containing N-fixer, P and K solubilizers and AM fungi on maize growth: a greenhouse trial. Geoderma 125:155–166. doi:10.1016/j.geoderma.2004.07.003

Zachow C, Müller H, Tilcher R, Berg G (2014) Differences between the rhizosphere microbiome of *Beta vulgaris* ssp. *maritima*—ancestor of all beet crops and modern sugar beets. Front Microbiol 5:415. doi:10.3389/fmicb.2014.00415

Zarraonaindia I, Owens SM, Weisenhorn P, West K, Hampton-Marcell J, Lax S, Bokulich NA, Mills DA, Martin G, Taghavi S, van der Lelie D, Gilberta JA (2015) The soil microbiome influences grapevine-associated microbiota. mBio 6:e02527-14. doi:10.1128/mBio.02527-14

Zolla G, Badri DV, Bakker MG, Manter DK, Vivanco JM (2013) Soil microbiomes vary in their ability to confer drought tolerance to *Arabidopsis*. Appl Soil Ecol 68:1–9. doi:10.1016/j.apsoil.2013.03.007

Plant-Associated Microbial Endophytes: Promising Source for Bioprospecting

15

Shipra Singh and Anita Pandey

Abstract

In the past few years, scientists have recognized various potential biotechnological applications of microorganisms that are useful for plants and also are of economical importance to humans. For such applications, endophytes ubiquitously present in plant species worldwide are becoming a primary target. These organisms can perform various tasks such as plant growth promotion and biological control of phytopathogens. Simultaneously, they may induce a range of novel compounds that could be potential candidates for use in pharma and agriculture industries. Besides, they enhance bioremediation to remove soil contaminants and may increase soil fertility through nutrient cycling. Secondary metabolites produced by these microorganisms have shown so much potential that they are being manipulated, both physicochemically and genetically, to multiply yields of desired compounds. In this chapter, we have tried to comprehend ecology along with the functional aspects of endophytes in plants with respect to their significance and impacts on man and environment.

Keywords

Endophytes • Natural products • Secondary metabolites • Bioprospecting

S. Singh • A. Pandey (✉)
Biotechnological Applications, G.B. Pant National Institute of Himalayan Environment and Sustainable Development, Kosi-Katarmal, Almora 263 643, Uttarakhand, India
e-mail: anita@gbpihed.nic.in, anitapandey333@gmail.com

V.C. Kalia et al. (eds.), *Mining of Microbial Wealth and MetaGenomics*,
DOI 10.1007/978-981-10-5708-3_15

15.1 Introduction

The microorganisms that inhabit inner tissues (intra- and/or intercellular) of higher plants, at least for one period of their life cycle, without causing any apparent symptoms are well accepted as endophytes (Bacon and White 2000). Endophytes are natural associates of plant-microbe ecology system and are precious gene pool (Zhang et al. 2006). Various groups of microorganisms, such as bacteria (Germaine et al. 2004), fungi (Molina et al. 2012) and actinobacteria (Yuan et al. 2008), are reported to colonize as microbial endophytes inside the plant tissues. They have been isolated from almost all the studied plants and are ubiquitous in distribution. In recent times, these organisms are gaining increased attention among scientists working in various fields owing to their unique ability to produce a range of secondary metabolites. Secondary metabolites produced by endophytes are shown to exhibit a promising range of biological activities as antibiotic, anticancerous agents, growth promoters of plants, biocontrol agents of phytopathogens, enzymes and many more on similar lines. The chances to recover new and novel microorganisms with potential bioactivity among the plants growing in different environmental conditions and ecological niches are excellent. In this chapter, we present status of endophytic ecology along with their significance as promising tools of biotechnological applications and, therefore, their importance for bioprospecting.

15.2 Ecology: Host-Endophyte Interaction

Probably, endophytes evolved along with their host plants millions of years ago when higher plants first appeared. Evidences for their long co-evolution with plants were discovered in fossilized tissues of temperate palm, *Trachycarpus fortunei* (Taylor et al. 1999). During this period they had adapted themselves to the niche and developed as gene regulated and completely compatible symbiont (Zhang et al. 2006). Exchange of genetic and/or chemical information among themselves and the host plant as well would have helped the endophytes in acclimatizing to the environmental conditions. The same would have been responsible for acquiring the compatibility to the host plant. Proposed ways of finding host by endophytes are chemotaxis and electrotaxis or by accidental encounter. Endophytes occupy plant's interior as a protected niche wherein they are benefited by little competition from other microorganisms for a constant and reliable nutritional source. Moreover, the internal colonization by endophytes provides an added advantage over epiphytes that they remain protected from various environmental conditions that otherwise may not be suitable for them such as extremes of temperature, osmotic potential and ultraviolet radiations, etc. In this microbiome, they are provided with photosynthetic products and minerals for their normal growth. In turn, endophytes promote the growth and chemical defence of host plants by directly providing with certain valuable metabolites or by transferring corresponding genes to the host genome (Wink 2008; Thongsandee et al. 2013). In this manner, they provide various ecological services,

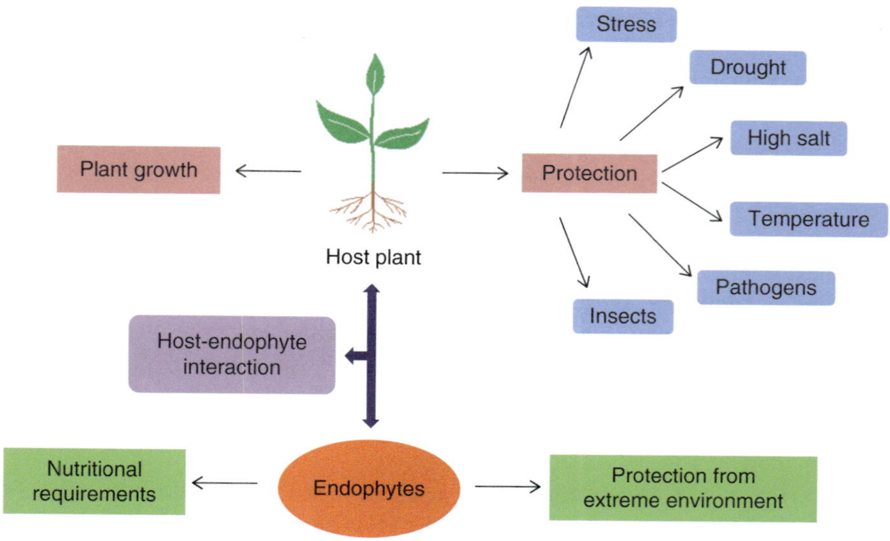

Fig. 15.1 Possible outcomes of host-endophyte interaction in terms of benefits to both the partners

such as protection against pathogens and enhanced tolerance against stressed environment like drought, high salt, improper temperature, etc. (Fig. 15.1).

More background knowledge on host-endophytes interaction is needed to elucidate the services provided by endophytes in nature and towards the search of natural active principles. Ability of an endophyte to produce bioactive compounds in its natural niche and under in vitro culture is highly uncertain. A particular endophyte may acquire dominance over others by producing certain bioactive compounds within its biological niche. This, in turn, may also provide resistance towards harmful pathogens. The case may be more appealing if the compound(s) is/are produced by the endophyte solely and not by the host plant. Such studies are keys in depicting the role of endophytes in the host plants. Several factors seem to involve in biological changes of endophytes in the host plant with respect to season, age, environment and location. Molecular approaches may unravel various facts underlying host-endophyte interactions and ecology of endophytes. An ecological awareness of the role of these microorganisms in natural niche will be helpful in targeting the particular endophytic bioactivity towards maximum possibility for bioprospecting.

15.3 Transmission of Endophytes

The main entry for endophytes is supposed to be horizontal transfer through wounds (Sprent and de Faria 1998) in root zone as well as in aerial parts (Sharrock et al. 1991). Stomata and lenticels may also serve as entry points for endophytes (Kluepfel 1993). There are studies showing that incidences of endophytes increase as leaves and seedling grow older whereas seed and seedling are virtually free of endophytes (Arnold et al. 2003; Gallery et al. 2007). In an asymptomatic host, it is difficult to understand the mechanism of horizontal transmission of endophytes. The question arises, how the inoculum for infection is produced; when and where? In case of latent saprophyte, it can be assumed that the inoculum which infects new host could be produced when the infected host tissue dies. There are reports showing saprophytes which produce fructification in dead plant parts are present as endophytes in healthy tissues (Sánchez et al. 2007). In other ways, endophytic inoculum may be produced in an inconspicuous manner in infected hosts. For example, the endophyte *Epichloë* develops a microscopic layer of hyphae and conidia on the surface of leaves of some grasses. It is thought that this inoculum might be responsible for horizontal transmission to occur (Tadych et al. 2007). According to Devarajan and Suryanarayana (2006), phytophagous insects may also take part in the spread of endophytes, since spores of some fungal species are resistant to gut digestion, and are present in their faecal pellets.

In another mode, the endophytes are vertically transmitted and are completely dependent on host reproduction for their own propagation and reproductive fitness. Studies on vertically transmitted endophytes are limited, and most of these are dependent on the study of seed transmitted fungi (Gallery et al. 2007). Some examples of vertical transmission to host progeny are *Neotyphodium* endophytes and some *Epichloë* species (*E. festucae*, *E. sylvatica*). In these cases, the seeds produced by fungal-infected plant showed the presence of fungal mycelium near the embryo and in the aleurone layer. Therefore, it can be assumed that endophytic species are vertically transmitted in a fashion similar to a maternally inherited character (Schardl et al. 2004). Incidences of these endophytes in natural populations of their hosts are very high (Arroyo et al. 2002) and are almost host specific (Schardl 2010). Vertically transmitted clavicipitaceous endophytes are associated with selected taxa of Poaceae and Convolvulaceae and are restricted to narrow host range (Clay and Schardl 2002; Steiner et al. 2011).

15.4 Biological and Chemical Diversity

Endophytic microorganisms are recovered from almost every studied plant species ranging from herbaceous plants such as tomato, rice, ferns and mosses and lichens (Li et al. 2007) to woody trees such as poplar, ginkgo, etc. A single plant species may harbour a diverse array of endophytic microorganisms. Classical methods to study endophytes involve isolation of endophytes from various plant parts following surface sterilization followed with phenetic and genetic characterization of recovered isolates (Miche and Balandreau 2001). These methods show cultural bias, and more often, species that grow very slowly or do not grow in culture media are overlooked. Environmental PCR, in which direct amplification of DNA from surface sterilized plant material is done, could have corrected this cultural bias. Use of these methods has shown that only 1% of species are culturable in case of bacteria. However, in case of fungi, relatively higher proportion can be cultured on media. It is difficult to compare culturing and direct amplification methods since these issues are based on methodology adopted. Consideration of morphotypes to calculate diversity is a preferred method, but this method may result in overestimation of the diversity. This was clear with the study on endophyte of *Pinus* sp., *Sphaeropsis sapinea* that showed the presence of more than five morphotypes on culture medium (Burgess et al. 2001). Other methodological factors such as sterilizing agent, size of the sample, season of collection, etc. are also known to collectively affect the number of endophytes isolated. Many of the studies are somewhat imprecise about the sampling strategy (Gamboa et al. 2002; Bayman 2006). For DNA-based methods, bias can be introduced by extraction method and primer. Still, there is no universal barcode region to study fungal diversity. Internal transcribed spacer (ITS) regions are being used to study fungal diversity, and other regions like cytochrome C oxidase subunit 1 (*CO1*) and multilocus barcodes are also getting attention these days (Seifert et al. 2007; Roe et al. 2010). Direct amplification methods may reveal new major lineages; however, coupling of both methods may provide best results (Bayman 2006).

Chemical diversity is another face of biological diversity. This is a fact for constant chemical changes that continuously take place in an ecosystem during evolutionary race. This condition is more remarkable in tropical rainforests where various conditions like competition, limitation for resources and selection pressure counteract simultaneously. Therefore, in such ecosystems, probability of finding new and novel molecular structures is relatively high (Redell and Gordon 2000). Bills et al. (2002) reported that number of bioactive compounds isolated from tropical endophytes was considerably higher as compared to compounds recovered from endophytes of temperate origin.

15.5 Groups of Endophytes

Based on the most popular classification methods, the microbial endophytes are mainly grouped into fungal, bacterial and actinobacterial endophytes.

15.5.1 Fungal Endophytes

As per available records, plants are supposed to be colonized with endophytic fungi for more than 400 million years ago. This was the time when plants started colonizing the land. Therefore, these interactions are considered to play an important role in driving the land colonization and evolution (Krings et al. 2007). Fungal endophytes are classified into four classes based on hosts they colonized. In Class I endophytes, *clavicipitaceous* species are limited to a small number of phylogenetically related *clavicipitaceous* species occurring in some cool and warm season grasses that exhibit fastidious growth in culture media. Primarily, they are transmitted vertically in which maternal plants pass on fungal endophyte on to offspring via producing seed infected with fungi. Plants associated with these endophytes are benefitted depending upon species and genotype of host and environmental conditions in which host is growing. Class 2 endophytes dwelling in individual host plants comprise a small group with members restricted to the *Dikarya* (*Ascomycota* or *Basidiomycota*). Major benefit they provide to their hosts is the tolerance towards various environmental stresses. Members of Class 3 endophytes exhibit great diversity within individual host belonging to vascular to non-vascular plants and woody to herbaceous angiosperms and in different ecoclimatic conditions like tropical forest to Antarctic communities. They also show horizontal transmission. Class 4 endophytes have colonization restricted to plant roots and are distinguished by the presence of darkly melanized septa. Most members of this class belong to *Ascomycetes* that form melanized structures within plant roots like inter and intracellular hyphae and microsclerotia (Rodriguez et al. 2009).

15.5.2 Bacterial Endophytes

Endophytic bacteria usually occupy intercellular spaces and also colonize vascular tissues of plants. They have been recovered from almost every higher plant that involve woody trees such as poplar and herbaceous plants such as sweet corn. During last few years, increasing tendency to report bacterial endophytes from medicinal plants is observed, e.g. *Codonopsis pilosula*, *Fructus forsythiae* (Ma et al. 2010), *Taxus chinensis* (Zhao et al. 2010a) and *Ginkgo biloba* (Kumar et al. 2009; Pandey et al. 2009). A diversified range of bacterial species are reported to be endophytic including *Acetobacter*, *Arthrobacter*, *Bacillus*, *Burkholderia*, *Pseudomonas*, etc. (Lodewyckx et al. 2002). Li et al. (2008) studied 150 soybean root nodules and could successfully recover a total of 98 non-symbiotic endophytic bacteria. More than 129 endophytic bacterial species that include both Gram-negative and Gram-positive bacteria belonging to 54 genera have so far been isolated from different crop plants. Most of these bacteria belong to the former pseudomonad group (*Pseudomonas*, *Burkholderia*, *Phyllobacterium*) and the family *Enterobacteriaceae*. Also, there are reports on endophytic bacteria colonizing plants that are growing in natural abominable conditions. For example, 61 endophytic bacteria were isolated from five plant species (*Vicia*, *Oxytropis*, *Medicago*, *Melilotus* and *Onobrychis*) growing in Qinghai-Tibet plateau and Loess plateau (Kan et al. 2007); many of these bacteria were shown for resistance to high alkaline conditions (pH 11) and high salt (NaCl) concentration (3–5%, w/v). In another study, community structure of endophytic bacteria showed dominance of *Bacillus* spp. in roots of the halophyte *Aster tripolium* (Szymanska et al. 2016).

15.5.3 Actinobacterial Endophytes

Actinobacteria have been found to produce more than 50,000 bioactive secondary metabolites, such as antibiotics, antitumor agents, immunosuppressive chemicals and enzymes. So, more attention is paid on isolating actinobacteria from different resources. *Streptomyces* species is the most commonly isolated endophytic actinobacteria. Medicinal plants have been of major interest for isolating actinobacterial endophytes that provided us several important secondary metabolites. Eighty-one endophytic actinobacteria belonging to eight genera were isolated from roots of Chinese cabbage (Lee et al. 2008). Wherein, *Microbispora* spp. were the most commonly isolated actinobacteria, followed by *Streptomyces* spp. and *Micromonospora* spp. Qin et al. (2011) isolated and characterized more than 40 new taxa employing polyphasic approaches that also included four new genera, i.e. *Plantactinospora*, *Actinophytocola*, *Phytohabitans* and *Jishengella*.

15.6 Applications

Endophytic microorganisms are excellent sources of bioactive natural products that can be used to satisfy demand of pharmaceutical, medical, agriculture and industry. The benefits provided by these organisms are discussed in the proceeding paras.

15.6.1 Plant Growth Promotion

Endophytes are well known for producing several important compounds, such as indoleacetic acid (IAA), indole acetonitrile, gibberellin and cytokinin. These compounds have an important role in root elongation and plant growth promotion. An endophytic *Pseudomonas* sp. was isolated from cortical cells of *Ginkgo biloba* roots and evaluated for its plant growth promoting abilities. This bacterium showed strong ability to solubilize tricalcium phosphate and enhanced plant biomass in barley and rice (Pandey et al. 2009). Improved survival of transplant stock stamped ability of this endophytic *Pseudomonas* sp. for conservation and propagation of *Ginkgo biloba* which is considered as a living fossil (Pandey et al. 2014, Fig. 15.2). Seed germination rate and seedlings length are also reported to be improved in rape and tomato (Nejad and Johnson 2000). An endophytic fungus, *Trichoderma gamsii*, was isolated from the lateral roots of lentil growing in mountain ecosystem of Indian Himalaya and is reported to possess plant growth promotion and biocontrol properties (Rinu et al. 2014).

Despite of this knowledge, there is yet to learn much more on the subject line, such as the role of seed endophytes. However, some of the seed endophytes have been investigated for enhancing plant growth on account of production of plant hormones and/or helping in improving the acquisition of plant nutrients (Johnston-Monje and Raizada 2011; Gagne-Bourgue et al. 2013; Xu et al. 2014). Herrera et al. (2016) demonstrated plant growth promotion in barley and wheat kernels through inoculation with endophytes isolated from wheat seeds.

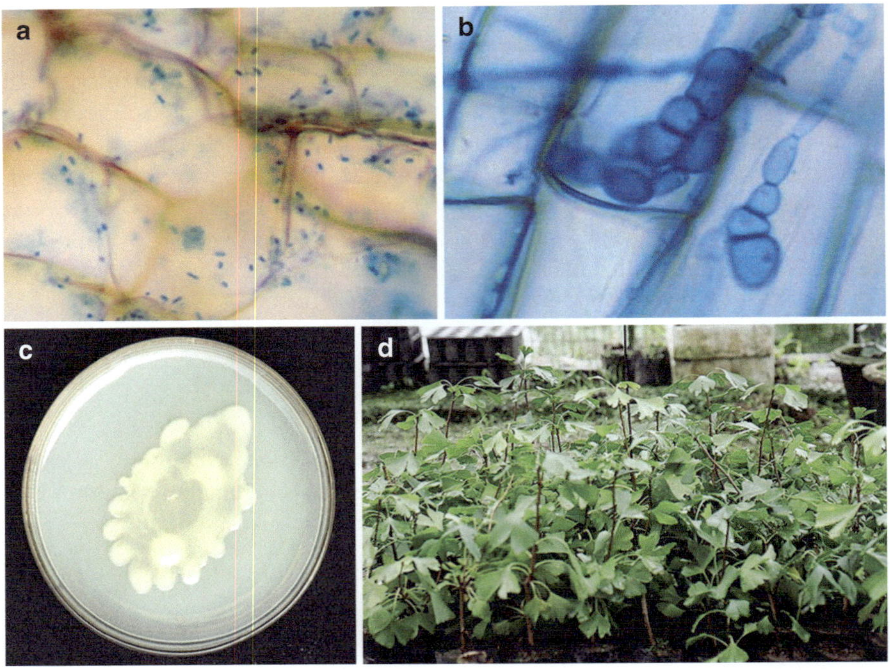

Fig. 15.2 Presence of bacterial (**a**) and fungal (**b**) endophytes in cortical cells of *Ginkgo biloba* roots (×400), pure culture of endophytic *Pseudomonas* sp. MTCC9476 growing on Tryptone Yeast extract agar (**c**) and use of endophytic bacterium in propagation of *G. biloba* through stem cuttings (**d**)

15.6.2 Biological Control

Endophytic microorganisms are regarded as an effective biocontrol agent, alternative to chemical control. They are known to provide protection against certain pathogenic organisms due to production of secondary metabolites and hydrolytic enzymes, and being competitive in nutrient acquisition. The mechanisms by which endobacteria benefit their host plants seem to be similar as described for rhizosphere bacteria. Compant et al. (2005) have already reviewed these mechanisms in a wide perspective. Reports are available for use of endophytes to provide protection against fungal, bacterial and viral diseases, and in some instances, harmful effects of insects and nematodes could also be reduced (Azevedo et al. 2000; Nejad and Johnson 2000; Sturz and Kimpinski 2004; Herrera et al. 2016). They can also trigger induced systemic resistance. Expression of defence-related genes responsible for triggering salicylic acid- or jasmonic acid-/ethylene-dependent signalling pathways in leaves following inoculation with *Streptomyces* sp. EN27 and *Micromonospora* sp. strain EN43 was observed by Conn et al. (2008). *Arabidopsis thaliana*, in turn, possessed resistance against phytopathogens, such as *Erwinia carotovora* and *Fusarium oxysporum*.

Not only naturally occurring endophytes are used as biocontrol agents but also they are genetically engineered to express antipest proteins, such as lectins. A fungal endophyte, *Chaetomium globosum* YY-11, isolated from rape seedlings and bacterial endophytes of *Enterobacter* sp. and *Bacillus subtilis* isolated from rice seedlings were used to express *Pinellia ternata* agglutinin (*PtA*) gene (Zhao et al. 2010b). These recombinant endophytes expressing *PtA* gene were found to effectively control the population of sap sucking pests in several crop seedlings. Similarly, in a different study, recombinant endophytic bacteria *Enterobacter cloacae* expressing *PtA* gene proved to be a bioinsecticide against white backed plant hopper, *Sogatella furcifera* (Zhang et al. 2011). Use of recombinant endophytes as biocontrol agents expressing different antipest proteins is a promising technique for control of plant pests.

15.6.3 Bioremediation/Biodegradation

Endophytes represent important resource for bioremediation. Probability of finding such endophytes is high in plants growing in soil contaminated with xenobiotics. These endophytes are shown to possess necessary contaminant-degrading genes required for degradation of contaminant xenobiotics (Siciliano et al. 2001). These bacteria are capable of improving biomass production under conditions of metal toxicity. This was shown in *Nicotiana tabaccum* where inoculation of plants with endophytes improved growth of plants as well as helped in accumulating more cadmium (Cd) in plants under conditions of Cd stress (Mastretta et al. 2009). To obtain plastic degrading endophytes, several endophytic fungi were screened under in vitro conditions using synthetic polymer polyester polyurethane (PUR) as substrate (Russell et al. 2011). Dominating genus for great PUR degradation activity in both solid and liquid suspensions was *Pestalotiopsis* with many other isolates showing moderate activity. Biodegradation of nitro-aromatic compounds such as 2,4,6-trinitrotoluene was shown by Van Aken et al. (2004) using an endophytic strain of *Methylobacterium* that was originally isolated from hybrid poplar trees (*Populus deltoides* x *nigra*). Nowadays, endophytic strains are being genetically engineered and tailor-made for obtaining desired activity. For example, in one such experiment, endophytes from yellow lupin were genetically modified for nickel resistance and successfully increased tolerance towards nickel toxicity and nickel accumulation in inoculated plants (Lodewyckx et al. 2001).

15.6.4 Natural Products: Bioactive and Novel Compounds

Endophytes naturally synthesize bioactive compounds that are useful in plant protection against diseases caused by pathogens. Besides, they have shown potential to produce compounds for novel drug discovery. A schematic representation showing how these endophytes can be studied in a systematic manner towards recovery of novel bioactive compounds is provided in Fig. 15.3. Although a large number of

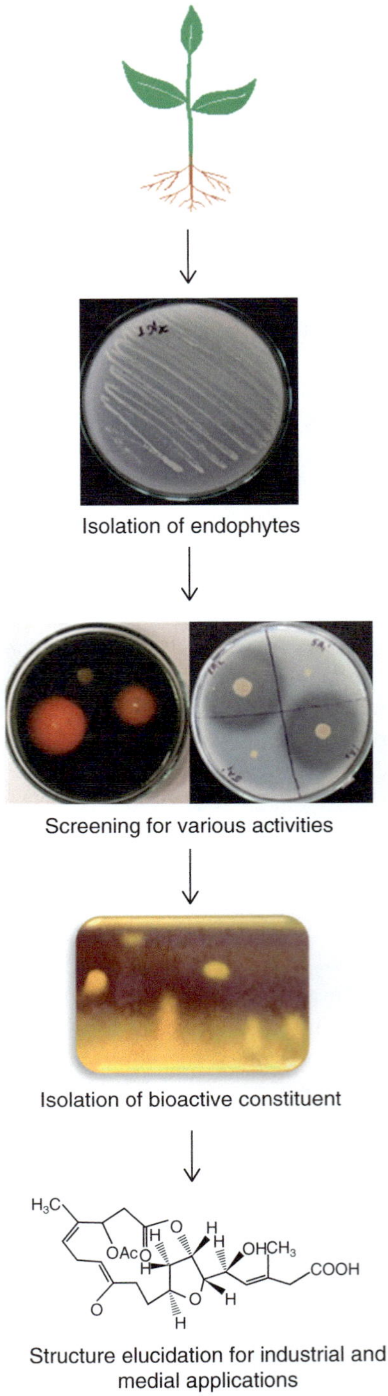

Fig. 15.3 Schematic representation of methodology to study bioactive constituents of endophytes

bioactive compounds have been isolated from endophytic organisms, there is still a lot of scope in this area since only a fraction of endophytes are utilized for obtaining novel natural products. Several natural products including alkaloids, terpenoids, flavonoids and steroids have been reported from endophytes till date. Most of the bioactive compounds isolated from endophytes are known to have functions of antibiotics, immunosuppressants, antidiabetic compounds, plant growth promoters and so forth (Joseph and Mini Priya 2011).

15.6.4.1 Antibiotics

Antibiotics are low-molecular weight organic compounds produced by microorganisms that have lethal effect on other microorganisms. Antibiotics have very specific mode of action, affecting the vital processes like DNA, RNA, protein and cell wall synthesis. To combat ever-increasing drug resistance of plant and human pathogens, new alternatives are urgently needed. Exploration of novel niche of endophytic biodiversity leads to discovery of several natural products of great importance. The endophytic fungi isolated from the *Dioscorea composita*, an important medicinal plant from northwest Himalaya, have recently been reported for their antimicrobial potential (Gupta et al. 2015). Antimicrobial agents produced by endophytic microorganisms fall in different structural classes such as alkaloids, peptides, steroids, terpenoids, phenols, quinones and flavonoids.

Alkaloids are well-known secondary metabolites produced by endophytes that may exhibit antimicrobial activities. Some examples of alkaloids are chaetoglobosins from *Chaetomium globosum*, an endophytic fungus, isolated from the leaves of *Ginkgo biloba* (Qin et al. 2009); pyrrocidines obtained from *Acremonium zeae*, an endophyte of maize (Wicklow et al. 2005); and phomoenamide synthesized by *Phomopsis* sp., an endophytic fungi isolated from *Garcinia dulcis* bolite (Rukachaisirikul et al. 2008).

Several peptides synthesized by endophytic microorganisms are also known for exhibiting significant activities against pathogenic microbes. For example, *Acremonium* sp., an endophyte of *Taxus baccata*, produced Leuesnostatin A that exhibited strong activity against pathogen *Pythium ultimum* (Strobel et al. 1997). The endophytic bacteria *Paenibacillus* sp. IIRAC-30, isolated from cassava, produced lipopeptide belonging to surfactin series (Canova et al. 2010). Several terpenoids, quinones, phenols, flavanoids, steroids and aliphatic compounds are already described and reviewed as antimicrobials from endophytes (Yu et al. 2010).

15.6.4.2 Antiviral Compounds

Research in the area of discovering potential antiviral compounds from endophytes is still in its infancy. The major factor limiting the discovery of such compound is probably the lack of proper antiviral screening systems. Cytonic acids A and B, two novel protease inhibitors of human cytomegalovirus, were isolated from the endophytic fungus *Cytonaema* sp. by solid-state fermentation (Guo et al. 2000). These results are encouraging enough for future studies.

15.6.4.3 Immunosuppressive Compounds

Immunosuppressive drugs are of key importance in transplant studies to prevent allograft rejection. Treatment of autoimmune diseases such as rheumatoid arthritis and insulin-dependent diabetes can be done using these compounds in the future. Immunosuppressive but noncytotoxic diterpene pyrones subglutinols A and B were isolated from *Fusarium subglutinans* (Lee et al. 1995). A promising beneficial immunosuppressant, cyclosporine, was isolated from fungus *Tolypocladium inflatum* using a computer aided programme for screening of fungi capable of producing bioactive compounds (Borel and Kis 1991).

15.6.4.4 Anticancerous Agents

First report of the production of paclitaxel from an endophyte of northwest Pacific yew was so encouraging that researchers took interest and successfully isolated several other important anticancer agents from endophytes (Stierle et al. 1993). The most common endophytes capable of producing anticancer agents are *Pestalotiopsis* spp. from world's yews; among these *P. microspora* is the most commonly isolated endophytic species. Endophytic *P. microspora*, isolated from *Taxus wallichiana*, was also shown to produce paclitaxel by monoclonal antibody test (Strobel et al. 1996; Strobel 2002).

15.6.4.5 Enzyme Production

The number of endophytic fungi capable of producing industrially important enzymes is high. These fungi are well reported; here are presented some examples. Fungal strains from the *Acremonium* endophyte species have been reported for production of hemicellulases and cellulases (de Almeida et al. 2012). Suto et al. (2002) recovered 155 endophytic fungal strains from 14 plant species that produced xylanases. A thermotolerant β-glucosidase was isolated and purified from an endophytic *Periconia* sp. (Harnpicharnchai et al. 2009). Robl et al. (2013) screened 110 endophytic fungi for the production of lignocellulosic biomass degrading hemicellulases and related enzymes. Out of these, six fungi, i.e. *Alternaria* sp. DR45, *Annulohypoxylon stygium* DR47, *Aspergillus niger* DR02, *Talaromyces wortmannii* DR49 and *Trichoderma atroviride* DR17 and DR19, were recognized as potent producers of hemicellulases and related hydrolytic enzymes.

15.6.4.6 Pigment Production

Endophytic fungus of *Ginkgo biloba* twigs, *Penicillium purpurogenum*, produced abundant red pigments that was soluble and could be used as natural food colourant (Qiu et al. 2010). Endophytic fungus *Monodictys castaneae* produced a pigment that was found to inhibit human pathogenic bacteria *Staphylococcus aureus*, *Klebsiella pneumonia*, *Salmonella typhi* and *Vibrio cholerae* and was proven to be more active than streptomycin (Visalakchi and Muthumary 2009).

15.6.4.7 Other Applications

Several other applications of the endophytes are well recognized and need to be worked out in an expanded way. Enhanced decomposition of ageing plants in the estuary or salt marshes by saprophytic bacteria was shown in the presence of endophytic fungi of marine plants (Cornick et al. 2005). Endophytes can also be applicable as a molecular tool in form of gene carrier. Exogenous genes can be expressed in the plants on successful colonization by the endophyte carrying that gene that in turn may provide protection to the host. For example, a heterologous gene was introduced into an endophytic microorganism; this gene responsible for the production of toxicant for insects was expressed in the host plant to provide protection from insects. This technique was applied commercially as well (Azevedo et al. 2000).

15.6.5 Future Prospects

In the past few years, there is significant increase pertaining to the field of endophytes, in terms of isolation of novel microbes as well as isolation of bioactive antimicrobial metabolites. Research for the exploitation of this excellent resource of poorly understood microorganisms got pace in the last two decades. It is now well established that these organisms hold an enormous potential for product and utilitarian discovery. Now, it is the time to review past research in the arena of natural products produced by endophytes and to scrutinize future prospects of endophytic biodiversity for various applications in biotechnology.

Plants are providing suitable environment for endophytes, which undoubtedly is storehouse of myriad of new microorganisms providing novel structurally diverse compounds that can be used in various industries. Thus, it is obvious that work in this field holds enormous promise. But the work related to endophytes is still in its juvenile stage. We can hunt for novel endophytes from plants that grow in extreme environments. To understand host-endophyte interactions in a better manner, molecular studies are required in this field. One area that is untouched in this review is that some of the endophytes are known human pathogens but also plant growth promoters.

Discovery of some potential antibiotics that can be effective against multidrug-resistant bacteria may be quite easier by improving some potent strains using genetic engineering and microbial fermentation processes. There is high probability for isolation of novel secondary metabolites because of boon of novel bioassay techniques and extensive chemical separation science. Thus, continuous investigations on endophytic bacteria from diverse and extreme habitats and application of new technologies are the key for future success. The mechanisms underlying existence of endophytes and their response to the surroundings are keys to predict a plant species with high probability to find potential microbes. Therefore, a better understanding is also required in this area to facilitate the processes in product discovery.

Endophyte-plant interactions can be explored to facilitate sustainable low-input agricultural system since they play important roles in plant growth promotion and plant protection. Both food and non-food crops can be benefitted by these interactions. Another promising area is development of endophytes for production of bioenergy crops along with bioremediation of toxic metals and biodegradation of xenobiotics.

Acknowledgements The editors of *Mining of Microbial Wealth and Metagenomics* are gratefully acknowledged for extending the invitation for sharing our views on one of the emerging research areas.

References

de Almeida MN, Guimarães VM, Bischoff KM, Falkoski DL, Pereira OL, Gonçalves DS, de Rezende ST (2012) Cellulases and hemicellulases from endophytic *Acremonium* species and its application on sugarcane bagasse hydrolysis. Appl Biochem Biotechnol 165:594–610

Arnold AE, Mejía LC, Kyllo D, Rojas EI, Maynard Z, Robbins N, Erre EA (2003) Fungal endophytes limit pathogen damage in a tropical tree. Proc Natl Acad Sci U S A 100:15649–15654

Arroyo GR, Martínez ZJM, García CB, Zabalgogeazcoa I (2002) Genetic structure of natural populations of the grass endophyte *Epichloë festucae* in semiarid grasslands. Mol Ecol 11:355–364

Azevedo J, Maccheroni W Jr, Pereira JO, Luiz de Arauzo W (2000) Endophytic microorganisms: a review on insect control and recent advances on tropical plants. Electron J Biotechnol 3:40–65

Bacon CW, White JF (2000) Microbial endophytes. Marcel Dekker Inc., New York, NY

Bayman P (2006) Diversity, scale and variation of endophytic fungi in leaves of tropical plants. In: Bailey MJ, Lilley AK, Timms-Wilson TM (eds) Microbial ecology of aerial plant surfaces. CABI, Oxfordshire, pp 37–50

Bills G, Dombrowski A, Pelaez F, Polishook J, An Z (2002) Recent and future discoveries of pharmacologically active metabolites from tropical fungi. In: Watling R, Frankland JC, Ainsworth AM, Issac S, Robinson CH (eds) Tropical mycology: micromycetes, vol 2. CABI Publishing, New York, NY, pp 165–194

Borel JF, Kis ZL (1991) The discovery and development of cyclosporine. Transplant Proc 23:1867–1874

Burgess T, Wingfield MJ, Wingfield BW (2001) Simple sequence repeat markers distinguish among morphotypes of *Sphaeropsis sapinea*. Appl Environ Microbiol 67:354–362

Canova S, Petta T, Reyes L, Zucchi T, Moraes L, Melo I (2010) Characterization of lipopeptides from *Paenibacillus* sp. (IIRAC30) suppressing *Rhizoctonia solani*. World J Microbiol Biotechnol 26:2241–2247

Clay K, Schardl C (2002) Evolutionary origins and ecological consequences of endophyte symbiosis with grasses. Am Nat 160:S99–S127

Compant S, Duffy B, Nowak J, Clément C, Brka EA (2005) Use of plant growth-promoting bacteria for biocontrol of plant diseases: principles, mechanisms of action, and future prospects. Appl Environ Microbiol 71:4951–4959

Conn VM, Walker AR, Franco CM (2008) Endophytic actinobacteria induce defense pathways in *Arabidopsis thaliana*. Mol Plant-Microbe Interact 21(2):208–218

Cornick J, Standwerth A, Fisher PJ (2005) A preliminary study of fungal endophyte diversity in a stable and declining bed of *Spartina anglica* Hubbard. Mycologist 19:24–29

Devarajan PT, Suryanarayana TS (2006) Evidence for the role of phytophagous insects in dispersal of non-grass fungal endophytes. Fungal Divers 23:111–119

Gagne-Bourgue F, Aliferis KA, Seguin P, Rani M, Samson R, Jabaji S (2013) Isolation and characterization of indigenous endophytic bacteria associated with leaves of switchgrass (*Panicum virgatum* L.) cultivars. J Appl Microbiol 114:836–853

Gallery RA, Dalling JW, Arnold AE (2007) Diversity, host affinity and distribution of seed-infecting fungi: a case study with *Cecropia*. Ecology 88:582–588

Gamboa MA, Laureano S, Bayman P (2002) Measuring diversity of endophytic fungi in leaf fragments: does size matter? Mycopathologia 156:41–45

Germaine K, Keogh E, Garcia-Cabellos G, Borremans B, van der Lelie D, Barac T, Oeyen L, Vangronsveld J, Moore FP, Moore ERB, Campbell CD, Ryan D, Dowling DN (2004) Colonization of poplar trees by gfp expressing bacterial endophytes. FEMS Microbiol Ecol 48:109–118

Guo B, Dai J, Ng S, Huang Y, Leong C, Ong W, Carte BK (2000) Cytonic acids A and B: novel tridepside inhibitors of hCMV protease from the endophytic fungus *Cytonaema* species. J Nat Prod 63:602–604

Gupta S, Kaul S, Dhar MK (2015) Exploring the antimicrobial potential of fungal endophytes of *Dioscorea composita*: a north-west Himalayan medicinal plant. ENVIS Bull Himalayan Ecol 23:113–118

Harnpicharnchai P, Champreda V, Sornlake W, Eurwilaichitr L (2009) A thermotolerant beta-glucosidase isolated from an endophytic fungi, *Periconia* sp., with a possible use for biomass conversion to sugars. Protein Expr Purif 67:61–69

Herrera SD, Grossia C, Zawoznika M, Groppaa MD (2016) Wheat seeds harbour bacterial endophytes with potential as plant growth promoters and biocontrol agents of *Fusarium graminearum*. Microbiol Res 186:37–43

Johnston-Monje D, Raizada MN (2011) Conservation and diversity of seed associated endophytes in *Zea* across boundaries of evolution, ethnography and ecology. PLoS One 6:e20396. doi:10.1371/journal.pone.0020396

Joseph B, Mini Priya R (2011) Bioactive compounds from endophytes and their potential in pharmaceutical effect: a review. Am J Biochem Mol Biol 1:291–309

Kan FL, Chen ZY, Wang ET, Tian CF, Sui XH, Chen WX (2007) Characterization of symbiotic and endophytic bacteria isolated from root nodules of herbaceous legumes grown in Qinghai–Tibet plateau and in other zones of China. Arch Microbiol 188:103–115

Kluepfel D (1993) The behavior and tracking of bacteria in the rhizosphere. Annu Rev Phytopathol 31:441–472

Krings M, Taylor TN, Hass H, Kerp H, Dotzler N, Hermsen EJ (2007) Fungal endophytes in a 400-million-yr-old land plant: infection pathways, spatial distribution, and host responses. New Phytol 174:648–657

Kumar A, Singh S, Pandey A (2009) General microflora, arbuscular mycorrhizal colonization and occurrence of endophytes in rhizosphere of two age groups of *Ginkgo biloba* L. of Indian central Himalaya. Indian J Microbiol 49:134–141

Lee J, Lobkovsky E, Pliam NB, Strobel GA, Clardy J (1995) Subglutinols a and B: immunosuppressive compounds from the endophytic fungus *Fusarium subglutinans*. J Org Chem 60:7076–7077

Lee SO, Choi GJ, Choi YH, Jang KS, Park DJ, Kim CJ, Kim JC (2008) Isolation and characterization of endophytic actinomycetes from Chinese cabbage roots as antagonists to *Plasmodiophora brassicae*. J Microbiol Biotechnol 18:1741–1746

Li WC, Zhou J, Guo SY, Guo LD (2007) Endophytic fungi associated with lichens in Baihua mountain of Beijing, China. Fungal Divers 25:69–80

Li JH, Wang ET, Chen WF, Chen WX (2008) Genetic diversity and potential for promotion of plant growth detected in nodule endophytic bacteria of soybean grown in Heilongjiang province of China. Soil Biol Biochem 40:238–246

Lodewyckx C, Taghavi S, Mergeay M, Vangronsveld J, Clijsters H, van der Lelie D (2001) The effect of recombinant heavy metal resistant endophytic bacteria in heavy metal uptake by their host plant. Int J Phytoremediation 3:173–187

Lodewyckx C, Vangronsveld J, Porteous F, Moore ERB, Taghavi S, van der Lelie D (2002) Endophytic bacteria and their potential applications. Crit Rev Plant Sci 6:583–606

Ma Y, Xie ZL, Li DY, Ma ZJ (2010) Research on biological diversity of endophytes from medicinal plants in Tibetan plateau. Biotechnol Bull 6:234–239

Mastretta C, Taghavi S, van der Lelie D, Mengoni A, Galardi F, Gonnelli C, Barac T, Boulet J, Weyens N, Vangronsveld J (2009) Endophytic bacteria from seeds of *Nicotiana tabacum* can reduce cadmium phytotoxicity. Int J Phytoremediation 11:251–267

Miche L, Balandreau J (2001) Effects of rice seed surface sterilization with hypochlorite on inoculated *Burkholderia vietnamiensis*. Appl Environ Microbiol 67:3046–3052

Molina G, Pimentel MR, Bertucci TCP, Pastore GM (2012) Application of fungal endophytes in biotechnological processes. Chem Eng Trans 27:289–294

Nejad P, Johnson PA (2000) Endophytic bacteria induce growth promotion and wilt disease suppression in oilseed rape and tomato. Biol Control 18(3):208–215

Pandey A, Singh S, Kumar A, Palni LMS, Malviya MK, Rinu K (2009) Isolation of an endophytic plant growth promoting *Pseudomonas* sp. strain gb3 (MTCC9476) from *Ginkgo biloba* L., growing in temperate Himalaya. Natl Acad Sci Lett 32:83–88

Pandey A, Sati P, Malviya MK, Singh S, Kumar A (2014) Use of endophytic bacterium (*Pseudomonas* sp., MTCC9476) in propagation and conservation of *Ginkgo biloba* L.: a living fossil. Curr Sci 106:1066–1067

Qin JC, Zhang YM, Gao JM, Bai MS, Yang SX, Laatsch H, Zhang AL (2009) Bioactive metabolites produced by *Chaetomium globosum*, an endophytic fungus isolated from *Ginkgo biloba*. Bioorg Med Chem Lett 19(6):1572–1574

Qin S, Xing K, Jiang XL, Li W (2011) Biodiversity, bioactive natural products and biotechnological potential of plant-associated endophytic actinobacteria. Appl Microbiol Biotechnol 89:457–473

Qiu M, Xie R, Shi Y, Chen H, Wen Y, Gao Y, Hu X (2010) Isolation and identification of endophytic fungus SX01, a red pigment producer from *Ginkgo Biloba* L. World J Microbiol Biotechnol 26:993–998

Redell P, Gordon V (2000) Lessons from nature: can ecology provide new leads in the search for novel bioactive chemicals from rainforests? In: Wrigley SK, Hayes MA, Thomas R, Chrystal EJT, Nicholson N (eds) Biodiversity: new leads for pharmaceutical and agrochemical industries. The Royal Society of Chemistry, Cambridge, United Kingdom, pp 205–212

Rinu K, Sati P, Pandey A (2014) *Trichoderma gamsii* (NFCCI 2177): a newly isolated endophytic, psychrotolerant, plant growth promoting, and antagonistic fungal strain. J Basic Microbiol 54(5):408–417

Robl D, Delabona PS, Mergel CM, Rojas JD, Costa PS, Pimentel IC, Vicente VA, Pradella JGC, Padilla G (2013) The capability of endophytic fungi for production of hemicellulases and related enzymes. BMC Biotechnol 13:94. doi:10.1186/1472-6750-13-94

Rodriguez RJ, White JF Jr, Arnold AE, Redman RS (2009) Fungal endophytes: diversity and functional roles. New Phytol 182:314–330

Roe AD, Rice AV, Bromilow SE, Cooke JEK, Sperling FAH (2010) Multilocus species identification and fungal DNA barcoding: insights from blue stain fungal symbionts of the mountain pine beetle. Mol Ecol Resour 10:946–959

Rukachaisirikul V, Sommart U, Phongpaichit S, Sakayaroj J, Kirtikara K (2008) Metabolites from the endophytic fungus *Phomopsis* sp. PSU-D15. Phytochemistry 69:783–787

Russell JR, Huang J, Anand P, Kucera K, Sandoval AG, Dantzler KW, Hickman DS, Jee J, Kimovec FM, Koppstein D, Marks DH, Mittermiller PA, Núñez SJ, Santiago M, Townes MA, Vishnevetsky M, Williams NE, Vargas MPN, Boulanger L, Bascom-Slack C, Strobel SA (2011) Biodegradation of polyester polyurethane by endophytic fungi. Appl Environ Microbiol 77:6076–6084

Sánchez MS, Bills GF, Zabalgogeazcoa I (2007) The endophytic mycobiota of the grass *Dactylis glomerata*. Fungal Divers 27:171–195

Schardl CL (2010) The epichloae, symbionts of the grass subfamily Poőideae. Ann Mo Bot Gard 97:646–665

Schardl CL, Leuchtmann A, Spiering MJ (2004) Symbioses of grasses with seedborne fungal endophytes. Annu Rev Plant Biol 55:315–340

Seifert KA, Samson RA, deWaard JR, Houbraken J, Levesque CA, Jean-Marc Moncalvo JM, Louis-Seize G, Hebert PDN (2007) Prospects for fungus identification using CO1 DNA barcodes, with *Penicillium* as a test case. Proc Natl Acad Sci U S A 104:3901–3906

Sharrock K, Parkes S, Jack H, Rees-George J, Hawthorne B (1991) Involvement of bacterial endophytes in storage rots of buttercup squash (*Cucurbita maxima* D. Hybrid '*Delicia*'). New Zealand J Crop Hort Sci 19:157–165

Siciliano S, Fortin N, Himoc N, Wisse G, Labelle S, Beaumier D, Ouellette D, Roy R, Whyte LG, Banks MK, Schwab P, Lee K, Greer CW (2001) Selection of specific endophytic bacterial genotypes by plants in response to soil contamination. Appl Environ Microbiol 67:2469–2475

Sprent J, de Faria S (1998) Mechanisms of infection of plants by nitrogen fixing organisms. Plant Soil 110:157–165

Steiner U, Leibner S, Schardl CL, Leuchtmann A, Leistner E (2011) *Periglandula*, a new fungal genus within the Clavicipitaceae and its association with Convolvulaceae. Mycologia 103:1133–1145

Stierle A, Strobel GA, Stierle D (1993) Taxol and taxane production by *Taxomyces andreanae*, an endophytic fungus of Pacific yew. Science 260:214–216

Strobel GA (2002) Microbial gifts from rain forests. Can J Plant Pathol 24:14–20

Strobel G, Yang X, Sears J, Kramer R, Sidhu RS, Hess WM (1996) Taxol from *Pestalotiopsis microspora*, an endophytic fungus of *Taxus wallichiana*. Microbiology 142:435–440

Strobel GA, Torzynski R, Bollon A (1997) *Acremonium* sp.-a leucinostatin A producing endophyte of European yew (*Taxus baccata*). Plant Sci 128:97–108

Sturz AV, Kimpinski J (2004) Endoroot bacteria derived from marigolds (*Tagetes* spp.) can decrease soil population densities of root-lesion nematodes in the potato root zone. Plant Soil 262:241–249

Suto M, Takebayashi M, Saito K, Tanaka M, Yokota A, Tomita F (2002) Endophytes as producers of xylanase. J Biosci Bioeng 93:88–90

Szymanska S, Płociniczakb T, Piotrowska-Segetb Z, Złocha M, Ruppelc S, Hrynkiewicza K (2016) Metabolic potential and community structure of endophytic and rhizosphere bacteria associated with the roots of the halophyte *Aster tripolium* L. Microbiol Res 182:68–79

Tadych M, Bergen M, Dugan FM, White JF (2007) Evaluation of the potential role of water in spread of conidia of the *Neotyphodium* endophyte of *Poa ampla*. Mycol Res 111:466–472

Taylor J, Hyde K, Jones E (1999) Endophytic fungi associated with the temperate palm, *Trachycarpus fortunei*, within and outside its natural geographic range. New Phytol 142:335–346

Thongsandee W, Matsuda Y, Shimizu M, Ehara H, Ito S (2013) Isolation of endophytic streptomycetes from above- and below-ground organs of *Quercus serrata*. J Forest Res 18:179–189

Van Aken B, Peres C, Doty S, Yoon J, Schnoor J (2004) *Methylobacterium populi* sp. nov., a novel aerobic, pink pigmented, facultatively methylotrophic, methane-ultilising bacterium isolated from poplar trees (*Populus deltoides* x *nigra* DN34). Int J Syst Evol Microbiol 54:1191–1196

Visalakchi S, Muthumary J (2009) Antimicrobial activity of the new endophytic *Monodictys castaneae* SVJM139 pigment and its optimization. Afr J Microbiol Res 3:550–556

Wicklow DT, Roth S, Deyrup ST, Gloer JB (2005) A protective endophyte of maize: *Acremonium zeae* antibiotics inhibitory to *Aspergillus flavus* and *Fusarium verticillioides*. Mycol Res 109:610–618

Wink M (2008) Plant secondary metabolism: diversity, function and its evolution. Nat Prod Commun 3:1205–1216

Xu M, Sheng J, Chen L, Men Y, Gan L, Guo S, Shen L (2014) Bacterial community compositions of tomato (*Lycopersicum esculentum* mill.) seeds and plant growth promoting activity of ACC deaminase producing *Bacillus subtilis* (HYT-12-1) on tomato seedlings. World J Microbiol Biotechnol 30:835–845

Yu H, Zhang L, Li L, Zheng C, Guo L, Li W, Sun P, Qin L (2010) Recent developments and future prospects of antimicrobial metabolites produced by endophytes. Microbiol Res 165:437–449

Yuan HM, Zhang XP, Zhao K, Zhong K, Gu YF, Lindstrom K (2008) Genetic characterisation of endophytic actinobacteria isolated from the medicinal plants in Sichuan. Ann Microbiol 58(4):597–604

Zhang HW, Song YC, Tan RX (2006) Biology and chemistry of endophytes. Nat Prod Rep 23:753–771

Zhang X, Li J, Qi G, Wen K, Lu J, Zhao X (2011) Insecticidal effect of recombinant endophytic bacterium containing *Pinellia ternata* agglutinin against white backed planthopper, *Sogatella furcifera*. Crop Prot 30:1478–1484

Zhao YX, Liu KH, Deng BW, Chen WQ, Geng Z, Li JH (2010a) Antibacterial activities of metabolites produced by two endophytic bacteria isolates in *Taxus chinensis*. J Food Sci Biotechnol 4:617–623

Zhao X, Qi G, Zhang X, Lan N and Ma X 2010b. Controlling sap sucking insect pests with recombinant endophytes expressing plant lectin. Nat Proc 21, article 21

Bifidobacterial Probiotics Through Fermented Foods

16

Sajan C. Achi and Prakash M. Halami

Abstract

The importance of bifidobacteria and its contribution to the health of human beings have made them attractive as a probiotic. Conventionally, fermented products served as a carrier for delivery of bifidobacteria. Little is known about the involvement of bifidobacteria in fermentation of such products. Foods fermented by bifidobacteria besides being a carrier may serve as a source of bioactive compounds, vitamins, amino acids, etc. However, maintaining viability of bifidobacteria in sufficient amount during storage has always been the concern for probiotic industry. The fermented foods associated with bifidobacteria and their roles in fermentation are described in this chapter. Furthermore, the significance of using fermented foods for delivering bifidobacteria and strategies employed to maintain the viability of bacteria are also discussed.

Keywords

Bifidobacteria • Health benefits • Fermented products • Probiotic delivery • Viability

16.1 Introduction

Among the diverse and complex ecosystems studied, human gastrointestinal tract (GIT) is one of the most important, comprising large numbers of bacteria of different phyla, class, family and species. These bacteria are involved in various physiological and metabolic processes that are responsible for the well-being of the host.

S.C. Achi • P.M. Halami (✉)
Academy of Scientific and Innovative Research (AcSIR), CSIR-Central Food Technological Research Institute (CSIR-CFTRI) Campus, Mysuru 570 020, India

Microbiology and Fermentation Technology Department, CSIR-Central Food Technological Research Institute, Mysuru 570 020, India
e-mail: prakashalami@cftri.res.in

V.C. Kalia et al. (eds.), *Mining of Microbial Wealth and MetaGenomics*,
DOI 10.1007/978-981-10-5708-3_16

Though there is a high amount of diversity and variation, the gut microbiota of an individual is considered to stabilise by time and will be unique (Martín et al. 2013). Microbes colonise the gut of infants immediately after birth, and their organisation is a lifelong process. Bifidobacteria belonging to the phylum of *Actinobacteria* are one of the dominant commensals present in the gut of vaginally delivered infants (Ventura et al. 2012). They are typical gut microbes and have been identified in various eukaryotic hosts such as mammals, birds and insects. They belong to the family *Bifidobacteriaceae*, and the genus currently includes 48 species (Bottacini et al. 2014).

Bifidobacteria was first reported by Henri Tisser in 1899, who isolated them from the faeces of breast-fed infants. They were initially named as *Bacillus bifidus* and were originally included in the genus of *Lactobacillus* but later found their separate genus *Bifidobacterium*. *Bifidobacterium* acquires its name from the bifid morphology (V- or Y-shaped morphology). Though the exact cause for the bifid morphology remains yet to be determined, low concentrations or the lack of *N*-acetylamino-sugar, Ca^{2+} ions or amino acids (alanine, aspartic acid, glutamic acid and serine) in growth media are reported to be responsible. Members of genus *Bifidobacterium* are Gram-positive, anaerobic, catalase-negative high G + C containing nonmotile and non-sporulating bacteria. Although bifidobacteria are generally portrayed as strict anaerobes, some strains can tolerate oxygen. The ability to tolerate oxygen may vary between species and also among strains of same species. They are chemoorganotrophs and metabolise hexoses through a peculiar pathway called bifid shunt pathway or fructose 6-phosphoketolase pathway (named on the basis of key enzyme involved: Fructose 6 phosphoketolase). This enzyme was considered as an important marker for identification of bifidobacteria (Lee and O'Sullivan 2010; Russell et al. 2011).

16.1.1 Bifidobacteria as a Commensal and Its Adaptation to Human Gut

Bifidobacteria have been isolated from various sources that include sewage, fermented milk products, anaerobic wastewater treatment facility, etc. But majority of them are linked with the alimentary tract of humans, animals and insects; other sources may be the result of direct or indirect contamination of their original source (Bottacini et al. 2014). Microbes present in the GIT are known to be involved in symbiotic interactions with its host, and these interactions include mutualism, commensalism and pathogenesis. Commensal relationships develop when there is a co-evolution of bacteria and their host. In such relationships, both the partners are benefited due to the unique metabolic and physiological processes shared by them, and neither of them is at a loss. Bifidobacteria are thought to colonise the gut along with other microflora from vagina and faeces of mother and environmental microbes, during the initial development of gut microbiota in infancy. In the case of breast-feeding infants, maternal milk may also act as a source of inoculum. In adults, bifidobacteria is considered to be one of the dominant gut microbiota and occupies

3% of the total gut microbiota. The association and ability to colonise GIT imply the adaptation of bifidobacteria to human gut and their further role as a commensal. Genomic studies including functional genome analysis, whole genome sequencing and comparative genomics have played a major role in revealing the mechanisms that are associated with survival, establishment and colonisation of these bacteria to human gut (Ventura et al. 2012).

Human GIT produces different enzymes that are involved in the hydrolysis of different disaccharides and polysaccharides. However, certain polysaccharides mainly dietary fibre-derived oligo and polysaccharides skip digestion as the GIT lacks a certain set of enzymes that are capable of hydrolysing these sugars. The ability to metabolise these sugars which reach the distal colon unutilised plays a significant role for the microbial colonisation (Ventura et al. 2007; Cronin et al. 2011). Eight percent of annotated genes in case of many bifidobacterial genomes encode enzymes involved in carbohydrate metabolism. Their genomes are predicted to encode various glycosyl hydrolases, essential for complex carbohydrate utilisation. These bacteria are also capable of utilising glycans and glycoproteins produced by the host, for instance, mucins produced by the host and released in the intestine. These mucins are comprised of acetyl glucosamine, acetyl galactosamine, fucose, galactose and sialic acid. Genome analysis revealed a set of genes in *B. bifidum* PRL2010 capable of metabolising mucin. Human milk oligosaccharides (HMOs) present in human milk that is underutilised by the host are also shown to be metabolised by bifidobacterial species. Gene cluster that encodes glycosidases and carbohydrate transporters involved in the metabolism of HMOs was identified in *B. longum* subsp. *infantis* ATCC 15697. The proficiency to utilise these sugars may vary between the species and within the species among different strains which subsequently determines their dominance, persistence and diversity in a particular ecological niche (Ventura et al. 2012; O'Sullivan and Halami 2013).

Bifidobacteria have to counteract the stressful environment of intestine for their successful colonisation. Bile salts, digestive enzymes and osmotic stress resulting from diet variation are some of the factors which these bacteria have to encounter in the intestinal environment. Alteration in the fatty acid composition of the cell wall, exopolysaccharide (EPS) synthesis, bile salt hydrolases (BSH), bile salt efflux transporter, multidrug-resistant ATP-binding cassette transporter and molecular chaperones are some of the mechanisms which enable bifidobacteria to tolerate bile stress. Production of serine protease inhibitors (serpin) offers protection from antitrypsin and neutrophil elastase present in the host (Ruiz et al. 2011; Ventura et al. 2012; O'Sullivan and Halami 2013).

Adhesion of bacteria to intestinal epithelial cells and mucin is one of the important features that facilitate colonisation of bacteria. The factors responsible for adhesion and further interaction of bifidobacteria with the host are subjects of investigation. Nevertheless, some of the factors identified include EPS, pili, fimbriae and some moonlighting proteins (Ventura et al. 2012).

Thus bifidobacteria are well-adapted commensals, and they are known to be involved in crosstalk with intestinal epithelial cells. These interactions and the metabolic activities of bifidobacteria are considered important for maintenance of

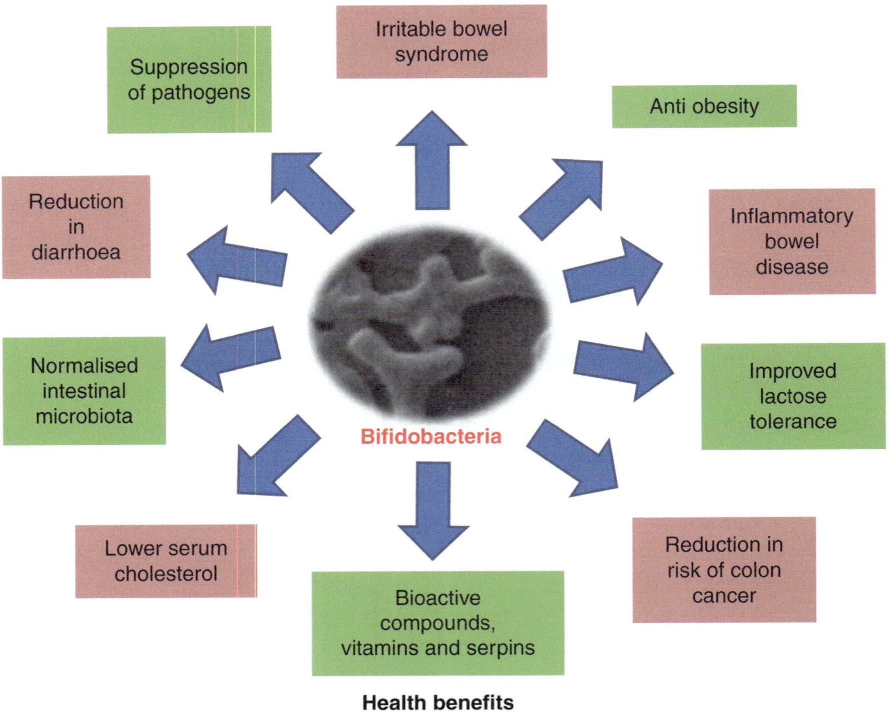

Fig. 16.1 Health benefits of bifidobacteria

health. Further, the inverse correlation between the presence of these bacteria and different diseases, as well as undesirable microbes, signifies the role of these bacteria in health. These positive attributes have made them attractive as a probiotic.

Probiotics are microorganisms that are proven to exert health-promoting effects on its hosts. The basic criteria for an organism to be considered as a probiotic include the ability to endure stress encountered during GI tract passage, attach epithelial cells of the intestine and colonise the tract, should be safe for consumption and free from negative effects, should help in gut homeostasis and have functional attributes beneficial for health. Since bifidobacteria are a natural inhabitant of the gut, these criteria are well satisfied by these organisms. Numerous in vitro, in vivo studies and clinical trials document the positive impact of bifidobacteria as a probiotic on human health is presented in Fig. 16.1 (Lee and O'Sullivan 2010).

16.2 Bifidobacteria and Fermented Foods

The concept of probiotics was introduced by Elie Metchnikoff when he found that microorganisms in fermented dairy products can prevent fouling of the intestine and later related that to longevity of life (Heller 2001). So, initially, the microorganisms considered for probiotics were lactic acid bacteria (LAB) from fermented foods; later on intestinal LAB garnered much attention as a probiotic. Bifidobacteria are heterofermenters and produce lactic acid and acetic acid. Very few bifidobacteria have been reported to be isolated from fermented food products which include *B. mongoliense* DSM 21395 from fermented milk, *B. crudilactis* LMG 23609 from raw milk cheese and *B. animalis* subsp. *lactis* DSM 10140 from fermented milk (Delcenserie et al. 2007, 2013; Bunesova et al. 2014). Most isolates of bifidobacteria are reported from non-food origin. However, fermented food products are often used as a carrier for delivery of these probiotic organisms.

16.2.1 Dairy Products

Bifidobacteria are used in a wide range of fermented dairy products which includes fermented milk, cheese, curd, yoghurt, etc. They may be used alone or in combination with other cultures for fermentation. In the case of yoghurt-like products, they may be added before or after fermentation. The poor proteolytic activity of bifidobacteria is presumed to be the cause of its slow growth in fermented milk products when compared with other LAB. Hence, bifidobacteria containing fermented dairy products are often prepared employing *Lactobacillus*, *Lactococcus* and *Streptococcus* as starter inoculum. However, these starters used for fermentation are generally known to result in slower growth of probiotic strains than if they were used alone. These may be due to various factors such as the production of organic acids, hydrogen peroxide, bacteriocins or other inhibitors or faster growth of the starter cultures resulting in less availability of nutrients. Hence, compatibility of the starter cultures and probiotic strains should be checked, and if required higher inoculum of bifidobacteria or growth promoting factors can be added to foster the growth and viability (Roy 2005). In cheese prepared with culture combination of *Lactococcus lactis* ssp. *lactis* and *Lc. lactis* ssp. *cremoris* along with *B. breve* and *B. longum*, survival of *B. breve* and *B. longum* was higher than 10^6 cfu g^{-1} for 15 days (Roy et al. 1997). Barona et al. (2000) noted that viability or growth of bifidobacteria was unaffected in the course of preparation of fermented milk when bifidobacterial cultures (*B. bifidum, B. breve, B. infantis* and *B. longum*) were inoculated in addition to a combination of lactococci, citrate-fermenting lactococci and leuconostoc cultures. Further, bifidobacteria were also found to have an influence on the final product characteristics. Milk fermented with bifidobacteria had lower residual lactose and higher acetaldehyde content than milk fermented without bifidobacteria. It was also noted that bifidobacteria in the presence of these mixed cultures increased the ethanol concentration (Barona et al. 2000). A bifidogenic factor was purified from *P. freudenreichii* which was used for fermentation of Swiss cheese. The bifidogenic factor provided an anaerobic environment which supported the growth of bifidobacteria. Such cultures compatible with bifidobacteria could be beneficial when considered for fermentation (Mori et al. 1997). The effect of probiotic cultures *Lactobacillus acidophilus* I and *B. bifidum* R and yoghurt and *dahi* starter culture on biochemical activity and viability was studied in buffalo skim milk. The study reported a log reduction in the mesophilic lactic count after 18 h of incubation at 30 °C. The study also concluded that the cultures did not affect the technological properties of *dahi* and yoghurt; hence, the inclusions of these cultures with starter cultures were recommended (Vijayendra and Gupta 2013). Bifidobacteria are often preferred to be used as an adjunct culture than a starter culture because of two main difficulties. The first is its low tolerance to oxygen and requires anaerobic conditions which result in slow growth or makes the process expensive for the industry. The second is a high amount of acetate production which is undesirable. These can be solved by selecting more stress-tolerant strains and strains producing less acetate, less ethanol and more volatile compounds desirable for the products, for example, *B. animalis* subsp. *lactis* CECT 7953 (Margolles and Sánchez 2012).

Compared to fermented milk and yoghurt, cheeses are proposed to provide a suitable environment for probiotic organisms improving its survival. Higher pH range, solid consistency, fat composition as well as buffering ability are some of the factors which make them a promising carrier for delivering probiotics (Boylston et al. 2004). During the ripening process of cheese, the metabolism of the lactic cultures develops an anaerobic environment which may be conducive for the growth of bifidobacteria. Bifidobacteria are known to be incorporated in different kinds of cheese. However, the survival of the bacteria depends on the stage of incorporation, the strain used for incorporation, the starter cultures present, the type of cheese and the process associated with its making (Boylston et al. 2004; Roy 2005). Several studies describe the incorporation of bifidobacteria as adjunct cultures or the use of cheese as a carrier for bifidobacterial culture and their viability. However, identification of new species of bifidobacteria, *B. crudilactis*, has been reported from cheese prepared using raw milk (Delcenserie et al. 2007). The possible contamination of raw milk with animal faeces was suspected to be the cause for the presence of these bacteria; however, the exact reason remains untraced. Similarly *B. crudilactis* and *B. animalis* subsp. *lactis* were also isolated from ovine cheese which was prepared by traditional methods devoid of starter cultures, from unprocessed raw milk (Bunesova et al. 2014). Probiotic properties of raw milk cheese isolates of *B. crudilactis* and *B. mongoliense* were studied and were reported to tolerate harsh conditions of acidic pH, bile salts and pancreatic juices. These organisms were also described to grow below 12 °C during cheese preparation. Subsequently, it was also stated that *B. crudilactis* was capable of tolerating heat treatment, and further investigations into these properties may help in exploiting such cultures for products where heat treatment procedures are present. The origin of these cultures remains untraced, but it was concluded that they may have a possible role in technological and sensory traits of cheeses (Delcenserie et al. 2013).

16.2.2 Soy Products

Apart from the dairy-associated products, bifidobacteria have been studied extensively for fermenting soy milk. Soybean is considered to be a good source of proteins, fibres, isoflavones and unsaturated fatty acids (Ghosh et al. 2013). As per

Protein Digestibility-Corrected Amino Acid Score (PDCAAS), they have the best protein value among vegetables and are also considered as equivalent to milk proteins and egg proteins (He and Chen 2013). The beany flavour of soya bean and the indigestible oligosaccharides were the major hurdle in the utilisation of soya bean. However, fermentation of soya beans has been reported to solve these problems (Ghosh et al. 2013). The ability of soya bean to support the growth of bacteria and its biochemical activities led to the employment of LAB for fermentation of soy milk. Products similar to cheese, yoghurt and sour milk beverages were developed from soy milk. The suitability of soy milk for bifidobacterial growth as a substrate alone as well as fortifying it with carbohydrates, protein hydrolysates and amino acids was assessed. The results of the study indicated that it could support the growth of bifidobacteria and, upon fortification, the growth was enhanced. The study also revealed the feasibility of developing yoghurt-like product by using reconstituted skimmed milk fortified with soy milk using bifidobacteria (Kamaly 1997). Several studies document the usage of soy yoghurt as carrier for bifidobacterial cultures. Bifidobacteria have been used alone or along with the yoghurt cultures for fermentation. The effect of these cultures on each other has also been reported. Studies have revealed that soy yoghurt fermented using mix cultures of *Streptococcus salivarius* ssp. *thermophilus*, *Lb. delbrueckii* ssp. *bulgaricus* and *Bifidobacterium* sp. was less acidic compared to that of cow milk yoghurt. It was also noted that the count of *Streptococci* was higher than that of *Lactobacillus* in the presence of bifidobacteria (Murti et al. 1992). *Streptococcus thermophilus* ATCC 4356, *Lb. delbrueckii* subsp. *bulgaricus* IM 025 and bifidobacteria of human origin were employed for fermentation, and it was demonstrated that the pH drop in soy yoghurt was faster than the milk yoghurt. However, the final pH was found to be similar. This study also interpreted that bifidobacteria and the starter cultures use different sugars for their growth, and thus the probiotic bacteria had no influence on the growth pattern of the starter cultures (Farnworth et al. 2007). Sensory and physicochemical characteristics and soya saponin metabolites were tested in soy yoghurt obtained after fermentation at 37 °C for 36 h using *B. breve* K-110, *Streptococcus thermophilus* 3781 or *Lb. acidophilus* Q509011. It was noted that only *B. breve* was able to produce soya sapogenol. Product employing *B. breve* had poor sensory score, but it was improved upon combining *B. breve, S. thermophilus* and *Lb. acidophilus* (Chang et al. 2010). Blending of soy milk with skim milk has also been considered for soy milk fermentation by bifidobacteria. Supplementation of soy milk with skim milk in a ratio of 75:25 was found to be better for fermentation by bifidobacteria. The fermentation parameters were also further optimised using response surface methodology (RSM). The effect of inoculum size, temperature and glucose on physicochemical properties and sensory score were also evaluated (Swathi 2015).

16.2.3 Cereal-Based Products

Cereals have been investigated for development of non-dairy probiotic products. Cereals may serve as fermentable substrate, dietary fibres, prebiotics or as an encapsulation material in the food formulation. Asia as well as Africa traditionally fermented cereals using LAB and prepared various beverages, gruels and porridge. Sourdough is prepared in western countries using cereals, and the fermentation is carried out by mixed cultures which have a higher population of LAB (Charalampopoulos et al. 2002). *B. longum* BB536 was used for the preparation of *medida* (Sudanese thin porridge prepared from fermented malted brown rice), and its suitability as probiotic dairy alternative was studied. It was noted that the total solids were high when compared to that of conventional *medida* and the viable count of the bacteria was also high. Further, the count remained stable under refrigerated conditions for a week (Kabeir et al. 2005). For developing a probiotic beverage, the suitability of malt hydrolysate as a substrate for fermentation by bifidobacteria was assessed. The study revealed that bifidobacteria were capable of growing in the fermentation medium but required the addition of growth factor (yeast extract) and the counts obtained indicated the possibility of developing a probiotic malt-based drink (Rozada-Sanchez et al. 2008). Another study evaluated the ability of *B. breve* NCIMB 702257 to survive in three malt-based media. Mathematical models were developed to investigate the impact of temperature during storage on total acidity and viability. Results indicated that survivability of bifidobacteria improved upon increasing the yeast extract concentration (Rozada et al. 2009). Analysis of malt beverage fermented with *B. breve* NCIMB 702257 demonstrated the presence of 12 volatile components; among them eight were due to the metabolic activities of bifidobacteria and others were due to Maillard reaction. The beverage had sour flavour with mild sweet and malty character (Salmerón et al. 2014). Presence of bifidobacteria has also been described in rice-based fermented beverage *haria* ethnic food in West Bengal, India. The origin of these bacteria in the beverage is thought to be from the starter *bakhar*. In the course of fermentation of *haria,* the LAB as well as bifidobacteria count increases, and yeast and mould count decreases (Ghosh et al. 2014; Ray et al. 2016). Conversions of starchy materials present in the rice to useful sugars are the result of the symbiotic interaction of yeast, mould, LAB and bifidobacteria. In addition to that, they are also known to supplement the product with antioxidants and other bioactive substances (Ghosh et al. 2015). A Chinese patent has been filed for bifidobeer wherein bifidobacteria have been used along with *Saccharomyces cerevisiae* for brewing (Zhang et al. 2010). The beer is considered to be enriched with functional factors secreted by bifidobacteria which include lactic acid, acetic acid, vitamins and bifidobacterial membrane components such as lipoteichoic acid (LTA), teichoic acids, etc. Antiobesity effects of bifidobeer have also been studied in obese rats (Shan et al. 2009). However, in another study when the suitability of beer was evaluated as a carrier for probiotic by adding *B. animalis* ssp. *lactis* Bb-12 to low-alcohol and nonalcoholic beer fermented by *Saccharomyces cerevisiae* 70,424 and *Saccharomyces rouxii* 2531, bifidobacteria were unable to retain the viability during the storage period of 20 days (Sohrabvandi et al. 2010).

Bifidobacteria have also been evaluated for bread making as potential starter cultures. Upon analysis of various fermentative parameters of dough making and technological parameters of bread making, it was known that bifidobacterial strains could substitute *Lactobacillus* strains, commercialised for bread making. The bread obtained had similar characteristics as that of from commercial *Lactobacillus* cultures with the additional advantage of softer crumbs. The study also concluded that the effects were strain and origin dependent (Palacios et al. 2006).

One of the major limitations of using cereal products like wheat is the presence of phytic acid which has chelating properties that decrease the mineral bioavailability. Sourdough fermentation reduces the phytate content (Raghavendra and Halami 2009). In another study, 23 strains of bifidobacteria were isolated and studied for higher phytase activity against inositol phosphate. They observed no differences in technological quality of bread with and without bifidobacteria. However, lower level of inositol phosphates was observed in the crumbs (Palacios et al. 2008a). Bifidobacteria have better adaptation to the dough ecosystem. The total titrable acidity and lactic acid production were found to be strain dependent. Moreover, phytic acid hydrolysis was higher, and inositol phosphate production was lower in species of *B. breve* and *B. longum.* Compared to commercial starter culture *Lb. plantarum*, the fermented dough had high pH and less acidity (Palacios et al. 2008b). Fermented dough and bread prepared using sourdough fermented with *B. pseudocatenulatum* ATCC 27919 had higher organic acid content. It was noted that phytate level was brought down by activation of endogenous phytase in cereals along with its own phytase. The formulation of whole wheat bread was also found to retard and decline amylopectin retrogradation (Sanz-Penella et al. 2012).

16.2.4 Meat Products

The inclusion of probiotic organisms in fermented meat products has been considered as a strategy for developing functional meat products. A salami product marketed by a German producer in 1998 which included three cultures *Lb. acidophilus*, *Lb. casei* and *Bifidobacterium* spp. with health claims is considered to be the first probiotic salami product (Arihara 2006). Dry fermented meat products which do not require any heat treatment or mild heat treatment can serve as a good carrier for probiotics (De Vuyst et al. 2008). In such products, the probiotic organisms are thought to be encapsulated in the matrix of meat and fat offering protection during intestinal transit (Rouhi et al. 2013). However, the viability of the organisms in the product is thought to be influenced by the high curing salt, low pH and water activity. Although this may be strain dependent, they are considered to be the limitations of using meat as a carrier for probiotic products (De Vuyst et al. 2008). Studies about utilisation of fermented sausages as a probiotic delivery medium are scarce. The improvement in protection against pathogens in Hungarian salami prepared by using three different strains of bifidobacteria as nonconventional starter cultures was evaluated. It was established that the cultures were able to survive the fermentation and maturation process and reduce the pathogen counts of *E. coli* O111 and

List. monocytogenes (Pidcock et al. 2002). Effect of encapsulated and nonencapsulated *B. longum* and *Lb. reuteri* on *E. coli* O157:H7, when used along with starter cultures in the preparation of dry fermented sausage, has been investigated. Nonencapsulated cultures had lower counts but had better antimicrobial activity against the pathogen compared to that of encapsulated ones. Comparison of technological properties of fermented mutton sausages prepared using traditional starter cultures and probiotic strains of *Lb. acidophilus* CCDM 476 and *B. animalis* 241a was found to be identical. The viable count of bifidobacteria at the end of fermentation was found to be low. However, technological properties of the product remained unaffected by the low counts (Holko et al. 2013). The positive effects of probiotic cultures *Lb. acidophilus* and *B. lactis* on the physicochemical, microbiological and sensory properties of Italian salami have been demonstrated. The microbial counts were within range, and the product had higher acceptance among the consumers (Ruiz et al. 2014).

16.2.5 Vegetable Products

Fermented vegetable juice has also been explored as a probiotic carrier for bifidobacteria. Fermented vegetables have high nutritional value along with several therapeutic and health-promoting effects (Vijayendra and Halami 2015). *Bifidobacterium* strains *B. animalis* ssp. *lactis* Bb-12 and *B. bifidum* B7.1 and B3.2 were used to determine the suitability of carrot juice for developing probiotic food. The study concluded that the strains were able to propagate in the juice without additional requirements and were viable until the end of fermentation with high volumetric productivities (Kun et al. 2008). Chemical properties of tomato juice fermented with three different bifidobacterial strains *B. breve*, *B. longum* and *B. infantis* were studied. Bifidobacterial fermentation did not have any effect on the lycopene content of the juice, and the addition of sugars like FOS improved the flavour and taste (Koh et al. 2010). The aptness of fermented noni juice as probiotic was tested, and it was revealed that *B. longum* had retained viability upon storage for a period of 4 weeks, and the product had higher antioxidant capacity compared to LAB-fermented juice (Wang et al. 2009). Supplementation of bifidobacteria to kimchi has been evaluated for improving the dietary value of the product. Among five different strains of bifidobacteria studied for viability in *mul*-kimchi, *B. longum* JK2 retained highest viability, and NaCl concentration above 3% was shown to reduce the viability (Lee et al. 1999). *B. longum* BO 11 fermented kimchi was shown to have higher taste preference compared with that of conventional kimchi without bifidobacteria (Chae et al. 2006). Moreover, the addition of fructooligosaccharides has also shown to improve the viability of *B. animalis* DY-64 in *baechu* kimchi fermented with *B. animalis* DY-64. Further, FOS was known to improve the organic content and the viability of other LAB compared to that of conventional and bifidobacteria-fermented kimchi (Chae and Jhon 2007).

16.3 Significance of Fermented Foods as a Probiotic Vehicle

Pharmaceutical formulation and food formulations are the widely used strategies for delivery of probiotics (Govender et al. 2014). The desire to have tangy food with additional health benefits has increased the popularity of functional foods among the consumers (Coman et al. 2012). The foods considered for probiotic delivery can be further grouped as fermented and non-fermented foods. Among them, the dairy-based products are the most popular probiotic carriers (Heller 2001; Kumar et al. 2015). The history of probiotic products explains the reason behind the popularity of dairy products. Other reasons for the popularity of fermented products are the records that they are healthful, and the consumers are aware of the fact that they contain viable organisms, facilitating the daily consumption of such products. The fermentation conditions for most of the products have already optimised for the survival of these organisms and thus allowing easy adaptation of technology for replacement of the probiotic bacteria (Heller 2001). Unlike non-fermented foods like chocolate, ice creams and fruit juices which merely serve as a carrier of probiotic bacteria, fermented foods allow interaction of bacteria with the food matrix which positively affects the food. Bifidobacteria-fermented foods are known to be an abundant source of vitamins and amino acids. Folate production was shown to be higher in yoghurt fermented with *S. thermophilus* and *B. animalis* (Crittenden et al. 2003). Bifidobacteria are also known to produce several of the water-soluble vitamins which may be released into the fermentation medium (Deguchi et al. 1985). They develop positive flavour texture and sensory characteristics. In the case of soy fermented products, bifidobacteria are known to reduce the beany flavour. Bifidobacteria-fermented kimchi was shown to have a better taste preference compared to that of kimchi without bifidobacteria. These foods have high antioxidant activity. The bioactive molecules released or the transformation of the substances occurring during fermentation may have additional health benefits. Bifidobacteria have been reported for the biotransformation of isoflavone glycosides to aflavones in fermented soy milk which has better absorption. This transformation can further be enhanced by supplementation with skim milk (Pham and Shah 2007). They may improve the digestibility or aid in the absorption of nutrients. Phytase-producing bifidobacteria are known to reduce the phytic acid content in the fermented product and can thus improve the bioavailability of minerals (Palacios et al. 2008a, b). Bifidobacteria are also known to metabolise and assimilate raffinose and stachyose sugars present in soybean responsible for flatulence (Farnworth et al. 2007). Fermented dairy products with bifidobacteria are known to contain less residual lactose than non-fermented milk (Prasanna et al. 2014). Thus bifidobacterial fermentation is known to improve the nutritional value of the product. Bifidobacterial fermented products have been established for different health beneficial effects which include lactose intolerance, irritable bowel syndrome, anti-mutagenic effects, anticancer, anti-inflammatory effects, cholesterol reduction, immunomodulation, inflammatory bowel disease, etc. (El-Gawad et al. 2005; Guyonnet et al. 2007; He et al.

Table 16.1 Some commercially available probiotic fermented products that contain bifidobacteria (Fasoli et al. 2003; Champagne et al. 2005 Masco et al. 2005; Siró et al. 2008)

Culture used	Product type	Commercial name	Producing country
B. animalis	Yoghurt	Activia	France
B. animalis subsp. *lactis*	Yoghurt	Biospega	Italy
B. longum	Yoghurt	B'A	France
B. animalis subsp. *lactis*	Yoghurt	Benecol	UK
B. animalis subsp. *lactis*	Yoghurt	Bifidus nature	France
B. animalis subsp. *lactis*	Yoghurt	Biobest	Germany
B. bifidum	Yoghurt	Biogarde	Germany
B. animalis subsp. *lactis* Bb12	Fermented milk	Crema actidrink	Germany
B. animalis subsp. *lactis* Bb12	Yoghurt/fermented milk	KYR	Italy
B. longum BB536	Yoghurt	Lactoferrin yoghurt	Japan
B. bifidum	Fermented milk	Mil-Mil	Japan
B. bifidum, *B. breve*	Fermented milk	Miru-Miru	Japan
B. animalis subsp. *lactis* Bb12	Yoghurt	Natural bio yoghurt	UK
Bifidobacterium	Cultured milk	Nutrigen	Malaysia
B. animalis subsp. *lactis* BB12	Yoghurt	Vitality	Germany
B. animalis subsp. *lactis* BB12	Yoghurt	Yogosan	Germany
B. breve Yakult strain	Fermented milk	Yakult bifiel	Japan
Bifidobacterium	Yoghurt	Yoghurt natural	Malaysia
B. animalis subsp. *lactis*	Yoghurt	Yomo	Italy
B. animalis subsp. *lactis*	Fermented oat-based product	Yosa Bioferme	Finland

2008; Vijayendra and Gupta 2012; Gomi et al. 2013; Lai et al. 2013; Gomi et al. 2015). Some of the commercially available fermented foods with bifidobacteria are presented in Table 16.1.

16.4 Viability and Strategies Adopted for Improving Viability

The definition of probiotic provided by the FAO states that the live organisms should be in adequate number to confer health benefits to the host. The efficacious dosage for probiotics is generally considered to be 10^8–10^9 cfu/day which is approximately 10^6–10^8 cfu/g of food as per the serving size. Achieving and maintaining this viability in the product till it reaches the consumers have always been the challenge for the probiotic industry. Processing conditions, kind of food matrix, pH, storage temperature, presence of oxygen and competing microflora are some of the factors held responsible for affecting the viability of probiotic bacteria. Besides these, the bacteria also have to survive the gastrointestinal transit and reach the gut in viable and sufficient numbers (Sanders and Marco 2010). Numerous studies report loss of viability of bifidobacteria in different products upon storage (Champagne et al. 2005). Several strategies have been adopted to address these problems. In some cases, the food matrix itself may play a role in maintaining the viability of bacteria. For instance, survival of bifidobacteria has been reported for a period of 15 days with higher counts in freshly prepared cheese upon storage at 12 °C (Roy et al. 1997). Cheddar cheese has been demonstrated as an excellent medium for bifidobacteria delivery where the cultures remained viable in adequate amount till 32 weeks (Phillips et al. 2006). The higher fat content of cheese may play a role in the viability of probiotic bacteria during storage at low temperature. Further ripening process in cheese also helps in achieving anaerobic conditions which may help in the survival of bifidobacteria. In another study, 24 commercially available probiotic fermented milk products were tested for viability, and in the majority of the products, a good number of bifidobacteria survived till the shelf life (Raeisi et al. 2013). Improvement in longevity of bifidobacteria in fermented products is also known to be achieved by the addition of prebiotics. Incorporation of prebiotics like FOS, inulin and raftilose in yoghurt has demonstrated to improve the viability of bifidobacteria for a period of 20–28 days under refrigerated storage conditions (Akalın et al. 2004; Capela et al. 2006; Varga et al. 2006; Akın et al. 2007). β-Glucan from oats has shown to improve the viability of bifidobacteria in yoghurts stored under refrigerated conditions (Rosburg et al. 2010). Coculturing of bifidobacteria with other lactic acid cultures also helps in extending the viability of the bacteria. The survivability of *B. breve* was found to be improved upon using *Lb. delbrueckii* subsp. *bulgaricus* with defective H + −ATPase activity as starter culture which prevents post-acidification of the yoghurt during storage (Ongol et al. 2007). Viability of bifidobacteria has also been reported to improve upon cocultivation with *Lc. lactis* subspecies *lactis* in fermented milk stored under refrigerated conditions by reducing the level of dissolved oxygen in the product (Odamaki et al. 2011). Encapsulation of probiotic bacteria has also been employed for improving survivability in the product. Encapsulation provides a microenvironment for the bacteria that protects it from harsh conditions of processing and storage and allows release at the desired site. The viability of the bacteria will depend upon the process employed for encapsulation, the material and its concentration used for encapsulation and the strain of the bacteria used (Chávarri et al. 2012; Corona-Hernandez et al. 2013). Microencapsulation protects the bacteria; at the same time it allows the diffusion of nutrients through the matrix (Talwalkar and Kailasapathy 2004). Addition of encapsulated *B. longum* ATCC 15696 to cheddar cheese during milling of the curd helped to maintain its viability during the entire period of ripening of 24 weeks, and the encapsulated organisms were metabolically inactive (Dinakar and Mistry 1994). Microencapsulation using k-carrageenan has demonstrated to preserve the viability of bifidobacteria in yoghurt during refrigerated storage conditions (Adhikari et al. 2000). Microencapsulation with alginate has been reported to improve the viability of bifidobacteria in freeze-dried yoghurt stored at 21 °C (Capela et al. 2006). Microcapsules of *B. longum* developed using *Eleutherine americana* extract and oligosaccharides extract improved survivability in food matrix during storage conditions and gastrointestinal transit (Phoem et al. 2015).

Conclusion

Health beneficial aspects of bifidobacteria are well documented, and its delivery as probiotic through fermented foods is of much interest. Bifidobacteria are mainly used as adjunct cultures in fermented foods. However, bifidobacterial fermentation is relatively unexplored when compared to that of LAB. Impact of fermentation by bifidobacteria on the product with respect to the physiochemical and sensory traits needs to be investigated. Reports of the presence of bifidobacteria in the fermented products imply its activity and the ability of these organisms to survive in such conditions. Such organisms should be additionally characterised for probiotic properties and technological properties which may further help in the development of better probiotic products with enhanced viability and shelf life. Dairy-based fermented foods have occupied a predominant place in the probiotic market. Nevertheless, the desire of consumers to have flavourful products can be explored by providing them with non-dairy fermented probiotic products. In addition to that, innovative ingredients and processes should be developed that can help in the prolonged survival of bifidobacteria in the product. Such fermented products having a proven health claim with an adequate number of bifidobacteria with prolonged shelf life and sustained viability will meet the consumer expectation.

Acknowledgements The authors are thankful to Director CSIR-Central Food Technological Research Institute (CFTRI), Mysuru, India, for providing necessary funds and facilities. SA would like to acknowledge CSIR for granting Senior Research Fellowship.

References

Adhikari K, Mustapha A, Grun IU, Fernando L (2000) Viability of microencapsulated bifidobacteria in set yogurt during refrigerated storage. J Dairy Sci 83:1946–1951. doi:10.3168/jds.S0022-0302(00)75070-3

Akalın AS, Fenderya S, Akbulut N (2004) Viability and activity of bifidobacteria in yoghurt containing fructooligosaccharide during refrigerated storage. Int J Food Sci Technol 39:613–621. doi:10.1111/j.1365-2621.2004.00829.x

Akın MB, Akın MS, Kırmac Z (2007) Effects of inulin and sugar levels on the viability of yogurt and probiotic bacteria and the physical and sensory characteristics in probiotic ice-cream. Food Chem 104:93–99. doi:10.1016/j.foodchem.2006.11.030

Arihara K (2006) Strategies for designing novel functional meat products. Meat Sci 74:219–229. doi:10.1016/j.meatsci.2006.04.028

Barona M, Roy D, Vuillemard JC (2000) Biochemical characteristics of fermented milk produced by mixed-cultures of lactic starters and bifidobacteria. Lait 80:465–478. doi:10.1051/lait:2000138

Bottacini F, Ventura M, van Sinderen D, Motherway MOC (2014) Diversity, ecology and intestinal function of bifidobacteria. Microb Cell Factories 13:S4. doi:10.1186/1475-2859-13-S1-S4

Boylston TD, Vinderola CG, Ghoddusi HB, Reinheimer JA (2004) Incorporation of bifidobacteria into cheeses: challenges and rewards. Int Dairy J 14:375–387. doi:10.1016/j.idairyj.2003.08.008

Bunesova V, Killer J, Vlkova E, Musilova S, Tomaska M, Rada V, Kmet V (2014) Isolation and characterization of bifidobacteria from ovine cheese. Int J Food Microbiol 188:26–30. doi:10.1016/j.ijfoodmicro.2014.07.001

Capela P, Hay TKC, Shah NP (2006) Effect of cryoprotectants, prebiotics and microencapsulation on survival of probiotic organisms in yoghurt and freeze-dried yoghurt. Food Res Int 39:203–211. doi:10.1016/j.foodres.2005.07.007

Chae MH, Jhon DY (2007) Effects of commercial fructooligosaccharides on bifidobacteria kimchi fermentation. Korean J Food Sci Technol 39:61–65

Chae MH, Park EJ, Oh TK, Jhon DY (2006) Preparation of kimchi containing *Bifidobacterium longum* BO-11. Korean J Food Sci Technol 38:232–236

Champagne CP, Gardner NJ, Roy D (2005) Challenges in the addition of probiotic cultures to foods. Crit Rev Food Sci Nutr 45:61–84. doi:10.1080/10408690590900144

Chang SY, Kim D, Han MJ (2010) Physicochemical and sensory characteristics of soy yogurt fermented with *Bifidobacterium breve* K-110, *Streptococcus thermophilus* 3781, or *Lactobacillus acidophilus* Q509011. Food Sci Biotechnol 19:107–113. doi:10.1007/s10068-010-0015-0

Charalampopoulos D, Wang R, Pandiella SS, Webb C (2002) Application of cereals and cereal components in functional foods: a review. Int J Food Microbiol 79:131–141. doi:10.1016/S0168-1605(02)00187-3

Chávarri M, Marañón I, Villarán MC (2012) Encapsulation technology to protect probiotic bacteria. In: Rigobelo E (ed) Probiotics. InTech Open Access Publisher, Rijeka, pp 501–539. doi:10.5772/50046

Coman MM, Cecchini C, Verdenelli MC, Silvi S, Orpianesi C, Cresci A (2012) Functional foods as carriers for SYNBIO®, a probiotic bacteria combination. Int J Food Microbiol 157:346–352. doi:10.1016/j.ijfoodmicro.2012.06.003

Corona-Hernandez RI, Alvarez-Parrilla E, Lizardi-Mendoza J, Islas-Rubio AR, Rosa L, Wall-Medrano A (2013) Structural stability and viability of microencapsulated probiotic bacteria: a review. Compr Rev Food Sci Food Saf 12:614–628. doi:10.1111/1541-4337.12030

Crittenden RG, Martinez NR, Playne MJ (2003) Synthesis and utilisation of folate by yoghurt starter cultures and probiotic bacteria. Int J Food Microbiol 80:217–222. doi:10.1016/S0168-1605(02)00170-8

Cronin M, Ventura M, Fitzgerald GF, Van Sinderen D (2011) Progress in genomics, metabolism and biotechnology of bifidobacteria. Int J Food Microbiol 149:4–18. doi:10.1016/j.ijfoodmicro.2011.01.019

De Vuyst L, Falony G, Leroy F (2008) Probiotics in fermented sausages. Meat Sci 80:75–78. doi:10.1016/j.meatsci.2008.05.038

Deguchi Y, Morishita T, Mutai M (1985) Comparative studies on synthesis of water-soluble vitamins among human species of bifidobacteria. Agric Biol Chem 49:13–19. doi:10.1080/00021369.1985.10866683

Delcenserie V, Gavini F, Beerens H, Tresse O, Franssen C, Daube G (2007) Description of a new species, *Bifidobacterium crudilactis* sp. nov., isolated from raw milk and raw milk cheeses. Syst Appl Microbiol 30:381–389. doi:10.1016/j.syapm.2007.01.004

Delcenserie V, Taminiau B, Gavini F, De Schaetzen MA, Cleenwerck I, Theves M, Mahieu M, Daube G (2013) Detection and characterization of *Bifidobacterium crudilactis* and *B. mongoliense* able to grow during the manufacturing process of French raw milk cheeses. BMC Micro 13:239. doi:10.1186/1471-2180-13-239

Dinakar P, Mistry VV (1994) Growth and viability of *Bifidobacterium longum* in cheddar cheese. J Dairy Sci 77:2854–2864. doi:10.3168/jds.S0022-0302(94)77225-8

El-Gawad IAA, El-Sayed EM, Hafez SA, El-Zeini HM, Saleh FA (2005) The hypocholesterolaemic effect of milk yoghurt and soy-yoghurt containing bifidobacteria in rats fed on a cholesterol-enriched diet. Int Dairy J 15:37–44. doi:10.1016/j.idairyj.2004.06.001

Farnworth ER, Mainville I, Desjardins MP, Gardner N, Fliss I, Champagne C (2007) Growth of probiotic bacteria and bifidobacteria in a soy yogurt formulation. Int J Food Microbiol 116:174–181. doi:10.1016/j.ijfoodmicro.2006.12.015

Fasoli S, Marzotto M, Rizzotti L, Rossi F, Dellaglio F, Torriani S (2003) Bacterial composition of commercial probiotic products as evaluated by PCR-DGGE analysis. Int J Food Microbiol 82:59–70. doi:10.1016/S0168-1605(02)00259-3

Ghosh D, Chattoraj DK, Chattopadhyay P (2013) Studies on changes in microstructure and proteolysis in cow and soy milk curd during fermentation using lactic cultures for improving protein bioavailability. J Food Sci Technol 50:979–985. doi:10.1007/s13197-011-0421-1

Ghosh K, Maity C, Adak A, Halder SK, Jana A, Das A, Parua (Mondal) S, PKD M, Pati BR, Mondal KC (2014) Ethnic preparation of *Haria*, a rice-based fermented beverage, in the Province of Lateritic West Bengal, India. Ethnobot Res Appl 12:039–049. doi:10.17348/era.12.0.039-049

Ghosh K, Ray M, Adak A, Dey P, Halder SK, Das A, Jana A, Parua (Mondal) S, PKD M, Pati BR, Mondal KC (2015) Microbial, saccharifying and antioxidant properties of an Indian rice based fermented beverage. Food Chem 168:196–202. doi:10.1016/j.foodchem.2014.07.042

Gomi A, Harima-Mizusawa N, Shibahara-Sone H, Kano M, Miyazaki K, Ishikawa F (2013) Effect of *Bifidobacterium bifidum BF-1* on gastric protection and mucin production in an acute gastric injury rat model. J Dairy Sci 96:832–837. doi:10.3168/jds.2012-5950

Gomi A, Iino T, Nonaka C, Miyazaki K, Ishikawa F (2015) Health benefits of fermented milk containing *Bifidobacterium bifidum* YIT 10347 on gastric symptoms in adults. J Dairy Sci 98:2277–2283. doi:10.3168/jds.2014-9158

Govender M, Choonara YE, Kumar P, du Toit LC, van Vuuren S, Pillay V (2014) A review of the advancements in probiotic delivery: conventional vs non-conventional formulations for intestinal flora supplementation. AAPS PharmSciTech 15:29–43. doi:10.1208/s12249-013-0027-1

Guyonnet D, Chassany O, Ducrotte P, Picard C, Mouret M, Mercier CH, Matuchansky C (2007) Effect of a fermented milk containing *Bifidobacterium animalis* DN-173 010 on the health-related quality of life and symptoms in irritable bowel syndrome in adults in primary care: a multicentre, randomized, double-blind, controlled trial. Aliment Pharmacol Ther 26:475–486. doi:10.1111/j.1365-2036.2007.03362.x

He FJ, Chen JQ (2013) Consumption of soybean, soy foods, soy isoflavones and breast cancer incidence: differences between Chinese women and women in Western countries and possible mechanisms. Food Sci Human Wellness 2:146–161. doi:10.1016/j.fshw.2013.08.002

He T, Priebe MG, Zhong Y, Huang C, Harmsen HJM, Raangs GC, Antoine JM, Welling GW, Vonk RJ (2008) Effects of yogurt and bifidobacteria supplementation on the colonic microbiota in lactose-intolerant subjects. J Appl Microbiol 104:595–604. doi:10.1111/j.1365-2672.2007.03579.x

Heller KJ (2001) Probiotic bacteria in fermented foods: product characteristics and starter organisms. Am J Clin Nutr 73:374S–379S

Holko I, Hrabě J, Šalaková A, Rada V (2013) The substitution of a traditional starter culture in mutton fermented sausages by *Lactobacillus acidophilus* and *Bifidobacterium animalis*. Meat Sci 94:275–279. doi:10.1016/j.meatsci.2013.03.005

Kabeir BM, Abd-Aziz S, Muhammad K, Shuhaimi M, Yazid AM (2005) Growth of *Bifidobacterium longum* BB536 in medida (fermented cereal porridge) and their survival during refrigerated storage. Lett Appl Microbiol 41:125–131. doi:10.1111/j.1472-765X.2005.01735.x

Kamaly KM (1997) Bifidobacteria fermentation of soybean milk. Food Res Int 30:675–682

Koh J, Kim Y, Oh JH (2010) Chemical characterization of tomato juice fermented with bifidobacteria. J Food Sci 75:C428–C432. doi:10.1111/j.1750-3841.2010.01632.x

Kumar BV, Vijayendra SVN, Reddy OVS (2015) Trends in dairy and non-dairy probiotic products - a review. J Food Sci Technol 52:6112–6124. doi:10.1007/s13197-015-1795-2

Kun S, Rezessy-Szabó JM, Nguyen QD, Hoschke A (2008) Changes of microbial population and some components in carrot juice during fermentation with selected *Bifidobacterium* strains. Process Biochem 43:816–821. doi:10.1016/j.procbio.2008.03.008

Lai LR, Hsieh SC, Huang HY, Chou CC (2013) Effect of lactic fermentation on the total phenolic, saponin and phytic acid contents as well as anti-colon cancer cell proliferation activity of soymilk. J Biosci Bioeng 115:552–556. doi:10.1016/j.jbiosc.2012.11.022

Lee JH, O'Sullivan DJ (2010) Genomic insights into bifidobacteria. Microbiol Mol Biol Rev 74:378–416. doi:10.1128/MMBR.00004-10

Lee SK, Ji GE, Park YH (1999) The viability of bifidobacteria introduced into kimchi. Lett Appl Microbiol 28:153–156. doi:10.1046/j.1365-2672.1999.00380.x

Margolles A, Sánchez B (2012) Selection of a *Bifidobacterium animalis* subsp. *lactis* strain with a decreased ability to produce acetic acid. Appl Environ Microbiol 78:3338–3342. doi:10.1128/AEM.00129-12

Martín R, Miquel S, Ulmer J, Kechaou N, Langella P, Bermúdez-Humarán LG (2013) Role of commensal and probiotic bacteria in human health: a focus on inflammatory bowel disease. Microb Cell Factories 12:71. doi:10.1186/1475-2859-12-71

Masco L, Huys G, De Brandt E, Temmerman R, Swings J (2005) Culture-dependent and culture-independent qualitative analysis of probiotic products claimed to contain bifidobacteria. Int J Food Microbiol 102:221–230. doi:10.1016/j.ijfoodmicro.2004.11.018

Mori H, Sato Y, Taketomo N, Kamiyama T, Yoshiyama Y, Meguro S, Sato H, Kaneko T (1997) Isolation and structural identification of bifidogenic growth stimulator produced by *Propionibacterium freudenreichii*. J Dairy Sci 80:1959–1964. doi:10.3168/jds.S0022-0302(97)76138-1

Murti TW, Bouillanne C, Landon M, De Smazeaud MJ (1992) Bacterial growth and volatile compounds in yoghurt-type products from soymilk containing *Bifidobacterium* ssp. J Food Sci 58:153–157. doi:10.1111/j.1365-2621.1993.tb03233.x

O'Sullivan DJ, Halami PM (2013) Genomics of prebiotics and probiotics. In: Otles S (ed) Probiotics and prebiotics in food, nutrition and health. CRC Press, Boca Raton, pp 446–463. doi:10.1201/b15561-23

Odamaki T, Xiao JZ, Yonezawa S, Yaeshima T, Iwatsuki K (2011) Improved viability of bifidobacteria in fermented milk by cocultivation with *Lactococcus lactis* subspecies *lactis*. J Dairy Sci 94:1112–1121. doi:10.3168/jds.2010-3286

Ongol MP, Sawatari Y, Ebina Y, Sone T, Tanaka M, Tomita F, Yokota A, Asano K (2007) Yoghurt fermented by *Lactobacillus delbrueckii* subsp. *bulgaricus* H+−ATPase-defective mutants exhibits enhanced viability of *Bifidobacterium breve* during storage. Int J Food Microbiol 116:358–366. doi:10.1016/j.ijfoodmicro.2007.02.019

Palacios MC, Sanz Y, Haros M, Rosell CM (2006) Application of *Bifidobacterium* strains to the breadmaking process. Process Biochem 41:2434–2440. doi:10.1016/j.procbio.2006.07.002

Palacios MC, Haros M, Rosella CM, Sanz Y (2008a) Selection of phytate-degrading human bifidobacteria and application in whole wheat dough fermentation. Food Microbiol 25:169–176. doi:10.1016/j.fm.2007.06.001

Palacios MC, Haros M, Sanz Y, Rosell CM (2008b) Phytate degradation by *Bifidobacterium* on whole wheat fermentation. Eur Food Res Technol 226:825–831. doi:10.1007/s00217-007-0602-3

Pham TT, Shah NP (2007) Biotransformation of isoflavone glycosides by *Bifidobacterium animalis* in soymilk supplemented with skim milk powder. J Food Sci 72:M316–M324. doi:10.1111/j.1750-3841.2007.00476.x

Phillips M, Kailasapathy K, Tran L (2006) Viability of commercial probiotic cultures (*L. acidophilus, Bifidobacterium* sp., *L. casei, L. paracasei* and *L. rhamnosus*) in cheddar cheese. Int J Food Microbiol 108:276–280. doi:10.1016/j.ijfoodmicro.2005.12.009

Phoem AN, Chanthachum S, Voravuthikunchai SP (2015) Applications of microencapsulated *Bifidobacterium longum* with *Eleutherine americana* in fresh milk tofu and pineapple juice. Forum Nutr 7:2469–2484. doi:10.3390/nu7042469

Pidcock K, Heard GM, Henriksson A (2002) Application of nontraditional meat starter cultures in production of Hungarian salami. Int J Food Microbiol 76:75–81. doi:10.1016/S0168-1605(02)00002-8

Prasanna PHP, Grandison AS, Charalampopoulos D (2014) Bifidobacteria in milk products: an overview of physiological and biochemical properties, exopolysaccharide production, selection criteria of milk products and health benefits. Food Res Int 55:247–262. doi:10.1016/j.foodres.2013.11.013

Raeisi SN, Ouoba LII, Farahmand N, Sutherland J, Ghoddusi HB (2013) Variation, viability and validity of bifidobacteria in fermented milk products. Food Cont 34:691–697. doi:10.1016/j.foodcont.2013.06.016

Raghavendra P, Halami PM (2009) Screening, selection and characterization of phytic acid degrading lactic acid bacteria from chicken intestine. Int J Food Microbiol 133:129–134. doi:10.1016/j.ijfoodmicro.2009.05.006

Ray M, Ghosh K, Singh S, Mondal KC (2016) Folk to functional: an explorative overview of rice-based fermented foods and beverages in India. J Ethn Foods 3:5–18. doi:10.1016/j.jef.2016.02.002

Rosburg V, Boylston T, White P (2010) Viability of bifidobacteria strains in yogurt with added oat beta-glucan and corn starch during cold storage. J Food Sci 75:439–444. doi:10.1111/j.1750-3841.2010.01620.x

Rouhi M, Sohrabvandi S, Mortazavian AM (2013) Probiotic fermented sausage: viability of probiotic microorganisms and sensory characteristics. Crit Rev Food Sci Nutr 53:331–348. doi:10.1080/10408398.2010.531407

Roy D (2005) Technological aspects related to use of bifidobacteria in dairy products. Lait 85:39–56. doi:10.1051/lait:2004026

Roy D, Mainville I, Mondou F (1997) Selective enumeration and survival of bifidobacteria in fresh cheese. Int Dairy J 7:785–793. doi:10.1016/S0958-6946(98)00012-0

Rozada R, Vázquez JA, Charalampopoulos D, Thomas K, Pandiella SS (2009) Effect of storage temperature and media composition on the survivability of *Bifidobacterium breve* NCIMB 702257 in a malt hydrolysate. Int J Food Microbiol 133:14–21. doi:10.1016/j.ijfoodmicro.2009.04.003

Rozada-Sanchez R, Sattur AP, Thomas K, Pandiella SS (2008) Evaluation of *Bifidobacterium* spp. for the production of a potentially probiotic malt-based beverage. Process Biochem 43:848–854. doi:10.1016/j.procbio.2008.04.002

Ruiz L, Ruas-Madiedo P, Gueimonde M, de Los Reyes-Gavilán CG, Margolles A, Sánchez B (2011) How do bifidobacteria counteract environmental challenges? Mechanisms involved and physiological consequences. Genes Nutr 6:307–318. doi:10.1007/s12263-010-0207-5

Ruiz JN, Villanueva NDM, Favaro-Trindade CS, Contreras-Castillo CJ (2014) Physicochemical, microbiological and sensory assessments of Italian salami. Sci Agric 71:204–211. doi:10.1590/S0103-90162014000300005

Russell DA, Ross RP, Fitzgerald GF, Stanton C (2011) Metabolic activities and probiotic potential of bifidobacteria. Int J Food Microbiol 149:88–105. doi:10.1016/j.ijfoodmicro.2011.06.003

Salmerón I, Rozada R, Thomas K, Ortega-Rivas E, Pandiella SS (2014) Sensory characteristics and volatile composition of a cereal beverage fermented with *Bifidobacterium breve* NCIMB702257. Food Sci Technol Int 20:205. doi:10.1177/1082013213481466

Sanders ME, Marco ML (2010) Food formats for effective delivery of probiotics. Annu Rev Food Sci Technol 1:65–85. doi:10.1146/annurev.food.080708.100743

Sanz-Penella JM, Tamayo-Ramos JA, Haros M (2012) Application of bifidobacteria as starter culture in whole wheat sourdough breadmaking. Food Bioprocess Technol 5:2370–2380. doi:10.1007/s11947-011-0547-1

Shan S, DeChun Z, QiaoFang C (2009) Antiobesity effect of bifidobeer on obese rats and its mechanism. Chin J Biologicals 22:671–675

Siró I, Kápolna E, Kápolna B, Lugasi A (2008) Functional food. Product development, marketing and consumer acceptance—a review. Appetite 51:456–467. doi:10.1016/j.appet.2008.05.060

Sohrabvandi S, Razavi SH, Mousavi MS, Mortazavian AM (2010) Viability of probiotic bacteria in low alcohol- and non-alcoholic beer during refrigerated storage. Philipp Agric Sci 93:24–28

Swathi Krishna E. (2015) Studies on soymilk fermentation by native isolates of *Bifidobacterium* species. Post graduate student (guide: Dr Prakash Halami) project report. http://ir.cftri.com / id/eprint/11950

Talwalkar A, Kailasapathy K (2004) A review of oxygen toxicity in probiotic yogurts: influence on the survival of probiotic bacteria and protective techniques. Compr Rev Food Sci Food Saf 3:117–124. doi:10.1111/j.1541-4337.2004.tb00061.x

Varga L, Szigeti J, Gyenis B (2006) Influence of chicory inulin on the survival of microbiota of a probiotic fermented milk during refrigerated storage. Ann Microbiol 56:139–141. doi:10.1007/BF03174995

Ventura M, O'Connell-Motherway M, Leahy S, Munoz JAM, Fitzgerald GF, Sinderen D (2007) From bacterial genome to functionality; case bifidobacteria. Int J Food Microbiol 120:2–12. doi:10.1016/j.ijfoodmicro.2007.06.011

Ventura M, Turroni F, Motherway MOC, MacSharry J, Sinderen D (2012) Host–microbe interactions that facilitate gut colonization by commensal bifidobacteria. Trends Microbiol 20:467–476. doi:10.1016/j.tim.2012.07.002

Vijayendra SVN, Gupta RC (2012) Assessment of probiotic and sensory properties of dahi and yoghurt prepared using bulk freeze-dried cultures in buffalo milk. Ann Microbiol 62:939–947. doi:10.1007/s13213-011-0331-5

Vijayendra SVN, Gupta RC (2013) Associative growth behavior of dahi and yoghurt starter cultures with *Bifidobacterium bifidum* and *Lactobacillus acidophilus* in buffalo skim milk. Ann Microbiol 63:461–469. doi:10.1007/s13213-012-0490-z

Vijayendra SVN, Halami PM (2015) Health benefits of fermented vegetable products. In: Tamang JP (ed) Health benefits of fermented foods and beverages. CRC Press, Boca Raton, pp 325–336. doi:10.1201/b18279-10

Wang C, Ng C, Su H, Tzeng W, Shyu YT (2009) Probiotic potential of noni juice fermented with lactic acid bacteria and bifidobacteria. Int J Food Sci Nutr 60:98–106. doi:10.1080/09637480902755095

Zhang P, Lee KG, Mu Xiaoping C (2010) 'Bifidus Factor Beer' Chinese Patent CN1970716B Filed: 29 Nov 2006, Published: 12 May 2010

Enterococci Prevalent in Processed Food Products: From Probiotics to Food Safety

17

S.R. Joshi and Koel Biswas

Abstract

Food processing for preservation using starter culture has always been a safer option in fermentation technology. However, spontaneous fermentation of food products using traditional technology encounters an array of organisms starting from yeasts to lactic acid bacteria. Although majority of LAB are designated GRAS strain, occurrence of *Enterococcus* in processed food products has been a subject of debate and concern. While *Enterococcus* are known to possess probiotic features, they are also known to carry virulence properties exposing the consumers of the processed products to unforeseeable danger and pathogenesis. Thus, *Enterococcus* spontaneously prevalent in traditional fermented food products raises food safety concern rather than their potentials in probiotics.

Keywords

Enterococcus • Food • Bacteriocins • Probiotics • Virulence factor • Antibiotic resistance

17.1 Introduction

Fermentation is a food processing and food preservation technology having its origin in the Middle East which dates as far back as 6000 BC, and it spread during the start of domestication of animals to the rest of the world (Ross et al. 2002).

S.R. Joshi (✉) • K. Biswas
Department of Biotechnology & Bioinformatics, North-Eastern Hill University, Shillong 793 022, Meghalaya, India
e-mail: srjoshi2006@yahoo.co.in

V.C. Kalia et al. (eds.), *Mining of Microbial Wealth and MetaGenomics*,
DOI 10.1007/978-981-10-5708-3_17

Fermentation of raw foods by the microbes gave rise to fermented food products, which became popular amongst the indigenous communities (Ross et al. 2002). Traditional fermentation methods were transmitted on from one generation to the next by using a small amount of the previously fermented product as a starter culture for the following fermentation (Sanni 1993). The latter process is also known as back slopping of the fermented product, and it results in the reduction of fermentation failure and conservation of the unique organoleptic properties (Leroy and De Vuyst 2004). Today most fermented products are produced on a commercial scale through highly developed equipment and industrial processes where fermentation is initiated by defined starter cultures.

The first 'pure' starter culture (*Lactococcus lactis*) was used in 1890 for producing fermented milk and cheese in Germany and Denmark (Holzapfel et al. 1995). A starter culture is a product with high viable microbial counts, and when added to certain foods, it accelerates fermentation leading to a final product with a desirable alteration in the aroma, texture and flavour profile (Holzapfel 2002). A better understanding of the metabolism and genomics of fermentation microbes led to improved strain selection to ensure product uniqueness (Leroy and De Vuyst 2004). Although commercial starter cultures can ensure end-product safety and quality, traditional starters result in a fermented product with diverse sensory attributes due to the wide variety of microbes present (Ross et al. 2002; Leroy and De Vuyst 2004). During fermentation, carbohydrates are oxidized (aerobically or anaerobically) by microbes, predominantly lactic acid bacteria (LAB), and produce end products like lactic acid, carbon dioxide and alcohol, organic acids, acetic, propionic, butyric and formic acids, as well as enzymes, bacteriocins, aroma compounds and exopolysaccharides (Ross et al. 2002; Leroy and De Vuyst 2004). As a result, raw materials are converted to a safe product with unique sensory characteristics. Due to the occurrence of partial oxidation, the fermented product still contains some carbohydrates and is, therefore, of nutritional value in the human diet. Such fermented foods are typical of a region and have unique flavour and texture (Sohliya et al. 2009). These products and the microbes have been studied, and these processed products have been classified into six main groups: (1) dairy, (2) fish, (3) soybean, (4) vegetable, (5) fish and meat and (6) alcoholic beverages.

17.2 *Enterococcus*: A Class of Lactic Acid Bacteria (LAB)

Bacteria that ferment carbohydrate and produce lactic acid are grouped under lactic acid bacteria. These bacteria aid in the flavour development and texture and enhance the nutritional value of food products; they are thus by and large employed in the fermentation industry (McKay and Baldwin 1990). Microbes belonging to the genus *Streptococcus*, *Lactobacillus*, *Lactococcus* and *Leuconostoc* are largely employed for producing fermented food products, like yogurt, sauerkraut, sausage, cheeses and other fermented foods. The major metabolic end product of these bacteria is lactic acid, and so this characteristic has linked LAB involved in food fermentation as food protectors as the acidification restrains the growth of spoilage agents like *Listeria monocytogenes, Staphylococcus aureus* and *Clostridium botulinum*. LAB are well documented for their beneficial role and their safety in food fermentation. Strains being employed as probiotics belong to *Enterococcus, Lactococcus, Lactobacillus* and *Bifidobacterium*.

Enterococci are coccus-shaped, Gram-positive, facultative anaerobic, oxidase-negative, non-endospore forming, catalase-negative bacteria that occur in chains, pairs or singly (Giraffa 2003; Foulquié Moreno et al. 2006). They are homofermenters and produce L(+) lactic acid from glucose and are also able to metabolize amino acids and citrate. Most enterococcal species can hydrolyze esculin and grow on azide medium: *Enterococcus asini, E. canis, E. avium, E. columbae, E. casseliflavus, E. cecorum, E. durans, E. faecium, E. faecalis, E. gallinarum, E. gilvus, E. haemoperoxidus, E. flavescens, E. hirae, E. moraviensis, E. pallens, E. mundtii, E. phoeniculicola, E. pseudoavium, E. ratti, E. raffinosus, E. malodoratus, E. saccharominimus, E. sulfureus, E. solitarius, E. dispar, E. saccharolyticus* and *E. villorum* have been listed to this genus (Foulquié Moreno et al. 2006).

Raw milk and dairy products have been a source of *Enterococcus faecium, E. durans, E. hirae, E. faecalis* and *E. casseliflavus* (Cortés et al. 2006; Ogier and Serror 2008). *E. faecalis* and *E. faecium* are generally isolated from faeces as they are normal flora of the human and animal gastrointestinal tracts. Enterococci enter the food and dairy environment from other primary habitats such as faeces, soil, plants and water (Giraffa 2003; Foulquié Moreno et al. 2006; Ogier and Serror 2008). These bacteria can grow in high salinity, extreme pH (4.0–9.6), temperatures between 10 and 45 °C and survive 30 min of heating at 60 °C and thus can adapt to various environment.

Bacteria belonging to the genus *Enterococcus* are, unlike most LAB, not GRAS. This is due to their association with faecal contamination of primarily water and their role as opportunistic pathogens. These are known to cause clinical human infections such as meningitis, endocarditis and bacteraemia. However, enterococci are also known as safe to use in food fermentations (Cortés et al. 2006; Foulquié Moreno et al. 2006; Ogier and Serror 2008).

17.3 Bacteriocins: A Class of Antibacterial Production by *Enterococcus*

Many bacteria produce bacteriocins (Riley and Wertz 2002) and have attracted enormous attention recently due to their prospects as natural preservatives in food industry (Ennahar et al. 1999). Small antimicrobial peptides and proteins produced by LAB, which possess activity towards Gram-positive bacteria, are categorized as bacteriocins (Klaenhammer 1988; Cotter et al. 2005).

LAB bacteriocins can be grouped as Class I which include the lantibiotics—heat-stable, lanthionine-containing, peptide bacteriocins, Class II bacteriocins which include non-lanthionine-containing bacteriocins, and Class III which include the bacteriolysins which are large, heat-labile, lytic proteins, often murein hydrolases (Klaenhammer 1988). Class II bacteriocins are classified as IIa, pediocin-like bacteriocins, IIb which include two-peptide bacteriocins, and IIc which include circular bacteriocins.

There are large reviews available on the application of enterococci in food, meat and dairy products and health (Foulquié Moreno et al. 2006; Giraffa 2003; Hugas et al. 2003). *Enterococcus faecium* linked to food ecosystems have been attributed to bacteriocin production (Giraffa 2003). These strains have been associated with dairy products (Foulquié Moreno et al. 2003; Leroy et al. 2003; Abriouel et al. 2005;), sausages and vegetable products (Floriano et al. 1998).

LAB strains that can participate as good candidates for starter cultures should possess the basic properties, which include acidification, absence of antibiotic resistance genes, tolerance to bile salts and also production of bacteriocins (Strompfova and Laukova 2007; Javed et al. 2011). The bacteriocins that have been best characterized and have originated from enterococci (enterocins) are associated to be originating from different foods (Table 17.1). These bacteriocins include the class II enterocins—P, B, A, bacteriocin 31 and CRL35 (Nami et al. 2015; Cintas et al. 2001).

17.4 Probiotic Properties of Enterococci

As stated by Havenar et al. (1992), probiotics can be described as 'single or mixed culture' of live microorganisms which exert helpful effect on humans or animals by enriching the properties of the native floras. The major function of probiotics includes growth inhibition of pathogens, lowering of blood cholesterol, anti-carcinogenic and anti-mutation activities and improvement of gut mucosal barrier and immune system (Holzapfel et al. 1998). Major requirements for a microorganism to be categorized as effective probiotic are adherence to cells; production of acids, H_2O_2 and bacteriocins, which are antagonistic to pathogens; reduction of pathogenic adherence; persistence and multiplicability; safety, non-invasiveness, noncarcinogenic and nonpathogenic; and having co-aggregation capacity to form normal balanced flora (Salminen et al. 1996).

Lactobacillus and *Bifidobacterium* are genera of bacteria that are commonly used as probiotics by humans. Other species of LAB have also been used as probiotics, and they include the strains belonging *Enterococcus faecium*, *E. faecalis*, *Lactococcus lactis* subsp. *lactis*, *L. mesenteroides*, *Pediococcus acidilactici*, etc. (Holzapfel et al. 1998).

Enterococci are used as probiotics in a lesser extent. *E. faecium* SF 68 is a well-established probiotic strain reported to be clinically efficient in prevention of diarrhoea associated with antibiotic therapy as well as in diarrhoea associated with children (Bhardwaj et al. 2008). Another strain *E. faecium* CRL 183 with probiotics potential has been employed for producing fermented soymilk products (Rossi et al. 1999).

The probiotic benefits of enterococcal strains have been well studied and widely reported, but the appearance of strains of enterococci which are antibiotic resistant and the reports of human diseases associated with these strains have raised concern

Table 17.1 List of some of the bacteriocins produced by *E faecium* and *E. faecalis*

Bacteriocin	Producer organism	Location of gene for production	Molecular mass	Post-translational modifications	Amino acid sequence of purified bacteriocin	References
Enterocin A	*E. faecium* CTC492	Chromosome	4833	Cleaves the leader peptide on export, aids in disulphide bridge formation	TTHSGKYYGNGVYCTKNKCTVDWAKATT-QCIAGMSIGGFLGGAIPGKC	Aymerich et al. (1996)
Enterocin B	*E. faecium* T136	Chromosome	5465	Cleaves the leader peptide on export, aids in disulphide bridge formation	ENDHRMPNELNRPNNLSKGGAKCGAAIAGG LFGIP-QKGPLAWAAGLANVYSKCN	Casaus et al. (1997)
Enterocin P	*E. faecium* P13	Chromosome	4630	Cleaves signal peptide on export, possible disulphide bridge formation	ATRSYGNGVYCNNSKCWVNWGEAKENIAGI-QVISGWASGLAGMGH	Cintas et al. (2001)
Enterocin AS-48	*E. faecalis* S-48	Pheromone responsive plasmid pMB2	7167	Cleaves the leader peptide and head-tail peptide bond formation	MAKEFGIPAAVAGTVLNVVEAGGWVTTIVSI LTAV-QGSGGLSLLAAAGRESIKAYLKKEIKK KGKRAVIAW	Martınez-Bueno et al. (1994)
Enterocin L50A	*E. faecium* L50	Plasmid pCIZ1	5190	No modification	MGAIAKLVAKFGWPIVKKYYKQIMQFIGEG WAIN-QKIIEWIKKHI	Cintas et al. (2001)
Enterocin L50B	*E. faecium* L50	Plasmid pCIZ1	5178	No modification	MGAIAKLVTKFGWPLIKKFYKQIMQFIGQG WTID-QQIEKWLKRH	Cintas et al. (2001)
Bacteriocin 31	*E. faecalis* YI717	Pheromone responsive plasmid pYI17	5009	Cleaves signal peptide on export, possible disulphide bridge formation	ATYYGNGLYCNKQKCWVDWNKASREIGKII-QVNGWVQHGPWAPR	Tomita et al. (1996)
Cyl L_L	*E. faecalis* DS16	n.r	n.r	Modification to form lanthionine-containing precursors, removes leader sequence on export and proteolytic activation	MENLSVVPSFEELSVEEMEAIQGSGDVQAET TPV-QCAVAATAAASSAACGWVGGGIFTGVT VVVSLKHC	Booth et al. (1996)
Cyl L_S	*E. faecalis* DS16	n.r	n.r		VLNKENQENYYSNKLELVGPSFEELSLEEME AIQG-QSGDVQAETTPACFTIGLGVGALFSAKFC	Gilmore et al. (1994)

n.r Not reported

on the subject of their use as probiotics. The greatest fear is that the virulence factors and genes encoding antimicrobial resistance may get transferred to other unwanted bacteria in the intestinal tract leading to larger dissemination of antibiotic resistance and development of drug-resistant strains.

17.5 Safety of Enterococci

Clear effects of enterococci promoting health have been illustrated, and some strains are safe to use, but these group of organisms are also recognized as major nosocomial pathogens that have been reported to cause bacteraemia, endocarditis and other infections related to the urinary tract, central nervous system and multi-antibiotic resistance (Morrison et al. 1997). These disadvantageous characteristics undoubtedly raise legitimate alarm for the safe application of enterococci as probiotics.

17.5.1 Antibiotic Resistance Amongst *Enterococcus*

A major cause of pathogenesis caused by enterococci is their resistance to a large number of antibiotics (Barbosa et al. 2009). Antibiotic resistance that has been found to occur in these strains belong to both natural or intrinsic resistance and transferable or acquired resistance. Enterococci are genetically resistant to antibiotics such as cephalosporins, aminoglycosides, monobactams and polymyxins. Acquired resistance is mediated by genes present on plasmid or transposons, and the examples include resistance to erythromycin, chloramphenicol, tetracycline, high level of clindamycin and aminoglycosides, β-lactams and glycopeptides like vancomycin (Clewell 1990; Leclercq 1997). The antibiotic vancomycin is considered the final shot for treating multiple-resistant enterococcal infections. Ampicillin, vancomycin and gentamicin are mostly used to cure infections caused by multiple antibiotic-resistant enterococci strains. Unrestricted use of vancomycin has elevated the number of vancomycin-resistant enterococci (VRE) in the recent past (Franz et al. 2003). This has also led to a situation where the VRE that have emerged in hospitals has not been able to be treated with conventional antibiotic treatments.

The virulence prevalent amongst enterococci cannot be explained based alone on antibiotic resistance. It is believed that enterococci that cause nosocomial infections have traits such as the colonization ability, adherence to host and tissue invasion which also add up to the pathogenesis of enterococci (Franz et al. 2003).

17.5.2 Enterococci and Virulence

Rampant and widespread use of antibiotics has always been a major contributor to virulence and drug-resistant factor amongst organism including *Enterococcus*. Besides this compatibility and resistance to desiccation has added to the resilience of enterococci becoming facultative parasites. Understanding their commensal relationships has been a focus while targeting treatment strategies of this group with new therapeutics. On the basis of the nature of the gastrointestinal tract or environments contaminated with human waste which are invariably inhabited with enterococci, the intrinsic resistance of the group to antimicrobials has resulted from their need to survive and compete the detrimental ecosystems such as GI tract (Gilmore et al. 2002).

Pathogenesis and virulence of enterococci cannot be justified primarily on antibiotic resistance. With various parameters and events such as colonization and adherence to host tissues, tissue invasion leads to pathogenesis of most infections. It is the effector molecule that supplements to the disease-causing ability of a microorganism which is defined as the virulence factor (Mundy et al. 2000). Enterococci harbour genes which can play a role in virulence directly or indirectly (Franz et al. 2001) (Table 17.2). Various genes that encode to the development of virulence in enterococci have been reported from foods and food products. These factors include properties such as aggregation, antigen production, gelatinase formation, surface protein formation and production of adherence protein especially the collagen.

17.5.2.1 Aggregation Substance

This is a pheromone-induced protein present on the surface of *Enterococcus faecalis* and is an aggregation substance (Agg). It is required during conjugation for cell-to-cell contact and also contributes to the adherence of the organism to eukaryotic organisms (Medeiros et al. 2014). Agg is an important constituent of the genetic exchange system of bacteria which are pheromone responsive, and it arbitrates effective contact that is established between enterococcal donor and recipient which facilitates the transfer of plasmid (Clewell 1993). This characteristic may be a factor for the pathogenesis of infections caused by enterococci. Agg can also result in superantigen activity by bonding and transferring to the surface of the organism the cognate ligand of enterococci (Foulquié Moreno et al. 2006). Agg is also predicted to increase the hydrophobicity of the enterococcal surface which leads to cholesterol localization to phagosomes preventing its fusion with lysosomal vesicles (Mundy et al. 2000). The prevalence of Agg in enterococci-encountered food is significantly high although it is reported to be exclusive to *Enterococcus faecalis*.

17.5.2.2 Gelatinase

Another important virulence factor responsible for pathogenesis of enterococcus is gelatinase (Gel) enzyme which cleaves fibrin and has an important contribution in virulence of enterococci (Rathnayake et al. 2012). Gelatinase is a zinc metalloendopeptidase which is found extracellularly and promotes hydrolysis of gelatin, haemoglobin, collagen and peptides (Su et al. 1991). This protein is encoded by the chromosomal *gel*E gene. Gel production is seen to be high in strains *E. faecalis* prevalent in foods (Eaton and Gasson 2001).

17.5.2.3 Extracellular Surface Protein

Shankar et al. (1999) initially reported the occurrence of extracellular surface protein (Esp) in a clinical isolate *E. faecalis* MMH594. *esp* is large consisting of 5622 nucleotides which encode 1873 amino acids as primary translation product with a

Table 17.2 List of identified *Enterococcus* virulence factors and antibiotic-resistant genes

Gene	Function	Primer sequence	Product size	Reference
esp	Cell wall-associated protein involved in immune evasion; may be associated with *cyl* genes on a pathogenicity island	TTACCAAGATGGTTCTGTAGGCACCCAAGTATACTTAGCATCTTTTGG	913	Shankar et al. (1999)
agg	Aggregation protein involved in adherence to eukaryotic cells; cell aggregation and conjugation	AAGAAAAAGAAGTAGACCAACAAACGGCAAGACAAGTAAATA	1553	Eaton and Gasson (2001)
gelE	Toxin; extracellular metalloendopeptidase hydrolyzes gelatin, collagen, haemoglobin and other bioactive compounds	ACCCCGTATCATTGGTTT ACGCATTGCTTTTCCATC	419	Eaton and Gasson (2001)
cylM	Post-translational modification of cytolysin	CTGATGGAAAGAAGATAGTATTGAGTTGGTCTGATTACATTT	742	Eaton and Gasson (2001)
cylB	Transport of cytolysin	ATTCCTACCTATGTTCTGTTA AATAAACTCTTCTTTTCCAAC	843	Eaton and Gasson (2001)
cylA	Activation of cytolysin	TGGATGATAGTGATAGGAAGTTCTACAGTAAATCTTTCGTCA	517	Eaton and Gasson (2001)
efaAfm	Cell wall adhesins expressed in serum by *E. faecium*, respectively	AACAGATCCGCATGAATA CATTTCATCATCTGATAGTA	735	Eaton and Gasson (2001)
efaAfms	Cell wall adhesins expressed in serum by *E. faecalis*, respectively	GACAGACCCTCACGAATAAGTTCATCATGCTGTAGTA	705	Eaton and Gasson (2001)
Cpd	Sex pheromones, chemotactic for human leukocytes; facilitate conjugation	TGGTGGGTTATTTTTCAATTC TACGGCTCTGGCTTACTA	782	Eaton and Gasson (2001)
cab		AACATTCAGCAAACAAAGC TTGTCATAAAGAGTGGTCAT	1405	Eaton and Gasson (2001)
ccf		GGGAATTGAGTAGTGAAGAAGAGCCGCTAAAATCGGTAAAAT	543	Eaton and Gasson (2001)
hyl	Hyaluronidase activity	ACAGAAGAGCTGCAGGAAATGGACTGACGTCCAAGTTTCCAA	276	Vankerckhoven et al. (2004)
ace	Adhesion of collagen	AAAGTAGAATTAGATCCACACTCTATCACATTCGGTTGCG	320	Dupre et al. (2003)
vanA	Vancomycin resistance	CCCCTTTAACGCTAATACGATCAACATGAATAGAATAAAAGTTGCAAT	1030	Clark et al. (1993)
vanB	Vancomycin resistance	GTGACAAACCGGAGGCGAGGACCGCCATCCTCCTGCAAAAAA	433	Clark et al. (1993)

Mol. mass of 202 kDa. Esp protein evades the immune system of the host and biofilm formation leading to resistance to environmental stresses (Borgmann et al. 2004). There is a difference in the frequency of Eps in food isolates of *E. faecium* and *E. faecalis*. The occurrence of this gene in *E. faecalis* is higher (Eaton and Gasson 2001; Franz et al. 2001).

17.5.2.4 Cytolysin

Cytolysin (Cyl) or β-haemolysin is a major virulence factor which is a cellular toxin found in enterococci and known to increase virulence in animals (Gilmore et al. 1994). Cytolysin permits the organism to evade the host immune response. It does so by destroying macrophages and neutrophils (Franz et al. 2001). Other potential virulence factors include cell wall adhesins from *E. faecium* (*efaAfm*) and *E. faecalis* (*efaAfs*). These proteins facilitate the adhesion of enterococci to diverse surfaces and evade the immune response (Shankar et al. 1999; Toledo-Arana et al. 2001). Production of the adhesin-like *E. faecalis* and *E. faecium* antigens is well intended to be probable virulence determinants.

17.6 Enterococci in Food and Their Safety for Use as Starter Cultures

Enterococci are frequently isolated from food, including traditionally fermented fish and meat products and also from a variety of milk products. Mammalian gastrointestinal tract comprises a large portion of enterococci. Once these bacteria are rejected in the environment through human and animal ejecta, they inhabit diverse niches as they can grow in extreme conditions, and so enterococci are also found in soil and water surfaces, plants and vegetables. They can also colonize raw foods such as milk and meat through environmental contamination.

Application of enterococci in food industry demands the basic characteristic such as production of bacteriocins (enterocin) and contribution to the organoleptic characteristics of fermented food materials (Foulquié Moreno et al. 2006). Fermented meat is produced around the world. Enterococci are thermotolerant (Magnus et al. 1988), and this trait can cause spoilage problem in meat products.

E. faecium strains are used as starter cultures for producing fermented meat. A few specific *E. faecium* strains RZS C13 and CCM 4213 were used as starters for producing fermented sausages (Callewaert et al. 2000). In addition, antilisterial efficiency of enterocins CRL 35 from *E. faecium* acts against meat infected with *L. innocua* and *L. monocytogenes* (Vignolo et al. 2000). Growth of enterococci in cheeses is quite enviable. They help in ripening and developing aroma in cheeses. The extent of milk contamination leads to varying levels of *E. faecium* and *E. faecalis* in different cheeses (Macedo et al. 1995). The positive role of enterococci in improving aroma of the cheese has led to addition of enterococcal strains in certain starters.

Enterococci also occur in many vegetables, plant material, olives, etc.; however limited literature data relates to their isolation and characterization (Giraffa 2002; Ben Omar et al. 2004). *E. faecalis* and *E. faecium* are frequently observed as contaminants in Spanish-style green olive fermentations (de Castro et al. 2002).

The assortment of vegetable products especially ready-to-eat products is gaining popularity. To enhance the safety of such products, bacteriocinogenic LAB or bacteriocins need to be used. Nonetheless, the usage of bacteriocins, as bio-preservatives in vegetables and olives, has also been reported (Floriano et al. 1998).

The pathogenic properties of enterococci because of horizontal gene transfer associated with virulence and antibiotic resistance are the major concern for enterococci present in food (Franz et al. 2003). *E. faecium* is mostly used in food, as probiotics and as starter cultures. *E. faecalis* is recognized to harbour higher number of virulence determinants than *E. faecium.* The transmission of plasmid in response to sex pheromones seems to be specific to *E. faecalis* (Franz et al. 2003). The occurrence of virulence determinants amongst food isolates seems to be strain specific (Eaton and Gasson 2001; Franz et al. 2001).

In order to narrow the risk of their usage, *Enterococcus* strain can be used as starter culture or probiotic. Each and every strain should be checked for any known virulence factor. Preferably such starter cultures of enterococci should not harbour any virulence factors and essentially be sensitive to antibiotics.

Conclusion

Enterococci have been gaining importance in terms of their use as probiotics in food. Because of their extensive prevalence in nature, critical analysis of their benefits and possible hazards calls for attention of food quality and safety managers. These bacteria are found to occur as spontaneous colonizers in traditional fermentation and contribute to the transformation of raw materials into finished products. Some of the isolates may be opportunist pathogens, but the virulence of enterococci is confined as well as strain specific, and hence all the opportunist pathogen may not cause disease in humans. Analysis of transferable and sustainable virulence properties should form the basis for widespread use of enterococci in food and food products to rule out the fact that the virulent genes are not disseminated randomly through the consumption of traditionally processed products. Thus, the knowledge domain of enterococci needs to be updated through modern systems of analysis to ensure that food safety of processed and unprocessed foods and their products are evaluated before releasing these foods for consumption.

Acknowledgements This chapter forms a part of the research study on probiotics of non-dairy origin funded by Department of Biotechnology, Govt. of India (Sanction No. BT/PR7343/FNS/20/738/2013), and the authors acknowledge the financial support received from the funding agency.

References

Abriouel H, Lucas R, Ben Omar N, Valdivia E, Maqueda M, Martínez-Cañamero M, Gálvez A (2005) Enterocin AS-48RJ: a variant of enterocin AS-48 chromosomally encoded by Enterococcus faecium RJ16 isolated from food. Syst Appl Microbiol 28:383–397. doi:10.1016/j.syapm.2005.01.007

Aymerich T, Holo H, Havarstein LS, Hugas M, Garriga M, Nes IF (1996) Biochemical and genetic characterization of Enterocin A from *Enterococcus faecium*, a new antilisterial bacteriocin in the pediocin family of bacteriocins. Appl Environ Microbiol 62:1676–1682

Barbosa J, Ferreira V, Teixeira P (2009) Antibiotic susceptibility of enterococci isolated from traditional meat products. Food Microbiol 26:527–532. doi:10.1016/j.fm.2009.03.005

Ben Omar N, Castro A, Lucas R, Abriouel H, Yousif NMK, Franz CMAP, Holzapfel WH, Perez-Pulido R, Martınez-Canamero M, Galvez A (2004) Functional and safety aspects of enterococci isolated from different Spanish foods. Syst Appl Microbiol 27:118–130. doi:10.1078/0723-2020-00248

Bhardwaj A, Malik RK, Chauhan P (2008) Functional and safety aspects of enterococci in dairy foods. Indian J Microbiol 48:317–325. doi:10.1007/s12088-008-0041-2

Booth MC, Bogie CP, Sahl HG, Siezen RJ, Hatter KL, Gilmore MS (1996) Structural analysis and proteolytic activation of *Enterococcus faecalis* cytolysin, a novel lantibiotic. Mol Microbiol 21:1175–1184. doi:10.1046/j.1365-2958.1996.831449.x

Borgmann S, Niklas DM, Klare I, Zabel LT, Buchenau P, Autenrieth IB, Heeg P (2004) Two episodes of vancomycin resistant *Enterococcus faecium* outbreaks caused by two genetically different clones in a newborn intensive care unit. Int J Hyg Environ Health 207:386–389. doi:10.1078/1438-4639-00304

Callewaert R, Hugas M, De Vuyst L (2000) Competitiveness and bacteriocin production of enterococci in the production of Spanish-style dry fermented sausages. Int J Food Microbiol 57:33–42. doi:10.1016/S0168-1605(00)00228-2

Casaus P, Nilsen T, Cintas LM, Nes IF, Hernandez PE, Holo H (1997) Enterocin B, a new bacteriocin from *Enterococcus faecium* T136 which can act synergistically with Enterocin A. Microbiology 143:2287–2294. doi:10.1099/00221287-143-7-2287

de Castro A, Montano A, Casado FJ, Sanchez AH, Rejano L (2002) Utilization of *Enterococcus casseliflavus* and *Lactobacillus pentosus* as starter cultures for Spanish-style green olive fermentation. Food Microbiol 19:637–644. doi:10.1006/fmic.2002.0466

Cintas LM, Casaus MP, Herranz C, Nes I, Hernandez PE (2001) Review: bacteriocins of lactic acid bacteria. Food Sci Technol Int 7:281–305. doi:10.1106/R8DE-P6HU-CLXP-5RYT

Clark N, Cooksey R, Hill B, Swenson J, Tenover F (1993) Characterization of glycopeptide-resistant enterococci from US hospitals. Antimicrob Agents Chemother 37(11):2311–2317. doi:10.1128/AAC.37.11.2311

Clewell DB (1990) Movable genetic elements and antibiotic resistance in enterococci. Eur J Clin Microbiol Infect Dis 9:90–102. doi:10.1007/BF01963632

Clewell DB (1993) Bacterial sex pheromone-induced plasmid transfer. Cell 73:9–12. doi:10.1016/0092-8674(93)90153-H

Cortés C, De la Fuente R, Contreras A, Sánchez A, Corrales JC, Ruiz-Santa-Quiteria JA, Orden JA (2006) Occurrence and preliminary study of antimicrobial resistance of enterococci isolated from dairy goats in Spain. Int J Food Microbiol 110:100–103. doi:10.1016/j.ijfoodmicro.2006.01.033

Cotter PD, Hill C, Ross RP (2005) Bacteriocins: developing innate immunity for food. Nat Rev Microbiol 3:777–788. doi:10.1038/nrmicro1273

Dupre I, Zanetti S, Schito AM, Fadda G, Sechi LA (2003) Incidence of virulence determinants in clinical *Enterococcus faecium* and *Enterococcus faecalis* isolates collected in Sardinia (Italy). J Med Microbiol 52(6):491–498. doi:10.1016/j.bjid.2015.11.011

Eaton TJ, Gasson MJ (2001) Molecular screening of *enterococcus* virulence determinants and potential for genetic exchange between food and medical isolates. Appl Environ Microbiol 67:1628–1635. doi:10.1128/AEM.67.4.1628-1635.2001

Ennahar S, Sonomoto K, Ishizaki A (1999) Class IIa bacteriocins from lactic acid bacteria: antibacterial activity and food preservation. J Biosci Bioeng 87:705–716. doi:10.1016/S1389-1723(99)80142-X

Floriano B, Ruiz-Barba JL, Jiménez-Díaz R (1998) Purification and genetic characterization of enterocin I from *Enterococcus faecium* 6T1a, a novel antilisterial plasmid-encoded bacteriocin which does not belong to the pediocin family of bacteriocins. Appl Environ Microbiol 64:4883–4890

Foulquié Moreno MR, Rea M, Cogan T, De Vuyst L (2003) Applicability of a bacteriocin-producing *Enterococcus faecium* as a coculture in cheddar cheese manufacture. Int J Food Microbiol 81:73–84. doi:10.1016/S0168-1605(02)00167-8

Foulquié Moreno MR, Sarantinopoulos P, Tsakalidou E, De Vuyst L (2006) The role and application of enterococci in food and health. Int J Food Microbiol 106:1–24. doi:10.1016/j.ijfoodmicro.2005.06.026

Franz CMAP, Muscholl-Silberhorn AB, Yousif NMK, Vancanneyt M, Swings J, Holzapfel W (2001) Incidence of virulence factors and antibiotic resistance among enterococci isolated from food. Appl Environ Microbiol 67:4385–4389. doi:10.1128/AEM.67.9.4385-4389.2001

Franz CMAP, Stiles ME, Schleiferc KH, Holzapfel WH (2003) Enterococci in foods a conundrum for food safety. Int J Food Microbiol 88:105–112. doi:10.1016/S0168-1605(03)00174-0

Gilmore MS, Segarra RA, Booth MC, Bogie CP, Hall LR, Clewell DB (1994) Genetic structure of the *Enterococcus faecalis* plasmid pAD1-encoded cytolytic toxin system and its relationship to lantibiotic determinants. J Bacteriol 176:7335–7344

Gilmore M, Coburn P, Nallapareddy S, Murry B (2002) Enterococcal virulence. In: Gilmore M, Clewell D, Courrallin P, Dunny G, Murry B, Rice L (eds) The enterococci. ASM Press, Washington DC, pp 301–354. doi:10.1128/9781555817923.ch8

Giraffa G (2002) Enterococci from foods. FEMS Microbiol Rev 26:163–171. doi:10.1111/j.1574-6976.2002.tb00608.x

Giraffa G (2003) Functionality of enterococci in dairy products. Int J Food Microbiol 88:215–222. doi:10.1016/S0168-1605(03)00183-1

Havenaar R, Brink BT, Huis In't Veld JHJ (1992) Selection of strains for probiotics use (Ed. R. Fuller). Probiotics: The Scientific Basis. pp. 209–222. Chapman and Hall, Landon

Holzapfel WH (2002) Appropriate starter culture technologies for small-scale fermentation in developing countries. Int J Food Microbiol 75:197–210. doi:10.1016/S0168-1605(01)00707-3

Holzapfel WH, Geisen R, Schillinger U (1995) Biological preservation of foods with reference to protective cultures, bacteriocins and food-grade enzymes. Int J Food Microbiol 24:343–362. doi:10.1016/0168-1605(94)00036-6

Holzapfel WH, Haberer P, Snel J, Schillinger U, JHJ H i't V (1998) Overview of gut flora and probiotics. Int J Food Microbiol 41:85–101. doi:10.1016/S0168-1605(98)00044-0

Hugas M, Garriga M, Aymerich MT (2003) Functionality of enterococci in meat products. Int J Food Microbiol 88:223–233. doi:10.1016/S0168-1605(03)00184-3

Javed I, Ali MI, Ahmed B, Ghumro PB, Abdul H, Safia A (2011) Optimization and partial purification of bacteriocins from *Enterococcus* spp indigenous to Pakistan. Food Biotechnol 25:130–139. doi:10.1080/08905436.2011.576559

Klaenhammer TR (1988) Bacteriocins of lactic acid bacteria. Biochimie 70:337–349. doi:10.1016/0300-9084(88)90206-4

Leclercq R (1997) Enterococci acquire new kinds of resistance. Clin Infect Dis 24(Suppl. 1):S80–S84. doi:10.1093/clinids/24.Supplement_1.S80

Leroy F, De Vuyst L (2004) Lactic acid bacteria as functional starter cultures for the food fermentation industry. Trends Food Sci Technol 15:67–78. doi:10.1016/j.tifs.2003.09.004

Leroy F, Foulquié-Moreno MR, De Vuyst L (2003) *Enterococcus faecium* RZS C5, an interesting bacteriocin producer to be used as a coculture in food fermentation. Int J Food Microbiol 88:235–240. doi:10.1016/S0168-1605(03)00185-5

Macedo AC, Malcata FX, Hogg TA (1995) Microbiological profile in Serra ewe's cheese during ripening. J Appl Bacteriol 79:1–11. doi:10.1111/j.1365-2672.1995.tb03117.x

Magnus CA, McCurdy AR, Ingledew WM (1988) Further studies on the thermal resistance of *Streptococcus faecium* and *Streptococcus faecalis* in pasteurized ham. J Can Inst Food Sci Technol 21:209–212. doi:10.1016/S0315-5463(88)70778-6

Martınez-Bueno M, Maqueda M, Galvez A, Samyn B, van Beeumen J, Coyette J (1994) Determination of the gene sequence and the molecular structure of the enterococcal peptide antibiotic AS-48. J Bacteriol 176:6334–6339

McKay LL, Baldwin KA (1990) Application for biotechnology: present and future improvements in lactic acid bacteria. FEMS Microbiol Rev 7:3–14. doi:10.1016/0378-1097(90)90694-L

Medeiros AW, Pereira RI, Oliveira DV, Martins PD, d'Azevedo PA, Van der Sand S, Frazzon J, Frazzon APG (2014) Molecular detection of virulence factors among food and clinical *Enterococcus faecalis* strains in South Brazil. Braz J Microbiol 45(1):327–332. doi:10.1590/S1517-83822014005000031

Morrison D, Woodford N, Cookson B (1997) Enterococci as emerging pathogens of humans. J Appl Microbiol Symp Supp 83:89S–99S. doi:10.1046/j.1365-2672.83.s1.10.x

Mundy LM, Sahm DF, Gilmore M (2000) Relationships between enterococcal virulence and antimicrobial resistance. Clin Microbiol Rev 13:513–522. doi:10.1128/CMR.13.4.513-522.2000

Nami Y, Haghshenas B, Haghshenas M, Khosroushahi AY (2015) Antimicrobial activity and presence of virulence factors and bacterial structural genes in *Enterococcus faecium* CM33 isolated from ewe colostrum. Front Microbiol 6:782. doi:10.3389/fmicb.2015.00782

Ogier J-C, Serror P (2008) Safety assessment of dairy microorganisms: the *Enterococcus* genus. Int J Food Microbiol 126:291–301. doi:10.1016/j.ijfoodmicro.2007.08.017

Rathnayake IU, Hargreaves M, Huygens F (2012) Antibiotic resistance and virulence traits in clinical and environmental *Enterococcus faecalis* and *Enterococcus faecium* isolates. Syst Appl Microbiol 35:326–333. doi:10.1016/j.syapm.2012.05.004

Riley MA, Wertz JE (2002) Bacteriocins: evolution, ecology, and application. Annu Rev Microbiol 56:117–137. doi:10.1146/annurev.micro.56.012302.161024

Ross RP, Morgan S, Hill C (2002) Preservation and fermentation: past, present and future. Int J Food Microbiol 79:3–16. doi:10.1016/S0168-1605(02)00174-5

Rossi EA, Vendramini RC, Carlos IZ, Pei YC, de Valdez GF (1999) Development of a novel fermented soymilk product with potential probiotic properties. Euro Food Res Technol 209:305–307. doi:10.1007/s002170050499

Salminen S, Isolauri E, Salminen E (1996) Clinical uses of probiotics for stabilizing the gut mucosal barrier: successful strains for future challenges. Antonie Van Leeuwenhoek 70:347–358. doi:10.1007/BF00395941

Sanni AI (1993) The need for process optimization of African fermented foods and beverages. Int J Food Microbiol 18:85–95. doi:10.1016/0168-1605(93)90213-Z

Shankar V, Baghdayan A, Huycke MM, Lindahl G, Gilmore MS (1999) Infection-derived *Enterococcus faecalis* strains are enriched in *esp*, a gene encoding a novel surface protein. Infect Immun 67:193–200

Sohliya I, Bhagobaty RK, Kumar R, Joshi SR (2009) *Tungrymbai*-traditional fermented soybean food of the ethnic tribes of Meghalaya. Ind J Trad Know 8(4):559–561

Strompfova V, Laukova A (2007) In vitro study on bacteriocin production of enterococci associated with chickens. Anaerobe 13:228–237. doi:10.1016/j.anaerobe.2007.07.002

Su YA, Sulavik MC, He P, Makinen KK, Makinen P, Fiedler S, Wirth R, Clewell DB (1991) Nucleotide sequence of the gelatinase gene (gelE) from *Enterococcus faecalis* subsp. *liquefaciens*. Infect Immun 59:415–420

Toledo-Arana A, Valle J, Solano C, Arrizubieta MJ, Cucarella C, Lamata M et al (2001) The enterococcal surface protein, esp, is involved in *Enterococcus faecalis* biofilm formation. Appl Environ Microbiol 67:4538–4545. doi:10.1128/AEM.67.10.4538-4545.2001

Tomita H, Fujimoto S, Tanimoto K, Ike Y (1996) Cloning and genetic organization of the bacteriocin 31 determinant encoded on the *Enterococcus faecalis* pheromone-responsive conjugative plasmid pYI17. J Bacteriol 178:3585–3593

Vankerckhoven V, Van Autgaerden T, Vael C, Lammens C, Chapelle S, Rossi R et al (2004) Development of a multiplex PCR for the detection of asa1, gelE, cylA, esp, and hyl genes in enterococci and survey for virulence determinants among European hospital isolates of *Enterococcus faecium*. J Clin Microbiol 42(10):4473–4479. doi:10.1128/JCM.42.10.4473-4479.2004

Vignolo G, Palacios J, Farias ME, Sesma F, Schillinger U, Holzapfel W, Oliver G (2000) Combined effect of bacteriocins on the survival of various *Listeria* species in broth and meat system. Curr Microbiol 41:410–416. doi:10.1007/s002840010159

Fermented Foods, Microbiota and Human Health

18

Ann Catherine Archer and Prakash M. Halami

Abstract

Fermentation is an age-old process aided by microorganisms that, directly or by the action of bioactive components produced by them, impart several health benefits. A plethora of fermented foods have opened new avenues of scientific research for tapping their various health-promoting properties. In addition, globalization has exposed consumers to ethnic fermented products once available only in a local community. Fermented foods are found to alleviate many diseases including gastrointestinal disorders, lactose intolerance, cancer, hypertension, allergy and metabolic syndromes and maintain digestive health and immune functions. Understanding of the complex interactions of diet, gut, microbiota and host in addition to microbiome studies of ethnic populations will reveal much about the mechanisms by which microbiota are associated with human health.

Keywords

Bioactive compound • Fermented foods • Fermented vegetables • Human health • Lactic acid bacteria • Probiotic

18.1 Introduction

Fermentation is an ancient preservation technique prevalent among various cultures and ethnic groups around the globe. The term fermentation is derived from *fermentum* (Latin) meaning 'to ferment'. Fermented foods play a large role in the cuisines of various cultures spanning countries across Asia, Africa and Europe. Be

A.C. Archer • P.M. Halami (✉)
Microbiology and Fermentation Technology, CSIR-Central Food Technological Research Institute, Mysore 570 020, India
e-mail: prakashalami@cftri.res.in

V.C. Kalia et al. (eds.), *Mining of Microbial Wealth and MetaGenomics*,
DOI 10.1007/978-981-10-5708-3_18

it fermented dairy products like *koumiss* and *kefir* from Russia and central Asia or *kimchi*, a famous Korean fermented cabbage product or *Natto* from Japan; all fermented foods are diverse in nature, origin, texture, flavour and microbiota. Milk and cultured dairy products are undoubtedly one of the oldest fermentation practices. Almost every continent in the world has formulated its own fermented dairy product given the wide consumption and nutritious properties of milk (Hutkins 2008). Ancient reference of Abraham serving milk and curds to guests in the Bible, preparation of alcoholic beverages by ancient Egyptians, preparation of yoghurt, *koumiss*, *kefir* by nomadics of central Asia, fermentation of olives by the Greeks and Romans, etc. are evidences of fermentation (Ouwehand and Röytiö 2014). However, in 1907 a Russian zoologist named Elie Metchnikoff attributed the long life of Bulgarian peasants to yoghurt consumption and advocated the theory of longevity owing to lactic acid bacteria present in sour milk (Praveesh et al. 2011), thus establishing the theory of well-being promoted by consumption of fermented foods.

Nutritive and nonnutritive components of food are known to modify specific target functions of our body, thereby sustaining good health or alleviation of chronic diseases like cancer, obesity, gastrointestinal illnesses, etc. (Alexandraki et al. 2013). Apart from preservation, fermentation of foods also leads to enhanced flavour, better digestibility and enhancement of nutritional and pharmacological properties (Jeyaram et al. 2009). Empiric knowledge tells us that fermented foods were used in traditional medicine (Todorov and Holzapfel 2014). Although fermented foods have been associated with human culinary culture, people ascribed their potential health benefits without knowing the scientific basis of these foods or the microbiota present in them. Hence, it is imperative that the scientific community has done extensive research on various fermented foods around the world and documented several health benefits related to intake of such foods. It is known that microorganisms play a large role in fermentation and play a vital role in preserving the well-being of consumers (Parvez et al. 2006). Fermented foods have recently gained much attention in the scientific community for their diverse microbiota which metabolize a wide range of substrates resulting in distinctive products with attractive consumer appeal and beneficial effects on human health. These beneficial bacteria termed 'probiotics' are defined as microorganisms that impart positive health effects on the host when supplied in sufficient amounts (FAO/WHO 2002). Probiotics have been widely exploited and commercially available in the form of supplements. Nonetheless, speculations on the viability and authenticity of the cultures still exist.

In view of green healthy living and shift of consumer's outlook to super foods, fermented foods have a great scope in fulfilling the needs of flavourful diet as well as welfare of the host. However, there is still a large gap in deciphering the health benefits acclaimed by consumption of fermented foods majorly due to lack of coordinated double blind, randomized, placebo-controlled scientific studies on such foods (Anukam and Reid 2009). This chapter deals with the microbiota found in fermented foods and the various health benefits imparted by components of the fermented product and/or microbiota present in them (Table 18.1).

Table 18.1 List of health benefits of some bioactive substances produced in fermented food products

Product	Substrate	Country	Microorganisms	Bioactive compound	Health benefits	References
Angkak	Red rice	China, Taiwan, Thailand, Phillipines	*Monascus purpureus*	Mevinolin citrinin	Inhibits HMG-CoA, a key enzyme in cholesterol synthesis	Pattanagul et al. (2008)
Chungkook-jang	Soybean	Korea	*B. subtilis, B. cereus, B. licheniformis, B. amyloliquefaciens, Enterococcus, Rhodococcus, Pantoea agglomerans, Pantoea ananatis* and *Pseudomonas* sp.	Fibrinolytic enzymes	Thrombolytic activity	Kim et al. (1996)
Doenjang	Soybean	Korea	*Lactobacillus sp., Enterococcus faecium, B. subtilis, B. pumilus, B. licheniformis, Mucor plumbeus, Aspergillus oryzae, Debaryomyces hansenii, Leuc. mesenteroides, Tetragenococcus halophilus*	Genistein	Reduces body weight	Kwak et al. (2012)
Fermented bamboo shoot	Bamboo shoot	India	*Lactococcus lactis, Lb. brevis, Lb. casei, Lb. plantarum, Lb. fermentum, Lb. curvatus, Lb. xylosus, T. halophilus, Leuconostoc* sp.	Crude fibre, phenolic compounds, flavonoids and tannins	Antioxidant, anti-ageing and anticancer activity	Tamang and Tamang (2009), Sonar and Halami (2014)
Gundruk	Leafy vegetable	India, Nepal, Bhutan	*Lb. casei, Lb. casei subsp. pseudoplantarum, Lb. plantarum, Lb. fermentum, Ped. Pentosaceus*	Organic acids	Appetizer	Tamang et al. (2009b)
Kanji	Carrot/ beet roots, Torani	India	*Pediococcus* sp., *Leuc. mesenteroides, Lb. pentosus, Lb. paraplantarum, Lb. dextranicum, Geotrichium candidum, Candida guilliermondii, C. tropicalis, Hansenlu anomala*	Minerals, betacyanin	Anticancer activity, prevents infection	Winkler et al. (2005)

(continued)

Table 18.1 (continued)

Product	Substrate	Country	Microorganisms	Bioactive compound	Health benefits	References
Kimchi	Cabbage, ginger, green onion and hot pepper	Korea	*Leuc. kimchii, Leuc. mesenteroides, Leuc. gasicomitatum, Leuc. citreum, Leuc. inhae, Lb. plantarum, Lb. fermentum, Lb. buchneri, Lb. brevis, Lb. sakei, Lb. delbrueckii, W. cibaria, W. koreensis, Ped. pentosaceus, Ped. acidilactici*	Isocyanate and sulfide Indole-3-carbinol	Anticancer and detoxification of heavy metals	Park and Kim (2010)
				Capsaicin, allicin	Cancer prevention, suppression of *H. pylori*	An et al. (2014)
				S-adenosyl-L-methionine (SAM)	Alleviates symptoms of depression	Lee and Lee (2009)
				β-Sitosterol	Anti-proliferative action	Choi et al. (2004)
Natto	Soybean	Japan	*B. subtilis (natto)*	Nattokinase, antibiotics, vitamin K	Immune modulation and antitumour activity	Nagai and Tamang (2010)
Nham	Pork meat, pork skin,	Thailand	*Lb. brevis, Lb. plantarum, Ped. cerevisiae*	GABA	Diuretic, anti-proliferative, prevents hypertension and diabetes	Ratanaburee et al. (2013)
Puer tea	Tea	China	*Asp. glaucus, Blastobotrys adeninivorans,* species of *Saccharomyces, Penicillium, Rhizopus, Actinoplanes, Streptomyces*	Bioactive substance	Prevents CVD	Mo et al. (2008)
Red wine	Grapes	Transcontinental	*Saccharomyces cerevisiae*, sp. of LAB	Resveratrol	Anti-inflammatory, anti-daibetic, antioxidative	Jeong et al. (2010), Ramadori et al. (2009), Corder et al. (2006)
				Phenolics and flavonoids	Prevention of CVD, digestion	Jackson (2008), Truelsen et al. (1998)

Sauerkraut	Cabbage	Europe, USA, Canada, Australia	*Leuc. mesenteroides, Lb. sakei, Ped. pentosaceus, Lb. plantarum, Lb. brevis*	Isothiocynate, glucosinolates	Prevention of cancer and antioxidant enzyme stimulation	Higdon et al. (2007), Martinez-Villaluenga et al. (2012)
Tempeh	Soybean	Indigenous of Indonesia, USA, Japan, Netherlands	*Asp. niger, Citrobacter freundii, Enterobacter cloacae, K. pneumoniae, K. pneumoniae subsp. ozaenae, Lb. lactis, Lb. plantarum, Lb. fermentum, Lb. reuteri, Pseudomonas fluorescens as vitamin B12-producing bacteria, Rhiz. oligosporus, Rhiz. stolonifer, Rhiz. oryzae, Rhiz. arrhizus*	Antioxidant genestein, daidzein, tocopherol, superoxide dismutase	Prevents oxidative stress and pancreatic β-cell damage, prevents dementia	Kiriakidis et al. (1997), Lu et al. (2009)
				γ-aminobutyric acid	Improves brain blood flow and retards high blood pressure	Aoki et al. (2003)

18.2 Microorganisms from Fermented Foods

There are about 5000 different types of fermented products available globally hitherto only a small percentage of them have been scientifically investigated (Tamang 2010b). It is well known that microorganisms are crucial for fermentation. Microorganisms in fermented foodstuffs govern the physiognomies of the product such as acidity, flavour and texture in addition to conferring various health benefits (Vogel et al. 2011). An understanding of the microorganisms involved, their diversity and properties could perhaps reveal more of the fermentation product and help in managing the fermentation process making it more reliable and predictable. Microorganisms in fermented foods may be present as native microbiome or may be externally added as starter cultures (Stevens and Nabors 2009) or to produce compounds such as enzymes, flavours and/or fragrances (Longo and Sanromán 2006). Lactic acid bacteria (LAB) and *Bacillus* are two main microbes engaged in fermentation quite often dominating the initial fermentation process followed by different species of yeasts (Tamang 2015). LAB produce lactic acid (inhibits pathogens and reduces sugar content) that prolongs the shelf life of the food whereas aroma and alcohols are generated by yeasts. Conversely, moulds produce proteolytic and lipolytic enzymes that govern sensory characteristics of the product (Tamang and Fleet 2009) (Table 18.2).

Fermented foods are classified based on the substrates into fermented milk foods, vegetable products, cereal products, meat products and fermented fish products, etc. Some of the milk products fermented naturally include *dahi*, *shrikhand*, *lassi*, *chhurpi*, *mishti doi*, *somar*, etc. (India, Bhutan, Nepal, Bangladesh and Pakistan); *koumiss* (Central Asia); *ergo*, *laban*, *zabady*, *mish amasi*, *kad*, *zeer*, etc. (Africa and Middle East); *kurut* (China); and *aaruul*, *byasulag*, *airag*, *khoormof*, *chigee*, etc. (Mongolia) (Harun-ur-Rashid et al. 2007; Sarkar 2008; Wu et al. 2009; Tamang 2010a; Sun et al. 2010; Akabanda et al. 2013; Tamang et al. 2015). *Lactococcus lactis* subsp. *lactis* and *Lc. Lactis* subsp. *cremoris* dominate the microbiota of fermented milks alongside *Lactobacillus casei*, *Lb. fermentum*, *Lb. helveticus*, *Lb. plantarum*, *Lb. acidophilus*, *Enterococcus faecium*, *Leuconostoc* and *Pediococcus* species (Tamang et al. 2000; Mathara et al. 2004; Patrignani et al. 2006; Yu et al. 2011; Akabanda et al. 2013). Yeasts include *Candida lusitaniae*, *C. tropicalis*, *C. parapsilosis*, *C. rugosa*, *Saccharomyces cerevisiae*, *Kluyveromyces marxianus*, *Pichia kudriavzevii*, *Galactomyces geotrichum*, etc. (Dewan and Tamang 2006). The most famous fermented cereal foods worldwide are sourdough (Australia, Europe and America) (De Vuyst et al. 2009); *idli* and *dosa* (India and Sri Lanka) (Sridevi et al. 2010); *ogi*, *kenkey*, *kunu-zaki*, *kisra*, *togwa* (Nigeria, Ghana, Sudan and Tanzania) (Oguntoyinbo et al. 2011; Mugula et al. 2003); and *tarhana* (Turkey, Cyprus and Greece) (Sengun et al. 2009). Species of *Enterococcus*, *Lc. lactis*, *Lactobacillus plantarum*, *Lb. sakei*, *Lb. panis*, *Lb. curvatus*, *Lb. sanfranciscensis*, *Lb. pontis*, *Lb. brevis*, *Lb. fructivorans*, *Lb. alimentarius*, *Lb. paralimentarius*, *Lb. crispatus*, *Lb. pentosus*, *Lb. spicheri*, *Lb. delbrueckii*, *Lb. reuteri*, *Lb. fermentum*, *Lb. acidophilus*, *Streptococcus*, *Ped. pentosaceus*, *Weissella confusa* and *Leuc. mesenteroides* and yeasts like *Sacch. cerevisiae*, *Candida*, *Trichosporon*,

Table 18.2 List of some probiotic bacterial cultures from fermented foods exhibiting health benefits

Probiotic culture	Product	Health benefit	Reference
Lactobacillus	*Dahi*	Anticancer effect, diarrhea	Mohania et al. (2013), Agarwal and Bhasin (2002)
LAB	Fermented vegetables	Appetizer, normal blood clotting	Breidt et al. (2013)
Lb. rhamnosus GG (Lipoteichoic acid)	Fermented milk	Protects against UV-induced carcinogenesis, reduction of cholesterol	Weill et al. (2013), Kiessling et al. (2002)
LAB	*Inziangsang*	Appetizer	Tamang et al. (2009b)
LAB	*Khalpi*	Biopreservative	Tamang et al. (2009b)
LAB	*Kimchi*	Inhibits pathogenic bacteria, stimulates beneficial bacteria,prevention of constipation and colon cancer, IBD, atopic dermatitis	Lim et al. (2011)
LAB	*Sinki*	Biopreservative, antidiarrhoeal and reduces stomach pain	Tamang et al. (2009b)
LAB	Yoghurt	Digestive health,suppression of pathogens, prevents irritable bowel syndrome, Crohn's disease, vaginitis, ulcerative colitis and infant gastroenteritis	Shah et al. (2013), Chandan and Kilara (2013)
LAB	Fermented sausages	Alleviation of gastric and intestinal illnesses	Marteau et al. (2002)
S. thermophilus and *Lb. delbrueckii* subsp. *bulgaricus*	Yoghurt	Ameliorates lactose intolerance, prevention of CVD	Shah (1993), Astrup (2014)
Lb. acidophilus	Acidophilus milk	Lactose intolerance	Shah (2004)
LAB	*Kefir*	Lactose intolerance, increase number of apoptotic cells and IgA cells in the mammary gland	Hertzler and Clancy (2003), de LeBlanc et al. (2007)
Lb. plantarum	*Kallappam*	Antioxidant activity, reduction of colon cancer	Kumar et al. (2012)
Lb. helveticus and *Sacch. cerevisiae*	Fermented milk	Reduction of systolic and diastolic blood pressure	Aihara et al. (2005)
B. subtilis	*Natto*	Thrombolytic activity	Singh et al. (2014)
Lb. kefiranofaciens M1	Kefir grains	Anti-allergic activity	Hong et al. (2010)
LAB	Fermented fish oil	Sensitization to allergens, asthma, atopic dermatitis, eczema	Han et al. (2012a, b)
Lb. acidophilus and *Lb. casei*	Probiotic *dahi*	Antidiabetic effect	Yadav et al. (2007)

(continued)

Table 18.2 (continued)

Probiotic culture	Product	Health benefit	Reference
LAB isolated from *kimchi*	Fermented soy milk	Inhibited transcription factors of adipocyte differentiation, antiobesity	Kim et al. (2008)
S. thermophilus and *E. faecium*	Yoghurt	Reduction of LDL cholesterol	Agerholm-Larsen et al. (2000)
B. subtilis	Commercial mouthwash	Reduces swelling in gums and eliminates microbes in plaque of periodontis patients	Tsubura (2012)
Lb. fermentum and *Lb. salivarius*	South Indian fermented foods and feces	Oxalate degradation	Gomathi et al. (2014)
Lb. fermentum	Indian *dahi*	Anti-inflammatory effect	Archer et al. (2015)
Bacillus spp.	*Kinema*	Antioxidant properties, possesses essential amino acids, vitamin and low cholesterol content	Sarkar et al. (2002)
Ent. faecium, Lb. fructosus, Lb. amylophilus, Lb. plantarum	*Ngari*	Antimicrobial activity	Thapa et al. (2004)
Lb. sanfrancisco, Lb. brevis	*Sourdough*	Acts as probiotic carrier	Tyopponen et al. (2003)

Pichia, *Yarrowia*, *Debaryomyces*, *Hansenula*, etc. are the most common microorganisms found in fermented cereals (Foschino et al. 2004; Veinocchi et al. 2006; Guyot 2010). Diverse microbiota have been reported from fermented vegetables worldwide. Species of LAB, namely, *Lc. lactis*, *Lb. plantarum*, *Lb. brevis*, *Lb. sakei* subsp. *sakei*, *Lb. curvatus*, *Ped. pentosaceus*, *Leuc. mesenteroides* subsp. *mesenteroides*, *Luec. citreum*, *Leuc. kimchii*, *Leuc. gasicomitatum*, *Leuc. gelidum*, *Weissella confusa*, *W. koreensis* and *W. kimchii*, and yeasts such as *Candida*, *Halococcus*, *Natronococcus*, *Haloterrigena*, *Pichia*, *Kluyveromyces*, *Saccharomyces*, *Lodderomyces*, *Natrialba*, *Sporisorium* and *Trichosporon* were isolated from Korean *Kimchi* (Chang et al. 2008; Park et al. 2010; Jung et al. 2011). Species of *Lb. plantarum*, *Lc. lactis* subsp. *lactis*, *Lb. brevis*, *Lb. curvatus*, *Lb. sakei*, *Leuc. fallax*, *Leuc. mesenteroides* and *Ped. Pentosaceus* are reported in *sauerkraut* (Plengvidhya et al. 2007; Johanningsmeier et al. 2007). Himalayan and Northeast Indian fermented vegetables such as *sinki*, *khalpi*, *gundruk*, *goyang* and *inziangsang* consists of *Lb. brevis*, *Lb. plantarum*, *Lb. fermentum*, *Lb. casei*, *Lb. casei* subsp. *pseudoplantarum*, *Ped. Pentosaceus* and *Leuc. fallax* species (Tamang and Tamang 2007, 2010). *Ent. durans*, *Lb. delbrueckii*, *Lb. brevis*, *Lb. curvatus*, *Lb. coryniformis*, *Lb. plantarum*, *Lb. xylosus*, *Leuc. citreum*, *Leuc. lactis*, *Leuc. mesenteroides*, *Leuc. fallax*, *Tetragenococcus halophilus* and *Ped. Pentosaceus* were found in fermented bamboo shoots of India (Tamang and Tamang 2009; Tamang et al. 2012; Sonar and Halami 2014). *Bacillus* species are mainly responsible for fermentation of fermented soybean products like *chungkokjang* (Korea); *natto* (Japan); *kinema*

(India, Bhutan and Nepal); *thua nao* (Thailand); *hawaijar, aakhune, tungrymbai, bekang* and *peruyaan* (India); *pepok* (Myanmar); and *sieng* (Cambodia and Laos) (Tamang 2010b). *B. cereus, B. amyloliquefaciens, B. circulans, B. licheniformis, B. sphaericus, B. subtilis, B. megaterium, B. thuringiensis* and *B. subtilis subsp. chungkokjang* were present in *kinema* and *chungkokjang* (Tamang 2003; Park et al. 2005; Kwon et al. 2009b). *B. natto* was isolated from *natto*. *Lb. curvatus, Lb. brevis, Lb. paraplantarum, Lb. plantarum, Lb. sakei, Lb. casei, Lb. carnis, Lb. sanfranciscensis, Lb. divergens, Leuc. mesenteroides, Leuc. pseudomesenteroides, Leuc. carnosum, Leuc. gelidium, Leuc. citreum, Ped. pentosaceus, Ped. acidilactici, E. durans, E. cecorum, E. faecalis, E. hirae, E. faecium, W. viridescens, W. cibaria, B. subtilis, B. lentus, B. licheniformis, B. mycoides* and *B. thuringiensis* are the principal LAB present in fermented meat products (Albano et al. 2009; Rai et al. 2010; Oki et al. 2011; Nguyen et al. 2013a). Species of *Lactobacillus, Bacillus, Micrococcus, Pediococcus, Candida, Saccharomyces and Haloanaerobium* are reportedly present in fermented fish (Kobayashi et al. 2000b; Thapa et al. 2007; Saithong et al. 2010).

18.3 Health Beneficial Properties

Different kinds of fermented foods are consumed by native population and other communities likewise due to their organoleptic, nutritional and health-promoting properties (Vijayendra and Halami 2015). Health benefits may be due to probiotics, antioxidants, antimicrobials, bioactive compounds, essential amino acids, etc. (Farhad et al. 2010). Some beneficial effects bestowed by these foods documented by in vitro, in vivo and human clinical trials are mentioned below.

18.3.1 Preventing Gastrointestinal Maladies and Inflammatory Bowel Disease

Lactic acid bacteria have been employed in both animals and humans to treat a wide array of gastrointestinal maladies as they restrict the growth of many enteric pathogens. LAB are purported to secrete several enzymes into the gut exerting synergistic effects on digestion thereby relieving intestinal malabsorption symptoms (Parvez et al. 2006). Randomised controlled trials have shown that probiotic bacteria could alleviate antibiotic-associated diarrhoea (Marteau et al. 2001). Probiotic therapy in children reportedly shortened the period of acute diarrhoea illness (Gill and Guarner 2004). Specific strains of LAB including *Lb. rhamnosus GG, Lb. reuteri, Sacc. boulardii* and *Bifidobacteria* have exhibited significant effects on diarrhoea, traveller's diarrhoea and rotavirus-infected diarrhoea in children (Guandalini et al. 2000; Benchimol and Mack 2004). Probiotics can assuage diarrhoea through effects of the immune system. In infants, they appear to reduce viral diarrhoea by increasing the secretion of IgA and decreasing viral shedding (Parvez et al. 2006). LAB might relieve constipation and ease intestinal mobility probably through intestinal pH reduction (Sanders and Klaenhammer 2001). Patients suffering from irritable bowel syndrome experienced reduced symptoms of bloating, stomach pain, flatulence and constipation on receiving a *Lb. plantarum* strain in a double-blind clinical study (MacFarlane and Cummings 2002). *Sacch. boulardii*, on the other hand, could relieve diarrhoea in patients with irritable bowel syndrome but was ineffective against other symptoms (Marteau et al. 2001). LAB isolated from *dahi* may be beneficial for the treatment of diarrhoea (Agarwal and Bhasin 2002). Live bacteria present in fermented milks may be beneficial by means of their immunomodulatory activity (Granier et al. 2013). LAB isolated from *kimchi* may be actively used for alleviating IBD and atopic dermatitis like inflammatory immune disorders (Lim et al. 2011). LAB present in fermented sausages reduced the period of gastric and intestinal illnesses in addition to restoring the microbial ecosystem of the gut (Marteau et al. 2002).

18.3.2 Improvement of Lactose Intolerance

Lactose intolerance (lactose malabsorption) is the condition where lactose present in milk is not entirely broken down into glucose and galactose. Deficiency of the enzyme β-D-galactosidase (lactase enzyme) which is responsible for the cleavage of lactose into its constituent monosaccharides results in lactose malabsorption (Lomer et al. 2008). Lactic acid in yoghurt and fermented milks produced by LAB present in them alleviates the symptoms of lactose intolerance by increasing lactase activity in the gut (Swallow 2003). Human studies have shown that ingesting fresh yoghurt containing live bacteria aided in better lactose digestion and absorption compared to intake of pasteurized milk containing heat killed bacteria (Marteau et al. 2002). *Kefir* consumption also minimizes symptoms of lactose intolerance (Hertzler and Clancy 2003). Due to its viscous nature, coagulated milk has slower oral-caecal transit time, and hence fermented acidophilus milk is well tolerated compared to unfermented sweet acidophilus milk (Shah 2004). Yoghurt bacteria or lactococci which possess probiotic properties are rapidly processed in the intestinal lumen and hence may be potential candidates as vectors for enzymes or oral immunization (Mercenier et al. 2000).

18.3.3 Synthesis of Nutrients and Bioavailability

Probiotics in fermented foods and gut microbiome are said to help in the digestibility, availability and enhance the quantity of some dietary nutrients (Parvez et al. 2006). A number of studies substantiate the ability of LAB to enhance the digestion of some nutrients like lactose by production of lactose hydrolysing lactase enzyme thereby alleviating the symptoms of lactose intolerance. LAB raise folic acid, niacin and riboflavin levels in *kefir*, *bifidus* milk and yoghurt (Parvez et al. 2006). LAB produce several enzymes and vitamins in the gut that exert synergistic effects on nutrient malabsorption and digestion; furthermore lactic acid released reduces the gut pH and helps in inhibition of pathogens (Mack et al. 1999). As a result of bacterial enzymatic hydrolysis of nutrients, there is enhanced bioavailability of protein and fat along with increased production of free amino acids, lactic acid, butyric acid, propionic acid and short chain fatty acids (SCFAs). SCFAs, upon absorption, add to the host's energy pool and also prevents any pathological fluctuations in the colonic mucosa (Rolfe 2000). Phytate-degrading LAB have been found to increase mineral bioavailability. Several LAB with phytate degrading property were obtained from fermented cereal and pulses products (Raghavendra et al. 2010). Leroy et al. (2006) reported that incorporation of probiotic bacteria in fermented meat products enhances its nutritional value as functional food. *Lb. fermentum* isolated from fermented pearl millet porridge (Kambu koozh) of India was found to produce feruloyl esterase enzyme that plays a vital role in ferulic acid bioavailability in cereal based foods (Palaniswamy and Govindaswamy 2016).

18.3.4 Prevention of Cardiovascular Disease

CVD is the primary cause of mortality among both males and females in the United States and ranks among the top five causes of death in the developing nations (Erdmann et al. 2008). Diets rich in high carbohydrates and low protein are known to reduce hypertension and low density lipoprotein (LDL) cholesterol (Svetkey et al. 1999). Intake of fermented whole grains seems to prevent incidence of diabetes and coronary heart disease (Anderson et al. 2000). Mediterranean-rich diet including olive oil, fruits, nuts, cereals and vegetables with fair intake of wine, fish, poultry and minimal consumption of meat and dairy products and sweets was found to reduce the occurrence of cardiovascular disease (Estruch et al. 2013). Numerous studies have reported the protective effects of wine consumption in reducing risk of CVD. Red wines are rich in polyphenols like gallic acid, *p*-coumaric acid, caffeic acid, flavonoids (catechin, quercetin, rutin, epicatechin, etc.), stilbenes (trans-resveratrol), etc. (Kammerer et al. 2004). Resveratrol among these is considered to be the key agent in wine that proved to prevent many illnesses including CVD (Bradamante et al. 2004). Long-term consumption (6 months) of fermented milk products was found to reduce LDL cholesterol and increased HDL cholesterol (Kiessling et al. 2002). Also, many species of *Bacillus* isolated from milk and dairy sources were found to possess potent cholesterol-reducing ability besides other probiotic properties (Shobharani and Halami 2015). Fibrinolytic enzymes present in fermented Asian foods like Japanese *natto*, fermented shrimp paste, *Kimchi* and *Chungkook-jang* (fermented soybean sauce) have been explored for application in thrombolytic therapy (Kim et al. 1996; Noh et al. 1999; Wong and Mine 2004). *Doenjang*, a fermented soybean paste, comprises isoflavones that enhance activation of LDL-C receptor and increase HDL levels, thus preventing risk of CVD (Kwak et al. 2012). Fermented *puer* tea (China) protects against development of CVD (Mo et al. 2008). Yoghurt consumption alongside a healthy diet helps prevent risk of CVD, a suggested clinical data (Astrup 2014; Marette and Picard-Deland 2014).

18.3.5 Prevention of Hepatic Disease

Liver disease encompasses a host of diseases that cause damage to the liver and can be life threatening (https://www.nlm.nih.gov/medlineplus/liverdiseases.html 2016). Probiotic bacteria (*Bifidobacteria, Strep. thermophilus, Lb. plantarum, Lb. acidophilus, Lb. delbrueckii bulgaricus, Lb. casei* and *E. faecium*) having beneficial properties disrupt the pathogenesis of hepatic encephalopathy by various mechanisms of action that lower portal pressure reducing the risk of haemorrhage contrary to conventional treatment (De Santis et al. 2000; Guslandi et al. 2000; Solga 2003).

18.3.6 Prevention of Cancer

Fermented foods and lactic acid bacteria present in them have shown tremendous potential in conferring nutritional and therapeutic health benefits including antimutagenic and anticarcinogenic activity (Lee et al. 2004). In vitro, in vivo and human studies demonstrated that fermented milks containing live probiotics played a protective role against colorectal cancer, responsible for highest mortality rates in the west (Rowland 2004; Saikali et al. 2004). Studies in Balb/c mice induced with sarcoma 180 cells revealed that *kimchi* might help detoxify xenotoxic compounds in the liver (Hur et al. 2000). It is stated that the anticancer effects of *kimchi* may be due to inhibition of inflammation and regulation of apoptosis of cancer cells (Park et al. 2014). *Kefir*, a fermented milk product consumed traditionally for therapeutic values exhibited antitumour activity in in vitro models as well as murine breast cancer models and colorectal cancer in patients (de LeBlanc et al. 2007; Topuz et al. 2008). Cruciferous vegetables like cabbage and broccoli are suggested to decrease the risk of various cancers. These vegetables are rich in glucosinolates, which are converted to isothiocyanates (known to block carcinogens) by the action of plant myrosinase and gut microbiota (Talahay and Fahey 2001). *Lb. plantarum* obtained from fermented South Indian product *kallappam* modulated the development of 1,2-dimethylhydrazine (DMH)-induced colon cancer through antioxidant-dependent mechanism (Kumar et al. 2012). Indian *dahi* possesses carcinopreventive properties (Arvind et al. 2010). *Sauerkraut* contains high levels of glucosinolate, which has anticancer activity (Martinez-Villaluenga et al. 2012). *Kochujang* showed high anticancer effect on stomach cancer cells MKN45 (Song et al. 2008).

18.3.7 Lowering High Blood Pressure

Hypertension is the main culprit for various CVD including stroke, myocardial infarction, heart failure and CHD (FitzGerald et al. 2004). Several animal and clinical studies have documented the role of probiotic microorganisms and their fermented food products in hypertension control. Hypertensive elderly subjects experienced reduced systolic and diastolic blood pressure post consumption of fermented milk containing *Lb. helveticus* and *Sacch. cerevisiae* starter cultures (Aihara et al. 2005). *Douchi* has shown to lower blood pressure by producing angiotensin I-converting enzyme inhibitors (Zhang et al. 2006). Soya-fermented food products especially from soybean, namely, *sufu*, *tempeh*, *doengjang*, *natto* and soy sauce (Gibbs et al. 2004; Rho et al. 2009), are known to possess antihypertensive properties. Given the current scenario with rise in risk of heart diseases, regular intake of probiotics and fermented food products may provide a prophylactic effect through their action on blood pressure and blood lipid levels.

18.3.8 Thrombosis

In clinical practices, patients with CVDs are administered with external fibrinolytic enzymes like streptokinase, urokinase, alteplase (t-PA), anistreplase, releplase (r-PA) and tenecteplase (TNK-t-PA) (Duggal and Harger 2011). WHO has estimated more than 23 million deaths due to CVDs by 2030 (WHO report 2013). Expensive cost, very short half-life, specificity and other complications like internal haemorrhage and allergic reactions have led to the search for safer, efficient and cheaper fibrinolytic enzymes (Blann et al. 2002). Recently many fibrinolytic enzymes from fermented foods were explored (Mine et al. 2005). *Nattokinase* from *B. subtilis* isolated from *natto* was the first microbial fibrinolytic enzyme from fermented food (Singh et al. 2014). Subsequently, *katsuwokinase* isolated from Japanese *shiokara* (Sumi et al. 1995), *myulchikinase* obtained from Korean *myulchi-jeot-gal* (Jeong et al. 2004), *subtilisin DFE* from *B. amyloliquefaciens* isolated from Chinese *douchi* (Peng et al. 2003), TPase of *B. subtilis* TP-6 (Indonesian *Tempeh*) and novel metalloprotease of *Bacillus* KA38 from Korean fermented fish *Jeot-Gal* are some of the fibrinolytic enzymes derived from fermented foods (Kotb 2012). Fibrinolytic enzymes from fermented fish and soybean products of Northeast India have also been explored, and various species of *Bacillus* and LAB were found to possess fibrinolytic activity (Singh et al. 2014). Fibrinolytic enzymes obtained from food-grade microbes in fermented foods can directly be consumed on a daily basis for the prevention of CVD (Singh et al. 2014; Mine et al. 2005). Recently, *Bacillus* species (*B. subtilis* and *B.* circulans) have been reported from our laboratory with unique fibrinolytic and fibrinogenolytic properties (Yogesh and Halami 2015a, b).

18.3.9 Allergy/Eczema

Fermented foods exert a variety of health effects on allergic reactions, in improving nutrient bioavailability, decreasing serum cholesterol, etc. (Galdeano and Perdigon 2006). Evidence from clinical reports suggests that dietary consumption of fermented foods, for instance, yoghurt, can prevent occurrence of hypersensitive reactions and ameliorate symptoms of atopic syndrome by regulation of immune status (Cross et al. 2001). Controlled studies have designated that LAB in fermented milks can augment systemic levels of type I and type II interferons. Breakdown of caseins with aging of fermented milks was found to stimulate gut tolerance by facilitating loss of allergic response during intestinal digestion (Alessandri et al. 2012). Soy sauce is reported to impart immunological functions pertaining to hypoallergenicity and antiallergic activity (Kobayashi 2005). Probiotics augment allergic reactions by improving barrier functions of gut mucosa and immune modulation (MacFarlane and Cummings 2002). Allergic response was attenuated in ovalbumin-sensitized BALBL/c mice when treated with heat-inactivated *Lb. kefiranofaciens* M1, an isolate obtained from *kefir* grains (Hong et al. 2010). *Lactobacillus* species obtained from *kimchi* alleviated atopic dermatitis and food allergy through modulation of Th1/Th2 balance as well as increasing secretion of IL-12 and IFN-γ (Won et al. 2011). With the rise in food allergy-related disease such as celiac disease due to gluten, alternative approaches to produce gluten-free bread are envisaged (Waldherr and Vogel 2009). Rühmkorf et al. (2012) reported that production of exopolysaccharides (EPS) in gluten-free sourdough could be enhanced by using specific LAB strains (*L. curvatus* TMW 1.624, *L. reuteri* TMW 1.106 and *L. animalis* TMW 1.971) under optimized conditions. EPS which is produced by a number of food bacteria (LAB) act as hydrocolloids and are used to enhance the quality of bread (especially gluten-free bread) and *dahi* (Galle et al. 2012; Vijayendra et al. 2008). Oral dosage of *Lb. sakei* proBio65 in mice induced with DNCB (1-chloro-2, 4-dinitro-benzene) reversed the effects of allergic dermatitis through regulation of IgE and IL-4 compared with the control group (Kim et al. 2013). Probiotic bacteria also act by upregulating IL-10, an anti-inflammatory cytokine in children suffering from atopy (Pessi et al. 2000). Apart from alleviating atopy, LAB in fermented foods may prevent the development of an allergic phenotype at infancy (Matricardi 2002). Natural and fermented fish oils are known to possess antiallergic activity ranging from improving sensitivity to allergens, asthma, improvement of eczema and atopic dermatitis (Han et al. 2012a, b).

18.3.10 Anti-ageing Effects

Chlorophyll, phenolic compounds, carotenoids, dietary fibres, LAB, etc. contained in fermented *kimchi* show antioxidative and antiaging properties (Park et al. 2014). The influence of *kimchi* intake on antiaging characteristics, production of free radicals and antioxidative enzyme activity in brains of senescence-accelerated mice showed that *baechu kimchi*-fed groups displayed better inhibition than the standard *kimchi* (Kim et al. 2002a). Antioxidant effects of *Baechu kimchi* on feeding hairless mice for 20 weeks were investigated by Ryu et al. (2004a, b). *Kimchi* seemed to promote healthy skin by increasing the epidermal thickness and greater free radical scavenging activity than the control groups. The levels of hydroxyl and free radicals in the plasma of aged subjects consuming over 112 g of *kimchi* per day were found to decrease compared to those who consumed less, indicating radical scavenging activity of *kimchi* (Kim et al. 2002b). Melatonin present in red wine has anti-ageing properties, regulates body clock and helps in better sleep (Corder et al. 2006).

18.3.11 Diabetes

Consumption of high fibre is known to prevent diabetes risk (Meyer et al. 2000). Clinical trials have shown that whole grain intake has a positive effect on insulin sensitivity, gut-hormone levels and insulin responses post meal time (Juntunen et al. 2000; Pereira et al. 2002). Possibly the high intake of whole grains improves the blood glucose levels of diabetic individuals and also controls the occurrence of diabetes (Anderson 2003). Consumption of *chungkokjang* improves insulin resistance thereby preventing and mitigating diabetes (Shin et al. 2011; Tolhurst et al. 2012). Yadav et al. (2007) reported a significant delay in occurrence of glucose tolerance, hyperglycaemia, dyslipidaemia, hyperinsulinemia and oxidative stress when administered with probiotic *dahi*-supplemented diet in diabetic rats induced with high fructose diet. Mothers supplemented with probiotic bacteria showed reduced risk of developing gestational diabetes (Luoto et al. 2010). Cani et al. (2007) observed an inverse correlation between serum concentrations of lipopolysaccharide (LPS), a bacterial cell wall component and numbers of colonic bifidobacteria in mice fed a high-fat diet. The same authors have proposed a theory of metabolic endotoxaemia stating that dysbiosis of the colonic microbiota caused by consumption of high fat diet leads to increased permeability of the intestine further promoting the translocation of pathogenic bacteria (causing increased LPS concentrations) and low-grade systemic inflammation. The key factor contributing to the development of metabolic syndrome and diabetes is said to be chronic systemic inflammation (Cani et al. 2008). These effects have also been correlated in humans (Pussinen et al. 2011). Antidiabetic effects of *Bifidobacteria animalis* subsp. *lactis* 420 and subsequent reduction in metabolic endotoxaemia have been demonstrated in mice maintained on high fat diet (Amar et al. 2011). Strains of *Lb. fermentum* have been investigated for reduction of metabolic endotoxaemia and potent anti-inflammatory activity associated with chronic inflammation in high fat diet and diabetic models (Archer and Halami, under communication). Several animal and few human studies have documented anti-diabetic effects of fermented soybeans (Kawakami et al. 2005; Kwon et al. 2007, 2009a; Taniguchi et al. 2008). Isoflavanoid glycones in soybeans are converted to isoflavanoid aglycones as a result of fermentation, and since isoflavanoid aglycones have greater activity than isoflavanoid glycones, fermented soybeans are more effective in controlling glucose metabolism (Kwon et al. 2010). *Kochujang* has been found to reduce body weight, serum leptin levels and visceral fat without modulating energy consumption in diabetic rats. It also improves tolerance to glucose by increasing insulin sensitivity (Kwon et al. 2009b).

18.3.12 Obesity

Obesity is a rapidly emerging metabolic disorder affecting humans all over the globe and poses a predisposition to diseases such as hypertension, atherosclerosis, metabolic syndrome and diabetes. The role of intestinal microbiota in weight gain has been demonstrated in which colonization of germ-free mice with microbiota taken from obese mice gained more weight compared to colonization of mice with microbiota obtained from lean mice (Backhed et al. 2004). Numerous reports have presented the antiobesity effects of *kimchi* and *doenjang* in clinical trials as well as in animal models (Kim et al. 2011; Moon et al. 2012; Park et al. 2012). *Doenjang* (fermented soybeans) is also known to reduce obesity as opposed to unfermented soybeans (Kwak et al. 2012). Fermented soymilk containing LAB isolated from *kimchi* (*Leu. kimchii*, *Leu. citreum* and *Lb. plantarum*) inhibited CCAAT/enhancer-binding protein-α, PPAR-γ expression and transcription factors of adipocyte differentiation. It also significantly decreased plasma LDL cholesterol in obese Sprague Dawley rats induced with high fat diet (Kim et al. 2011). *Weissella koreensis* OK 1–6 showed antiobesity activity in groups that received the culture by significantly decreasing the accumulation of intracellular and lipid triglyceride concentration compared to control group (Park et al. 2012). There are limited studies that report the beneficial role of probiotics on adiposity. Intake of fermented milk containing probiotic *Lb. gasseri* SBT2055 for 3 months resulted in substantial reduction of subcutaneous and visceral fat in the abdomen, weight and body mass index (BMI) in subjects with elevated BMI at the start of the trial (Kadooka et al. 2010). However, more definite research is needed to claim the protective mechanisms of probiotics against obesity (Sanz et al. 2013). Intake of *kochujang* presented antiobesity effect when compared to groups that consumed red pepper and non-fermented *kochujang* (Rhee et al. 2003).

18.3.13 Hypercholesterolaemia

Elevated levels of serum cholesterol act as major risk factor for CHD and atherosclerosis (Ouwehand and Röytiö 2014). The cholesterol-lowering properties of LAB have been recognised, and lactobacilli are known to possess enzymes that help in the deconjugation of bile salts, thus inhibiting cholesterol absorption from the intestine and the enterohepatic circulation of cholesterol. In addition, end products of nutrient digestion like SCFAs are also known to impact cholesterol synthesis and lipid metabolism (Aggarwal et al. 2013). Several animal studies have documented cholesterol-reduction activity by probiotic microbes (Huang et al. 2013a, b); however, evidences from human studies are found to be inconsistent (Aggarwal et al. 2013). Daily consumption of yoghurt containing *S. thermophilus* and *E. faecium* for 8 weeks in a controlled clinical trial resulted in drop of LDL up to 8.4% and elevated levels of fibrinogen (Agerholm-Larsen et al. 2000). *Bacillus* sp. investigated for their probiotic properties were also found to have hypocholesterolaemic effect when fed to hypercholesterolaemic mice (Shobharani and Halami, unpublished data). *Kimchi* consumption is known to elevate HDL cholesterol and decrease LDL cholesterol as evident by human studies (Lee et al. 2012). *Kefir* helps against risks of CVD by managing elevated cholesterol levels (Otes and Cagindi 2003).

18.3.14 Immune Functions

The human intestinal microbiome is a complex organ, viz., also critically involved in the modulation of the immune system via cross talk and maintenance of immune homeostasis through a healthy balance of beneficial and harmful microbes. A number of gut-related inflammatory conditions are attributed to shift in microecology and inflammation of the gut along with dysbiosis of the gut microbiota (Gill and Guarner 2004). Therefore, the immune-stimulating effect of beneficial/probiotic bacteria has been investigated in various aspects of the immune system function and disease models (Purchiaroni et al. 2013). Probiotics alter both specific and non-specific immune responses via activation of macrophages, natural killer cell activity and stimulation of cytokine and immunoglobulins production (Ouwehand et al. 2002). In vitro and in vivo studies show that *Kimchi* has positive effect on immune function (Kim et al. 1997; Kim et al. 2001). The cell wall fraction of *Lb. plantarum* isolated from *kimchi* enhanced the production of antibodies, TNF-α and IL-6, in murine macrophage cell line RAW 264.7 (Lee et al. 2006). *Lb. plantarum* strain YU isolated from fermented foods exhibited immune stimulation via activation of Th1 immune responses in ovalbumin (OVA)-immunized mice and influenza A virus-infected model (Kawashima et al. 2011).

18.3.15 Other Therapeutic Effects of Fermented Foods

Traditionally ethnic people around the globe consume fermented foods for therapeutic values and to prevent illness besides nutritional benefits. Probiotics, essential amino acids, antimicrobials, antioxidants and bioactive compounds are some of the health beneficial elements ascribed to ethnic fermented foods (Farhad et al. 2010). Himalayan-fermented milk products are proposed to possess immune-stimulating properties and cure stomach-related problems. Mild alcoholic beverages that include *poko*, *bhaati jaanr*, *kodo ko jaanr*/*chyang*, etc. of Himalayan region possess high calorie content (Tamang 2010a) and are consumed by ailing patients and postnatal women for strength (Tamang and Thapa 2006). *Gundruk* and *sinki* are good appetizers and relieve indigestion (Tamang 2010a). *Kinema* is consumed for its high nutritive and health-promoting properties (Omizu et al. 2011). *Natto* consumption in Japan helps prevent haemorrhage in vitamin K-deficient infants and also improves serum MK-7 and γ-carboxylated osteocalcin levels in healthy persons (Tsukamoto et al. 2000). *Kimchi* among several other health benefits mentioned earlier possesses antistress principles beneficial for treating depression, hepatic disease, osteoarthritis and irregular bowel movements (Lee et al. 2008; Lee and Lee 2009). *Kombucha* (a sweetened tea beverage) contains high concentrations of gluconic, acetic and lactic acid along with several other metabolites that promote wellness (Schillinger et al. 2010). T*empeh* contains γ-aminobutyric acid which improves bloodflow to the brain and regulates hypertension (Aoki et al. 2003). Indian *dahi* has anticholesterolaemic and antidiarrhoeal remedial effect (Agarwal and Bhasin 2002). Commercially available mouthwash containing an extract of *natto* isolate *B. subtilis* reduced swelling in the gums and eliminated microbes in plaque of periodontitis patients after 1 month of usage (Tsubura 2012). *Natto*, *chungkokjang*, *douche*, *kinema*, *bekang*, *tempeh*, *kimchi*, *kefir*, yoghurt and fermented whole grains are known to possess antioxidant activities (Iwai et al. 2002; Shon et al. 2007; Wang et al. 2007a; Horii 2008; Moktan et al. 2008; Sim and Han 2008; Sun et al. 2009; Sabeena et al. 2010; Park et al. 2011; Chettri and Tamang 2014). Native isolates of LAB from various fermented vegetables displayed strong antimicrobial properties (Rubia-Soria et al. 2006; Tamang et al. 2009b; Lee et al. 2011; Jiang et al. 2012). Cheese is considered a cariostatic food as it helps stimulate rate of saliva flow and sugar clearance, increases the pH and inhibits plaque bacteria and deposition of excess calcium to plaque (Kashket and DePaola 2002). Yoghurt consumption seems to strengthen the bones preventing possibility of fractures due to old age (Morelli 2014; Prentice 2014). It also helps fight common cold and infections of the upper respiratory tract (King et al. 2014). Isoflavones (estrogenic compound) present in *tempeh* may play a role in the prevention of menopausal symptoms in women and also have neuroprotective effects (Sapbamrer et al. 2013; Ahmad et al. 2014). Oxalate-degrading bacteria, *Lb. fermentum* and *Lb. salivarius*, isolated from human faeces and south Indian fermented foods were found to degrade oxalate in vitro between 40 and 62% in addition to possessing other probiotic properties, thus suggesting their potentiality for preventing hyperoxaluria (Gomathi et al. 2014). *Lb. plantarum* C88 isolated from traditional Chinese tofu exhibited antioxidant activity in vitro, and when administered to mice suffering from induced oxidative stress, it was found to alleviate liver glutathione peroxidase activity, malondialdehyde, superoxide dismutase activity, and total antioxidant capacity of the liver was elevated significantly (Li et al. 2012).

18.4 New Beneficial Microorganisms from Fermented Foods

Lb. brevis UN isolated from Dhulliachar, a salted pickle of Northeast India, was found to possess probiotic properties and a potent bacteriocin active against several food-borne pathogens (Gautam et al. 2014). It could be a potent probiotic for addressing food-borne illnesses and for preservation. Species of *Lb. plantarum*, *Lb. pobuzihii* and *Lb. rossiae* possessing several probiotic properties like acid and bile tolerance, adhesion to hydrocarbons, antimicrobial activity and cholesterol utilization were obtained from traditional fermented fish *Tungtap* (North East India) (Rapsang and Joshi 2015). *Lb. spicheri* G2 was another novel probiotic isolated from *Gundruk* and showed probiotic potential in vitro (Gautam and Sharma 2015). Probiotic application of strains of *Bacillus licheniformis* that produce high amounts of EPS which have antioxidant and anti-ageing properties has been proposed by Song et al. (2011). *W. cibaria* and *Lb. plantarum* isolated from *kimchi* displayed anticancer and immunomodulating effects in vitro (Kwak et al. 2014). In vitro inhibition of *Helicobacter pylori* by *Lb. plantarum* KNUC25 from *kimchi* has been reported (Ki et al. 2010). *Lb. plantarum* DM9218-A was able to reduce serum uric acid concentration in hyperuricaemic rats and said to be a promising candidate as adjunct therapy for hyperuricemia (Li et al. 2014). *Lb. plantarum* JSA22 from *sokseongjang*, a Korean fermented soybean food was found to possess high enzymatic, fibrinolytic, antimicrobial activity and also suppressed *S. typhimurium* infection upon coincubation in intestinal cells (Eom et al. 2015). A novel bacteriocin, paracin 1.7, produced by *Lb. paracasei* HD1–7 isolated from Chinese sauerkraut juice was studied extensively for its application in the food preservation industry (Ge et al. 2016). In our laboratory, LAB have been isolated from native fermented dairy sources and have exhibited probiotic properties (Archer and Halami 2015; Devi et al. 2015). In addition, *Lb. fermentum* from dairy and infant faecal origin displayed anti-inflammatory effect in Wistar rat carrageenan induced paw oedema model (Archer et al. 2015).

18.5 Opinion

Fermented foods are a rich and diverse source of microorganisms that can be exploited to study microbial diversity, harvest potent indigenous probiotic cultures, develop starter cultures and formulate functional fermented product with positive health-promoting effects. Knowledge of microbial niche of fermented food would help in large-scale production and optimization of product for the larger benefit of the consumer. Microorganisms in the process of fermentation also produce bioactive compounds that not only increase nutritional value but also impart health benefits. Insights into the diet-gut-microbe-host relationship and studies on microbiome changes of people consuming ethnic fermented foods will give better understanding of health benefits of microbes or the metabolites produced by them. Saying this, detailed and rigorous clinical studies are required to validate health claims. At last, there is still an umpteen wealth of ethnic fermented foods that entails scientific attention and validation.

Acknowledgements The authors gratefully acknowledge the director, CSIR-Central Food Technological Research Institute, Mysore, for providing necessary facilities. ACA wishes to acknowledge University Grants Commission for grant of Maulana Azad National Fellowship.

References

Aihara K, Kajimoto O, Hirata H, Takahashi R, Nakamura Y (2005) Effect of powdered fermented milk with Lactobacillus helveticus on subjects with high-normal blood pressure or mild hypertension. J Am Coll Nutr 24(4):257–265

An J, Jung SM, Chan K, Tam CF (2014) Development of a 28-day Kimchi Cyclic menu for Health. J Culinary Sci Technol 12(1):43–66

Anukam KC, Reid G (2009) African traditional fermented foods and probiotics. J Med Food 12(6):1177–1184

Aoki H, Uda I, Tagami K, Furuta Y, Endo Y, Fujimoto K (2003) The production of a new *tempeh*-like fermented soybean containing a high level of γ-aminobutyric acid by anaerobic incubation with Rhizopus. Biosci Biotechnol Biochem 67:1018–1023. doi:10.1271/bbb.67.1018

Archer AC, Halami PM (2015) Probiotic attributes of *Lactobacillus fermentum* isolated from human feces and dairy products. Appl Microbiol Biotechnol 99(19):8113–8123. doi:10.1007/s00253-015-6679-x

Archer AC, Muthukumar SP, Halami PM (2015) Anti-inflammatory potential of probiotic *Lactobacillus* spp. on carrageenan induced paw edema in Wistar rats. Int J Biol Macromol 81:530–537. doi:10.1016/j.ijbiomac.2015.08.044

Arvind K, Nikhlesh KS, Pushpalata RS (2010) Inhibition of 1,2-dimethylhydrazine induced colon genotoxicity in rats by the administration of probiotic curd. Mol Biol Rep 37:1373–1376. doi:10.1007/s11033-009-9519-1

Astrup A (2014) Yogurt and dairy product consumption to prevent cardiometabolic diseases: epidemiologic and experimental studies. Am J Clin Nutr 99(5):1235S–1242S. doi:10.3945/ajcn.113.073015

Backhed F, Ding H, Wang T, Hooper LV, Koh GY, Nagy A, Semenkovich CF, Gordon JI (2004) The gut microbiota as an environmental factor that regulates fat storage. Proc Natl Acad Sci USA 101(44):15718–15723

Benchimol EI, Mack DR (2004) Probiotics in relapsing and chronic diarrhea. J Pediatr Hematol Oncol 26:515–517

Blann AD, Landray MJ, Lip GY (2002) An overview of antithrombotic therapy. BMJ 325(7367):762–765. doi:10.1136/bmj.325.7367.762

Bradamante S, Barenghi L, Villa A (2004) Cardiovascular protective effects of resveratrol. Cardiovasc Drug Rev 22(3):169–188

Breidt F, McFeeters RF, Perez-Diaz I, Lee CH (2013) Fermented vegetables. In: Doyle MP, Buchanan RL (eds) Food microbiology: fundamentals and frontiers, 4th edn. ASM Press, Washington, pp 841–855. doi:10.1128/9781555818463.ch33

Cani PD, Neyrinck AM, Fava F, Knauf C, Burcelin RG, Tuohy KM et al (2007) Selective increases of bifidobacteria in gut microflora improve high-fat-diet-induced diabetes in mice through a mechanism associated with endotoxaemia. Diabetologia 50(11):2374–2383

Cani PD, Bibiloni R, Knauf C, Waget A, Neyrinck AM, Delzenne NM et al (2008) Changes in gut microbiota control metabolic endotoxemia-induced inflammation in high-fat diet-induced obesity and diabetes in mice. Diabetes 57(6):1470–1481

Chandan RC, Kilara A (eds) (2013) Manufacturing yogurt and fermented milks, 2nd edn. Wiley, Chichester, 477pp

Chang HW, Kim KH, Nam YD, Roh SW, Kim MS, Jeon CO, Oh HM, Bae JW (2008) Analysis of yeast and archaeal population dynamics in kimchi using denaturing gradient gel electrophoresis. Int J Food Microbiol 126:159–166

Chettri R, Tamang JP (2014) Functional properties of *tungrymbai* and *bekang*, naturally fermented soybean foods of North East India. Int J Ferment Foods 3:87–103

Choi YH, Kim YA, Park C, Choi BT, Lee WH, Hwang KMH, Jung KO, Park KY (2004) β-sitosterol induced growth inhibition is associated with up-regulation of Cdk inhibitor p21WAF1/CIP1 in human colon cancer cells. J Kor Soc Food Sci Nutr 33:1–6

Corder R, Mullen W, Khan NQ, Marks SC, Wood EG, Carrier MJ, Crozier A (2006) Oenology: red wine procyanidins and vascular health. Nature 444:566. doi:10.1038/444566a

Cross ML, Stevenson LM, Gill HS (2001) Anti-allergy properties of fermented foods: an important immunoregulatory mechanism of lactic acid bacteria? Int Immunopharmacol 1(5):891–901

De Santis A, Famularo G, De Simone C (2000) Probiotics for the hemodynamic alterations of patients with liver cirrhosis. Am J Gastroenterol 95:323–324. doi:10.1111/j.1572-0241.2000.01726.x

De Vuyst L, Vrancken G, Ravyts F, Rimaux T, Weckx S (2009) Biodiversity, ecological determinants, and metabolic exploitation of sourdough microbiota. Food Microbiol 26:666–675. doi:10.1016/j.fm.2009.07.012

Devi SM, Archer AC, Halami PM (2015) Screening, characterization and in vitro evaluation of probiotic properties among lactic acid bacteria through comparative analysis. Probiotics Antimicrob Proteins 7(3):181–192. doi:10.1007/s12602-015-9195-5

Dewan S, Tamang JP (2006) Microbial and analytical characterization of *Chhu*, a traditional fermented milk product of the Sikkim Himalayas. J Sci Ind Res 65:747–752

Duggal RW, Harger NJ (2011) The safe and appropriate use of thrombolytics in the emergency department. US Pharm 36:11–16
Eom JS, Song J, Choi HS (2015) Protective Effects of a Novel Probiotic Strain of *Lactobacillus plantarum* JSA22 from Traditional Fermented Soybean Food Against Infection by *Salmonella enterica* Serovar *typhimurium*. J Microbiol Biotechnol 25:479–491. doi:10.4014/jmb.1501.01006
Erdmann K, Cheung BWY, Schroder H (2008) The possible role of food derived bioactive peptides in reducing the risk of cardiovascular diseases. J Nutr Biochem 19:643–654. doi:10.1016/j.jnutbio.2007.11.010
Estruch R, Ros E, Salas-Salvadó J, Covas MI, Corella D, Arós F, Gómez-Gracia E, Ruiz-Gutiérrez V, Fiol M, Lapetra J, Lamuela-Raventos RM (2013) Primary prevention of cardiovascular disease with a Mediterranean diet. N Engl J Med 368(14):1279–1290. doi:10.1056/NEJMoa1200303
FAO/WHO (2002) Joint FAO/WHO (Food and Agriculture Organization/World Health Organization) working group report on drafting guidelines for the evaluation of probiotics in food. London, Ontario, Canada
Farhad M, Kailasapathy K, Tamang JP (2010) Health aspects of fermented foods. In: Tamang JP, Kailasapathy K (eds) Fermented foods and beverages of the world. CRC Press, Taylor & Francis Group, New York, pp 391–414
FitzGerald RJ, Murray BA, Walsh DJ (2004) Hypotensive peptides from milk protein. J Nutr 34:9805–9885
Foschino R, Gallina S, Andrighetto C, Rossetti L, Galli A (2004) Comparisons of cultural methods for the identification and molecular investigations of yeasts from sourdoughs for Italian sweet baked products. FEMS Yeasts Res 4:609–618. doi:10.1016/j.femsyr.2003.12.006
Galdeano CM, Perdigon G (2006) Probiotic bacterium *Lactobacillus casei* induces activation of the gut mucosal immune system through innate immunity. Clin Vaccine Immunol 13:219–226. doi:10.1128/CVI.13.2.219-226.2006
Galle S, Schwab C, Dal Bello F, Coffey A, Ganzle MG, Arendt EK (2012) Influence of *in-situ* synthesized exopolysaccharides on the quality of gluten-free sorghum sourdough bread. Int J Food Microbiol 155:105–112
Gautam N, Sharma N (2015) Evaluation of probiotic potential of new bacterial strain, *Lactobacillus spicheri* G2 isolated from *Gundruk*. Proc Natl Acad Sci India Sect B Biol Sci 85(4):979–986. doi:10.1007/s40011-014-0458-9
Gautam N, Sharma N, Ahlawat OP (2014) Purification and characterization of bacteriocin produced by *Lactobacillus brevis* UN isolated from Dhulliachar: a traditional food product of north east India. Ind J Microbiol 54(2):185–189. doi:10.1007/s12088-013-0427-7
Ge J, Sun Y, Xin X, Wang Y, Ping W (2016) Purification and partial characterization of a novel bacteriocin synthesized by *Lactobacillus paracasei* HD1-7 isolated from Chinese sauerkraut juice. Sci Rep 6:19366. doi:10.1038/srep19366
Gibbs BF, Zougman A, Masse R, Mulligan C (2004) Production and characterization of bioactive peptides from soy hydrolysate and soy-fermented food. Food Res Int 37:123–131. doi:10.1016/j.foodres.2003.09.010
Gill HS, Guarner F (2004) Probiotics and human health: a clinical perspective. Postgrad Med J 80:516–526. doi:10.1136/pgmj.2003.008664
Gomathi S, Sasikumar P, Anbazhagan K, Sasikumar S, Kavitha M, Selvi MS, Selvam GS (2014) Screening of indigenous oxalate degrading lactic acid bacteria from human faeces and South Indian fermented foods: assessment of probiotic potential. Sci World J. doi:10.1155/2014/648059
Granier A, Goulet O, Hoarau C (2013) Fermentation products: immunological effects on human and animal models. Pediatr Res 74:238–244. doi:10.1038/pr.2013.76
Guandalini S, Pensabene L, Zikri MA, Dias JA, Casali LG, Hoekstra H, Kolacek S, Massar K et al (2000) *Lactobacillus* GG administered in oral rehydration solution to children with acute diarrhea: a multicenter European trial. J Pediatr Gastroenterol Nutr 30:54–60
Guslandi M, Mezzi G, Sorghi M, Testoni PAL (2000) *Saccharomyces boulardii* in maintenance treatment of Crohn's disease. Dig Dis Sci 45:1462–1464

Guyot JP (2010) Fermented cereal products. In: Tamang JP, Kailasapathy K (eds) Fermented foods and beverages of the world. CRC Press, Taylor & Francis Group, New York, pp 247–261. doi:10.1201/EBK1420094954-c8

Han S, Kang G, Ko Y, Kang H, Moon S, Ann Y, Yoo E (2012a) Fermented fish oil suppresses T helper 1/2 cell response in a mouse model of atopic dermatitis via generation of CD4+CD25+Foxp3+ T cells. BMC Immunol 13:44. doi:10.1186/1471-2172-13-44

Han S, Kang G, Ko Y, Kang H, Moon S, Ann Y, Yoo E (2012b) External application of fermented olive flounder (*Paralicthys olivaceus*) oil alleviates inflammatory responses in 2,4-dinitrochlorobenzene-induced atopic dermatitis mouse model. Toxicol Res 28(3):159–164. doi:10.5487/TR.2012.28.3.159

Harun-ur-Rashid M, Togo K, Useda M, Miyamoto T (2007) Probiotic characteristics of lactic acid bacteria isolated from traditional fermented milk "Dahi" in Bangladesh. Pak J Nutr 6:647–652

Hertzler SR, Clancy SM (2003) Kefir improves lactose digestion and tolerance in adults with lactose maldigestion. J Am Diet Assoc 103:582–587

Higdon JV, Delage B, Williams DE, Dashwood RH (2007) Cruciferous vegetables and human cancer risk: epidemiologic evidence and mechanistic basis. Pharmacol Res 55:224–236. doi:10.1016/j.phrs.2007.01.009

Hong W, Chen Y, Chen M (2010) The antiallergic effect of kefir Lactobacilli. J Food Sci 75(8):H244–H253

Horii M (2008) Tempeh. In: Kiuchi K, Nagai T, Kimura K (eds) Advanced science on natto. Kenpakusha, Tokyo, pp 234–237. (in Japanese)

https://www.nlm.nih.gov/medlineplus/liverdiseases.html "Liver Diseases: MedlinePlus". www.nlm.nih.gov. Retrieved 2016 May 3

Huang Y, Wang X, Wang J, Wu F, Sui Y, Yang L et al (2013a) *Lactobacillus plantarum* strains as potential probiotic cultures with cholesterol-lowering activity. J Dairy Sci 95(5):2746–2753. doi:10.3168/jds.2012-6123

Huang Y, Wu F, Wang X, Sui Y, Yang L, Wang J (2013b) Characterization of *Lactobacillus plantarum* Lp27 isolated from Tibetan kefir grains: a potential probiotic bacterium with cholesterol-lowering effects. J Dairy Sci 96(5):2816–2825. doi:10.3168/jds.2012-6371

Hur YM, Kim SH, Choi YW, Park KY (2000) Inhibition of tumor formation and changes in hepatic enzyme activities by kimchi extracts in sarcoma 180 cell transplanted mice. J Food Sci Nutr 5:48–53

Hutkins RW (2008) Microbiology and technology of fermented foods, vol 22. Wiley, Hoboken

Iwai K, Nakaya N, Kawasaki Y, Matsue H (2002) Antioxidative function of *Natto*, a kind of fermented soybeans: effect on LDL oxidation and lipid metabolism in cholesterol-fed rats. J Agric Food Chem 50:3597–3601. doi:10.1021/jf0117199

Jackson RS (2008) Wine science: principles and applications, 3rd edn. Academic Press, San Diego, pp 686–706

Jeong J, Junga H, Leea S, Leea H, Hwanga KT, Kimb T (2010) Anti-oxidant, anti-proliferative and anti-inflammatory activities of the extracts from black raspberry fruits and wine. Food Chem 123:338–344. doi:10.1016/j.foodchem.2010.04.040

Jeong YK, Yang WS, Kim KH, Chung KT, Joo WH, Kim JH, Kim DE, Park JU (2004) Purification of a fibrinolytic enzyme (myulchikinase) from pickled anchovy and its cytotoxicity to the tumor cell lines. Biotechnology letters 26(5):393–397

Jeyaram J, Anand ST, Romi W, Ranjita DA, Mohendro SW, Dayanidhi H, Rajmuhon SN, Tamang JP (2009) Traditional fermented foods of Manipur. Indian J Tradit Knowl 8(1):115–121

Jiang J, Shi B, Zhu D, Cai Q, Chen Y, Li J, Qi K, Zhang M (2012) Characterization of a novel bacteriocin produced by *Lactobacillus sakei* LSJ618 isolated from traditional Chinese fermented radish. Food Control 23:338–344. doi:10.1016/j.foodcont.2011.07.027

Johanningsmeier S, McFeeters RF, Fleming HP, Thompson RL (2007) Effects of *Leuconostoc mesenteroides* starter culture on fermentation of cabbage with reduced salt concentrations. J Food Sci 72:M166–M172. doi:10.1111/j.1750-3841.2007.00372.x

Jung JY, Lee SH, Kim JM, Park MS, Bae JW, Hahn Y, Madsen EL, Jeon CO (2011) Metagenomic analysis of *kimchi*, a traditional Korean fermented food. Appl Environ Microbiol 77:2264–2274. doi:10.1128/AEM.02157-10

Juntunen KS, Mazur WM, Liukkonen KH, Uehara M, Poutanen KS, Adlercreutz HC, Mykkanen HM (2000) Consumption of wholemeal rye bread increases serum concentrations and urinary excretion of enterolactone compared with consumption of white wheat bread in healthy Finnish men and women. Br J Nutr 84:839–846

Kadooka Y, Sato M, Imaizumi K, Ogawa A, Ikuyama K, Akai Y et al (2010) Regulation of abdominal adiposity by probiotics (*Lactobacillus gasseri* SBT2055) in adults with obese tendencies in a randomized controlled trial. Eur J Clin Nutr 64(6):636–643

Kammerer D, Claus A, Carle R, Schieber A (2004) Polyphenol screening of pomace from red and white grape varieties (*Vitis vinifera* L.) by HPLC-DAD-MS/MS. J Agric Food Chem 52(14):4360–4367

Kashket S, DePaola DP (2002) Cheese consumption and the development and progression of dental caries. Nutr Rev 60(4):97–103. doi:10.1301/00296640260085822

Kawakami Y, Tsurugasaki W, Nakamura S, Osada K (2005) Comparison of regulative functions between dietary soy isoflavones aglycone and glucoside on lipid metabolism in rats fed cholesterol. J Nutr Biochem 16(4):205–212. doi:10.1016/j.jnutbio.2004.11.005

Kawashima T, Hayashi K, Kosaka A, Kawashima M, Igarashi T, Tsutsui H, Tsuji NM, Nishimura I, Hayashi T, Obata A (2011) *Lactobacillus plantarum* strain YU from fermented foods activates Th1 and protective immune responses. Int Immunopharmacol 11(12):2017–2024. doi:10.1016/j.intimp.2011.08.013

Ki MR, Ghim SY, Hong IH, Park JK, Hong KS, Ji AR, Jeong KS (2010) *In vitro* inhibition of *Helicobacter pylori* growth and of adherence of cagA-positive strains to gastric epithelial cells by *Lactobacillus paraplantarum* KNUC25 isolated from kimchi. J Med Food 13(3):629–634. doi:10.1089/jmf.2009.1265

Kim W, Choi K, Kim Y (1996) Purification and characterization of a fibrinolytic enzyme produced from *Bacillus* sp. strain CK 11-4 screened from Chungkook-Jang. Appl Environ Microbiol 62:2482–2488

Kim MJ, Kwon MJ, Song YO, Lee EK, Yoon HJ, Song YS (1997) The effects of kimchi on hematological and immunological parameters *in vivo* and *in vitro*. J Food Sci Nutr 26:1208–1214

Kim KH, Kim SH, Park KY (2001) Effects of kimchi extracts on production of nitric oxide by activated macrophages, transforming growth factor b1 of tumor cells and interleukin-6 in splenocytes. J Food Sci Nutr 6:126–132

Kim JH, Ryu JD, Lee HG, Park JH, Moon GS, Cheigh HS, Song YO (2002a) The effect of kimchi on production of free radicals and anti-oxidative enzyme activities in the brain of SAM. J Kor Soc Food Sci Nutr 31:117–123

Kim JH, Ryu JD, Song YO (2002b) The effect of kimchi intake on free radical production and the inhibition of oxidation in young adults and the elderly people. Korean J Community Nutr 7:257–265

Kim NH, Moon PD, Kim SJ, Choi IY, An HJ, Myung NY, Jeong HJ, Um JY, Hong SH, Kim HM (2008) Lipid profile lowering effect of Soypro™ fermented with lactic acid bacteria isolated from Kimchi in high-fat diet-induced obese rats. Biofactors 33(1):49–60

Kim EK, An SY, Lee MS, Kim TH, Lee HK, Hwang WS et al (2011) Fermented kimchi reduces body weight and improves metabolic parameters in overweight and obese patients. Nutr Res 31:436–443. doi:10.1016/j.nutres.2011.05.011

Kim JY, Park BK, Park HJ, Park YH, Kim BO, Pyo S (2013) Atopic dermatitis-mitigating effects of new *Lactobacillus* strain, *Lactobacillus sakei* probio 65 isolated from kimchi. J Appl Microbiol 115:517–526

King S, Glanville J, Sanders ME, Fitzgerald A, Varley D (2014) Effectiveness of probiotics on the duration of illness in healthy children and adults who develop common acute respiratory infectious conditions: a systematic review and meta-analysis. Br J Nutr 112(01):41–54. doi:10.1017/S0007114514000075

Kiessling G, Schneider J, Jahreis G (2002) Long-term consumption of fermented dairy products over 6 months increases HDL cholesterol. Eur J Clin Nutr 56(9):843–849

Kiriakidis S, Stathi S, Jha HC, Hartmann R, Egge H (1997) Fatty acid esters of sitosterol 3β-glucoside from soybeans and tempe (fermented soybeans) as antiproliferative substances. J Clin Biochem Nutr 22:139–147. doi:10.3164/jcbn.22.139

Kobayashi M (2005) Immunological functions of soy sauce: Hypoallergenicity and antiallergic activity of soy sauce. J Biosci Bioeng 100:144–151. doi:10.1263/jbb.100.144

Kobayashi T, Kimura B, Fujii T (2000b) *Haloanaerobium fermentans* sp. nov., a strictly anaerobic, fermentative halophile isolated from fermented puffer fish ovaries. Int J Syst Evol Microbiol 50:1621–1627. doi:10.1099/00207713-50-4-1621

Kotb E (2012) Fibrinolytic bacterial enzymes with thrombolytic activity. Springer, Berlin, pp 1–74. doi:10.1007/978-3-642-24980-8_1

Kumar RS, Kanmani P, Yuvaraj N, Paari KA, Pattukumar V, Thirunavukkarasu C, Arul V (2012) *Lactobacillus plantarum* AS1 isolated from south Indian fermented food Kallappam suppress 1, 2-dimethyl hydrazine (DMH)-induced colorectal cancer in male Wistar rats. Appl Biochem Biotechnol 166(3):620–631. doi:10.1007/s12010-011-9453-2

Kumar M, Verma V, Nagpal R, Kumar A, Behare PV, Singh B, Aggarwal PK (2012) Anticarcinogenic effect of probiotic fermented milk and chlorophyllin on aflatoxin-B1-induced liver carcinogenesis in rats. Br J Nutr 107:1006–1016. doi:10.1017/S0007114511003953

Kwak CS, Park S, Song KY (2012) Doenjang, a fermented soybean paste, decreased visceral fat accumulation and adipocyte size in rats fed with high fat diet more effectively than nonfermented soybeans. J Med Food 15:1–9. doi:10.1089/jmf.2010.1224

Kwak SH, Cho YM, Noh GM, Om AS (2014) Cancer preventive potential of kimchi lactic acid bacteria (*Weissella cibaria*, *Lactobacillus plantarum*). J Cancer Prev 19(4):253–258. doi:10.15430/JCP.2014.19.4.253

Kwon DY, Jang JS, Hong SM, Lee JE, Sung SR, Park HR, Park S (2007) Long-term consumption of fermented soybean-derived Chungkookjang enhances insulinotropic action unlike soybeans in 90% pancreatectomized diabetic rats. Eur J Nutr 46(1):44–52. doi:10.1007/s00394-006-0630-y

Kwon DY, Hong SM, Ahn IS, Kim YS, Shin DW, Park S (2009a) Kochujang, a Korean fermented red pepper plus soybean paste, improves glucose homeostasis in 90% pancreatectomized diabetic rats. Nutrition 25(7):790–799. doi:10.1016/j.nut.2008.12.006

Kwon GH, Lee HA, Park JY, Kim JS, Lim J, Park CS, Kwon DY, Kim JH (2009b) Development of a RAPD-PCR method for identification of *Bacillus* species isolated from Cheonggukjang. Int J Food Microbiol 129:282–287. doi:10.1016/j.ijfoodmicro.2008.12.013

Kwon DY, Daily JW, Kim HJ, Park S (2010) Antidiabetic effects of fermented soybean products on type 2 diabetes. Nutr Res 30(1):1–13. doi:10.1016/j.nutres.2009.11.004

de LeBlanc ADM, Matar C, Farnworth E, Perdigon G (2007) Study of immune cells involved in the antitumor effect of *kefir* in a murine breast cancer model. J Dairy Sci 90(4):1920–1928. doi:10.3168/jds.2006-079

Lee HR, Lee JM (2009) Anti-stress effects of kimchi. Food Sci Biotechnol 18(1):25–30

Lee JW, Shin JG, Kim EH, Kang HE, Yim IB, Kim JY, Joo HG, Woo HJ (2004) Immunomodulatory and antitumor effects *in vivo* by the cytoplasmic fraction of *Lactobacillus casei* and *Bifidobacterium longum*. J Vet Sci 5:41–48

Lee JH, Kweon DH, Lee SC (2006) Isolation and characterization of an immunopotentiating factor from *Lactobacillus plantarum* in kimchi: assessment of immunostimulatory activities. Food Sci Biotechnol 15:877–883

Lee Y, Lee HJ, Lee H-S, Jang Y-A, Kim C-I (2008) Analytical dietary fiber database for the National Health and Nutrition Survey in Korea. J Food Compos Anal 21:S35–S42. doi:10.1016/j.jfca.2007.07.008

Lee H, Yoon H, Ji Y, Kim H, Park H, Lee J, Shin H, Holzapfel WH (2011) Functional properties of *Lactobacillus* strains isolated from kimchi. Int J Food Microbiol 145:155–161. doi:10.1016/j.ijfoodmicro.2010.12.003

Lee SY, Song YO, Han ES, Han JS (2012) Comparative study on dietary habits, food intakes, and serum lipid levels according to kimchi consumption in college students. J Korean Soc Food Sci Nutr 41:351–361. doi:10.3746/jkfn.2012.41.3.351

Leroy F, Verluyten J, De Vuyst L (2006) Functional meat starter cultures for improved sausage fermentation. Int J Food Microbiol 106:270–285. doi:10.1016/j.ijfoodmicro.2005.06.027

Li S, Zhao Y, Zhang L, Zhang X, Huang L, Li D, Niu C, Yang Z, Wang Q (2012) Antioxidant activity of *Lactobacillus plantarum* strains isolated from traditional Chinese fermented foods. Food Chem 135(3):1914–1919. doi:10.1016/j.foodchem.2012.06.048

Li M, Yang D, Mei L, Yuan L, Xie A, Yuan J (2014) Screening and characterization of purine nucleoside degrading lactic acid bacteria isolated from Chinese sauerkraut and evaluation of the serum uric acid lowering effect in hyperuricemic rats. PLoS One 9(9):e105577. doi:10.1371/journal.pone.0105577

Lim J, Seo BJ, Kim JE, Chae CS, Im SH, Hahn YS, Park YH (2011) Characteristics of immunomodulation by a *Lactobacillus sakei* proBio65 isolated from Kimchi. Korean J Microbiol Biotechnol 39:313–316

Lomer MCE, Parkes GC, Sanderson JD (2008) Review article: lactose intolerance in clinical practice–myths and realities. Aliment Pharmacol Ther 27(2):93–103

Longo MA, Sanromán MA (2006) Production of food aroma compounds: microbial and enzymatic methodologies. Food Technol Biotechnol 44:335–353

Lu Y, Wang W, Shan Y, Zhiqiang E, Wang L (2009) Study on the inhibition of fermented soybean to cancer cells. J Northeast Agric Univ 16(1):25–28

Luoto RM, Kinnunen TI, Aittasalo M, Ojala K, Mansikkamäki K, Toropainen E, Kolu P, Vasankari T (2010) Prevention of gestational diabetes: design of a cluster-randomized controlled trial and one-year follow-up. BMC Pregnancy Childbirth 10(1):1. doi:10.1186/1471-2393-10-39

MacFarlane GT, Cummings JH (2002) Probiotics, infection and immunity. Curr Opin Infect Dis 15:501–506

Mack DR, Michail S, Wet S (1999) Probiotics inhibit enteropathogenic *E. coli* adherence *in vitro* by inducing intestinal mucin gene expression. Am J Phys 276:G941–G950

Marette A, Picard-Deland E (2014) Yogurt consumption and impact on health: focus on children and cardiometabolic risk. Am J Clin Nutr 99(5):1243S–1247S. doi:10.3945/ajcn.113.073379

Marteau P, deVrese M, Cellier CJ, Schrezenmeir J (2001) Protection from gastrointestinal diseases with the use of probiotics. Am J Clin Nutr 73:430S–436S

Marteau P, Seksik P, Jian R (2002) Probiotics and intestinal health effects: A clinical perspective. Br J Nutr 88:51–57. doi:10.1079/BJN2002629

Martinez-Villaluenga C, Peñas E, Sidro B, Ullate M, Frias J, Vidal-Valverde C (2012) White cabbage fermentation improves ascorbigen content, antioxidant and nitric oxide production inhibitory activity in LPS-induced macrophages. LWT Food Sci Technol 46:77–83. doi:10.1016/j.lwt.2011.10.023

Mathara JM, Schillinger U, Kutima PM, Mbugua SK, Holzapfel WH (2004) Isolation, identification and characterisation of the dominant microorganisms of kule naoto: the Maasai traditional fermented milk in Kenya. Int J Food Microbiol 94:269–278. doi:10.1016/j.ijfoodmicro.2004.01.008

Matricardi PM (2002) Probiotics against allergy: data, doubts, and perspectives. Allergy 57(3):185–187. doi:10.1034/j.1398-9995.2002.1a3299.x

Mercenier A, Muller-Alouf H, Grangette C (2000) Lactic acid bacteria as live vaccines. Curr Issues Mol Biol 2:17–25

Meyer K, Kushi L, Jacobs D, Slavin J, Sellers T, Folsom A (2000) Carbohydrates, dietary fiber, and incidence of type 2 diabetes in older women. Am J Clin Nutr 71:921–930

Mine Y, Wong AHK, Jiang B (2005) Fibrinolytic enzymes in Asian traditional fermented foods. Food Res Int 38:243–250

Mo H, Zhu Y, Chen Z (2008) Microbial fermented tea—a potential source of natural food preservatives. Trends Food Sci Technol 19:124–130

Mohania D, Kansal VK, Sagwal R, Shah D (2013) Anticarcinogenic effect of probiotic dahi and piroxicam on DMH-induced colorectal carcinogenesis in Wistar rats. Am J Cancer Ther Pharmacol 1(1):8–24

Moktan B, Saha J, Sarkar PK (2008) Antioxidant activities of soybean as affected by *Bacillus*-fermentation to Kinema. Food Res Int 4(6):586–593

Moon YJ, Soh JR, Yu JJ, Sohn HS, Cha YS, Oh SH (2012) Intracellular lipid accumulation inhibitory effect of *Weissella koreensis* OK1-6 isolated from Kimchi on differentiating adipocyte. J Appl Microbiol 113:652–658. doi:10.1111/j.1365-2672.2012.05348.x

Morelli L (2014) Yogurt, living cultures, and gut health. Am J Clin Nutr 99(5):1248S–1250S. doi:10.3945/ajcn.113.073072

Mugula JK, Ninko SAM, Narvhus JA, Sorhaug T (2003) Microbiological and fermentation characteristics of togwa, a Tanzanian fermented food. Int J Food Microbiol 80:187–199

Nagai T, Tamang JP (2010) Fermented soybeans and non-soybeans legume foods. In: Tamang JP, Kailasapathy K (eds) Fermented Foods and Beverages of the World. CRC Press, Taylor & Francis Group, New York, pp 191–224

Nguyen DTL, Van Hoorde K, Cnockaert M, de Brandt E, de Bruyne K, Le BT, Vandamme P (2013a) A culture-dependent and -independent approach for the identification of lactic acid bacteria associated with the production of nem chua, a Vietnamese fermented meat product. Food Res Int 50(1):232–240

Noh KA, Kim DH, Choi NS, Kim SH (1999) Isolation of fibrinolytic enzyme producing strains from *kimchi*. Korean J Food Sci Technol 31:219–223

Oguntoyinbo FA, Tourlomousis P, Gasson MJ, Narbad A (2011) Analysis of bacterial communities of traditional fermented West African cereal foods using culture independent methods. Int J Food Microbiol 145:205–210. doi:10.1016/j.ijfoodmicro.2010.12.025

Oki K, Rai AK, Sato S, Watanabe K, Tamang JP (2011) Lactic acid bacteria isolated from ethnic preserved meat products of the Western Himalayas. Food Microbiol 28:1308–1315. doi:10.1016/j.fm.2011.06.001

Omizu Y, Tsukamoto C, Chettri R, Tamang JP (2011) Determination of saponin contents in raw soybean and fermented soybean foods of India. J Sci Ind Res 70:533–538

Otes S, Cagindi O (2003) Kefir: a probiotic dairy-composition, nutritional and therapeutic aspects. Pak J Nutr 2:54–59

Ouwehand AC, Röytiö H (2014) Probiotic fermented foods and health promotion. In: Holzapfel W (ed) Advances in fermented foods and beverages: improving quality, technologies and health benefits. Elsevier, Amsterdam

Ouwehand AC, Salminen S, Isolauri E (2002) Probiotics: an overview of beneficial effects. Antonie Van Leeuwenhoek 82:279–289. doi:10.1023/A:1020620607611

Palaniswamy SK, Govindaswamy V (2016) In-vitro probiotic characteristics assessment of feruloyl esterase and glutamate decarboxylase producing *Lactobacillus* spp. isolated from traditional fermented millet porridge (kambu koozh). LWT Food Sci Technol 68:208–216. doi:10.1016/j.lwt.2015.12.024

Park C, Choi JC, Choi YH, Nakamura H, Shimanouchi K, Horiuchi T, Misono H, Sewaki T, Soda K, Ashiuchi M, Sung MH (2005) Synthesis of super-high-molecular-weight poly-γ-glutamic acid by *Bacillus subtilis* subsp. chungkookjang. J Mol Catal B Enzym 35:128–133. doi:10.1016/j.molcatb.2005.06.007

Park KY, Kim BK (2010) Kimchi lactic acid bacteria and health benefits. FASEB J 24: (Meeting Abstract Supplement) No. 340.6

Park JM, Shin JH, Lee DW, Song JC, Suh HJ, Chang UJ, Kim JM (2010) Identification of the lactic acid bacteria in kimchi according to initial and over-ripened fermentation using PCR and 16S rRNA gene sequence analysis. Food Sci Biotechnol 19:541–546. doi:10.1007/s10068-010-0075-1

Park JM, Shin JH, Gu JG, Yoon SJ, Song JC, Jeon WM, Suh HJ, Chang UJ, Yang CY, Kim JM (2011) Effect of antioxidant activity in kimchi during a short-term and over-ripening fermentation period. J Biosci Bioeng 112:356–359. doi:10.1016/j.jbiosc.2011.06.003

Park JA, TirupathiPichiah PB, Yu JJ, Oh SH, Daily JW 3rd, Cha YS (2012) Anti-obesity effect of kimchi fermented with *Weissella koreensis* OK1–6 as starter in high-fat diet-induced obese C57BL/6 J mice. J Appl Microbiol 113:1507–1516. doi:10.1111/jam.12017

Park KY, Jeong JK, Lee YE, Daily JW 3rd (2014) Health benefits of kimchi (Korean fermented vegetables) as a probiotic food. J Med Food 17(1):6–20. doi:10.1089/jmf.2013.3083

Parvez S, Malik KA, Ah Kang S, Kim HY (2006) Probiotics and their fermented food products are beneficial for health. J Appl Microbiol 100:1171–1185. doi:10.1111/j.1365-2672.2006.02963.x

Patrignani F, Lanciotti R, Mathara JM, Guerzoni ME, Holzapfel WH (2006) Potential of functional strains, isolated from traditional Maasai milk, as starters for the production of fermented milks. Int J Food Microbiol 107:1–11. doi:10.1016/j.ijfoodmicro.2005.08.004

Pattanagul P, Pinthong R, Phianmongkhol A, Tharatha S (2008) Mevinolin, citrinin and pigments of adlay angkak fermented by *Monascus* sp. Int J Food Microbiol 126(1):20–23. doi:10.1016/j.ijfoodmicro.2008.04.019

Peng Y, Huang Q, Zhang R, Zhang Y (2003) Purification and characterization of a fibrinolytic enzyme produced by *Bacillus amyloliquefaciens* DC-4 screened from douchi, a traditional Chinese soybean food. Comp Biochem Physiol B Biochem Mol Biol 134:45–52. doi:10.1016/S1096-4959(02)00183-5

Pereira MA, Jacobs DR Jr, Pins JJ, Raatz SK, Gross MD (2002) Effect of whole grains on insulin sensitivity in overweight hyper insulinemic adults. Am J Clin Nutr 75:848–855

Pessi T, Sutas Y, Hurme M, Isolauri E (2000) Interleukin-10 generation in atopic children following oral *Lactobacillus rhamnosus* GG. Clin Exp Allergy 30:1804–1808. doi:10.1046/j.1365-2222.2000.00948.x

Plengvidhya V, Breidt F, Fleming HP (2007) Use of RAPD-PCR as a method to follow the progress of starter cultures in sauerkraut fermentation. Int J Food Microbiol 93:287–296. doi:10.1016/j.ijfoodmicro.2003.11.010

Praveesh BV, Angayarkanni J, Palaniswamyint M (2011) Antihypertensive and anticancer effect of cow milk fermented by *Lactobacillus plantarum* and *Lactobacillus casei*. J Pharm Pharm Sci 3(5):452–456

Prentice AM (2014) Dairy products in global public health. Am J Clin Nutr 99(5):1212S–1216S. doi:10.3945/ajcn.113.073437

Purchiaroni F, Tortora A, Gabrielli M, Bertucci F, Gigante G, Ianiro G et al (2013) The role of intestinal microbiota and the immune system. Eur Rev Med Pharmacol Sci 17(3):323–333

Pussinen PJ, Havulinna AS, Lehto M, Sundvall J, Salomaa V (2011) Endotoxemia is associated with an increased risk of incident diabetes. Diabetes Care 34(2):392–397. doi:10.2337/dc10-1676

Raghavendra P, Rao T, Halami P (2010) Evaluation of beneficial attributes for phytate-degrading Pediococcus pentosaceus CFR R123. Benef Microbes 1(3):259–264. doi:10.3920/BM2009.0042

Rai AK, Palni U, Tamang JP (2010) Microbiological studies of ethnic meat products of the Eastern Himalayas. Meat Sci 85:560–567. doi:10.1016/j.meatsci.2010.03.006

Ramadori G, Gautron L, Fujikawa T, Claudia R, Vianna J, Elmquist E, Coppari R (2009) Central administration of resveratrol improves diet-induced diabetes. Endocrinology 150:5326–5333. doi:10.1210/en.2009-0528

Rapsang GF, Joshi SR (2015) Molecular and probiotic functional characterization of *Lactobacillus* spp. associated with traditionally fermented fish, Tungtap of Meghalaya in Northeast India. Proc Natl Acad Sci India Sect B Biol Sci 85(4):923–933. doi:10.1007/s40011-013-0234-2

Ratanaburee A, Kantachote D, Charernjiratrakul W, Sukhoom A (2013) Selection of γ-aminobutyric acid-producing lactic acid bacteria and their potential as probiotics for use as starter cultures in Thai fermented sausages (Nham). Int J Food Sci Technol 48(7):1371–1382. doi:10.1111/ijfs.12098

Rhee SH, Kong KR, Jung KO, Park KY (2003) Decreasing effect of Kochujang on body weight and lipid levels of adipose tissues and serum in rats fed a high-fat diet. J Korean Soc Food Sci Nutr 32:882–886

Rho SJ, Lee JS, Chung YL, Kim YW, Lee HG (2009) Purification and identification of an angiotensing I-converting enzyme inhibitory peptide from fermented soybean extract. Process Biochem 44:490–493. doi:10.1016/j.procbio.2008.12.017

Rolfe RD (2000) The role of probiotic cultures in the control of gastrointestinal health. J Nutr 130:396S–402S

Rowland I (2004) Probiotics and colorectal cancer risk. Br J Nutr 91:805–807. doi:10.1079/BJN20041161

Rubia-Soria A, Abriouel H, Lucas R, Omar NB, Martinez-Caõamero M, Gálvez A (2006) Production of antimicrobial substances by bacteria isolated from fermented table olives. World J Microbiol Biotechnol 22(7):765–768. doi:10.1007/s11274-005-9101-5

Rühmkorf C, Jungkunz S, Wagner M, Vogel RF (2012) Optimization of homoexopolysaccharide formation by lactobacilli in gluten-free sourdoughs. Food Microbiol 32:286–294. doi:10.1016/j.fm.2012.07.002

Ryu BM, Ryu SH, Jeon YS, Lee YS, Moon GS (2004a) Inhibitory effect of solvent fraction of various kinds of kimchi on ultraviolet B induced oxidation and erythema formation of hairless mice skin. J Kor Soc Food Sci Nutr 33:785–790

Ryu BM, Ryu SH, Lee YS, Jeon YS, Moon GS (2004b) Effect of different kimchi diets on oxidation and photooxidation in liver and skin of hairless mice. J Korean Soc Food Sci Nutri 33:291–298

Sabeena FKH, Baron CP, Nielsen NS, Jacobsen C (2010) Antioxidant activity of yoghurt peptides: Part 1-*in vitro* assays and evaluation in ω-3 enriched milk. Food Chem 123:1081–1089. doi:10.1016/j.foodchem.2010.05.067

Saikali J, Picard V, Freitas M, Holt P (2004) Fermented milks, probiotic cultures, and colon cancer. Nutr Cancer 49:14–24. doi:10.1207/s15327914nc4901_3

Saithong P, Panthavee W, Boonyaratanakornkit M, Sikkhamondhol C (2010) Use of a starter culture of lactic acid bacteria in plaa-som, a Thai fermented fish. J Biosci Bioeng 110:553–557. doi:10.1016/j.jbiosc.2010.06.004

Sanders ME, Klaenhammer TR (2001) Invited review: the scientific basis of *Lactobacillus acidophilus* NCFM functionality as a probiotic. J Dairy Sci 84:319–331. doi:10.3168/jds.S0022-0302(01)74481-5

Sanz Y, Rastmanesh R, Agostonic C (2013) Understanding the role of gut microbes and probiotics in obesity: how far are we? Pharm Res 69(1):144–155. doi:10.1016/j.phrs.2012.10.021

Sapbamrer R, Visavarungroj N, Suttajit M (2013) Effects of dietary traditional fermented soybean on reproductive hormones, lipids, and glucose among postmenopausal women in northern Thailand. Asia Pac J Clin Nutr 22(2):222

Sarkar S (2008) Innovations in Indian fermented milk products—a review. Food Biotechnol 22(1):78–97. doi:10.1080/08905430701864025

Sarkar PK, Hasenack B, Nout MJR (2002) Diversity and functionality of *Bacillus* and related genera isolated from spontaneously fermented soybeans (Indian Kinema) and locust beans (African Soumbala). Int J Food Microbiol 77:175–186. doi:10.1016/S0168-1605(02)00124-1

Schillinger U, Ban-Koffi L, Franz CMAP (2010) Tea, coffee and cacao. In: Tamang JP, Kailasapathy K (eds) Fermented foods and beverages of the world. CRC Press, Taylor & Francis Group, New York, pp 353–375

Sengun IY, Nielsen DS, Karapinar M, Jakobsen M (2009) Identification of lactic acid bacteria isolated from Tarhana, a traditional Turkish fermented food. Int J Food Microbiol 135:105–111. doi:10.1016/j.ijfoodmicro.2009.07.033

Shah NP (1993) Effectiveness of dairy products in alleviation of lactose intolerance. Food Australia 45:268–271

Shah NP (2004) Probiotics and prebiotics. Agro Food Industry HiTech 15(1):13–16

Shah NP, da Cruz AG, Faria JAF (eds) (2013) Probiotics and probiotic foods: technology, stability and benefits to human health. Nova Science Publishers, New York

Shin SK, Kwon JH, Jeon M, Choi J, Choi MS (2011) Supplementation of Cheonggukjang and Red Ginseng Cheonggukjang can improve plasma lipid profile and fasting blood glucose concentration in subjects with impaired fasting glucose. J Med Food 14:108–113. doi:10.1089/jmf.2009.1366

Shobharani P, Halami PM (2015) In vitro evaluation of the cholesterol-reducing ability of a potential probiotic *Bacillus* spp. Ann Microbiol:1–9. doi:10.1007/s13213-015-1146-6

Shon M-Y, Lee J, Choi J-H, Choi S-Y, Nam S-H, Seo K-I, Lee S-W, Sung N-J, Park S-K (2007) Antioxidant and free radical scavenging activity of methanol extract of chungkukjang. J Food Compos Anal 20:113–118. doi:10.1016/j.jfca.2006.08.003

Sim KH, Han YS (2008) Effect of red pepper seed on kimchi antioxidant activity during fermentation. Food Sci Biotechnol 17(2):295–301

Singh TA, Devi KR, Ahmed G, Jeyaram K (2014) Microbial and endogenous origin of fibrinolytic activity in traditional fermented foods of Northeast India. Food Res Int 55:356–362. doi:10.1016/j.foodres.2013.11.028

Solga SF (2003) Probiotics can treat hepatic encephalopathy. Med Hypotheses 61:307–313. doi:10.1016/S0306-9877(03)00192-0

Sonar RN, Halami PM (2014) Phenotypic identification and technological attributes of native lactic acid bacteria present in fermented bamboo shoot products from North-East India. J Food Sci Technol. doi:10.1007/s13197-014-1456-x

Song HS, Kim YM, Lee KT (2008) Antioxidant and anticancer activities of traditional Kochujang added with garlic porridge. J Life Sci 18(8):1140–1146. doi:10.5352/JLS.2008.18.8.1140

Song YR, Song NE, Kim JH, Nho YC, Baik SH (2011) Exopolysaccharide produced by *Bacillus licheniformis* strains isolated from Kimchi. J Gen Appl Microbiol 57(3):69–175. 10.2323/jgam.57.169

Sridevi J, Halami PM, Vijayendra SVN (2010) Selection of starter cultures for idli batter fermentation and their effect on quality of idli. J Food Sci Technol 47:557–563. doi:10.1007/s13197-010-0101-6

Stevens HC, Nabors L (2009) Microbial food cultures: a regulatory update. Food Technol (Chicago) 63:36–41

Sumi H, Nakajima N, Yatagai C (1995) A unique strong fibrinolytic enzyme (katsuwokinase) in skipjack "shiokara", a Japanese traditional fermented food. Comp Biochem Physiol Biochem Mol Biol 112:543–547. doi:10.1016/0305-0491(95)00100-X

Sun YP, Chou CC, Yu RC (2009) Antioxidant activity of lactic-fermented Chinese cabbage. Food Chem 115(3):912–917. doi:10.1016/j.foodchem.2008.12.097

Sun Z, Liu W, Gao W, Yang M, Zhang J, Wang J, Menghe B, Sun T, Zhang H (2010) Identification and characterization of the dominant lactic acid bacteria from kurut: The naturally fermented yak milk in Qinghai, China. J Gen Appl Microbiol 56:1–10. doi:10.2323/jgam.56.1

Svetkey LP, Simons-Morton D, Vollmer WM, Appel LJ, Conlin PR, Ryan DH, Ard J, Kennedy BM (1999) Effects of dietary patterns on blood pressure: subgroup analysis of the Dietary Approaches to Stop Hypertension (DASH) randomized clinical trial. Arch Intern Med 159(3):285–293. doi:10.1001/archinte.159.3.285

Swallow DM (2003) Genetics of lactase persistence and lactose intolerance. Annu Rev Genet 37(1):197–219

Tamang JP, Thapa S (2006) Fermentation dynamics during production of bhaati jaanr, a traditional fermented rice beverage of the Eastern Himalayas. Food Biotechnol 20(3):251–261

Talahay P, Fahey JW (2001) Phytochemicals from cruciferous plants protect against cancer by modulating carcinogen metabolism. J Nutr 131(11):3027S–3033S

Tamang JP (2003) Native microorganisms in fermentation of kinema. Ind J Microbiol 43:127–130

Tamang JP (2010a) Himalayan fermented foods: microbiology, nutrition, and ethnic values. CRC Press, Taylor & Francis Group, New York

Tamang JP (2010b) Diversity of fermented foods. In: Tamang JP, Kailasapathy K (eds) Fermented foods and beverages of the world. CRC Press, Taylor & Francis Group, New York, pp 41–84

Tamang JP (ed) (2015) Health benefits of fermented foods and beverages. CRC Press, New York

Tamang JP, Fleet GH (2009) Yeasts diversity in fermented foods and beverages. In: Satyanarayana T, Kunze G (eds) Yeasts biotechnology: diversity and applications. Springer, New York, pp 169–198. doi:10.1007/978-1-4020-8292-4_9

Tamang B, Tamang JP (2007) Role of lactic acid bacteria and their functional properties in Goyang, a fermented leafy vegetable product of the Sherpas. J Hill Res 20:53–61

Tamang B, Tamang JP (2009) Lactic acid bacteria isolated from indigenous fermented bamboo products of Arunachal Pradesh in India and their functionality. Food Biotechnol 23:133–147. doi:10.1080/08905430902875945

Tamang B, Tamang JP (2010) In situ fermentation dynamics during production of gundruk and khalpi, ethnic fermented vegetables products of the Himalayas. Ind J Microbiol 50(Suppl 1):93–98. doi:10.1007/s12088-010-0058-1

Tamang JP, Dewan S, Thapa S, Olasupo NA, Schillinger U, Wijaya A, Holzapfel WH (2000) Identification and enzymatic profiles of predominant lactic acid bacteria isolated from soft-variety chhurpi, a traditional cheese typical of the Sikkim Himalayas. Food Biotechnol 14:99–112. doi:10.1080/08905430009549982

Tamang JP, Tamang B, Schillinger U, Guigas C, Holzapfel WH (2009b) Functional properties of lactic acid bacteria isolated from ethnic fermented vegetables of the Himalayas. Int J Food Microbiol 135:28–33. doi:10.1016/j.ijfoodmicro.2009.07.016

Tamang JP, Tamang N, Thapa S, Dewan S, Tamang BM, Yonzan H, Rai AK, Chettri R, Chakrabarty J, Kharel N (2012) Microorganisms and nutritional value of ethnic fermented foods and alcoholic beverages of North East India. Indian J Tradit Knowl 11(1):7–25

Tamang JP, Thapa N, Tamang BM, Rai AK, Chettri R (2015) Microorganisms in Fermented Foods and Beverages. In: Tamang JP (ed) Health benefits of fermented foods and beverages. CRC Press, New York, pp 2–87

Taniguchi M, Nagao K, Inoue K, Imaizumi K (2008) Cholesterol lowering effect of sulfur-containing amino acids added to a soybean protein diet in rats. J Nutr Sci Vitaminol 54(6):448–453. 10.3177/jnsv.54.448

Thapa N, Pal J, Tamang JP (2004) Microbial diversity in ngari, hentak and tungtap, fermented fish products of Northeast India. World J Microbiol Biotechnol 20:599–607. doi:10.1023/B:WIBI.0000043171.91027.7e

Thapa N, Pal J, Tamang JP (2007) Microbiological profile of dried fish products of Assam. Ind J Fish 54:121–125

Todorov SD, Holzapfel W (2014) Traditional cereal fermented foods as sources of functional microorganisms. In: Holzapfel W (ed) Advances in fermented foods and beverages: Improving quality, technologies and health benefits. Elsevier, Amsterdam, pp 123–153

Tolhurst G, Heffron H, Lam YS, Parker HE, Habib AM, Diakogiannaki E, Cameron J, Grosse J, Reimann F, Gribble FM (2012) Short-chain fatty acids stimulate glucagon-like peptide-1 secretion via the g-protein-coupled receptor FFAR2. Diabetes 61:364–371. doi:10.2337/db11-1019

Topuz E, Derin D, Can G, Kürklü E, Çınar S, Aykan F, Çevikbaş A, Dişçi R, Durna Z, Şakar B, Saglam S (2008) Effect of oral administration of kefir on serum proinflammatory cytokines on 5-FU induced oral mucositis in patients with colorectal cancer. Investig New Drugs 26(6):567–572. doi:10.1007/s10637-008-9171-y

Truelsen T, Gronbaek M, Schnohr P, Boysen G (1998) Intake of beer, wine, and spirits and risk of stroke: the Copenhagen city heart study. Stroke 29:2467–2472. doi:10.1161/01.STR.29.12.2467

Tsubura S (2012) Anti-periodontitis effect of *Bacillus subtilis* (natto). Shigaku (Odontology) 99:160–164

Tsukamoto Y, Ichise H, Kakuda H, Yamaguchi M (2000) Intake of fermented soybean (natto) increases circulating vitamin K_2 (menaquinone-7) and γ-carboxylated osteocalcin concentration in normal individuals. J Bone Miner Metab 18:216–222. doi:10.1007/s007740070023

Tyopponen S, Petaja E, Mattila-Sandholm T (2003) Bioprotectives and probiotics for dry sausages. Int J Food Microbiol 83:233–244. doi:10.1016/S0168-1605(02)00379-3

Veinocchi P, Valmossi S, Dalai I, Torrsani S, et Gianotti A, Suzzi G, Guerzoni ME, Mastrocola D, Gardini F (2006) Characterization of the yeast populations involved in the production of a typical Italian bread. J Food Sci 69:M182–M186.doi: 10.1111/j.1365-2621.2004.tb13618.x

Vijayendra SVN, Halami PM (2015) Health benefits of fermented vegetable products. In: Tamang JP (ed) Health benefits of fermented foods and beverages. CRC Press, New York, pp 325–336

Vijayendra SVN, Palanivel G, Mahadevamma S, Tharanathan RN (2008) Physico-chemical characterization of an exopolysaccharide produced by a non-ropy strain of *Leuconostoc* sp. CFR 2181 isolated from dahi, an Indian traditional lactic fermented milk product. Carbohydr Polym 72:300–307. doi:10.1016/j.carbpol.2007.08.016

Vogel RF, Hammes WP, Habermeyer M, Engel KH, Knorr D, Eisenbrand G (2011) Microbial food cultures—opinion of the Senate Commission on Food Safety (SKLM) of the German Research Foundation (DFG). Mol Nutr Food Res 55:654–662. doi:10.1002/mnfr.201100010

Waldherr FW, Vogel RF (2009) Commercial exploitation of homo-exopolysaccharides in non-dairy food systems. In: Ullrich M (ed) Bacterial polysaccharides: current innovations and future trends. Caister Academic Press, Norfolk, pp 313–329

Wang L-J, Li D, Zou L, Chen XD, Cheng Y-Q, Yamaki K, Li L-T (2007a) Antioxidative activity of douchi (a Chinese traditional salt-fermented soybean food) extracts during its processing. Int J Food Prop 10:1–12. doi:10.1080/10942910601052715

Weill FS, Cela EM, Paz ML, Ferrari A, Leoni J, Gonzalez Maglio DH (2013) Lipoteichoic acid from *Lactobacillus rhamnosus GG* as an oral photoprotective agent against UV-induced carcinogenesis. Br J Nutr 109:457–466. doi:10.1017/S0007114512001225

Winkler C, Wirleitner B, Schroecksnadel K, Schennach H, Fuchs D (2005) *In vitro* effects of beetroot juice on stimulated and unstimulated peripheral blood mononuclearcells. Am J Biochem Biotechnol 1:180–185

Won TJ, Kim B, Song DS, Lim YT, Oh ES, Lee DI, Park ES, Min H, Park S, Hwang KW (2011) Modulation of Th1/Th2 balance by *Lactobacillus* strains isolated from kimchi via stimulation of macrophage cell line J774A.1 *in vitro*. J Food Sci 76(2):H55–H61. doi:10.1111/j.1750-3841.2010.02031.x

Wong AHK, Mine Y (2004) A novel fibrinolytic enzyme in fermented shrimp paste, a traditional Asian fermented seasoning. J Agric Food Chem 52:980–986. doi:10.1021/jf034535y

World Health Organization (2013) Global action plan for the prevention and control of noncommunicable diseases 2013–2020

Wu R, Wang L, Wang J, Li H, Menghe B, Wu J, Guo M, Zhang H (2009) Isolation and preliminary probiotic selection of lactobacilli from Koumiss in Inner Mongolia. J Basic Microbiol 49:318–326. doi:10.1002/jobm.200800047

Yadav H, Jain S, Sinha PR (2007) Antidiabetic effect of probiotic dahi containing *Lactobacillus acidophilus* and *Lactobacillus casei* in high fructose fed rats. Nutrition 23(1):62–68. doi:10.1016/j.nut.2006.09.002

Yogesh D, Halami PM (2015a) Evidence that multiple proteases of *Bacillus subtilis* can degrade fibrin and fibrinogen. Int Food Res J 22(4):1662–1667

Yogesh D, Halami PM (2015b) A fibrin degrading serine metallo protease of *Bacillus circulans* with α-chain specificity. Food Biosci 11:72–78. doi:10.1016/j.fbio.2015.04.007

Yu J, Wang WH, Menghe BL, Jiri MT, Wang HM, Liu WJ, Bao QH et al (2011) Diversity of lactic acid bacteria associated with traditional fermented dairy products in Mongolia. J Dairy Sci 94:3229–3241. doi:10.3168/jds.2010-3727

Zhang J-H, Tatsumi E, Ding C-H, Li L-T (2006) Angiotensin I- converting enzyme inhibitory peptides in douche, a Chinese traditional fermented soybean product. Food Chem 98:551–557. doi:10.1016/j.foodchem.2005.06.024

Metagenomics of Fermented Foods: Implications on Probiotic Development

19

Shiwani Guleria, Anil Kumar, Samriti Sharma, Saurabh Kulshrestha, and Anjali Chauhan

Abstract

Fermented foods act as delivery vehicles of probiotic cells in human body. These food products boost human health through enhanced nutrition content, digestibility, microbial stability, and detoxification. The importance of fermented foods as probiotics is increasing continuously as they play significant roles in regulating and balancing intestinal microflora. Therefore, efforts are needed toward mining and characterization of microbial communities of fermented foods. The application of metagenomics techniques provides a right way to explore and

S. Guleria
Department of Microbiology, Lovely Professional University, Phagwara, Punjab 144 411, India

DR YS Parmar University of Horticulture and Forestry, Nauni, Solan, Himachal Pradesh 173 230, India

A. Kumar
Jaypee University of Information Technology, Waknaghat, Solan, Himachal Pradesh 173 234, India

DR YS Parmar University of Horticulture and Forestry, Nauni, Solan, Himachal Pradesh 173 230, India

S. Sharma
DR YS Parmar University of Horticulture and Forestry, Nauni, Solan, Himachal Pradesh 173 230, India

S. Kulshrestha
Faculty of Applied Sciences and Biotechnology, Shoolini University, Bajol, Solan, Himachal Pradesh 173 229, India

A. Chauhan (✉)
Department of Soil Science and Water Management, College of Forestry, DR YS Parmar University of Horticulture and Forestry, Nauni, Solan, Himachal Pradesh 173 230, India
e-mail: anjali_chauhan22@yahoo.co.in

V.C. Kalia et al. (eds.), *Mining of Microbial Wealth and MetaGenomics*,
DOI 10.1007/978-981-10-5708-3_19

characterize the unexplored beneficial microbial flora. Furthermore, for commercial development of a probiotic from fermented food, there are some prerequisite, which need to be fulfilled. Here we present the benefits, pitfall, and development of fermented food as probiotic.

Keywords
Probiotic • Human health • Microbiome • Fermented foods

19.1 Introduction

For thousands of years, our ancestors are using beverages and fermented foods as established part of their diet with proper preservation techniques without any skilled knowledge of microbes (Selhub et al. 2014). Good microbes are frequently being used for their applications in improving human health. A living organism that may provide health benefits beyond its nutritional value when ingested is *probiotics*. Most predominant beneficial microflora of the gastrointestinal tract and small and large intestine are *Lactobacillus reuteri*, *Streptococcus thermophilus* (gastrointestinal tract), *Lactobacillus acidophilus* (small intestine), *Bifidobacterium bifidum* (large intestine), *Lactobacillus delbrueckii*, and *Lactobacillus bulgaricus*. Most common are the strains of *Lactobacillus and Bifidobacterium* which are used in rapidly developing area of probiotics.

Before going further, let us remind *what is fermentation and fermented food*. An energy-yielding process in which an organic molecule is oxidized in the absence of exogenous electron acceptor is *fermentation*. Usually pyruvate or a pyruvate derivative serves as an electron acceptor. Many fungi, protists, and some bacteria ferment sugar to ethanol and CO_2 called alcoholic fermentation. This process has been used by humans for years for alcohol production using *Saccharomyces cerevisiae*. Even more common is the reduction of pyruvate to lactate, i.e., lactic acid fermentation. Depending upon the end product, lactic acid fermenters can be divided into homolactic (end product, lactate) and heterolactic fermenters (end product, lactate, ethanol, and CO_2). Representative homolactic LAB (lactic acid bacteria) include *Lactobacillus*, *Streptococcus*, *Pediococcus*, *Enterococcus*, and *Lactococcus* spp., and heterofermentative LAB include *Leuconostoc*, *Weissella*, *Oenococcus*, and certain lactobacilli. This fermentation is used in commercial production of kimchi, poi, sauerkraut, kefir, and yogurt.

Fermented food is produced and/or preserved by action of microorganisms. Growth of natural or inoculated population of microorganism causes chemical and/or textural changes to form a product with pleasing flavors and odors, which can be stored for longer periods. Fermented food is an easily digestible part of a meal, which is enzyme and nutrient rich and healthy to eat. All fermented foods are not probiotics as the ability to colonize the intestine is not many (Selhub et al. 2014). Foods like sauerkraut and coconut kefir are with active bacterial activity and therefore considered extremely beneficial to the colon and the entire digestive tract.

Lactic acid fermentation is the most important and common type being done for milk, vegetable, and fruits. Lacto-fermented foods should be part of a meal, because it can boost up the overall nutrient level of our meal. Traditionally the use of salt and sugar for preservation in homemade fermented food is done because salt and sugar play an important role for growth of beneficial bacteria and prevention of pathogens.

19.2 Probiotics from Fermented Foods

Various food products have been proposed as viable probiotic cell carrier for consumers. Fermented food is the main driving force for boosting immunity in the human body as they provide probiotic cells in the human body. Among different fermented products, dairy products are considered the most important probiotic cell delivery vehicle. Other than dairy food products, many fermented meats, vegetable, fruit, and cereals are also used as probiotic cell carriers which are healthy and according to the consumers expectancy (Reid 2008; Granato et al. 2010; Pereira et al. 2011) (Fig. 19.1).

19.2.1 Probiotics from Dairy Products

19.2.1.1 Yogurt

Yogurt is a well-liked food prepared by bacterial fermentation of milk. It is also an important food carrier of probiotic cell delivery. Milk, milk powder, standard pure microbial cultures of LAB sugar, fruit, flavors, stabilizers, and emulsifiers are constituents of yogurt. It is generally produced under well-controlled processes carried out at 42–45 °C. *Lactobacillus bulgaricus* and *Streptococcus thermophilus* are commonly employed fermenters for this process. The pH of finally produced commercial yogurt is 3.7–4.3, and it is a commercial probiotic with several nutritional benefits and well accepted by the consumers around the world.

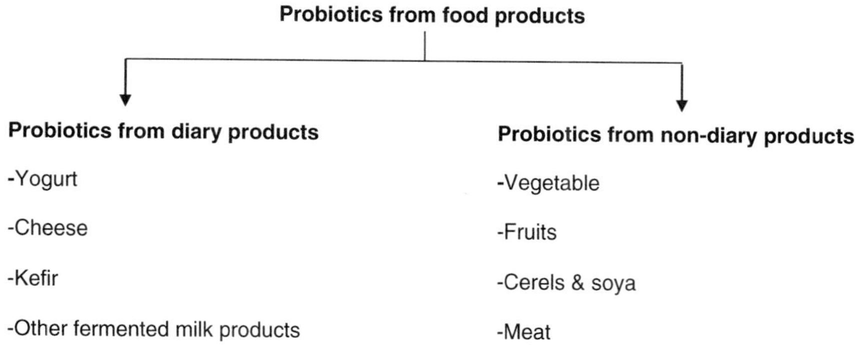

Fig. 19.1 Probiotics from various fermented foods

Incorporation of furthermore probiotic bacteria into yogurt can additionally increase its nutritional status. The addition of bacterial strains such as *B. bifidum* and *L. acidophilus* enhances the nutritional and physiological conditions of yogurt (Vandenplas et al. 2014; Wedajo 2015). A new yogurt product named as "bio-yogurt" contains probiotic cells of different strains like *Lactobacillus* and *Bifidobacteria* spp. (Hekmat et al. 2009; Cruz et al. 2009; Awaisheh 2011; Awaisheh et al. 2012).

19.2.1.2 Cheese

Cheese is consumed throughout the world in various textures and flavors. Production of cheese includes curd formation, followed by pressing, salting, and ripening. In curd formation, liquid milk is converted into casein and fat-containing solid mass. This coagulation of casein is achieved by either acid addition of starter cultures or by rennet. Thus, cheese is named as fermented/non-fermented product. According to their moisture content, cheese can be categorized as soft, semisoft, hard, and very hard. The most suitable and desired form of cheese as probiotic food carrier is fresh and soft cheese as it is unripened with limited self-life (Health Canada 2007). Some examples of soft, semisoft, and hard cheese developed and marketed as food vehicle of probiotic bacteria are shown in Table 19.1.

The benefits of cheese as food carrier of probiotic delivery are (Health Canada 2007):

- Protection of probiotic bacteria due to high protein content of cheese from the gastrointestinal tract's acidic environment.
- Protection of probiotic bacteria in the stomach due to dense matrix of cheese.
- Protection of probiotic bacteria in the stomach due to high fat content of cheese.
- In rennet cheese, high pH and the absence of antagonistic effects of starter cultures are important factors.

19.2.1.3 Kefir

Kefir is a fermented milk material having ethanol concentration of 2%. Traditionally, kefir was produced by kefir grain addition into the milk in leather sacks. These leather sacks were hung by the front door during the day, so that passerby pushed it

Table 19.1 Examples of commercialized probiotic cheese

Commercialized probiotic cheese	Probiotic bacteria
Jordanian probiotic soft cheese	*L. acidophilus* and *L. Reuteri* (Health Canada 2007)
Cheddar-like cheese	*B. infantis* (Ross et al. 2002a, b)
Cheddar-like cheese	*L. acidophilus*, *L. casei*, *L. paracasei*, and *Bifidobacterium* spp. (Ong et al. 2006)
Argentinian Fresco cheese	*Bifidobacterium*, *L. acidophilus and L. casei*, and *L. paracasei A13* (Heller 2001; Vinderola et al. 2009)

and stimulates the fermentation. Kefir grains were used as inoculum of milk fermentation. These grains have complex mixture of LAB (*Lactobacillus*, *Leuconostoc*, *Lactococcus*), acetic acid bacteria, and yeast mixture (Guzel-Seydim et al. 2005). Generally, mixture of yeast includes lactose- as well as non-lactose-fermenting ones including *Saccharomyces cerevisiae* (Farnworth 2005). In fermentation process, boiled milk is inoculated with kefir grain (2–10%) at 20–25 °C for 18–24 h. After incubation, grains are separated from the final product. The final product is foamy and frothy due to carbon dioxide produced by yeast. There are numerous health benefits of having kefir in daily diet. It is a rich source of vitamins such as K, B1, and B12, biotin, folic acid, amino acids such as tryptophan, and ions like calcium, magnesium, and phosphorus. Many reports had demonstrated the role of kefir products in antibacterial, antitumor, and immunological effects (Lavermicocca 2006; Sheehan et al. 2007).

19.2.2 Probiotics from Nondairy Products

19.2.2.1 Fruit and Vegetable Products

Many fruits and vegetable products especially juices including apple, orange, tomato, pineapple, beet, and carrot juices are examined for their suitability as probiotic carriers. Fruits and vegetable products are also tested for their acceptance by the consumer (Sheehan et al. 2007; Pereira et al. 2011). For the development of vegetable and fruit products mainly following probiotic strains, *Lactobacillus* and *Bifidobacteria* spp. are used. However, *L. plantarum* inoculated juices were observed with unpleasant dairy aromas and sour flavors. Most frequently faced constraints while processing and commercial development of probiotic juices include (1) exposure of probiotic bacteria to acidic and other unfavorable conditions and (2) aroma and flavor of final product. In a study on fresh apple juices done by Roble et al. (2010), they had reported the fresh apple slices inoculated with *L. rhamnosus* GG as a good probiotic vehicle with a viability of probiotic bacteria for 10 days at 2–4 °C.

19.2.2.2 Cereals and Soya Probiotic Products

Many soya products can be fermented and used as probiotic carrier such as:

Ingredient	Fermenting microorganism	Food
Soya bean	*Aspergillus oryzae*, *Zygosaccharomyces rouxii*	Miso
Soya bean	*A. oryzae*, *Z. rouxii*, *Lactobacillus delbrueckii*	Soy sauce
Soya bean	*Actinimucor elegans*, *Mucor* sp.	Sufu
Soya bean	*Rhizopus oligosporus*, *R. oryzae*	Tempeh
Soya bean	*A. oryzae*	Tao-si

Soybean being rich in protein, its products also play very important role in preventing diseases such as atherosclerosis, cancer, osteoporosis, and menopausal disorder (Liu et al. 2006). Very complex nutritional requirements have been observed for growth of probiotic strains such as *Lactobacillus* and *Bifidobacteria*

(Schoenlechner et al. 2008). Fermentation products of cereals provide complex nutrition to probiotic bacteria such as improved protein; fermentable carbohydrates; B vitamins; minerals such as manganese, iron, zinc, and calcium; and some nondigestible poly- and oligosaccharides. Cereal grains can stimulate the growth of *Lactobacilli* and *Bifidobacteria* bacteria in the colon. Therefore, cereal grains, which are good source of nondigestible carbohydrates, can act as probiotics (Reid 2008; Schoenlechner et al. 2008). Thus, cereals make a suitable medium for probiotic bacteria growth. However, there are some challenges in commercial development of probiotic cereal products, which include antimicrobial activity of LAB against added probiotic bacteria (Farnworth et al. 2007).

19.3 Development of Probiotics by Exploring Microbiome of Fermented Foods

There are some prerequisite for commercial development of a probiotics, which need to be fulfilled before the marketing process of final product. The most essential prerequisite is evaluating and approving the efficacy and safety of developing probiotic. Confirmation regarding safe use of microbes in food has been obtained by safe and long experience of starter culture use such as *Lactobacillus* and *Leuconostoc* in food (Huys et al. 2013). There was no harmful effect reported until now from LAB consumption except rare cases of bacteremia (Sanders et al. 2010); therefore, LAB are classified as GRAS (Widyastuti and Rohmatussolihat 2014) ("GRAS"—*G*enerally *R*ecognized *A*s *S*afe).

The use of intestinal bacterial isolates in high numbers has raised the concern regarding safety of probiotic bacteria. Therefore, it is essential to evaluate the acute, subacute, and chronic toxicity of all potential probiotic strains. However, such evaluation may not be necessary for strains with established long safe use. Characterization of probiotic strains is based on a large number of tests in order to assure their safety: antibiotic resistance, metabolic activities, side effects, epidemiological surveillance, etc.

19.3.1 FAO/WHO Regulation of Probiotics Products

Markets of probiotic products are expanding globally. Therefore, regulation and guidelines for development of probiotic products at national and international level are becoming extremely important. Until now, there are no international regulatory standards for probiotic products, and the national probiotic standards differ among different countries. The various categories of probiotics according to their use are a dietary supplement, food/food ingredient, and/or a drug (Venugopalan et al. 2010). The regulatory standards also vary according to intended use of a probiotic. Probiotics, which are used as dietary supplements and foods additives, may apply for general health claims (Pineiro and Stanton 2007).

19.3.1.1 Guidelines for Probiotics

Identification of Genus/Species/Strain

Clarification regarding identity of probiotic used in the product is essential, and this can be done by speciation of the probiotic bacteria on the product. For this, most current and valid methodology should be used. Combination of phenotypic and genotypic identification is recommended. Product labels with older and misleading nomenclature are not acceptable. Current and scientifically recognized names of bacteria must be used and can be retrieved as follows:

- http://www.bacterio.cict.fr/ having list of Bacterial Names
- International Journal of Systematic Bacteriology having published list of Bacterial Names

In Vitro Tests to Screen Potential Probiotics

There are number of recommended target-specific in vitro tests to screen potential probiotics. Results of these in vitro tests should correlate with in vivo tests, such as correlation between in vitro bile salt resistance with in vivo gastric survival (Conway et al. 1987). For screening of potential probiotics, the list of current in vitro tests according to Joint FAO/WHO working group report (2002) was as follows: (1) resistance to gastric acidity, bile acid, (2) ability to adhere human epithelial cells, (3) decreased pathogen adhesion, (4) antimicrobial activity, and (5) safety.

Very few cases of systemic infections were reported due to probiotic consumption. All cases evidenced in patients with underlying medical conditions. Already documented infections with consumption of commercial product are (although necessarily not proven):

- Consumption of *L. rhamnosus* probiotic product resulted to two cases (Mackay et al. 1999; Rautio et al. 1999).
- Consumption of *Saccharomyces fungemia* probiotic product resulted to 13 cases due to vascular catheter contamination (Hennequin et al. 2000).
- Consumption of probiotic product lead to *Bacillus* infections which include three reports (Richard et al. 1988; Oggioni et al. 1998; Spinosa et al. 2000).
- There is no report of *Bifidobacterium* infection.

In Vivo Studies Using Animals and Humans

Impact of potential probiotics has been evaluated on different clinical conditions of animal and human. Many centers have recommended repeated human trials for confirmation. Cases of adverse effects should be carefully reported and monitored. There should be no report against considered probiotic food. Standard methods of clinical evaluation of probiotic are as follows:

Phase 1 (safety)
Phase 2 (efficacy): measure adverse effects and efficacy
Phase 3 (effectiveness): not applicable for food vehicles of probiotic
Phase 4 (surveillance)

Health Claims and Labeling

Requisite appropriate information regarding the product must be available on the label for the choice of consumer. Product manufacturers are mainly responsible for truthful health claims, and these claims should not be misleading. The following information must be described on the label according to Joint FAO/WHO working group report (2002):

- Designation of genus, species, and strain on the label
- Minimum viable cells
- Serving size
- Health claim

19.4 Mining Fermented Food Microbiome for Probiotics

Fermented food products are widely produced and consumed as essential components of human diet globally. These food products boost human health through enhanced nutritional content, digestibility, and detoxification (Hayes et al. 2007; Ray and Sivakumar 2009) and are the oldest means of preservation to the current day (Bokulich and Mills 2012). The most important fermentations typically used in food microbiology are lactic acid (fruits, vegetables, cassava, meat, and milk), propionic acid, high salt (fish and soy sauce), and alcoholic fermentation (rice and cassava), which are carried out by variety of microorganisms including bacteria, yeast, and mold. However, in some fermented product like soy sauce, fermentation is carried by mold followed by brine fermentation in which LAB and yeasts are involved (Röling et al. 1996). Fermentation changes organoleptic properties of food, thereby developing diversity of flavors, textures, and aroma (Nuraida 2015). Beside nutritional needs and sensory indulgence, fermentation microorganisms produce bioactive metabolites and impart health-promoting characteristics like protection against degenerative diseases and various other gastrointestinal disorders (gastrointestinal tract infection, lactose digestion, colon cancer, and inflammatory bowel disease) (Peres et al. 2012).

Fermented foods with health-promoting characteristics, imparted by fermentative microbes, belong to class probiotics (Metchnikoff 2004) and have gained attraction worldwide. Presently, bacteria used as commercial probiotics products are the members of genus *Bifidobacterium* and *Lactobacillus* and strains of *Bacillus* and *Enterococcus* and some yeasts (Heller 2001). Of these genera, bacterial strain with probiotic potential is listed in Table 19.2. The recent medical evidences suggested enterococci as commercial probiotic strains for food formulation (Franz et al. 2011); but remain controversial as few isolates appeared to be opportunistic pathogen, hence not GRAS (Ogier and Serror 2008; Peres et al. 2012). Given the continued importance for probiotics of fermented foods, efforts are needed toward mining and characterization of microbial communities of fermented foods. The microbial mining techniques include (Ercolini 2013):

1. Culture-dependent approaches
2. Culture-independent approaches

Table 19.2 Commercially used probiotic bacterial strain (Heller 2001; Soccol et al. 2010; Davis 2014)

Sr. no.	Genus	Species and strain
1.	*Lactobacillus*	*Lactobacillus acidophilus*
		Lactobacillus gasseri
		Lactobacillus rhamnosus
		Lactobacillus johnsonii
		Lactobacillus casei Shirota
		Lactobacillus reuteri SD2112/MM2
		Lactobacillus lactis L1A
		Lactobacillus bulgaricus Lb12
2.	*Bifidobacterium*	*Bifidobacterium bifidum* Bb-11
		Bifidobacterium longum UCC 35624
		Bifidobacterium infantis
		Bifidobacterium adolescentis ATCC 15703
		Bifidobacterium animalis Bb-12
		Bifidobacterium breve
		Bifidobacterium essencis
3.	*Bacillus*	*Bacillus lactis* DR10
4.	*Enterococcus*	*Enterococcus faecium*
5.	*Saccharomyces*	*Saccharomyces boulardi*
6.	*Pediococcus*	*Pediococcus acidilactici*

1. Culture-dependent approaches. This is a traditional approach to cultivate microorganisms from given source.
   (a) Sampling of source (fermented foods in our case), homogenized representative samples were taken.
   (b) Cultivation on medium. The medium choice depends on group of bacteria you want to isolate, e.g., PCA medium, KAA medium, MSE medium, FH medium, and MRS medium for total aerobic count, enterococci, leuconostocs, mesophilic lactobacillus, and thermophilic lactobacillus, respectively (Giraffa 2004).
   (c) Bacterial strains were differentiated on the bases of morphological and biochemical characteristics (Phenotypic basis). Enumeration should follow acceptable rage between 25–250 and 30–300 of colonies on agar plate (Davis 2014).
   (d) Bacterial enumeration selectivity can be improved by altering incubation time, temperature, pH, and nutrient composition of the medium. (Nowadays, the medium for separate enumeration of bacterial species (*Lactobacillus* and *Bifidobacterium* species) is also available.)
   (e) Pure culture of bacterial species was obtained using standard streak plate method followed by DNA extraction, PCR amplification using 16S rRNA gene primers, and sequencing to confirm the identity of bacterial species (Kumar et al. 2015).

Yes, a culture-dependent technique is commonly employed to enumerate microbial population, but only accounts for metabolically active cells and requires experienced persons for reliable results. Additionally, there is limited number of selective differential medium to enumerate bacterial species of probiotic interest. To date, no single culture-dependent approach is available to cultivate all microorganisms of probiotic interest due to significant variation in species and even in strain (Davis 2014), which remains a challenge to scientific community.

2. Culture-independent methods. To define microbial diversity in given environment, culture-independent methods were developed, which exclude the isolation of microbial colonies (Giraffa 2004). These techniques include (Davis 2014):
   (a) Direct imaging and visual enumeration—FISH
   (b) Flow cytometry (FC)
   (c) Nucleic acid-based enumeration methods
   (d) Next-generation sequencing (metagenomics—the recent advancement in technology)

   Although culture-dependent techniques are widely used to isolate microorganisms, it however faces challenges of cultivability and reproducibility and may not account for true microbial population. To circumvent this, various culture-independent techniques like denaturing gradient gel electrophoresis (DGGE) and restriction analysis of amplified ribosomal DNA, followed by sequencing of 16S rRNA genes, were additionally employed to enumerate microbial community. However these approaches are unable to predict true genotypic diversity, because of no correlation between diversity of 16S rRNA gene sequences and genotypic and phenotypic diversity (Jung et al. 2011). In this context, metagenomics approach could serve as a powerful tool by providing access to the phylogenetic diversity of complex microbial assemblages, plentiful information (free of PCR bias) about metabolic potential, gene content, and the function of microbial communities (Jung et al. 2011; Almeida et al. 2014). Recently developed 454-pyrosequencing metagenomics technique has enabled the mining of microbial communities from environmental habitat such as soil, coastal water, mines, and fermented foods (cheese and kimchi) (Jung et al. 2011; Lessard et al. 2014). Metagenomics makes it possible to access large biodiversity beyond and within the species level.

19.5 Exploiting Microbes and Metagenome of Fermented Foods for Probiotics

To identify microbial communities associated with fermented foods through metagenomics (massive parallel pyrosequencing method) (Nam et al. 2012; Leite et al. 2012):

- The first very step is to collect homogenized samples.
- The collected sample representative of all the cells is subjected for environmental/total DNA isolation followed by PCR amplification of 16S rRNA regions.
- PCR amplicon with sample-specific barcode sequences is pooled equally and further assessed for quality and quantity of DNA to ensure the removal of smaller fragments prior to emulsion PCR.

DNA (equally pooled) is amplified using emulsion PCR and subjected for pyrosequencing. These high-throughput sequencing techniques like pyrosequencing and illumina sequencing generate shorter DNA fragments, approx. 25–500 bp, and exclude cloning of DNA prior to sequencing, a prerequisite step in shotgun sequencing method. Moreover, these (NGS) techniques produce high volume of data in very short period of time.

- The sequences obtained are further analyzed through application of bioinformatics tool such as to produce high quality sequences (by excluding low quality sequences), assignment of operational taxonomic units (OTU), and comparing against the database.

Then after functionalities of bacteria strains can be predicted (Alkema et al. 2015). After knowing the function of novel bacterial strains, specific culture conditions are required to isolate the bacteria to know the probiotic potential of that particular type of strain. Although metagenomics describes overall bacterial diversity, cultural methods provide proof.

Researchers reported bacterial diversity associated with different fermented foods using metagenomics chiefly pyrosequencing method. The bacterial community of *cheonggukjang* (Korean fermented food derived from soybean) was predominant by *Bacillus* species; however, unclassified *Bacillus* and LAB (lactic acid bacteria) were also present (Nam et al. 2012). The microbial diversity of Brazilian kefir grains was dominated by 96% of *Lactobacillus*, while *Leuconostoc*, *Lactococcus*, *Acetobacter*, and *Streptococcus* were at low levels (Leite et al. 2012). Further, metagenome analysis of cocoa bean fermentation revealed dominance by *Hanseniaspora uvarum*, *Hansenias poraopuntiae*, *Saccharomyces cerevisiae*, *Lactobacillus fermentum*, and *Acetobacter pasteurianus*. Other genera such as *Erwinia*, *Lactobacillus*, *Lactococcus*, *Leuconostoc*, and *Oenococcus* were also revealed (Illeghems et al. 2012). Jung et al. studied the metagenome of kimchi (traditional Korean fermented food) which showed the presence of *Leuconostoc* and *Lactobacillus* genomes which were highly represented (Jung et al. 2011). The metagenome of fermented foods confirmed the presence of *Lactobacillus* in majority, proving their importance in fermentation as well as probiotics.

LAB strains from fermented foods are tolerant to bile salt and low pH and inhibit foodborne pathogens. These are similar to gut microorganisms *Lactobacillus acidophilus* and *Lactobacillus casei*, commonly used as probiotics (Nuraida 2015). The bacterial strains *Lactococcus lactis* IS-16183 and *Lactobacillus rhamnosus* IS-7257 from fermented food Dadiah showed adherence to mucus layer, caco-2 cells, and inhibited *Escherichia coli* O157:H7 adhesion and are important probiotic properties similar to commercial probiotics strains (Dharmawan et al. 2006). Further promising health-promoting effects like stimulation of the immune system, diarrhea prevention, removal of microcystin-LR a cyclic heptapeptide hepatotoxin, and anticancer and hypocholesterolemic effects have been observed by bacterial strains isolated from fermented foods (Leite et al. 2012). Human-based studies are needed to check the efficacy of probiotics, which are still limited. Therefore, fermented foods are rich source of efficient probiotic strains which need to be explored. The application of metagenomics techniques provides a right way to explore the unexplored.

19.6 Fermented Foods, Microbiota, and Human Health

The word *fermentation* has its origin from a Latin word *fermentare*, which means "to leaven." Fermentation is a microbial-mediated metabolic process, which converts carbohydrate molecules to alcohol, gases, and organic acids. It is a type of preservation, allowing a food to be shelf stable without rotting or degrading. Different types of mold (including *Penicillium* species used for cheeses), bacteria (especially *Lactobacillus* and *Acetobacter* species), and yeast (including the *Saccharomyces* family) are all used to produce different types of fermented foods. The fermented food provides palatability and nutritional value, preserves food, and has medicinal value (Selhub et al. 2014).

Fermentation involves the action of desirable microorganisms, or their enzymes such as amylases, proteases, and lipases (Steinkraus 2002) which hydrolyze the lipid, polysaccharides, and proteins to nontoxic products. Fermentation brings significant biochemical changes along with improvement in flavor, texture, and aroma of the food, which makes the food more attractive for human consumption. However, product originated as a result of enzyme activities is not considered safe for consumption, if it is disease causing/toxic or contains unpleasant flavor, texture, and aroma (Steinkraus 1996).

19.6.1 What Does Fermentation Do?

- *Preserves food*: Fermentation is an energy efficient and cheap means of preserving raw materials. Preservation is achieved by inhibiting growth of food-deteriorating microorganisms like bacteria and fungi (mold/yeast). The process of fermentation allows foods to preserve for longer time due to the production of alcohol and organic acids like lactic acid, acetic acid, etc. (Breidt et al. 2013). Examples include wine, cheese, pickled vegetables, and salami (Porto-Fett et al. 2010; Magalhaes et al. 2010; Perez-Diaz et al. 2013).
- *Supplements microorganism to the gut*: Fermented foods are good/rich source of live "beneficial" bacteria (Kirov et al. 2010). Some of the examples are yogurt, sauerkraut, kimchi, kombucha tea, and kefir (Farnworth 2005; Goh et al. 2012; Sfakianakis and Tzia 2014; Enwa 2014; Wadamori et al. 2014).
- *Enhances nutrient (micronutrients) supply*: Microorganisms hydrolyze polymeric compounds (polysaccharide) and nutritional by-products, e.g., vitamins and organic acids, which improve digestibility. The quantity of vitamin B12 is high in palm wine in West Africa, which is very important for people primarily dependent on a vegetarian diet or people with low meat intake. Other than this southern African sorghum beer and *Idli* (fermented product of Indian origin) contain higher concentration of riboflavin, nicotinic acid and (Ghosh and Chattopadhyay 2011; Ratnavathi et al. 2013; Sathe and Mandal 2015).
- *Makes food more digestible*: Microorganisms produce certain kind of enzymes, such as cellulases, which are not normally synthesized by humans. Cellulose is hydrolyzed into sugars by the action of microbial cellulases, which are not nor-

mally digestible by humans. Pectinase is another enzyme that softens the food texture and releases sugars for digestion. Unfermented foods are hard to digest as compared to fermented foods.

- *Changes taste*: Fermented food is pleasantly sour or tangy and develops flavor (fermentation of sugars produces acids (lactic/acetic acids)), which makes food more acidic. Whereas in some cases, bitterness is reduced by enzymatic action. Moreover, the characteristic flavor of vanilla and chocolate is also due to fermentation (Camu et al. 2008; Hansen et al. 2014; Gu et al. 2015).
- *Anti-nutrient removal*: Fermentation eliminates/destroys anti-nutrient compounds. The adsorption of iron and zinc when eaten is reduced by binding of phytic acid, which is present in seeds and legumes. However, the minerals become available only after the breakdown of phytic acid during fermentation. Examples of fermented legumes are miso and tempeh (Salem et al. 2014).
- *Decreases cooking times*: Proteins and carbohydrate food are soft because of complex changes. The fermented foods reduce the need for cooking and fuel as compared to unfermented foods that are tough, difficult to digest (Sekar and Mariappan 2007).
- *Produces carbon dioxide*: The carbon dioxide is produced during the formation of ethanol by fermented yeast. The carbon dioxide is used for the production of carbonated drinks (champagne and beer) leavening bread (Conlon and Bird 2015).
- *Increases aroma and color*: Fermentation enhances acidity and reduces sweetness and bitterness. Fermented food products are rich in protein, essential amino acids, as well as fatty acids. Proteolytic activity, degradation of chlorophyll, and enzymatic browning may produce brown pigments (Ba et al. 2012).

The use of fermented food products is ascertained back to 10,000 years ago. According to history of fermented products (Table 19.3), beer and fermented milk products are being utilized in Babylon since 3000 BC. However, market for fermented products was established in 1941–1946 such as antibiotics and germ warfare. After that, there is nonstop increase in industrial utilization of microorganisms and in market of fermented food products till now (Ross et al. 2002a, b).

19.7 Microbiota of Human

The human body is inhabited with large number of viruses, bacteria, unicellular eukaryotes, and archaea. The microorganisms that are normally coexistence within their hosts are known as microbiota, microflora, or normal flora (Kirov et al. 2010). If we measured the entire DNA that makes "us," the majority of it would not be our own DNA, but rather that of a wide variety of bacteria. The number of beneficial microorganism can be increased by ingestion of fermented food containing live microorganisms, when administered in adequate quantities; deliberate health benefits to the host is known as "probiotic" (Table 19.4).

Table 19.3 History of the use of fermentation product

Time	Fermentation product	Place
Antiquity	Bread, vinegar, soy sauce, wine, beer	
7000 B.C.	Beer and wine	Assyria, Caucasia, Mesopotamia, Sumer
6000 B.C.	Winemaking	Georgia
5000 B.C.	Wine jars	Zagros Mountains, Iran
	Fermented beverages	Babylon
3000 B.C.	Beer and fermented milk products	Babylon
2600 B.C.	Bread	Egypt
1000 B.C.	Soy sauce and miso	China
600 B.C.	Cheese	Asia
500 B.C.	Preservation of fish and meat	
100 B.C.	Bread	Ancient Rome
Modern times		
1700s	Vinegar from fruit pomace	
	Gallic acid	
1800s	Yeast induce fermentation	Erxleben, Germany
1850s	1. Bacteria produce lactic acid which conserve food 2. Pasteurization—heat treatment to prevent unwanted fermentation 3. Yeast + grape juice to wine—beginning of the science of food fermentation	Louis Pasteur, France
1907	Publication of book *Prolongation of Life* by Eli Metchnikoff describing therapeutic benefits of fermented milks	
1900–1930	Application of microbiology to fermentation, use of defined cultures	
1970–present	Development of products containing probiotic cultures or friendly intestinal bacteria	

1. *Flatulence-reducing effect* (Hesseltine 1983): For the preparation of tempeh from bean fermentation, inactivation of trypsin inhibitor is important. This result in reduction of several oligosacharides which usually cause flatulence.
2. *Anticholesterolemic effect* (Ooi and Liong 2010): The fermented yogurt has the ability to lower cholesterol levels.
3. *Effect on transit time, bowel function, and glycemic index* (Strandhagen et al. 1994).
4. *Anticarcinogenic effect* (Nami et al. 2014) and *immunoactive effects* (Yan and Polk 2011; Wedajo 2015).
5. *Balance the production of stomach acid*: Main property of fermented foods is to remove digestive discomfort due to stomach acid. Fermented food improves acidity of gastric juices even in low concentration of HCl. They also protect the stomach if it produces too much acid (Tamime 2002).
6. *Foods help the body produce acetylcholine* (Jeudi 2015).

7. *Beneficial for people with diabetes* (Selhub et al. 2014; Chilton et al. 2015).
8. *Produce unknown compounds*: In the early 1950s, fresh sauerkraut has been proved as an effective agent for killing pathogenic bacteria during typhoid fever epidemics in Europe (Rattanachaikunsopon and Phumkhachorn 2010). Because, maximum number of pathogenic bacteria are sensitive to acidic environment.

Table 19.4 Advantages of probiotic bacterial strains

Microflora	Associated actions	Reference
Bifidobacteria species	*Bifidobacterium* used as probiotic because of resistant to variety of bile salts. This bacteria is mainly used for the treatment of constipation travelers' diarrhea, antibiotic-associated diarrhea, maintaining remission of disease activity of gut inflammation and moderate ulcerative colitis, prevention as well as treatment of necrotizing enterocolitis in newborns, reduction of radiation-induced diarrhea, reducing the development of disease risk for eczema, treatment of food allergies, and cholesterol lowering capacities	Fijan (2014)
Enterococcus faecium	Duration of acute diarrhea is decreased from gastroenteritis.	Caramia et al. (2015), Riddle et al. (2016)
Lactobacillus strains	• *Lactobacillus* generally present in the gastrointestinal tract and along with *Bifidobacterium* are the first bacteria to be colonized into the gut of new born baby	Matamoros et al. (2013)
	• Lactose digestion improved/decreased diarrhea and symptoms of intolerance in lactose-intolerant individuals, children with diarrhea, and individuals with short bowel syndrome	Marteau et al. (2001), Parvez et al. (2006)
	• Some lactobacilli are used for the production of yogurt, cheese, sauerkraut, pickles, sourdough, wine, and other fermented products. In all cases, sugars are metabolized into lactic acid, thus creating a hostile environment for spoilage microorganisms and enabling food preservation	
	• Microbial interference therapy—the use of nonpathogenic bacteria to eliminate pathogens and as an adjunct to antibiotics	Parvez et al. (2006)
	• Improved mucosal immune function, mucin secretion, and prevention of disease	Hardy et al. (2013)
Lactobacillus acidophilus	• Significant decrease of diarrhea in patients receiving pelvic irradiation. Decreased polyps, adenomas, and colon cancer in experimental animals. Prevented urogenital infection with subsequent exposure to three uropathogens: *Escherichia coli*, *Klebsiella pneumoniae*, *Pseudomonas aeruginosa*	Marteau et al. (2001), Abdulamir et al. (2011)
	• Lowered serum cholesterol levels	Ooi and Liong (2010)

(continued)

Table 19.4 (continued)

Microflora	Associated actions	Reference
Lactobacillus plantarum	• Reduced incidence of diarrhea in daycare centers when administered to only half of the children	Riddle et al. (2016)
	• Especially effective in reducing inflammation in inflammatory bowel, e.g., enterocolitis in rats, small bowel bacterial overgrowth in children, and pouchitis	Parvez et al. (2006), Marteau et al. (2001)
	• Reduced pain and constipation of irritable bowel syndrome	
	• Reduced bloating, flatulence, and pain in irritable bowel syndrome in controlled trial	Korpela and Niittynen (2012)
	• Positive effect on immunity in HIV+ children	Monachese et al. (2011)
	• It has the ability to destroy pathogens and to preserve critical nutrients, vitamins, and antioxidant. It has also shown the rare ability to produce L. lysine, a beneficial amino acid	
Lactobacillus casei	• *Lactobacillus casei* is a beneficial bacterium that is found naturally in both the mouth and intestine of human being • It is able to improve and promote digestion, control diarrhea, has anti-inflammatory effect on the gut, reduces lactose intolerance, alleviates constipation, and even modulates the immune system	Wedajo (2015)
Lactobacillus reuteri	• Shortened the duration of acute gastroenteritis by reducing the growth of harmful *E.coli* bacteria	Marteau et al. (2001)
	• Shortened acute diarrhea	Lin et al. (2008)
Lactobacillus rhamnosus	• Enhanced cellular immunity in healthy adults in controlled trial	Wen et al. 2014
	• It is also used as a natural preservative in yogurt-based products, where the bacterium attaches to the lining of the intestine, where it encourages the growth of helpful organism that aid indigestion	Yan and Polk (2011)
Lactobacillus salivarius	Suppressed and eradicated *Helicobacter pylori* in tissue cultures and animal models by lactic acid secretion	Parvez et al. (2006)
Bacteroides species	Chronic colitis, gastritis, arthritis (increased bacterial urease activity in chronic juvenile arthritis)	Schaubeck et al. (2016)
Saccharomyces boulardii (yeast)	Reduced recurrence of *Clostridium difficile* diarrhea	Tung et al. (2009)
	Effects on *C. difficile* and *Klebsiella oxytoca* resulted in decreased risk and/or shortened duration of antibiotic-associated diarrhea	Marteau et al. (2001)
	Shortened the duration of acute gastroenteritis	Marteau et al. (2001)
	Decreased only functional diarrhea, but not any other symptoms of irritable bowel syndrome	Marteau et al. (2001)

Probiotic microbes of fermented foods include bacteria, yeast, and mold (Penner et al. 2005; Parvez et al. 2006; Sullivan and Nord 2005; Singh et al. 2014; Mummah et al. 2014). The benefits and pitfalls of fermented foods have been well described in literature (Rhee et al. 2011; Selhub et al. 2014; Salem et al. 2014; Chilton et al. 2015).

Benefits/pitfalls	Description
Benefits	
General advantages	• Unique texture and flavors of food • Low operating costs and energy consumption • Easy and simple technologies
Inhibition of pathogens and spoilage organisms	• Mostly food is fermented by lactic acid resulting in lowering of pH to 4 • Bacteriocins production • Hydrogen peroxide production • Ethanol and diacetyl production
Detoxification and softening	Reduce the natural toxins in plant food
Digestibility of oligosaccharides and dietary fibers is enhanced	• Foods having amylase-rich flour and LAB • Phytate transformation in fermented plant food. This increases severalfold the bioavailability of iron • Increases in protein digestibility and minerals absorption of grains due to lactic acid fermentation
Beneficial health effects	Consumption of probiotic results: • Protection from *E. coli* and other pathogens • Enhance hypocholesterolemic and anticarcinogenic effects • Enhance food safety
Pitfalls	
Complex and sensitive	• Quality and safety of raw materials • Environment hygiene • Metabolites • Processing
Risk of contamination	Final product formation depends upon the condition of fermentation. If conditions are not properly maintained, there are chances of pathogen growth. It may also result to foul smell and bad taste in final product
Risk of intoxication	Several cases of botulism were found in 1980s when fermentation began to carry out in plastic containers

Conclusion

The importance of fermented food to humankind is well known worldwide. Diverse fermented products are in use since antiquity because of high nutritional, therapeutic, and antimicrobial values, imparted by fermentative microbes, so-called probiotics. Fermented food and their products with promising probiotic effects offer an alternative solution to fight against emerging antibiotic-resistant microorganism. However, the most frequently used probiotic microbes are limited to genera; *Lactobacillus* and *Bifidobacterium* warrant further research to explore more probiotic bacterial strains. In this context, next-generation sequencing (pyrosequencing) can be promising in exploring the novel microbial community of fermented foods, thereby investigation of its probiotic potential. Research needs to be focused on identification of traditional fermented food of every region of the country/world, investigating its microbial community, antibiotic resistance, metabolic properties, and beneficial health impacts associated. The identified novel probiotic bacteria will be accessed for its genes, structure, and function of the product through metagenomic analysis.

Acknowledgments The authors wish to thank Dr. Yashwant Singh Parmar University of Horticulture and Forestry, for providing necessary funds and facilities.

References

Abdulamir AS, Hafidh RR, Bakar FA (2011) The association of *Streptococcus bovis/gallolyticus* with colorectal tumors: the nature and the underlying mechanisms of its etiological role. J Exp Clin Cancer Res 30(1):11. doi:10.1186/1756-9966-30-11

Alkema W, Boekhorst J, Wels M, Van Hijum SA (2015) Microbial bioinformatics for food safety and production. Brief Bioinform 34. doi:10.1093/bib/bbv034

Almeida M, Hébert A, Abraham AL, Rasmussen S, Monnet C, Pons N et al (2014) Construction of a dairy microbial genome catalog opens new perspectives for the metagenomic analysis of dairy fermented products. BMC Genomics 15:1. doi:10.1186/1471-2164-15-1101

Awaisheh SS (2011) Development of probiotic soft cheese manufactured using goat's milk with the addition of thyme. Milchwissenschaft 66:51–54

Awaisheh SS, Al-Dmoor HM, Omar SS, Hawari A, Al-Rwaily MM (2012) Impact of selected nutraceuticals on viability of probiotic strains in milk during refrigerated storage at 4°C for 15 days. Int J Dairy Technol 65:268–273. doi:10.1111/j.1471-0307.2011.00817.x

Ba HV, Hwang I, Jeong D, Touseef A (2012) Principle of meat aroma flavors and future prospect. doi: 10.5772/51110

Bokulich NA, Mills DA (2012) Next-generation approaches to the microbial ecology of food fermentations. BMB Rep 45:377–389. doi:10.5483/BMBRep.2012.45.7.148

Breidt F, McFeeters RF, Diaz IP, Lee CH (2013) Fermented vegetables. Food Microbiol 4:841–855. doi:10.1128/9781555818463

Camu N, Winter TD, Addo SK, Takrama JS, Bernaert H, Vuyst LD (2008) Fermentation of cocoa beans: influence of microbial activities and polyphenol concentrations on the flavour of chocolate. J Sci Food Agric 88:2288–2297. doi:10.1002/jsfa.3349

Caramia G, Silvi S, Verdenelli MC, Coman MM (2015) Treatment of acute diarrhoea: past and now. Int J Enteric Pathog 3(4):e28612. doi:10.17795/ijep28612

Chilton SN, Burton JP, Reid G (2015) Inclusion of fermented foods in food guides around the world. Nutrients 7:390–404. doi:10.3390/nu7010390

Conlon MA, Bird AR (2015) The impact of diet and lifestyle on gut microbiota and human health. Nutrients 7:17–44. doi:10.3390/nu7010017

Conway PL, Gorbach SL, Goldin BR (1987) Survival of lactic acid bacteria in the human stomach and adhesion to intestinal cells. J Dairy Sci 70:1–12. doi:10.3168/jds.S0022-0302(87)79974-3

Cruz AG, Antunes AE, Sousa ALOP, Faria JAF, Saad SMI (2009) Ice-cream as a probiotic food carrier. Food Res Int 42:1233–1239. doi:10.1016/j.foodres.2009.03.020

Davis C (2014) Enumeration of probiotic strains: review of culture-dependent and alternative techniques to quantify viable bacteria. J Microbiol Methods 103:9–17. doi:10.1016/j.mimet.2014.04.012

Dharmawan J, Surono IS, Kun LY (2006) Adhesion properties of indigenous dadih lactic acid bacteria on human intestinal mucosal surface. Asian Aust J Anim Sci 19:751. doi:10.5713/ajas.2006.751

Enwa FO (2014) A mini review on the microbiochemical properties of sauerkraut. Afr J Sci Res 3:15–16

Ercolini D (2013) High-throughput sequencing and metagenomics: moving forward in the culture-independent analysis of food microbial ecology. Appl Environ Microbiol 79:3148–3155. doi:10.1128/AEM.00256-13

Farnworth ER (2005) Kefir-a complex probiotic. Food Sci Technol Bull Funct Foods 2:1–17. doi:10.1616/1476-2137.13938

Farnworth ER, Mainville I, Desjardins MP, Gardner N, Fliss I, Champagne C (2007) Growth of probiotic bacteria and *Bifidobacteria* in a soy yogurt fomulation. Int J Food Microbiol 116:174–181. doi:10.1016/j.ijfoodmicro.2006.12.015

Fijan S (2014) Microorganisms with claimed probiotic properties: an overview of recent literature. Int J Environ Res Public Health 11(5):4745–4767. doi:10.3390/ijerph110504745

Franz CM, Huch M, Abriouel H, Holzapfel W, Gálvez A (2011) Enterococci as probiotics and their implications in food safety. Int J Food Microbiol 151:125–140. doi:10.1016/j.ijfoodmicro.2011.08.014

Ghosh D, Chattopadhyay P (2011) Preparation of idli batter, its properties and nutritional improvement during fermentation. J Food Sci Technol 48:610–615. doi:10.1007/s13197-010-0148-4

Giraffa G (2004) Studying the dynamics of microbial populations during food fermentation. FEMS Microbiol Rev 28:251–260. doi:10.1016/j.femsre.2003.10.005

Goh WN, Rosma A, Kaur B, Fazilah A, Karim AA, Bhat R (2012) Fermentation of black tea broth (Kombucha): I. Effects of sucrose concentration and fermentation time on the yield of microbial cellulose. Int Food Res J19:109–117

Granato D, Branco GF, Nazzaro F, Cruz AG, Faria JAF (2010) Functional foods and nondairy probiotic food development: trends, concepts, and products. Compr Rev Food Sci Food Saf 9:293–302. doi:10.1111/j.1541-4337.2010.00110.x

Gu F, Chen Y, Fang Y, Wu G, Tan L (2015) Contribution of *Bacillus* isolates to the flavor profiles of vanilla beans assessed through aroma analysis and chemometrics. Molecules 20:18422–18436. doi:10.3390/molecules201018422

Guzel-Seydim Z, Twyffels J, Seydim C, Greene K (2005) Turkish kefir and kefir grains: microbial enumeration and electron microscopic observation. Int J Dairy Technol 58:25–29. doi:10.1111/j.1471-0307.2005.00177.x

Hansen AMS, Fromberg A, Frandsen HL (2014) Authenticity and traceability of vanilla flavour by analysis of stable isotopes. J Agric Food Chem 62(42):10326–10331. doi:10.1021/jf503055k

Hardy H, Harris J, Lyon E, Beal J, Foey AD (2013) Probiotics, prebiotics and immunomodulation of gut mucosal defences: homeostasis and immunopathology. Nutrients 5:1869–1912. doi:10.3390/nu5061869

Hayes M, Ross RP, Fitzgerald GF, Stanton C (2007) Putting microbes to work: dairy fermentation, cell factories and bioactive peptides. Part I: Overview. Biotechnol J 2:426–434. doi:10.1002/biot.200600246

Health Canada (2007) List of licensed natural health products. Available at http://www.hc-sc.gc.ca/dhp-mps/prodnatur/applications/licen-prod/lists/listapprnhp-listeapprpsn_e.html

Hekmat S, Soltani H, Reid G (2009) Growth and survival of *Lactobacillus reuteri* RC-14 and *Lactobacillus rhamnosus* GR-1 in yogurt for use as a functional food. Innov Food Sci Emerg Technol 10:293–296. doi:10.1016/j.ifset.2008.10.007

Heller KJ (2001) Probiotic bacteria in fermented foods: product characteristics and starter organisms. J Clin Nutr 73:374S–379S

Hennequin C, Kauffmann-Lacroix C, Jobert A, Viard JP, Ricour C, Jacquemin JL, Berche P (2000) Possible role of catheters in *Saccharomyces boulardiifungemia*. Eur J Clin Microbiol Infect Dis 19:16–20. doi:10.1007/s100960050003

Hesseltine CW (1983) The future of fermented foods. Nutr Rev 41:293–301. doi:10.1111/j.1753-4887.1983.tb07122.x

Huys G, Botteldoorn N, Delvigne F, De Vuyst L, Heyndrickx M, Pot B, Dubois JJ, Daube G (2013) Microbial characterization of probiotics–advisory report of the Working Group “8651 Probiotics” of the Belgian Superior Health Council (SHC). Mol Nutr Food Res 57:1479–1504. doi:10.1002/mnfr.201300065

Illeghems K, DeVuyst L, Papalexandratou Z, Weckx S (2012) Phylogenetic analysis of a spontaneous cocoa bean fermentation metagenome reveals new insights into its bacterial and fungal community diversity. PLoS One 7:e38040. doi:10.1371/journal.pone.0038040

Jeudi (2015) Fermented foods can dramatically improve your digestive health and immunity. Ecrit par Cybermed 1–2

Jung JY, Lee SH, Kim JM, Park MS, Bae JW, Hahn Y, Madsen EL, Jeon CO (2011) Metagenomic analysis of kimchi, a traditional Korean fermented food. Appl Environ Microbiol 77:2264–2274. doi:10.1128/AEM.02157-10

Kirov I, Russell SL, Antunes LCM, Finlay BB (2010) Gut microbiota in health and disease. Am Physiol Soc 90:859–904. doi:10.1152/physrev.00045.2009

Korpela R, Niittynen L (2012) Probiotics and irritable bowel syndrome. Microb Ecol Health Dis 23:18573. doi:10.3402/mehd.v23i0.18573

Kumar A, Guleria S, Mehta P, Walia A, Chauhan A, Shirkot CK (2015) Plant growth-promoting traits of phosphate solubilizing bacteria isolated from *Hippophae rhamnoides* L. (Seabuckthorn) growing in cold desert Trans-Himalayan Lahul and Spiti regions of India. Acta Physiol Plant 37:1–12. doi:10.1007/s11738-015-1793-z

Lavermicocca P (2006) Highlights on new food research. Dig Liver Dis 38:S295–S299. doi:10.1016/S1590-8658(07)60014-0

Leite A, Mayo B, Rachid C, Peixoto R, Silva J, Paschoalin VM, Delgado S (2012) Assessment of the microbial diversity of Brazilian kefir grains by PCR-DGGE and pyrosequencing analysis. Food Microbiol 31:215–221. doi:10.1016/j.fm.2012.03.011

Lessard MH, Viel C, Boyle B, St-Gelais D, Labrie S (2014) Metatranscriptome analysis of fungal strains *Penicillium camemberti* and *Geotrichum candidum* reveal cheese matrix breakdown and potential development of sensory properties of ripened Camembert-type cheese. BMC Genomics 15:235. doi:10.1186/1471-2164-15-235

Lin YP, Thibodeaux CH, Peña JA, Ferry GD, Versalovic J (2008) Probiotic *Lactobacillus reuteri* suppress proinflammatory cytokines via c-Jun. Inflamm Bowel Dis 14:1068–1083. doi:10.1002/ibd.20448

Liu DM, Li L, Yang XQ, Liang SZ, Wang JS (2006) Survivability of *Lactobacillus rhamnosus* during the preparation of soy cheese. Food Technol Biotechnol 44:417–422

Mackay AD, Taylor MB, Kibbler CC, Hamilton-Miller JMT (1999) *Lactobacillus endocarditis* caused by a probiotic organism. Clin Microbiol Infect 5:290–292. doi:10.1111/j.1469-0691.1999.tb00144.x

Magalhaes KT, Pereira MA, Nicolau A, Dragone G, Domingues L, Teixeira JA, Silva JBDA, Schwan RF (2010) Production of fermented cheese whey-based beverage using kefir grains as starter culture: evaluation of morphological and microbial variations. Bioresour Technol 101:8843–8850. doi:10.1016/j.biortech.2010.06.083

Marteau P, de Vrese M, Cellier CJ, Schrezenmeir J (2001) Protection from gastrointestinal diseases with the use of probiotics. Am J Clin Nutr 73:430–436

Matamoros S, Gras-Leguen C, LeVacon F, Potel G, deLaCochetiere MF (2013) Development of intestinal microbiota in infants and its impact on health. Trends Microbiol 21:167–173. doi:10.1016/j.tim.2012.12.001

Metchnikoff II (2004) Introduction to reprinted edition, 1977. In: Metchnikoff II (ed) The prolongation of life: optimistic studies. Springer Publishing Company, New York, p xxvi

Monachese M, Cunningham-Rundles S, Diaz M, Guerrant R, Hummelen R, Kemperman R, Kerac M, Kort R, Merenstein DJ, Panigrahi P, Ramakrishna B (2011) Probiotics and prebiotics to combat enteric infections and HIV in the developing world: a consensus report. Gut Microbes 2:198–207. doi:10.4161/gmic.2.3.16106

Mummah S, Oelrich B, Hope J, Vu Q, Gardner CD (2014) Effect of raw milk on lactose intolerance: a randomized controlled pilot study. Ann Fam Med 12:134–141. doi:10.1370/afm.1618

Nam YD, Yi SH, Lim SI (2012) Bacterial diversity of cheonggukjang, a traditional Korean fermented food, analyzed by barcoded pyrosequencing. Food Control 28:135–142. doi:10.1016/j.foodcont.2012.04.028

Nami Y, Abdullah N, Haghshenas B, Radiah D, Rosli R, Khosroushahi AY (2014) Assessment of probiotic potential and anticancer activity of newly isolated vaginal bacterium *Lactobacillus plantarum* 5BL. Microbiol Immunol 58:492–502. doi:10.1111/1348-0421.12175

Nuraida L (2015) A review: health promoting lactic acid bacteria in traditional Indonesian fermented foods. Food Sci Human Wellness 4:47–55. doi:10.1016/j.fshw.2015.06.001

Oggioni MR, Pozzi G, Balensin PE, Galieni P, Bigazzi C (1998) Recurrentsepticemia in an immunocompromised patient due to probiotic strains of *Bacillus subtilis*. J Clin Microbiol 36:325–326

Ogier JC, Serror P (2008) Safety assessment of dairy microorganisms: the *Enterococcus* genus. Int J Food Microbiol 126:291–301. doi:10.1016/j.ijfoodmicro.2007.08.017

Ong L, Henriksson A, Shah NP (2006) Development of probiotic cheddar cheese containing *Lactobacillus acidophilus*, *Lactobacillus casei*, *Lactobacillus paracasei* and *Bifidobacterium* spp. and the influence of these bacteria on proteolytic patterns and production of organic acid. Int Dairy J 16:446–456. doi:10.1155/2014/801918

Ooi LG, Liong MT (2010) Cholesterol-lowering effects of probiotics and prebiotics: a review of *in vivo* and *in vitro* findings. Int J Mol Sci 11:2499–2522. doi:10.3390/ijms11062499

Parvez S, Malik KA, Ah Kang S, Kim HY (2006) Probiotics and their fermented food products are beneficial for health. J Appl Microbiol 100:1171–1185. doi:10.1111/j.1365-2672.2006.02963

Penner R, Fedorak RN, Madsen KL (2005) Probiotics and nutraceuticals: non-medicinal treatments of gastrointestinal diseases. Curr Opin Pharmacol 5:596–603. doi:10.1016/j.coph.2005.06.009

Pereira ALF, Maciel TC, Rodrigues S (2011) Probiotic beverage from cashew apple juice fermented with *Lactobacillus casei*. Food Res Int 44:1276–1283. doi:10.1016/j.foodres.2010.11.035

Peres CM, Peres C, Hernández-Mendoza A, Malcata FX (2012) Review on fermented plant materials as carriers and sources of potentially probiotic lactic acid bacteria–with an emphasis on table olives. Trends Food Sci Technol 26:31–42. doi:10.1016/j.tifs.2012.01.006

Perez-Diaz IM, Breidt F, Buescher RW, Arroyo-Lopez FN, Jimenez-Diaz R, Fernandez AG, Gallego JB, Yoon SS, Johanningsmeier SD (2013) Fermented and acidified vegetables. In: Salfinger Y, Tortorello ML (eds) Compendium of methods for the microbiological examination of foods, vol 1. APHA Press, Washington, DC, pp 1–22. doi:10.2105/MBEF.0222.056

Pineiro M, Stanton C (2007) Probiotic bacteria: legislative framework requirements to evidence basis. J Nutr 137:850S–853S

Porto-Fett ACS, Call JE, Shoyer BE, Hill DE, Pshebniski C, Cocoma GJ, Luchansky JB (2010) Evaluation of fermentation, drying, and/or high pressure processing on viability of *Listeria monocytogenes, Escherichia coli* O157:H7, *Salmonella spp,* and *Trichinella piralis* in raw pork and Genoa salami. Int J Food Microbiol 140:61–75. doi:10.1016/j.ijfoodmicro.2010.02.008

Ratnavathi CV, Ranga B, Mahender A, Patila JV (2013) Nutritional analysis of semi processed and processed foods of sorghum. J Nutr Food Sci 3:4. doi:10.4172/2155-9600.1000247

Rattanachaikunsopon P, Phumkhachorn P (2010) Lactic acid bacteria: their antimicrobial compounds and their uses in food production. Ann Biol Res 1:218–228

Rautio M, Jousimies-Somer H, Kauma H, Pietarinen I, Saxelin M, Tynkkynen S, Koskela M (1999) Liver abscess due to a *Lactobacillus rhamnosus* strain indistinguishable from *L. rhamnosus* strain GG. Clin Infect Dis 28:1159–1160. doi:10.1086/514766

Ray RC, Sivakumar PS (2009) Traditional and novel fermented foods and beverages from tropical root and tuber crops: review. Int J Food Sci Technol 44:1073–1087. doi:10.1111/j.1365-2621.2009.01933.x

Reid G (2008) probiotics and prebiotics-progress and challenges. Int Dairy J 18:969–975. doi:10.1016/j.idairyj.2007.11.025

Rhee SJ, Lee JE, Lee CH (2011) Importance of lactic acid bacteria in Asian fermented foods. Microb Cell Fact 10:1–13. doi:10.1186/1475-2859-10-S1-S5

Richard V, Auwera P, Snoeck R, Daneau D, Meunier F (1988) Nosocomial bacteremia caused by *Bacillus* species. Eur J Clin Microbiol Infect Dis 7:783–785. doi:10.1007/BF01975049

Riddle MS, DuPont HL, Connor BA (2016) ACG clinical guideline: diagnosis, treatment, and prevention of acute diarrheal infections in adults. Am J Gastroenterol 111:602–622. doi:10.1038/ajg.2016.126

Roble C, Auty MAE, Brunton N, Gormley RT, Butler F (2010) Evaluation of fresh-cut apple slices enriched with probiotic bacteria. Innov Food Sci Emerg Technol 11:203–209. doi:10.1016/j.ifset.2009.08.016

Röling WF, Apriyantono A, Van Verseveld HW (1996) Comparison between traditional and industrial soy sauce (kecap) fermentation in Indonesia. J Ferment Bioeng 81:275–278. doi:10.1016/0922-338X(96)82223-9

Ross RP, Fitzgerald G, Collins K, Stanton C (2002a) Cheese delivering biocultures-probiotic cheese. Aust J Dairy Technol 57:71–91

Ross RP, Morgan S, Hill C (2002b) Preservation and fermentation: past, present and future. Int J Food Microbiol 79:3–16. doi:10.1016/S0168-1605(02)00174-5

Salem FMA, Mohamed RK, Gibriel AY, Rasmy NMH (2014) Levels of some antinutritional factors in tempeh produced from some legumes and jojobas seeds. Int J Biol Biomol Agric Food Biotechnol Eng 8:296–301

Sanders ME, Akkermans LM, Haller D, Hammerman C, Heimbach JT, Hörmannsperger G, Huys G et al (2010) Safety assessment of probiotics for human use. Gut Microbes 1:164–185. doi:10.4161/gmic.1.3.12127

Sathe GB, Mandal S (2015) Fermented products of India and its implication. Asian J Dairy Food Res 35:1–9. doi:10.18805/ajdfr.v35i1.9244

Schaubeck M, Clavel T, Calasan J, Lagkouvardos I, Haange SB, Jehmlich N, Basic M, Dupont A, Hornef M, von Bergen M, Bleich A (2016) Dysbiotic gut microbiota causes transmissible Crohn's disease-like ileitis independent of failure in antimicrobial defence. Gut 65:225–237. doi:10.1136/gutjnl-2015-309333

Schoenlechner R, Siebenhandl S, Berghofer E (2008) Pseudocereals. In: Arendt EK, Bello FD (eds) Gluten-free cereal products and beverages. Academic, New York, pp 149–190

Sekar S, Mariappan S (2007) Usage of traditional fermented products by Indian rural folks and IPR. Ind J Trad Knowl 6:111–120. http://hdl.handle.net/123456789/840

Selhub EM, Logan AC, Bested AC (2014) Fermented foods, microbiota, and mental health: ancient practice meets nutritional psychiatry. J Physiol Anthropol 33:1–12. doi:10.1186/1880-6805-33-2

Sfakianakis P, Tzia C (2014) Conventional and innovative processing of milk for yogurt manufacture; development of texture and flavor. Foods 3:176–193. doi:10.3390/foods3010176

Sheehan VM, Ross P, Fitzgerald GF (2007) Assessing the acid tolerance and the technological robustness of probiotic cultures for fortification in fruit juices. Innov Food Sci Emerg Technol 8:279–284. doi:10.1186/1880-6805-33-2

Singh S, Kotla NG, Lal UR (2014) Omnipresence of probiotics in diversified clinical practices. J Prob Health 3:1–9. doi:10.4172/2329-8901.1000125

Soccol CR, Vandenberghe LPS, Spier MR, Medeiros ABP, Yamaguishi CT, Lindner JDD et al (2010) The potential of probiotics: a review. Food Technol Biotechnol 48:413–434

Spinosa MR, Wallet F, Courcol RJ, Oggioni MR (2000) The trouble in tracing opportunistic pathogens: cholangitis due to *Bacillus* in a French hospital caused by a strain related to an Italian probiotic? Microb Ecol Health Dis 12:99–101. doi:10.3402/mehd.v12i2.8115

Steinkraus KH (1996) Handbook of indigenous fermented foods. Marcel Decker, New York

Steinkraus KH (2002) Fermentations in world food processing. Compr Rev Food Sci Food Saf 1:23–32. doi:10.1111/j.1541-4337.2002.tb00004.x

Strandhagen E, Lia A, Lindstrand S, Bergstroem P, Lundstroem A, Fonden R, Andersson H (1994) Fermented milk (ropy milk) replacing regular milk reduces glycemic response and gastric emptying in healthy subjects. Scand J Nutr 38:117–121

Sullivan A, Nord CE (2005) Probiotics and gastrointestinal diseases. J Int Med 257:78–92. doi:10.1111/j.1365-2796.2004.01410.x

Tamime AY (2002) Fermented milks: a historical food with modern applications-a review. Eur J Clin Nutr 56:2–15. doi:10.1038/sj.ejcn.1601657

Tung JM, Dolovich LR, Lee CH (2009) Prevention of *Clostridium difficile* infection with Saccharomyces boulardii: a systematic review. Can J Gastroenterol 23:817–821

Vandenplas Y, Huys G, Daube G (2014) Probiotics: an update. J Pediatr (Rio J) 91:6–21. doi:10.1016/j.jped.2014.08.005

Venugopalan V, Shriner KA, Wong BA (2010) Regulatory oversight and safety of probiotic use. Emerg Infect Dis 6:1661–1665. doi:10.3201/eid1611.100574

Vinderola G, Prosello W, Molinari F, Ghiberto D, Reinheimer J (2009) Growth of *Lactobacillus paracasei* A13 in argentinian probiotic cheese and its impact on the characteristics of the product. Int J Food Microbiol 135:171–174. doi:10.1016/j.ijfoodmicro.2009.08.021

Wadamori Y, Vanhanen L, Savage GP (2014) Effect of kimchi fermentation on oxalate levels in silver beet (*Beta vulgaris* var. cicla). Foods 3:269–278. doi:10.3390/foods3020269

Wedajo B (2015) Lactic acid bacteria: benefits, selection criteria and probiotic potential in fermented food. J Prob Health 3:1–9. doi:10.4172/2329-8901.1000129

Wen K, Tin C, Wang H, Yang X, Li G, Giri-Rachman E, Kocher J, Bui T, Clark-Deener S, Yuan L (2014) Probiotic *Lactobacillus rhamnosus* GG enhanced Th1 cellular immunity but did not affect antibody responses in a human gut microbiota transplanted neonatal gnotobiotic pig model. PLoS One 9:e94504. doi:10.1371/journal.pone.0094504

Widyastuti Y, Rohmatussolihat FA (2014) The role of lactic acid bacteria in milk fermentation. Food Nutr Sci 5:435–442. doi:10.4236/fns.2014.54051

Yan F, Polk DB (2011) Probiotics and immune health. Curr Opin Gastroenterol 27:496–501. doi:10.1097/MOG.0b013e32834baa4d

Probiotics from Fermented Foods

20

Sundru Manjulata Devi and Prakash M. Halami

Abstract

Probiotic food products and the beneficial microflora display several nutritional and health-promoting factors. The probiotic microorganisms are found to maintain a balance between the host immunity and gastrointestinal tract. The extensive use of probiotic food products has gained attraction in the global market, leading to an increase of a million dollar annually. Fermented vegetables, meat, plants, and dairy products are widely explored and commercialized in different parts of the world, spreading their proven benefits and a healthy inner ecosystem. The present chapter focuses on different fermented foods and the microbes involved and highlights about the functional properties exerted by probiotic microbes. The fermented foods with identified probiotic strains are found to show antiallergic, antimutagenic, anticarcinogenic, antimicrobial, and other properties. Apart from these, the intake of fermented foods by the consumers has established to stimulate host immunity and is described in this chapter. More industrial and clinical research activities would lead to a mechanism of probiotic strain interaction to host intestinal layer. Presently, the consumption of fermented foods has increased, and understanding the probiotic bacteria and their functionalities will open up the discovery of novel foods with traditional practices and health-stimulating/health-promoting benefits.

Keywords

Probiotics • fermented foods • *Lactobacillus* species • nutrition • food preservation • health benefits

---

S.M. Devi • P.M. Halami (✉)
Microbiology and Fermentation Technology Department, CSIR-Central Food Technological Research Institute, Mysore 570020, Karnataka, India
e-mail: prakashalami@cftri.res.in

V.C. Kalia et al. (eds.), *Mining of Microbial Wealth and MetaGenomics*,
DOI 10.1007/978-981-10-5708-3_20

20.1 Introduction

From ancient Greece to date, the fermented foods have become a part of human nutrition and nourishment, imparting a major role in fermentation, food preservation, and alcohol production (Chilton et al. 2015). Over 5000 different fermented foods have been developed all over the world, either naturally or by the incorporation of starter cultures. Several traditional foods that were fermented naturally such as *kimchi*, *sauerkraut*, Northeast Indian fermented foods, and *cortido* have gained importance because of their nutritional benefits (Tamang et al. 2015). Depending on the bacterial culture presence and production of end products, different fermentation processes have been proposed, which include nonalcoholic, alcoholic, lactic/acetic acid, amino acid/peptide sauce, and, finally, alkaline fermentation (Blandino et al. 2003; Anukam and Reid 2009). It is noted that the natural fermentation of different food products is due to the presence of mixed colonies of various microorganisms such as bacteria, yeast, and molds (Antony and Chandra 1997). However, lactic acid bacteria (LAB) which include *Lactobacillus*, *Streptococcus*, *Leuconostoc*, *Pediococcus*, *Lactococcus*, etc. contribute majorly in the acidic fermented foods. The release of certain organoleptic compounds and enzymes such as proteases, lipases, and amylases by certain LAB and *Bifidobacteria* makes the food palatable by improving the flavor and aroma (Blandino et al. 2003).

The fermented foods are also found in improving the nutritional quality, by releasing the antinutrients such as phytic acid and tannins, thereby increasing the bioavailability of minerals (Santos et al. 2008; Chelule et al. 2010). It was suggested that, the fermented foods are found to manage cardiovascular risk factors and drug side effects; prevent insulin sensitivity; reduce inflammation, swellings, stomach discomfort, and inhibition of pathogens; reduce toxic compounds; etc. (Chilton et al. 2015; Tamang et al. 2016). The interaction of microbes with the gut microbiota and their adherence to host intestinal tract are quite complex, and an in-depth study is required. The fermentation of fruits and vegetables by LAB or *Bifidobacteria* might enhance the vitamins and minerals availability and promote the bacterial growth (Swain et al. 2014). The awareness of health concerns among the consumers has enhanced the formulation of natural or traditional fermented foods without the addition of chemical preservatives. However, the protection to fermented foods is regulated by the production of organic acids, antimicrobial or antifungal compounds, hydrogen peroxide, and ethanol (Settanni and Corsetti 2008). The present article highlights about the beneficial properties exerted by the microbes in the fermented foods and discusses the functional properties of fermented foods.

20.2 Antiquity of Fermented Foods

The fermented foods were found to be the oldest food preservation method with a lot in antiquity. The fermentation is considered as an art and originated in Indian subcontinent during Indus Valley Civilization. The fermentation process was at first evolved during pre-Aryans or Middle Age (around 10,000 BC) from surplus milk and later in ca. 7000 BC. It was observed in cheese and bread making (Prajapati and Nair 2003). About 7000 years ago, the Neolithic people first discovered the curdling of milk after a few hours of milking from animals. Over time, they could observe rich flavors by the addition of vegetable juices to milk ingredients. By the end of the Paleolithic Age (12,000 years ago), the discovery of mead wine (fermentation of honey) has spread from Spain to South Africa, leading to practical practices in food science field (Katz 2003).

The knowledge of microorganisms during the process of fermentation was first witnessed in 1680 by Antony Van Leeuwenhoek and later by Louis Pasteur where the role of microorganisms in different fermentation processes was perceived. After the discovery of microbes, the prevention and treatment of infections by the live beneficial microbes came into limelight by Elie Metchnikoff in 1907. He was referred as "grandfather of modern probiotics" due to the landmark discovery of lactic acid bacteria in fermented yogurt (dairy product). The detection of intestinal *Lactobacillus acidophilus* by Hennerberg, referred as *Acidophilusmilch*, or reform yogurt, was successful in Western European countries in the early 1980s. Later on, peculiar, Y-shaped bacteria called as "bifid" bacteria were discovered in the stool sample of a diarrhea patient by Henry Tissier. The administration of "bifid bacteria" in diarrhea patients resulted in a healthy gut and restored the gut flora (Prajapati and Nair 2003; Tamang et al. 2015).

The concept of Metchnikoff was taken seriously, and the term "probiotics" was first mentioned by Lilly and Stillwell in 1965, and later on many definitions were proposed in scientific community. In 1998, Salminen redefined probiotics as "foods containing live bacteria which are helpful to health" (Salminen et al. 1998). Finally, the FAO/WHO in 2001 defined probiotics as "live microorganisms which when administered in adequate amounts confer a health benefit on the host" (FAO/WHO 2001). The word fermentation was originated from a Latin term *fermentum* (to ferment), which can be defined as "slow decomposition process of organic substances induced by microorganisms or enzymes that convert carbohydrates to alcohols or organic acids" (Tamang et al. 2016). Different fermentation processes are practiced throughout the world and have been used for centuries (Hansen 2004). High heterogeneity was observed in fermented foods because of the traditional and cultural practices followed in different geographical areas. For example, fermented milk products like yogurt, cheese, and paneer have been a staple diet in India, other Asian countries, Australia, Europe, North America, and New Zealand. Also, the ethnic fermented and/or non-fermented foods made of maize/millets/sorghum, legume seeds, milk, meat, and cassava are found mostly in South America and Africa. In Asian countries, the non-fermented legume products and pickles and fermented fish and meat products, vegetables, traditional boiled rice, and milk products have become famous because of the abundant presence of minerals, nutrients, and proteins (Tamang et al. 2015).

20.3 Microorganisms in Fermented Foods

Many microorganisms are involved in the fermentation of milk, plant products, meat products, etc. worldwide. Presently, LAB microorganisms were found abundantly in the fermentation foods and beverages (Tamang 2010b). The microorganisms used in any food impact many characteristics beyond nutrition which include texture, enzymes, acidity, flavor, fragrances, etc. (Devi and Halami 2015). Till date over 5000 varieties of fermented foods are available with different starter cultures; however, in many of these commercialized food products, the exact mechanism against the fermentation processes was poorly understood. The discovery of microorganisms and well-defined starter cultures has given a helping hand in understanding the food fermentation processes. Many genera of LAB (*Lactobacillus*, *Pediococcus*, *Enterococcus*, *Leuconostoc*, etc.) were globally used in the fermentation of food and beverages (Holzapfel and Wood 2014). Also, the non-LAB, i.e., *Bacillus* species, are also used in the fermentation of legume-based foods, dairy products, etc. (Kubo et al. 2011).

Several species of *Bifidobacterium*, *Brevibacterium*, *Propionibacterium*, etc. come under non-LAB category and are usually found associated with fermented cheese and milk products (Coton et al. 2010). Few of the yeasts are also reported to be involved in the fermentation of vegetables, alcoholic beverages, and fish and meat products (Lv et al. 2013). Fungi like *Penicillium*, *Aspergillus*, *Mucor*, *Rhizopus*, etc. are found in few fermented foods and used as amylolytic starter cultures (Nout and Aidoo 2002). In many of the fermented meat and fish products, the LAB are mostly involved, and some of the coagulase-negative *Staphylococcus* and *Micrococcus* species are also found (Marty et al. 2011). Table 20.1 enlists some of the microorganisms involved in fermented foods from different parts of the world.

The food and beverages containing the probiotic beneficial microbes are found to impart their functionalities and can be delivered as functional food to the consumers. Moreover, the viability of the probiotics microbes is considered as an important trait (Homayouni et al. 2008). According to Italian law, the minimum concentration required for live probiotic bacteria in any food product must be $>10^6$ cfu/mL (Fortina 2007). Several fruit juices were found contaminated with spore-forming microorganisms, molds, and yeasts. The addition of 8 log CFU/mL of *L. plantarum* c19 and *Bifidobacterium animalis* ssp. *lactis* DSMC 10140 has improved the shelf life of apple juice and red-fruit juice at 4 °C and 37 °C for 26 days and found to inhibit *Zygosaccharomyces bailii* (Bevilacqua et al. 2013). In another study, the incorporation of *L. plantarum* and *L. delbrueckii* ssp. *bulgaricus* with 10^6 and 10^{10} cfu/mL, respectively, in pasteurized milk and skim milk has efficiently improved the shelf life for 21 days at

Table 20.1 Microorganisms involved in fermented products

Product name (raw material used)	Country	Microorganisms	References
Fermented vegetables			
Eup/ekung/hiring (bamboo shoot)	India	*Lactobacillus plantarum, L. brevis, L. casei, L. curvatus, Leuconostoc mesenteroides, Leu. fallax, Enterococcus durans, Pediococcus pentosaceus*	Tamang and Tamang (2009)
Hom-dong (fermented red onion)	Thailand	*L. plantarum, L. fermentum, L. buchneri, Leu. mesenteroides, Ped. cerevisiae*	Phithakpol et al. (1995)
Kimchi (cabbage, green onion, ginger)	Korea	*Leu. mesenteroides, Leu. Kimchi, L. delbrueckii, L. buchneri, L. plantarum, L. fermentum, L. sakei, Ped. acidilactici, Ped. pentosaceus, Weissella koreensis, W. Kimchi, W. cibaria, Lactococcus lactis, Candida* sp., *Pichia* sp., *Saccharomyces* sp.	Jung et al. (2011)
Jeruk	Malaysia	*L. fermentum, L. plantarum, L. rhamnosus, L. casei, E. faecium, P. acidilactici, P. pentosaceus*	Tamang et al. (2015)
Khalpi (cucumber)	Nepal, India	*L. brevis, L. plantarum, Leu. fallax, P. acidilactici, P. pentosaceus*	Tamang and Tamang (2010)
Olives (olive)	Spain, USA, Peru, Chile, Portugal	*L. pentosus, L. plantarum, L. suebicus, L. vaccinostercus, Pediococcus* sp., *Pichia* sp., *Saccharomyces* sp., *Candida apicola*	Abriouel et al. 2011
Pak-gard-dong (leafy vegetables, boiled rice, and salt)	Thailand	*L. plantarum, L. brevis, Ped. cerevisiae*	Tamang et al. (2016)
Yan-jiang (ginger)	Taiwan	LAB	Chang et al. (2011)
Fermented dairy products			
Gariss (camel milk)	Sudan	LAB	Akabanda et al. (2013)
Dahi	Asia	LAB	Tamang 2010b
Cheese and butter (animal milk)	Worldwide	LAB, *Brevibacterium, Propionibacterium, Penicillium*	Quigley et al. (2011)
Kurut (yak milk)	China	LAB	Sun et al. 2010
Sua chua	Vietnam	*L. bulgaricus, Streptococcus thermophilus*	Alexandraki et al. (2013)
Yogurt (animal milk)	Worldwide	LAB	Tamime and Robinson (2007)
Fermented cereals			
Busa (maize, sorghum, millet)	Kenya, Africa	LAB, *Sacch. cerevisiae, Schizosaccharomyces pombe*	Kolawole et al. (2013)

(continued)

Table 20.1 (continued)

Product name (raw material used)	Country	Microorganisms	References
Boza (cereals)	Bulgaria	LAB	Blandino et al. (2003)
Dosa (black gram and rice)	India	*Leu. mesenteroides, E. faecalis, E. faecium, L. plantarum, Trichosporon pullulans*	Tamang et al. (2015)
Kenkey (maize)	Ghana	*L. plantarum, L. brevis, L. fermentum, Cand. mycoderma, Sacch. cerevisiae*	Oguntoyinbo et al. (2011)
Gowe (maize)	Benin	LAB	Greppi et al. (2013)
Lao-chao (rice)	China	*Rhizospora oryza, Rhiz chinensis, Sachhromyces* sp.	Blandino et al. (2003)
Me (rice)	Vietnam	LAB	Alexandraki et al. (2013)
Mung bean starch (mung bean)	Thailand	*Leuc. mesenteroides*	Alexandraki et al. (2013)
Perkarnaya (rye)	Russia	LAB, *yeast*	Alexandraki et al. (2013)
Natto (soybean fermented)	Japan	*Bacillus subtilis*	Tamang et al. (2015)
Togwa (cassava, maize, millets, sorghum)	Tanzania	LAB, *Candida* sp., *Issatchenkia orientalis*	Mugula et al. (2003)
Fermented meat/fish products			
Chartayshya (chevon)	India	*Ent. faecium, Ent. faecalis, Leu. mesenteroides, Ped. pentosaceus, Weissella cibaria*	Oki et al. (2012)
Nem-chua (pork)	Vietnam	LAB	Nguyen et al. (2011)
Peperoni (pork, beef)	USA, Europe, Australia	LAB, *Micrococcus* sp.	Adams (2010)
Hentak (fingered fish)	India	LAB	Thapa et al. (2004)
Joet kal (fish)	Korea	LAB	Guan et al. (2011)
Kusaya (mackerel)	Japan	*Spirillum* sp., *Corynebacterium* sp.	Alexandraki et al. (2013)
Surstromming (fish)	Sweden	*Haloanaerobium praevalens*	Kobayashi et al. (2000)
Yu lu (sardine)	China	*LAB, Micrococcus*	Jiang et al. (2007)

refrigeration temperature and had acceptable sensory properties (Mirlohi et al. 2014). Nithya et al. (2013) have observed a rapid reduction of *Listeria monocytogenes* ScottA in milk samples packed in a film coated with antimicrobial peptide with 6400 AU/mL from *Bacillus licheniformis* Me1; this has aided in the application of the probiotic *Bacillus* sp. from dairy sources for bio-preservation. Recently, in our laboratory, we have isolated several species of *L. plantarum*-group (LPG) isolates with good adherence ability, antimicrobial activity, and several other probiotic properties (Devi et al. 2016). The food products of vegetables, fruits, meat, dairy, and cereals are found rich in vitamins, dietary fibers, minerals, and antioxidants and give a platform for delivering the probiotics through fermented functional foods to consumers.

20.4 Functionality of Microbes in Fermented Foods

After the discovery of probiotic beneficial microbes with several health benefits, the probiotic market has gained 10% annual profit in a year (UBIC-consulting 2008). The beneficial properties of most of the probiotic products includes supply of vitamins, producing antimicrobial compounds; competitive exclusion of pathogens, binding to intestinal mucosal layer, strengthening intestinal barrier function, and enhancing immunity; and adaptation to the host gut (Kleerebezem and Vaughan 2009). The fermented foods or dairy products act as major carriers of probiotics to the host and deliver several health benefits (Divya et al. 2012). The FAO/WHO in 2001 has provided certain guidelines for the selection of probiotic bacteria for promoting health benefits, which includes:

1. Ability to tolerate the intestinal epithelial cell lining.
2. Retain high viability in a foodstuff throughout the shelf life of the product.
3. Produce antimicrobial substance toward pathogens.
4. Exert beneficial properties like anticholesterol, antioxidant, and anticarcinogen.
5. Must resist to the acid and bile conditions of the host intestine, etc. (Parvez et al. 2006).

Today, the commercialized fermented foods are marketed as functional, organic health drinks, bio-foods, medico-foods, etc. globally (Badis et al. 2006). Several LAB species and *Bifidobacteria* are found to play a crucial role in the fermentation of milk, vegetables, meat, and wine products. During fermentation, several *Lactobacillus*, *Pediococcus*, *Leuconostoc*, and *Bifidobacteria* spp. produce lactic acid and acetic acid, thereby providing significant acidification of substrates and improving the existence and life of fermented foods (Tamang et al. 2008; Ammor and Mayo 2007). Some of the bioactive peptides produced by proteolytic microbes in *kimchi*, *natto*, and *kinema* have shown a positive impact on functional aspects which include immunomodulation, antioxidant, anticholesterol, antimicrobial, antithrombic, and antihypertensive properties (Tamang et al. 2015, 2016). The *Bacillus subtilis*, *Bacillus licheniformis*, *Saccharomycopsis fibuligera*, *Pichia burtonii*, *Mucor* sp., and *Rhizopus* sp. are found to produce α-amylase, amyloglucosidase, cellulose, lipases, etc. in fermented vegetables (*natto*, *kinema*) and soybean products (*tungrymbai*, *bekang*) (Tamang 2010a, b, 2014; Rai et al. 2010; Nithya and Halami 2013; Shobharani and Halami 2014). Reduction of serum cholesterol by *Lactobacillus* sp. was also reported in *kefir*, *kinema*, and *tempeh* food products and observed to control diabetes, obesity, hypertriacylglycerolemia, and insulin resistance, improving antioxidant status, etc. (Otes and Cagindi 2003; Anderson 2003).

The microbes in fermented foods act as protective cultures, inhibiting the growth of undesirable microbes. The fermentation process was found to improve the viability of beneficial microbes by allowing the production of antimicrobial agents such as organic acids, bacteriocins, diacetyl, and hydrogen peroxide (EFFCA 2011; Liong 2008). Gaggia et al. (2011) have also suggested that these compounds contribute in improving the texture and flavor of the end product. The subtilin-like peptide produced by *Bacillus licheniformis* was found to suppress the growth of food-borne pathogens (Shobharani et al. 2015). The production of pediocin PA-1 like bacteriocins from *Pediococcus acidilactici*, *Lactobacillus plantarum*, and *Enterococcus faecium* during soy milk fermentation has efficiently reduced the growth of *Listeria monocytogenes* (Devi et al. 2014). The sporulation of *Clostridium* and *Bacillus* sp. was prevented after heat treatment and incorporation of nisin (de Arauz et al. 2009). The application of *B. licheniformis* and *B. subtilis* in the aquatic field, i.e., shrimp and mollusk production, has resulted in the reduction of *Vibrio* and *Aeromonas* spp. and improvement of water quality and gut flora (Decamp and Moriarty 2006; Kesarcodi-Watson et al. 2008).

Several LAB (like *Lactobacillus brevis*, *Enterococcus durans*, *Enterococcus faecium*, etc.) in soybean foods were found to degrade undesirable compounds like phytic acid and found to increase the mineral bioavailability (Raghavendra et al. 2011; Chettri and Tamang 2014). It was noted that the trypsin inhibitor present in *tempeh* product was found to be inactivated by *Rhizopus oligosporus* during fermentation (Hesseltine 1983). Also, the microorganisms during *idli* fermentation are also found to reduce phytic acid, thereby improving the mineral bioavailability to the consumers (Reddy and Salunkhe 1980). The fermented foods like *gundruk*, *sinki*, *kinema*, *poko*, *douche*, and *kimchi* and fermented bamboo shoots have many health benefits which include antimicrobial, probiotic, anticholesterol, production of vitamins, folic acid, and amino acids and are used to stimulate immune system, curing stomach-related illness, reducing vitamin K deficiency in infants, antistress stabilizing blood sugar, etc. (Park et al. 2006; Lee and Lee 2009; Tamang 2010a, Omizu et al. 2011; Kim et al. 2011; Syal and Vohra 2013; Russo et al. 2014). Similarly, several functional properties like cholesterol removal, antimutagenic activity, and anti-oxidative effect have been observed in *Bifidobacterium* strains isolated from human origin (Awasti et al. 2016).

The polyphenols present in wine and isoflavones of *doenjang* are found to activate the LDL-C receptor and HDL-C level and thereby prevent certain cardio-related diseases (Wallerath et al. 2005; Kwak et al. 2012). Some of the fermented foods like *kefir*, *soybean sauce*, dahi, etc. are used in the treatment of cancer, tuberculosis, and tumorigenesis in the stomach, showing their therapeutic and anticarcinogenic activity (Otes and Cagindi 2003; Arvind et al. 2010). The microorganisms like *Lactobacillus acidophilus*, *Lactobacillus plantarum*, and *Bifidobacterium breve* are found associated with the anticarcinogenic activity (Rhee and Park 2001). The probiotic *Lactobacillus rhamnosus* GG strain is widely used in treating acute diarrhea (Szajewska et al. 2007). Some of the yeast strains, i.e., *Saccharomyces boulardii*, are effective in decreasing irritable bowel syndrome, flatulence, constipation, etc. (MacFarlane and Cummings 2002). Granier et al. (2013) have reported immunomodulation capacity by the consumption of fermented milk. The LAB of *kimchi* were found to control atopic dermatitis (Lim et al. 2011). The *Lactobacillus helveticus* and *Sacch. cerevisiae* in fermented milk, *Lactobacillus plantarum* in *kimchi*, and *Bacillus subtilis* in *natto* are found to possess the antihypertensive property and used as a weapon for cardiovascular problems (Mine et al. 2005; Aihara et al. 2005). The modulation of allergic reactions was improved by *Lactobacillus kefiranofaciens* M1 and found to contain certain probiotic functionalities like anticarcinogenic and antiplatelet activities (Ando et al. 2003; Hong et al. 2010). Due to the continuous emergence of antibiotic-resistant bacteria in poultry, meat, and aquaculture, the application of probiotic bacteria has shown a promising beneficial effect toward competitive elimination of pathogens, enhancing immunity, etc. (Tan et al. 2016). Table 20.2 enlists some of the beneficial microbes with functional properties provided in the fermented foods.

Many commercial probiotic cultures were already applied in fermented foods worldwide (Tamime et al. 2005). For the implementation of a culture in any food product, certain properties are required, which include safety, technological, functional, and desirable physiological criteria. In 1992, the United Kingdom Advisory Committee and in 1996 the Japanese functional food authorities have tested the functional, safety, and efficacy of *L. rhamnosus* GG extensively (Seppo and Donohue 1996). Later on, Zago et al. (2011) suggested that *L. plantarum* and *L. paracasei* present abundantly in many varieties of cheese are used in commercial probiotic products as a potential probiotic candidate. The absence of β-hemolytic activity, antibiotic resistance genes, *N*-acetyl-β-glucosaminidase, β-glucosidase, and β-glucuronidase is a major criterion required for a probiotic starter cultures, as the presence of these genes leads to carcinogenic and mutagenic activity on the host (Gill and Rowland 2002; Giraffa 2009). The viability of *Bifidobacteria* and LAB was found to be maintained in commercial yogurt samples of Michigan, USA, for 3 weeks with 10^6 CFU/mL at refrigerated temperatures (Shin et al. 2000). In another study, the commercialized local yogurt products of Greensboro, North Carolina, use *Bifidobacteria*, *L. delbrueckii*, and *Streptococcus thermophilus* and offered the beneficial consumption of yogurt foods at 4 °C for 3 weeks with a bacterial count ranging from 5.5 log to 8.94 log CFU/mL (Ibrahim and Carr 2006). At industrial level, the probiotic microorganisms are inoculated into food products for fermentation either in the form of lyophilized powder or in freeze-dried form (Sandine 1996). The presence of certain bioactive compounds like ornithine,

Table 20.2 Functionalities of the microbes exhibited in the fermented foods

Microorganism	Food product	Pronounced health benefit	References
Lactobacillus acidophilus KFR 1342	*Kimchi*	Reduced the growth of cancer cells SNU-C4	Chang et al. (2010)
Bifidobacterium longum and *B. infantis*	Fermented milk	Produce effective antitumor agents	Mitsuoka et al. (2014)
Lactobacillus casei Shirota	Fermented milk	Exerts anticancer activity	Shida and Nomato (2013)
Lactobacillus acidophilus, L. casei, L. paracasei	*Yogurt*	Produce strong antimutagenicity	Sah et al. (2014)
Bacillus sp.	Dairy milk and cereals	Exhibits anticholesterol, antioxidant, potential tolerance to acid and exerts probiotic properties	Shobharani and Halami (2016)
L. lactis subsp. *lactis*	Fermented vegetables	Inhibited the proliferation of SNUC2 human colon cancer cell line	Kim et al. (2003)
Penicillium roqueforti	Blue cheese	Produce andrastin A, a secondary metabolite used in the anticancer treatments	Fernandez-Bodega et al. (2009)
L. jensenii TL2937	–	Interact with immune cells and maintain balance between tolerance and inflammation	Villena and Kitazawa (2014)
L. rhamnosus CRL1505, *L. casei* CRL431	–	Stimulates immune system in respiratory tract	Marranzino et al. (2012)
L. helveticus, Bifidobacterium sp., *L. rhamnosus*	Fermented milk	Production of folic acid and biotin	Strozzi and Mogna 2008; D'Aimmo et al. (2012)
L. kefiranofaciens M1	*Kefir* grains	Suppressed mast cells degranulation and cytokine production and reduced inflammation associated with allergy	Hong et al. (2010)
L. rhamnosus GG, *L. gasseri* TMC0356	Fermented milk	Significantly inhibited IL-4 and IL-5 production and effective on nasal blockage	Kawase et al. (2009)
L. lactis biovar *diacetylactis*	*Dahi*	Reduced allergy by production of T-helper cytokines, i.e., IFN-γ and IL2	Jain et al. (2010)
L. lactis A17	Fermented cabbage	Showed significant immunomodulation by increasing IFN-γ and decreasing IL-4 production	Mei et al. (2013)
Streptococcus bacillaris and *Strep. cinereus*	Fermented green tea	Increased the proinflammatory cytokines, IL-12, and reduced allergy	Jeng et al. (2007); Hou et al. (2010)

Table 20.2 (continued)

Microorganism	Food product	Pronounced health benefit	References
L. plantarum	*Kimchi*	Inhibit the growth and adherence of *Helicobacter pylori* in MKN-45 cell line with the production of bacteriocin peptides	Lee and Lee (2006)
Pediococcus pentosaceus	Wine	Has high anti-adhesion activity against *Escherichia coli* CIAL-153	García-Ruiz et al. (2014)
L. fermentum	Sourdough	Production of riboflavin	Russo et al. (2014)
L. sakei	*Kimchi*	Folate production	Jung et al. (2013)
Lactobacillus, Pediococcus, Streptococcus spp.	Human feces and dairy products	Had antimicrobial property, acid and bile tolerance ability, adherence ability	Devi et al. (2015)
Pediococcus acidilactici, Enterococcus faecium, Lactobacillus plantarum	Fermented soy milk	Enhanced the production of pediocin PA-1 bacteriocin and improved the antioxidant activity. These cultures proficiently inhibited the growth of *Listeria monocytogenes* ScottA during soy milk fermentation	Devi et al. (2014)
Saccharomyces cerevisiae, Candida tropicalis, Pichia manshurica	Indian fermented foods like *idli, jalebi*	Production of vitamin B12	Syal and Vohra (2013)
L. fermentum, L. delbrueckii	Human feces and breast milk	Had probiotic properties with anti-inflammatory, adhesion ability	Archer and Halami (2015); Archer et al. (2015)
Pediococcus sp., *Lactobacillus* sp.,	*Sauerkraut*	Reduces tumorigenesis of the stomach	Kris-Etherton et al. (2002)
L. acidophilus	Fermented milk	Decreases azoreductase and nitroreductase and found to remove procarcinogens and activates the immunity	Macouzet et al. (2009)
L. plantarum, W. cibaria	*Kimchi*	Used for curing cancer	Kwak et al. (2014)
L. lactis, L. plantarum, L. fermentum	Probiotic *dahi*	Found to reduce diabetes	Yadav et al. (2007)
L. kefiranofaciens	*Kefir*	Found to produce antiallergic properties	Hong et al. (2010)
L. plantarum	*Doenjang*	Prevents cardiovascular diseases	Shin et al. (2015)
Oenococcus sp.	Red wine	Produces polyphenols, vitamin E, and ascorbic acid and inhibits lipid peroxidation	Feher et al. (2007)
Monascus purpureus	*Angkak* (fermented red rice)	Reduces cholesterol levels	Pattanagul et al. (2008)

Not available

nattokinase, phenolics, genistein, etc. in the synthesized fermented foods is found to provide certain health benefits like prevention of cancer; treatment of diabetes; detoxification of the liver, kidney, and intestine; and immunomodulation (Tamang et al. 2016) (Table 20.2).

20.5 Conclusion

The continuous demand for organic foods has made the fermented foods as an important food commodity. The probiotic fermented foods like *natto*, *kimchi*, fermented dairy products, soybean products, etc. are a good example of functional foods and have become popular after addition of probiotic bacteria. The incorporation of a known viable starter culture/strain in any food commodity would provide the claimed health benefits. The proper identification of probiotic bacteria and their commercialization would improve the functionality of the fermented foods. The presence of bioactive peptides; anticholesterol, antithrombotic, immunomodulatory, anti-oxidative, antiallergic, and anticarcinogenic properties; vitamins production; and mineral binding efficiency would make these fermented foods as natural with added nutritional values. Much literature is not available on the interaction and contribution of health benefits of the microbes to the host cells. Research on human microbiome and genome sequencing of beneficial microbes would open an exciting field in biology with fascinating insights on metabolomics, meta-transcriptomics, evolution, microbial interaction, etc. An investigation of the microbial interaction with the host would reveal certain microbiome-based biomarkers providing an alternative strategy for improving and immunomodulating human health.

20.6 Opinion

The research activities on probiotic microbes have been doubled in the past few years, where the articles emphasized on the characterization of probiotic strains, studying their clinical findings, preventing the growth of pathogenic microbes, and exerting health benefits. Depending on the consumer's requirement, several methodologies have been followed to extend its application in the food industries. A future research on the interaction of microbes to the host epithelium and its mechanism of action is required to expand its commercialization. Different combinations of prebiotics and probiotics might give an insight in the removal of microbial disorders.

Acknowledgments The authors acknowledge Prof. Ram Rajasekharan, Director of CSIR-CFTRI for the encouragement. SMD extends her thanks to SERB-DST, New Delhi, for endorsing the study under start-up grant scheme (Project No. SB/YS/LS-353/2013).

References

Abriouel H, Benomar N, Lucas R, Gálvez A (2011) Culture-independent study of the diversity of microbial populations in brines during fermentation of naturally-fermented Aloreña green table olives. Int J Food Microbiol 144:487–496. doi:10.1016/j.ijfoodmicro.2010.11.006

Adams MR (2010) Fermented meat products. In: Tamang JP, Kailasapathy K (eds) Fermented foods and beverages of the world. CRC Press, Taylor & Francis Group, New York, pp 309–322

Aihara K, Kajimoto O, Hirata H, Takahashi R, Nakamura Y (2005) Effect of powdered fermented milk with *Lactobacillus helveticus* on subjects with high-normal blood pressure or mild hypertension. J Am Coll Nutr 24:257–265. doi:10.1080/07315724.2005.10719473

Akabanda F, Owusu-Kwarteng J, Tano-Debrah K, Glover RLK, Nielsen DS, Jespersen L (2013) Taxonomic and molecular characterization of lactic acid bacteria and yeasts in nunu, a Ghanaian fermented milk product. Food Microbiol 34(2):277–283. doi:10.1016/j.fm.2012.09.025

Alexandraki V, Tsakalidou E, Papadimitriou K, Holzapfel WH (2013) Status and trends of the conservation and sustainable use of microorganisms in food processes. Commission on Genetic Resources for Food and Agriculture. FAO Background Study Paper No. 65

Ammor MS, Mayo B (2007) Selection criteria for lactic acid bacteria to be used as functional starters in dry sausage production: an update. Meat Sci 76(1):138–146. doi:10.1016/j.meatsci.2006.10.022

Anderson JW (2003) Whole grains protect against atherosclerotic cardiovascular disease. Proc Nutr Soc 62:135–142. doi:10.1079/PNS2002222

Ando M, Harada K, Kitao S, Kobayashi M, Tamura Y (2003) Relationship between peroxyl radical scavenging capability measured by the chemiluminescence method and an aminocarbonyl reaction product in soy sauce. Int J Mol Med 12:1018–1023. doi:10.3892/mmr00000313

Antony U, Chandra TS (1997) Microbial population and biochemical changes in fermenting finger millet (Eleusine coracana). World J Microbiol Biotechnol 13(5):533–537. doi:10.1023/A:1018561224777

Anukam KC, Reid G (2009) African traditional fermented foods and probiotics. J Med Food 12(6):1177–1184. doi:10.1089/jmf.2008.0163

Archer AC, Halami PM (2015) Probiotic attributes of *Lactobacillus fermentum* isolated from human feces and dairy products. Appl Microbiol Biotechnol 99:8113–8123. doi:10.1007/s00253-015-6679-x

Archer AC, Muthukumar SP, Halami PM (2015) Anti-inflammatory potential of probiotic *Lactobacillus* spp. on carrageenan induced paw edema in Wistar rats. Int J Biol Macromol 81:530–537. doi:10.1016/j.ijbiomac.2015.08.044

Arvind K, Nikhlesh KS, Pushpalata RS (2010) Inhibition of 1, 2-dimethylhydrazine induced colon genotoxicity in rats by the administration of probiotic curd. Mol Biol Rep 37:1373–1376. doi:10.1007/s11033-009-9519-1

Awasti N, Tomar SK, Pophaly SD, Poonam, Lule VK, Singh TP, Anand S (2016) Probiotic and functional characterization of bifidobacteria of Indian human origin. J Appl Microbiol 120:1021–1032. doi:10.1111/jam.13086

Badis A, Guetarni D, Moussa-Boujemaa B, Henni DE, Tornadijo ME, Kihal M (2006) Identification of cultivable lactic acid bacteria isolated from Algerian raw goat's milk and evaluation of their technological properties. Food Microbiol 21:343–349. doi:10.1016/S0740-0020(03)00072-8

Bevilacqua A, Campaniello D, Corbo MR, Maddealena L, Sinigaglia M (2013) Suitability of *Bifidobacterium* spp. and *Lactobacillus plantarum* as probiotics intended for fruit juices containing citrus extracts. J Food Sci 78(11):1764–1771. doi:10.1111/1750-3841.12280

Blandino A, Al-Aseeri ME, Pandiella SS, Cantero D, Webb C (2003) Cereal-based fermented foods and beverages. Food Res Int 36(6):527–543. doi:10.1016/S0963-9969(03)00009-7

Chang JH, Shim YY, Cha SK, Chee KM (2010) Probiotic characteristics of lactic acid bacteria from *Kimchi*. J Appl Microbiol 109:220–230. doi:10.1111/j.1365-2672.2009.04648.x

Chang CH, Chen YS, Yanagida F (2011) Isolation and characterisation of lactic acid bacteria from yan-jiang (fermented ginger), a traditional fermented food in Taiwan. J Sci Food Agric 91:1746–1750. doi:10.1002/jsfa.4364

Chelule PK, Mbongwa HP, Carries S, Gqaleni N (2010) Lactic acid fermentation improves the quality of amahewu, a traditional South African maize-based porridge. Food Chem 122(3):656–661. doi:10.1016/j.foodchem.2010.03.026

Chettri R, Tamang JP (2014) Functional properties of Tungrymbai and Bekang, naturally fermented soybean foods of North East India. Int J Ferment Foods 3:87–103. doi:10.5958/2321-712X.2014.01311.8

Chilton NS, Burton JP, Reid G (2015) Inclusion of fermented foods in food guides around the world. Nutrients 7:390–404. doi:10.3390/nu7010390

Coton E, Desmonts MH, Leroy S, Coton M, Jamet E, Christieans S, Donnio PY, Lebert I, Talon R (2010) Biodiversity of coagulase-negative staphylococci in French cheeses, dry fermented sausages, processing environments and clinical samples. Int J Food Microbiol 137:221–229. doi:10.1016/j.ijfoodmicro.2009.11.023

D'Aimmo MR, Mattarelli P, Biavati B, Carlsson NG, Andlid T (2012) The potential of bifidobacteria as a source of natural folate. J Appl Microbiol 112:975–984. doi:10.1111/j.1365-2672.2012.05261.x

de Arauz LJ, Jozala AF, Mazzola PG, Penna TCV (2009) Nisin biotechnological production and application. A review. Trends Food Sci Technol 20:146–154. doi:10.1016/j.tifs.2009.01.056

Decamp O, Moriarty DJW (2006) Probiotics as alternative to antimicrobials: limitations and potential. World Aquacult 37:60–62

Devi SM, Halami PM (2015) Metabolic characteristics of lactic starters. In: Fermented milk and dairy products. CRC Press Taylor and Francis Group, New York, pp 109–127. isbn:9781466577978

Devi SM, Ramaswamy AM, Halami PM (2014) In situ production of pediocin PA-1-like bacteriocin by lactic acid bacteria in soymilk fermentation and evaluation of sensory properties. J Food Sci Technol 51:3325–3332. doi:10.1007/s13197-012-0870-1

Devi SM, Archer AC, Halami PM (2015) Screening, characterization and in vitro evaluation of probiotic properties among lactic acid bacteria through comparative analysis. Probiotics Antimicrob Proteins 7:181–192. doi:10.1007/s12602-015-9195-5

Devi SM, Aishwarya S, Halami PM (2016) Discrimination and divergence among *Lactobacillus plantarum*-group (LPG) isolates with reference to their probiotic functionalities from vegetable origin. Syst Appl Microbiol 39(8):562–570. doi:10.1016/j.syapm.2016.09.005

Divya JB, Varsha KK, Nampoothiri KM, Ismail B, Pandey A (2012) Probiotic fermented foods for health benefits. Eng Life Sci 12:377–390. doi:10.1002/elsc.201100179

EFFCA (2011) Protective cultures. European Food and Feed cultures Association/2011/52. Brussels, 17 Oct 2011

FAO/WHO (2001) http://www.fao.org/es/ESN/probio/probio.htm. Accessed 28 Apr 2007

Feher J, Lengyel G, Lugasi A (2007) The cultural history of wine—theoretical background to wine therapy. Cent Eur J Med 2:379–391. doi:10.2478/s11536-007-0048-9

Fernandez-Bodega MA, Mauriz E, Gomez A, Martin JF (2009) Proteolytic activity, mycotoxins and andrastin A in *Penicillium roqueforti* strains isolated from Cabrales, Valdeón and Bejes-Tresviso local varieties of blue-veined cheeses. Int J Food Microbiol 136:18–25. doi:10.1016/j.ijfoodmicro.2009.09.014

Fortina MG (2007) I prodotti lattiero-caseari. In: Cocolin LS, Comi G (eds) Lamicrobiologia applicata alle industrie alimentari. Aracne Editrice, Roma, pp 289–336

Gaggia F, Gioia D, Baffoni L, Biavati B (2011) The role of protective and probiotic cultures in food and feed and their impact in food safety. Trends Food Sci Technol 22(1):58–66. doi:10.1016/j.tifs.2011.03.003

García-Ruiz A, Esteban-Fernández DGA, Requena T, Bartolomé B, Moreno-Arribas MV (2014) Assessment of probiotics properties in lactic acid bacteria isolated from wine. Food Microbiol 44:220–225. doi:10.1016/j.fm.2014.06.015

Gill CI, Rowland IR (2002) Diet and cancer: assessing the risk. Br J Nutr 88:S73–S87

Giraffa G (2009) Antibiotic resistance in food lactic acid bacteria. In: Bonilla AR, Muniz KP (eds) Antibiotic resistance: causes and risk factors, mechanisms and alternatives. Nova Science Publishers, New York, pp 371–378

Granier A, Goulet O, Hoarau C (2013) Fermentation products: immunological effects on human and animal models. Pediatr Res 74:238–244. doi:10.1038/pr.2013.76

Greppi A, Rantsiou K, Padonou W, Hounhouigan J, Jespersen L, Jakobsen M et al (2013) Determination of yeast diversity *in ogi*, *mawè*, *gowé* and *tchoukoutou* by using culture-dependent and-independent methods. Int J Food Microbiol 165:84–88. doi:10.1016/j.ijfoodmicro.2013.05.005

Guan L, Cho KH, Lee JH (2011) Analysis of the cultivable bacterial community in jeotgal, a Korean salted and fermented seafood, and identification of its dominant bacteria. Food Microbiol 28:101–113. doi:10.1016/j.fm.2010.09.001

Hansen EB (2004) Micro-organisms. In: Hui YH, Meunier-Goddik L, Hansen ÅS, Josephsen J, Nip WK, Stanfield PS, Toldrá F (eds) Handbook of food and beverage fermentation technology, vol 2. Marcel Dekker, Inc., 270 Madison Avenue, New York, pp 9–21. doi:10.1201/9780203913550.ch2

Hesseltine CW (1983) Microbiology of oriental fermented foods. Annual Review of Microbiology 37:575–601

Holzapfel WH, Wood BJB (2014) Lactic acid bacteria: biodiversity and taxonomy. Wiley-Blackwell, New York, p 632. doi:10.1002/9781118655252

Homayouni A, Azizi A, Ehsani MR, Yarmand MS, Razavi SH (2008) Effect of microencapsulation and resistant starch on the probiotic survival and sensory properties of symbiotic ice cream. Food Chem 111:50–55. doi:10.1016/j.foodchem.2008.03.036

Hong W, Chen Y, Chen M (2010) The anti-allergic effect of kefir Lactobacilli. J Food Sci 75(8):H244–H253. doi:10.1111/j.1750-3841.2010

Hou CW, Heng KC, Chen YS (2010) Enhancement of fermentation process in pu-erh tea leaf extract. J Food Sci 75:44–48. doi:10.1111/j.1750-3841.2009.01441.x

Ibrahim SA, Carr JP (2006) Viability of bifidobacteria in commercial yogurt products in North Caroline during refrigerated storage. Int J Dairy Technol 59(4):272–277. doi:10.1111/j.1471-0307.2006.00282.x

Jain S, Yadav H, Sinha PR, Kapila S, Naito Y, Marotta F (2010) Anti-allergic effects of probiotic *dahi* through modulation of the gut immune system. Turk J Gastroenterol 21:244–250. doi:10.4318/tjg.2010.0095

Jeng K-C, Chen C-S, Fang Y-P, Hou RC-W, Chen Y-S (2007) Effect of microbial fermentation on content of statin, GABA, and polyphenols in Pu-Erh Tea. Journal of Agricultural Food Chemistry 55: 8787–8792. doi:10.1021/jf071629p

Jiang JJ, Zeng QX, Zhu ZW, Zhang LY (2007) Chemical and sensory changes associated Yu-lu fermentation process-A traditional Chinese fish sauce. Food Chem 104:1629–1634. doi:10.1016/j.foodchem.2007.03.024

Jung JY, Lee SH, Kim JM, Park MS, Bae JW, Hahn Y et al (2011) Metagenomic analysis of Kimchi, a traditional Korean fermented food. Appl Environ Microbiol 77:2264–2274. doi:10.1128/AEM.02157-10

Jung JY, Lee SH, Jin HM, Hahn Y, Madsen EL, Joen CO (2013) Metatranscriptomic analysis of lactic acid bacterial gene expression during Kimchi fermentation. Int J Food Microbiol 163:171–179. doi:10.1016/j.ijfoodmicro.2013.02.022

Katz ES (2003) Wild fermentation—the flavour, nutrition and craft of live-culture foods. In: Food and drink, health and wellness. Chelsea Green Publishing Company

Kawase M, He F, Kubota A, Hiramatsu M, Saito H, Ishii T, Yasueda H, Akiyama K (2009) Effect of fermented milk prepared with two probiotic strains on Japanese cedar pollinosis in a double-blind placebo-controlled clinical study. Int J Food Microbiol 128:429–434. doi:10.1016/j/ijfoodmicro.2008.09.17

Kesarcodi-Watson A, Kaspar H, Lategan MJ, Gibson L (2008) Probiotics in aquaculture: the need, principles and mechanisms of action and screening processes. Aquaculture 274:1–14. doi:10.1016/j.aquaculture.2007.11.019

Kim MJ, Koh E, Surh J, Kim YKL, Kwon H (2003) Distribution of isoflavones and coumestrol in legumes and their products consumed in Korea. Food Sci Biotechnol 12:278–284. doi:10.1080/09637480600802348

Kim EK, An SY, Lee MS, Kim TH, Lee HK, Hwang WS et al (2011) Fermented Kimchi reduces body weight and improves metabolic parameters in overweight and obese patients. Nutr Res 31:436–443. doi:10.1016/j.nutres.2011.05.011

Kleerebezem M, Vaughan EE (2009) Probiotic and gut lactobacilli and bifidobacteria: molecular approaches to study diversity and activity. Annu Rev Microbiol 63:269–290. doi:10.1146/annurev.micro.091208.073341

Kobayashi T, Kimura B, Fujii T (2000) Strictly anaerobic halophiles isolated from canned Swedish fermented herrings (Surströmming). Int J Food Microbiol 54:81–89

Kolawole OM, Kayode RMO, Akinduyo B (2013) Proximate and microbial analyses of burukutu and pito produced in Ilorin, Nigeria. Afr J Microbiol 1(1):15–17

Kris-Etherton PM, Hecker KD, Bonanome A, Coval SM, Binkoski AE, Hilpert KF et al (2002) Bioactive compounds in foods: their role in the prevention of cardiovascular disease and cancer. Am J Med 113:71S–88S. doi:10.1016/S0002-9343(01)00995-0

Kubo Y, Rooney AP, Tsukakoshi Y, Nakagawa R, Hasegawa H, Kimura K (2011) Phylogenetic analysis of *Bacillus subtilis* strains applicable to natto (fermented soybean) production. Appl Environ Microbiol 77:6463–6469. doi:10.1128/AEM.00448-11

Kwak CS, Park S, Song KY (2012) Doenjang, a fermented soybean paste, decreased visceral fat accumulation and adipocyte size in rats fed with high fat diet more effectively than nonfermented soybeans. J Med Food 15:1–9. doi:10.1089/jmf.2010.1224

Kwak SH, Cho YM, Noh GM, Om AS (2014) Cancer preventive potential of Kimchi lactic acid bacteria (*Weissella cibaria*, *Lactobacillus plantarum*). J Cancer Prev 19:253–258. doi:10.15430/JCP.2014.19.4.253

Lee HM, Lee Y (2006) Isolation of *Lactobacillus plantarum* from kimchi and its inhibitory activity on the adherence and growth of *Helicobacter pylori*. J Microbiol Biotechnol 16(10):1513–1517. doi:10.1089/jmf.2009.1265

Lee HR, Lee JM (2009) Anti-stress effects of Kimchi. Food Sci Biotechnol 18(1):25–30. http://doi.org/10.3389/fmicb.2016.00578

Lim J, Seo BJ, Kim JE, Chae CS, Im SH, Hahn YS, Park YH (2011) Characteristics of immunomodulation by a *Lactobacillus sakei* proBio65 isolated from Kimchi. Korean J Microbiol Biotechnol 39:313–316

Liong MT (2008) Safety of probiotics: translocation and infection. Nutr Rev 66:192–202. doi:10.1111/j.1753-4887.2008.00024.x

Lv XC, Huang XL, Zhang W, Rao PF, Ni L (2013) Yeast diversity of traditional alcohol fermentation starters for Hong Qu glutinous rice wine brewing, revealed by culture-dependent and culture-independent methods. Food Control 34:183–190. doi:10.1016/j.foodcont.2013.04.020

MacFarlane GT, Cummings JH (2002) Probiotics, infection and immunity. Curr Opin Infect Dis 15:501–506

Macouzet M, Lee BH, Robert N (2009) Production of conjugated linoleic acid by probiotic *Lactobacillus acidophilus* La-5. J Appl Microbiol 106:1886–1891. doi:10.1111/j.1365-2672.2009.04164.x

Marranzino G, Villena J, Salva S, Alvarez S (2012) Stimulation of macrophages by immunobiotic *Lactobacillus* strains: influence beyond intestinal tract. Microbiol Immunol 56:771–781. doi:10.1111/j.1348-0421.2012.00495.x

Marty E, Buchs J, Eugster-Meier E, Lacroix C, Meile L (2011) Identification of staphylococci and dominant lactic acid bacteria in spontaneously fermented Swiss meat products using PCR–RFLP. Food Microbiol 29:157–166. doi:10.1016/j.fm.2011.09.011

Mei H-C, Liu Y-W, Chiang Y-C, Chao S-H, Mei N-W, Liu Y-W, Tsai Y-C (2013) Immunomodulatory activity of Lactococcus lactis A17 from Taiwan Fermented cabbage in OVA-sensitized BALB/c mice. Evid Based Complement Alternat Med 2013:287803. doi:10.1155/2013/287803

Mine Y, Wong AHK, Jiang B (2005) Fibrinolytic enzymes in Asian traditional fermented foods. Food Res Int 38:243–250. doi:10.1016/j.foodres.2004.04.008

Mirlohi M, Soleimanian-Zad S, Dokhani S, Sheikh-Zeinodin M (2014) Microbial and physio-chemical changes in yoghurts containing different *Lactobacillus delbrueckii* subsp. *bulgaricus* strains in association with *Lactobacillus plantarum* as an adjunct culture. Int J Dairy Technol 67:246–254. doi:10.1111/1417-0307.12122

Mitsuoka T (2014) Establishment of intestinal bacteriology. Bioscience Microbiota Food Health 33(3):99–116. doi:10.12938/bmfh.33.99

Mugula JK, Ninko SAM, Narvhus JA, Sorhaug T (2003) Microbiological and fermentation characteristics of togwa, a Tanzanian fermented food. Int J Food Microbiol 80:187–199. doi:10.1016/S0168-1605(02)00141-1

Nguyen H, Elegado F, Librojo-Basilio N, Mabesa R, Dizon E (2011) Isolation and characterisation of selected lactic acid bacteria for improved processing of nem chua, a traditional fermented meat from Vietnam. Benef Microbes 1(1):67–74. doi:10.3920/BM2009.0001

Nithya V, Halami PM (2013) Evaluation of the probiotic characteristics of *Bacillus* species isolated from different food sources. Ann Microbiol 63:129–137. doi:10.1007/s13213-012-0453-4

Nithya V, Murthy PSK, Halami PM (2013) Development and application of active films for food packaging using antibacterial peptide of *Bacillus licheniformis* Me1. J Appl Microbiol 115:475–483. doi:10.1111/jam.12258

Nout MJR, Aidoo KE (2002) Asian fungal fermented food. In: Osiewacz HD (ed) The mycota. Springer, New York, pp 23–47

Oguntoyinbo FA, Tourlomousis P, Gasson MJ, Narbad A (2011) Analysis of bacterial communities of traditional fermented West African cereal foods using culture independent methods. Int J Food Microbiol 145:205–210. doi:10.1016/j.ijfoodmicro.2010.12.025

Oki K, Kudo Y, Watanabe K (2012) *Lactobacillus saniviri* sp. nov. and *Lactobacillus senioris* sp. nov., isolated from human faeces. Int J Syst Evol Microbiol 62:601–607. doi:10.1099/ijs.0.031658-0

Omizu Y, Tsukamoto C, Chettri R, Tamang JP (2011) Determination of Saponin contents in raw soybean and fermented soybean foods in India. J Sci Ind Res 70:533–538

Otes S, Cagindi O (2003) Kefir: a probiotic dairy-composition, nutritional and therapeutic aspects. Pak J Nutr 2:54–59

Park KY, Kil JH, Jung KO, Kong CS, Lee J (2006) Functional properties of Kimchi (Korean fermented vegetables). Acta Horticult 706:167–172. doi:10.17660/ActaHortic.2006.706

Parvez S, Malik KA, Ah Khang S, Kim HY (2006) Probiotics and their fermented food products are beneficial for health. J Appl Microbiol 100:1171–1185. doi:10.1111/j.1365-2672.2006.02963.x

Pattanagul P, Pinthong R, Phianmongkhol A, Tharatha S (2008) Mevinolin, citrinin and pigments of adlay angkak fermented by Monascus sp. Int J Food Microbiol 126:20–23. doi:10.1016/j.ijfoodmicro.2008. 04.019

Phithakpol B, Varanyanond W, Reungmaneepaitoon S, Wood H (1995) The Traditional Fermented Foods of Thailand. Kuala Lumpu: ASEAN Food Handling Bureau

Prajapati JB, Nair BM (2003) The history of fermented foods. In: Farnworth R (ed) Handbook of fermented functional foods. CRC Press, New York, pp 1–25. Metchnikoff E (1908) Optimistic studies. Putman's Sons, New York, pp 161–183

Quigley L, O'Sullivan O, Beresford TP, Ross RP, Fitzgerald GF, Cotter PD (2011) Molecular approaches to analysing the microbial composition of raw milk and raw milk cheese. Int J Food Microbiol 150:81–94. doi:10.1016/j.ijfoodmicro.2011.08.001

Raghavendra P, UshaKumari SR, Halami PM (2011) Phytate degrading Pediococcus pentosaceus CFR R123 for application in functional foods. Benefic Microbes 2:57–61. doi:10.3920/BM2010.0031

Rai AK, Palni U, Tamang JP (2010) Microbiological studies of ethnic meat products of the Eastern Himalayas. Meat Sci 85:560–567. doi:10.1016/j.meatsci.2010.03

Reddy NR, Salunkhe DK (1980) Effect of fermentation on phytate phosphorus, and mineral content in black gram, rice, and black gram and rice blends. J Food Sci 45:1708–1712. doi:10.1111/j.1365-2621.1980.tb07594.x

Rhee CH, Park HD (2001) Three glycoproteins with mutagenic activity identified in *Lactobacillus plantarum* KLAB21. Appl Environ Microbiol 67:3445–3449. doi:10.1128/AEM.67.8.3445-3449.2001

Russo P, Capozzi V, Arena MOP, Spadaccino G, Duenas MT, Lopez P, Fiocco D, Spano G (2014) Riboflavin-overproducing strains of *Lactobacillus fermentum* for riboflavin-enriched bread. Appl Microbiol Biotechnol 98:3691–3700. doi:10.1007/s00253-013-5484-7

Sah BN, Vasilijevic T, McKechnic S, Donkar ON (2014) Effect of probiotics on antioxidant and antimutagenic activities of crude peptides extracted from yogurt. Food Chem 156:264–270. doi:10.1016/j.foodchem.2014.01.105

Salminen S, von Wright A, Morelli L, Marteau P, Brassart D, de Vos WM, Fonden R, Saxelin M et al (1998) Demonstration of safety of probiotics—a review. Int J Food Microbiol 44:93. doi:10.1016/S0168-1605(98)00128-7

Sandine WE (1996) Commercial production of starter cultures. In: Cogan TM, Accolas J-P (eds) Dairy starter cultures. VCH Publishers, New York, pp 191–206

Santos F, Wegkamp A, de Vos WM, Smid EJ, Hugenholtz J (2008) High-level folate production in fermented foods by the B12 producer *Lactobacillus reuteri* JCM1112. Appl Environ Microbiol 74(10):3291–3294. doi:10.1128/AEM.02719-07

Seppo J, Donohue DC (1996) Safety assessment of *Lactobacillus* strain GG (ATCC 53103). Nutr Today 31:12S–15S

Settanni L, Corsetti A (2008) Application of bacteriocins in vegetable food biopreservation. Int J Food Microbiol 121(2):123–138. doi:10.1016/j.ijfoodmicro.2007.09.001

Shida K, Nomato K (2013) Probiotics as efficient immunopotentiators: translational role in cancer prevention. Indian J Med Res 138:808–814

Shin HS, Lee JH, Pestka JJ, Ustunol Z (2000) Viability of bifidobacteria in commercial dairy products during refrigerated storage. J Food Prot 63(3):327–331. doi:10.1111/j.1750-3841.2010.01620.x

Shin DH, Jung SJ, Chae SW (2015) Health benefits of Korean fermented soybean products. In: Tamang JP (ed) Health benefits of fermented foods. CRC Press, New York, pp 395–431

Shobharani P, Halami PM (2014) Cellular fatty acid profile and H+-ATPase activity to assess acid tolerance of Bacillus sp. for potential probiotic functional attributes. Appl Microbiol Biotechnol 98:9045–9058. doi:10.1007/s00253-014-5981-3

Shobharani P, Halami PM (2016) *In vitro* evaluation of the cholesterol-reducing ability of a potential probiotic *Bacillus* sp. Ann Microbiol 66:643–651. doi:10.1007/s13213-015-1146-6

Shobharani P, Padmaja RJ, Halami PM (2015) Diversity in the antibacterial potential of probiotic cultures Bacillus licheniformis MCC2514 and Bacillus licheniformis MCC2512. Res Microbiol 166(6):546–554. doi:10.1016/j.resmic.2015.06.003

Strozzi GP, Mogna L (2008) Quantification of folic acid in human feces after administration of *Bifidobacterium* probiotic strains. J Clin Gastroenterol 42:S179–S184

Sun Z, Liu W, Gao W, Yang M, Zhang J, Wang J, Menghe B, Sun T, Zhang H (2010) Identification and characterization of the dominant lactic acid bacteria from kurut: the naturally fermented yak milk in Qinghai, China. J Gen Appl Microbiol 56:1–10. doi:10.2323/jgam.56.1

Swain MR, Anandharaj M, Ray RC, Rani RP (2014) Fermented fruits and vegetables of Asia: a potential source of probiotics. Biotechnol Res Int 2014:1–19. doi:10.1155/2014/250424

Syal P, Vohra A (2013) Probiotic potential of Yeasts isolated from traditional Indian fermented foods. Int J Microbiol Res 5:390–398. doi:10.9735/0975-5276.5.2.390-398

Szajewska H, Skorka A, Ruszcynski M, Gieruszczak-Bialek D (2007) Meta-analysis: *Lactobacillus* GG for treating acute diarrhoea in children. Aliment Pharmacol Ther 25:871–881. doi:10.1111/j.1365-2036.2007.03282.x

Tamang JP (2010a) Himalayan fermented foods: microbiology, nutrition, and ethnic values. CRC Press, Taylor and Francis Group, New York. isbn:9781420093247

Tamang JP (2010b) Diversity of fermented foods. In: Tamang JP, Kailasapathy K (eds) Fermented foods and beverages of the world. CRC Press, Taylor and Francis Group, New York, pp 41–84. doi:10.1201/ebk1420094954-c2

Tamang JP (2014) Biochemical and modern identification techniques—microfloras of fermented foods. In: Batt C, Tortorello MA (eds) Encyclopaedia of food microbiology, 2nd edn. Elsevier, Oxford, pp 250–258

Tamang B, Tamang JP (2009) Lactic acid bacteria isolated from indigenous fermented bamboo products of Arunachal Pradesh in India and their functionality. Food Biotechnol 23:133–147. doi:10.1080/08905430902875945

Tamang B, Tamang JP (2010) *In situ* fermentation dynamics during production of gundruk and *khalpi*, ethnic fermented vegetables products of the Himalayas. Indian J Microbiol 50(Suppl 1):93–98. doi:10.1007/s12088-010-0058-1

Tamang B, Tamang JP, Schillinger U, Franz CMAP, Gores M, Holzapfel WH (2008) Phenotypic and genotypic identification of lactic acid bacteria isolated from ethnic fermented tender bamboo shoots of North East India. Int J Food Microbiol 121:35–40. doi:10.1016/j.ijfoodmicro.2007.10.009

Tamang JP, Thapa N, Tamang B, Rai A, Chettri R (2015) Microorganisms in fermented foods and beverages. In: Tamang JP (ed) Health benefits of fermented foods. CRC Press, Taylor and Francis Group, New York, pp 1–110. doi:10.1201/b18279-2

Tamang JP, Watanabe K, Holzapfel WH (2016) Review: diversity of microorganisms in global fermented foods and beverages. Front Microbiol 7:377. doi:10.3389/fmicb.2016.0037

Tamime AY, Robinson RK (2007) Yoghurt science and technology. Woodhead Publishing, Cambridge

Tamime AY, Saarela M, Korslund SA, Mistry VV et al (2005) Production and maintenance of viability of probiotic micro-organisms in dairy products. In: Tamime AY (ed) Probiotic dairy products. Blackwell Publishing, Oxford, pp 39–72

Tan LT, Chan K-G, Lee L-H, Goh B-H (2016) Streptomyces bacteria as potential probiotics in aquaculture. Front Microbiol 7:79. doi:10.3389/fmicb.2016.00079

Thapa N, Pal J, Tamang JP (2004) Microbial diversity in ngari, hentak and tungtap, fermented fish products of Northeast India. World J Microbiol Biotechnol 20:599–607. doi:10.1023/B:WIBI.0000043171.91027.7e

UBIC-Consulting (2008) The world probiotic ingredient market 2005–2007 [WWW document]. http://www.ubic-consulting.com/template/fs/documents/Nutraceuticals/Probiotic-synbiotic-Ingredient-Market

Villena J, Kitazawa H (2014) Modulation of intestinal TLR4-inflammatory signaling pathways by probiotic microorganisms: lessons learned from *Lactobacillus jensenii* TL2937. Front Immunol 4:512. doi:10.3389/fimmu.2013.00512

Wallerath T, Li H, Godtel-Ambrust U, Schwarz PM, Forstermann U (2005) A blend of polyphenolic compounds explains the stimulatory effect of red wine on human endothelial NO synthase. Nitric Oxide 12(2):97–104

Yadav H, Jain S, Sinha PR (2007) Antidiabetic effect of probiotic *dahi* containing *Lactobacillus acidophilus* and *Lactobacillus casei* in high fructose fed rats. Nutrition 23:62–68. doi:10.1016/j.nut.2006.09.002

Zago M, Fornasari ME, Carminati D, Burns P, Suarez V et al (2011) Characterization and probiotic potential of *Lactobacillus plantarum* strains isolated from cheeses. Food Microbiol 28:1033–1040. doi:10.1016/j.fm.2011.02.009

Dietary Impacts on the Composition of Microbiota in Human Health and Disease

21

Anil Kumar Verma, Reena Kumari, Alok Bhattacharya, and Jaishree Paul

Abstract

More and more metabolic diseases, chronic inflammatory diseases, and cancers are being linked with the alteration in gut microbiota. Host genetics and environment are some of the factors that are thought to contribute in shaping gut flora. Recent research has suggested that development of gut microbial consortia and thereby host-microbe interaction is essentially guided by the early colonizers and the diet. Important metabolites derived from the gut bacteria and diet help in the development and maintenance of a healthy gut and consequently the immune system. In the present review, we have examined the impact of diet on evolution, stabilisation and dysbiosis of human gut microbiota and how diet induced subtle changes in the microbiota lead to the disease state. Diet, an important environmental factor, plays crucial role in tilting this fine balance in either directions. The importance of dietary pattern in regulating growth of beneficial bacteria has been discussed. We have also examined different intervention strategies that affect microbiota and consequently the metabolite profiles resulting in a diseased state. Various high-throughput techniques that are used in the studies described here have also been discussed.

Keywords

Diet • Dietary pattern • Diseases • Gut microbiota • Health • Metabolism

---

A.K. Verma • A. Bhattacharya • J. Paul (✉)
School of Life Sciences, Jawaharlal Nehru University, New Delhi 110 007, Delhi, India
e-mail: jpaul33@hotmail.com

R. Kumari
Division of Hematology and Blood and Marrow Transplant, Markey Cancer Center, Internal Medicine, University of Kentucky, Lexington 40536, KY, USA

V.C. Kalia et al. (eds.), *Mining of Microbial Wealth and MetaGenomics*,
DOI 10.1007/978-981-10-5708-3_21

21.1 Gut Flora: From Oblivion to Limelight

We have witnessed a gigantic leap in the past one and half decade in the field of microbiome research that has helped us to realize the critical role of gut bacteria in shaping our health through manipulation of overall human physiology. The dynamic changes in the gut microbiome (GM) can now be evaluated with new technologies that allow us to interrogate nucleic acid sequence, metabolome, and proteome at an unprecedented depth. This helps in analyzing complex microbial communities, such as GM of healthy and diseased individuals. As early as 2001, Hooper and Gordon had described that human lives are affected by the microbes, both positively and negatively from birth to death, and human body surfaces are colonized by vast, diverse, and dynamic microbial consortia. The structure, variability, and functionality of the microbial consortium vary depending on the body environment. The human gut is densely populated in a nutrient-rich environment housing trillions of microbes of diverse microbial species. The entire gut from oral cavity to rectum is populated by bacteria, but the majority lives in our colon with maximum densities reaching to 10^{11}–10^{12} cells/mL (Whitman et al. 1998). Each individual harbors about 1000 or more phylotypes of gut bacteria, and the combined gene content of GM exceeds that of human by about 100-fold (Qin et al. 2010). Moreover, there is tremendous amount of diversity in terms of different bacteria within and between different persons and can fluctuate at different time points especially during disease (Lozupone et al. 2012). The gut bacteria broadly include the members of phyla *Bacteroides*, *Firmicutes*, *Actinobacteria*, *Proteobacteria*, *Verrucomicrobia*, *Cyanobacteria* (*Melainabacteria*), and *Archaea* (Eckburg et al. 2005).

One of the most important metabolic functions carried out by gut microbiota (GMt) is fermentation of non-digested food components in the large intestine and extraction of extra energy in the form of short-chain fatty acids (SCFAs), i.e., acetate, propionate, and butyrate. GMt also produces vitamins and helps in absorption of minerals by the host gut. Gut flora maintains the equilibrium between the enteric nervous system and gut immune system of the host. The alterations in configuration of gut flora may be potentially harmful to the host. Several recent research reports have indicated the involvement of GMt in the progression of several chronic diseases, such as obesity, diabetes mellitus, rheumatoid arthritis, inflammatory bowel disease (IBD), heart diseases, allergy, and cancer. Alterations in GMt can be attributed to the use of antibiotics, age, disease, stress, dietary habits, socioeconomic status, and lifestyle (Flint et al. 2012). The diet has been considered to be a universal external factor that can influence the gut microbial composition of the host. Preliminary evidence suggests that the dietary patterns are associated with specific composition of the microbiota, and it is now possible to link various constituents of food with respect to growth of specific bacteria. In this chapter, we have focused on summarizing current understanding on various effects of diet on gut flora, optimizing gut microbiota composition by dietary means and various challenges faced in designing and interpreting the results from human studies.

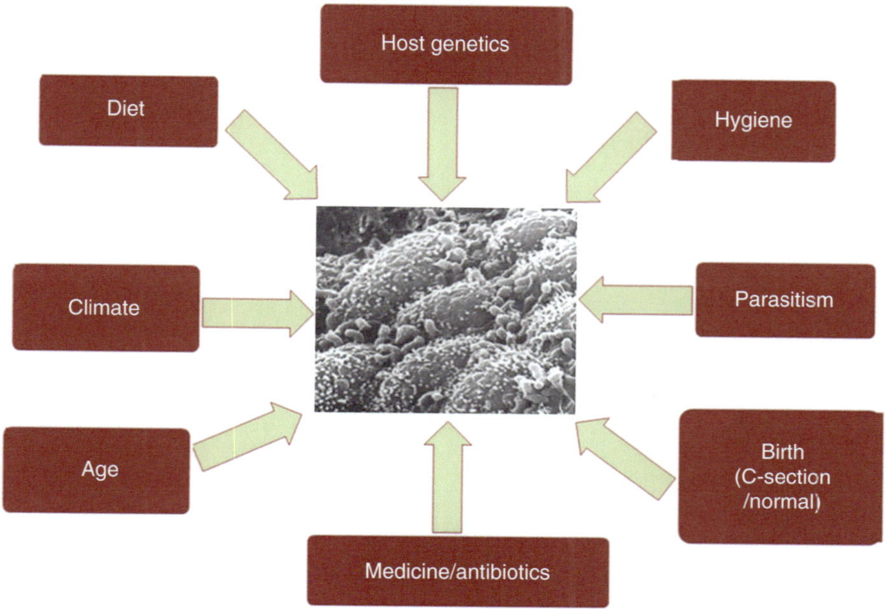

Fig. 21.1 Schematic representation of factors affecting the gut flora of an individual

21.2 Factors Affecting the Composition of Gut Microbiota

The composition of gut bacterial communities is generally stable in adult individuals and yet quite dynamic as it varies from person to person. In the uterus, fetus encounters very few microbes and microbial components maintaining near sterility in the gut. The colonization of infant gut begins during the birth and continues after the birth. The diversity, richness and the stability of gut microbiota are influenced by several factors, such as mode of delivery, age, diet, host genetics, hygiene, sanitation and pharmacological exposure to antibiotics (Fig. 21.1). Broadly the factors can be defined as follows.

21.2.1 Age

The development of human fetus takes place in almost sterile ambience, and the colonization of gut begins immediately after the birth of an infant. Firstly, the mode of delivery (vaginal birth vs. cesarean section) affects the early colonization of gut. Infants born through vaginal route predominantly harbor *Lactobacillus* and *Prevotella* (bacteria found in vagina), whereas infants delivered through cesarean section are likely to harbor bacteria found on mother's skin and in the hospital environment (Dominguez-Bello et al. 2010). Babies born through cesarean section are

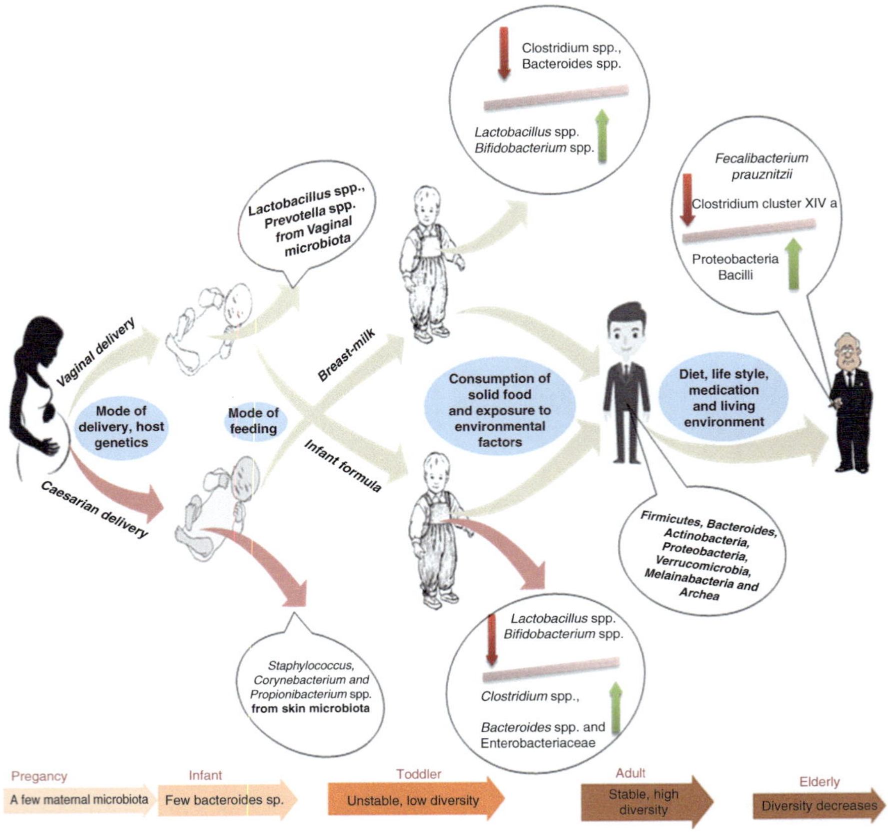

Fig. 21.2 Evolution of gut microbial diversity in humans

exposed to antibiotics early in life which may influence the gut microbiota. The gut flora of infants is initially shaped by mother's milk or formula milk and later by solid food at the postweaning stage. In about 2–5 years, the gut flora of child matures and resembles with that of the adult in terms of richness and diversity (Yatsunenko et al. 2012). The gut flora of elderly persons depends on the lifestyle, diet, health status, and exposure to antibiotics (Claesson et al. 2012; Jeffery et al. 2016). The diversity and stability of gut microbiota vary with age from infancy to adulthood and depend on interaction with host genetics and environmental factors as depicted in Fig. 21.2.

21.2.2 Diet

The word "diet" originated from the Greek word "diaita" which means balanced way of life. Diet plays very critical role in configuring the gut microbial composition and the effect of diet on gut flora begins very early in the life with very first meal of the infant in the form of mother's milk. Gut flora of breast-fed infants is different from that of the formula-fed infants (Harmsen et al. 2000). Dietary interventions with high-fat and

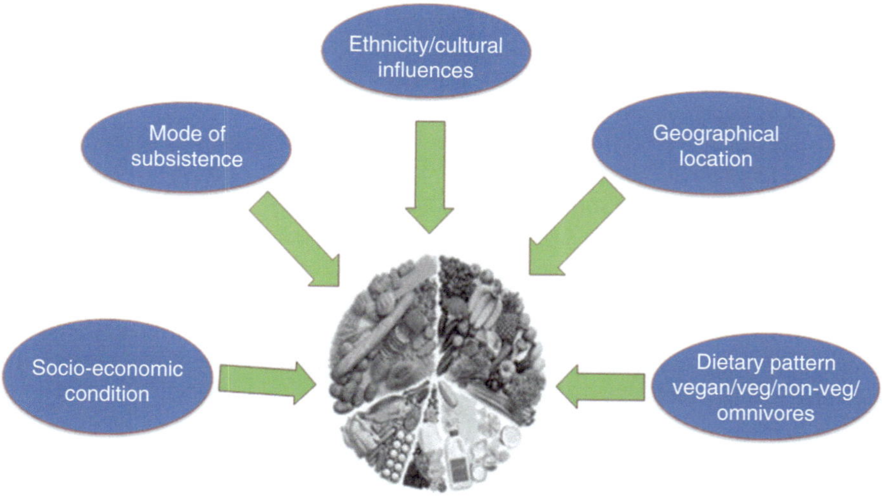

Fig. 21.3 Schematic representation of factors affecting the diet of an individual

high-sugar (HFHS) diet and plant polysaccharide-based diet in mice of different genetic background have shown that the effect of diet dominates the genetic background in configuring the gut microbiota (Kashyap et al. 2013; Carmody et al. 2015). Diet not only affects the composition of gut bacterial flora of host but it can also affect gene expression of gut microbes. The gut bacteria *Bacteroides thetaiotaomicron* changes gene expression depending on the availability of dietary fiber (plant glycan) in food. If diet is low in dietary fiber, the expression of gene for fermentation of host mucus glycans is upregulated, and reverse phenomenon is seen when dietary fiber is high in food (Kashyap et al. 2013). The diet of an individual depends on several factors, such as geographical location, climate, socioeconomic status, ethnicity, mode of subsistence and health status (Fig. 21.3). The composition of major food components (proportion of carbohydrate, protein and fat), source/type (vegan, vegetarian, nonvegetarian, or omnivorous), and pattern (long term or short term) of diet affects gut microbiota of individuals.

21.2.3 Host Genetics

The composition of gut bacterial flora is the result of coevolution between host and gut bacteria (Ley et al. 2006). The gut microbiota of twins is more similar than that of non-twins and parents depending on relatedness (Zoetendal et al. 2001). The host genotype can affect the gut microbial diversity in number of ways. For example, the absence of α-1-2-fucosyltransferase activity in non-secretor (FUT2 mutants) individuals led to altered mucus structures in the host intestine and eventually deceased α-diversity of gut microbiota (Kashyap et al. 2013). Humans harboring mutation in FUT2 gene are more susceptible to gut microbiota-associated chronic inflammatory diseases, such as Crohn's disease (McGovern et al. 2010) and primary sclerosing cholangitis (Folseraas et al. 2012). The host genotype is responsible for lactose digestion in the small intestine of individuals. Two SNPs ($C/T_{13,910}$ and $G/A_{22,018}$) located in the 14 and 22 kb upstream of the 5′-end of lactase gene have been associated with lactase persistence. Lactose-intolerant individuals are unable to digest it in the small intestine and slowly move to the colon, and subsequently colonic flora metabolizes them into unwanted by-products like large quantity of H_2 and methane gas (Enattah et al. 2002).

Even, the presence of some microbial taxa has been linked with host genetics. The members of the family *Christensenellaceae*, methanogenic archaea, and few other undefined bacteria coexist in human gut in the form of a network that is inherited. The consortium of *Christensenellaceae* along with the partners is associated with low body mass index (BMI) individuals indicating the role of host genetics in shaping the gut microbial composition and consequently the host metabolism (Goodrich et al. 2014).

21.2.4 Hygiene

Several animal studies indicate that gut microbial richness and diversity play an important role in the development and progression of noncommunicable diseases like IBD, diabetes, autoimmune disorders, colon cancer, and cardiovascular diseases. Due to poor hygienic practices, infectious diseases are highly prevalent whereas noncommunicable diseases are very less prevalent in nonindustrialized countries such as Papua New Guinea, Burkina Faso, Malawi, and Tanzania. Various studies on the fecal samples of children and adults of non-industrialized countries (Burkina Faso (De Filippo et al. 2010), children and adults in Malawi and Amazonian Amerindians (Yatsunenko et al. 2012), and adult Hadza hunter-gatherers in Tanzania (Schnorr et al. 2014), adults in Papua New Guinea) have shown high α-diversity and lower β-diversity in gut microbiota in comparison to US and European population (Clemente et al. 2015; Martinez et al. 2015). Progressive shrinkage in gut microbial diversity is a cumulative result of changes in lifestyle, diet, hygienic practices, and exposure to antibiotics over the years in Western countries. According to microflora hypothesis, "the low microbial exposure of individuals following western lifestyle affects colonization of the infant's gut and disturbs the immune system development leading to chronic inflammatory diseases" (Wold 1998). Several reports do indicate and validate the microflora hypothesis in humans. Recently Zhou et al. reported that gut microbial diversity and serum IgE levels in mice vary with exposure to antigen or bacteria from housing environment such as soil, house dust, and decaying plants (Zhou et al. 2016). Excessive hygienic and sanitation practices affect the diversity of gut microbiota and lead to poor development of immune system.

21.2.5 Antibiotics

The use of antibiotics in control, management, and treatment of bacterial infections has been a milestone in public health and has saved millions of lives worldwide. Though antibiotics have saved the lives, the collateral damage they cause to indigenous host flora is now being evaluated to fine-tune its use in medicine (Modi et al. 2014). Exposure to antibiotics may start very early in life of C-section babies and other children depending on health status. In mice, pulsed antibiotic treatment in early life has resulted in perturbed gut flora with altered metabolic functions, altered gene expression in intestinal epithelial cells, and recruitment of different T-cells in the lamina propria. These non-obese diabetic mice developed type 1 diabetes early in life than controls (Livanos et al. 2016). The use of broad spectrum antibiotic results in depletion of commensal gut flora diversity, and colonization and proliferation of antibiotic-associated diarrhea (AAD) associated pathogens such as *Salmonella typhimurium* and *Clostridium difficile* (Hogenauer et al. 1998). AAD is prevalent in 5–25% of antibiotic-treated patients (Bergogne-Berezin 2000).

The growth of *S. typhimurium* and *Clostridium difficile* in lumen of antibiotic-treated patients is facilitated by microbiota-generated mucosal carbohydrates, fucose and sialic acid. *S. typhimurium* catabolizes fucose and sialic acid, while *C. difficile* uses sialic acid in vivo. The impact of antibiotic treatment on human gut flora can be detected as early as 3–4 days of treatment initiation and may persist for more than 2 years. Frequent antibiotic courses may result in alternative and stable microbial configuration enriching antibiotic resistance gene in gut metagenome. The recovery to pre-antibiotic treatment state varies across individuals and remains often incomplete (Dethlefsen and Relman 2011). The unnecessary, excessive, incomplete dose compliance has resulted in multidrug resistance in pathogens. Interestingly, functional antibiotic resistance genes have been reported in gut flora of Yanomami tribe, Venezuela, which has never been exposed to commercial antibiotics. However, the source of antibiotic resistance genes for Yanomami tribe may be from antibiotic-producing soil bacteria or human commensal ancestors found in soil (Clemente et al. 2015). Antibiotic exposure leads to enrichment of antibiotic resistance genes and transfers of these genes to other bacteria via horizontal gene transfer. Antibiotic exposure along with other environmental factors, such as high-fat diet, enhances the propensity of development of a number of metabolic diseases mainly due to alteration in GM (Mahana et al. 2016).

21.2.6 Parasitism

It has been hypothesized that the gut bacterial components from the intestinal microbiota function as a stimulatory factor in the pathogenicity of the parasite giardia (Torres et al. 2000). *Entamoeba histolytica*, an intestinal protozoan parasite infection, has been shown to alter the gut flora (Verma et al. 2012). A recent study showed that colonization of gut by commensal segmented filamentous bacteria protects from *E. histolytica* infection in a murine model (Burgess et al. 2014). The higher parasitic load (*E. histolytica*) and significantly increased population of *Prevotella copri* in symptomatic cases have been found to be associated with diarrhea in Bangladesh children below 2 years of age (Gilchrist et al. 2016). Further, a recently conducted metagenomic study in African population stated that colonization by *E. histolytica* can be estimated with ~80% accuracy on the basis of gut microflora of an individual. Individuals with entamoeba infection show high alpha diversity and low interindividual variation. However, it is important to note that these subjects were asymptomatic at the time of this study (Morton et al. 2015). Furtther studies are required to understand if the observed relationship of parasitic infection and gut flora is cause or effect.

21.3 Diet as a Modulating Factor of Gut Bacterial Flora

Diet is considered a crucial factor that shapes gut flora, and the population of each microbial community can also be linked with diet. Diet-induced modifications in gut microbiota have been implicated in high incidence/prevalence of diseases of affluence such as obesity, cardiovascular disease, diabetes and IBD in industrialized USA and other Western countries (Hou et al. 2011; Moschen et al. 2012). Population can be separated from each other on the basis of characteristic differences in their gut flora associated with dietary pattern. The ratio of dominant gut flora like *Bacteroides* and *Prevotella* shows good correlation with overall diversity and dietary pattern. High concentration of *Prevotella* has been associated with consumption of high-fiber diet in children of African villages (agrarian diet), whereas high concentration of *Bacteroides* has been linked with diet rich in animal protein, saturated fat, and refined carbohydrates typical of Western diet (Wu et al. 2011; Yatsunenko et al. 2012).

The categorization of microbial configuration of individuals into distinct groups is called enterotyping, and a particular discrete microbial configuration is called enterotype. The concept of enterotypes links gut microbial configurations to diet. It is clear from multiple studies that evolution of specific enterotypes happens due to long-term dietary pattern. Specific genera that constitute an enterotype remain stable unaffected by age, sex, BMI, and nationality of an individual (Arumugam et al. 2011). The predominant enterotypes can be categorized as *Bacteroides*, *Prevotella*, and *Ruminococcus*. *Bacteroides* enterotype dominated by members of phylum *Bacteroidetes* (especially by *Alistipes* and *Parabacteroides*) is associated with the consumption of high animal fat and high-protein diet. Similarly *Prevotella* enterotype dominated by *Paraprevotella* (phylum *Bacteroidetes*) and *Catenibacterium* (phylum *Firmicutes*) is associated with high-fiber/high-carbohydrate diet. The *Ruminococcus* enterotype partially overlaps with that of *Bacteroides* enterotype and cannot be classified as a separate enterotype (Wu et al. 2011; Tremaroli and Backhed 2012; Graf et al. 2015).

21.4 Effect of Dietary Constituents on Gut Flora

Relative composition of major diet constituents, such as lipids, proteins, and carbohydrates, affects gut microbiota and human health differently. Gut microbes use different mechanisms to collect energy from lipids, proteins, and carbohydrates.

21.4.1 Carbohydrates

A large fraction of our diet is made up of different types of carbohydrates, a main source of energy and nutrient for both host and microbes. Many components of diet including digestible carbohydrates like sucrose, lactose, starch, etc. are digested and assimilated in the small intestine by human enzymes except undigested dietary complex carbohydrates. The undigested products are subsequently passed into the large intestine for fermentation by gut bacteria. Human gut microbiome (a model microbiome of 177sp.) is very rich in glycoside hydrolases (>9000) and polysaccharide lyases (~300) in comparison to human genome which barely encodes ~17 glycoside hydrolases and no polysaccharide lyases (Cantarel et al. 2012; El Kaoutari et al. 2013). According to the "American Society of Cereal Chemists" (2001), dietary fiber can be defined as "the edible parts of plants or analogous carbohydrates that are resistant to digestion and absorption in human small intestine with complete or partial fermentation in the large intestine." The carbohydrates which are metabolically available for fermentation by gut bacteria are collectively referred as microbiota-accessible carbohydrates (MACs) (Sonnenburg and Sonnenburg 2014). The main sources of dietary MACs are plant glycans (e.g., dietary fiber), glycans of animal origin, host-produced glycans (e.g., mucin), and polysaccharides produced by food-associated microbes. The quantity of MAC present in a food source depends on members of gut microbiota of an individual. In Japan, a popular food sushi is prepared using a nutritional seaweed *Porphyra* spp. (nori). The gut bacterium *Bacteroides plebeius* which can digest complex algal polysaccharides porphyran and agarose is found only in Japanese individuals but not in Westerners. Thus, algal polysaccharide porphyran is MAC for Japanese but not for North Americans and Europeans (Hehemann et al. 2010). However, it is important to note that the genes for digestion of algal polysaccharides have been acquired by gut bacterium *Bacteroides plebeius* from marine bacteria associated with seafood. This indicates the source of novel genes and how gut flora evolves with diet in individuals of a geographic location.

Consumption of fiber-rich diet has been associated with high microbial diversity with predominance of *Prevotella* enterotype. The microbial fermentation of dietary fibers produces acetate, propionate, and butyrate commonly known as short-chain fatty acids (SCFAs). Acetate, propionate, and butyrate are found at a molar ratio of 60:20:20 in colonic lumen. About 80–90% of SCFAs are absorbed in the colon and have profound impact on gut and host physiology. The acetate and propionate are sources of energy in peripheral tissues whereas butyrate for colonocytes. Both of these SCFAs are used as substrate for lipogenesis and gluconeogenesis by the liver, respectively (Rombeau and Kripke 1990). Propionate is reported to enhance intestinal gluconeogenesis via afferent nervous system (De Vadder et al. 2014). Acetate is used as substrate for production of butyrate by *Faecalibacterium prausnitzii* and *Roseburia* spp. (Duncan et al. 2004) and protects from enteric infections (Fukuda et al. 2011). Butyrate (Davie 2003) and acetate (Thorburn et al. 2015) can inhibit histone deacetylase inhibitors and consequently affect gene expression in colonic epithelial cells and peripheral tissues, respectively. Butyrate has shown anti-inflammatory and anticarcinogenic activity in cell lines, but studies on human subjects are limited (Hamer et al. 2008). Butyrate has also been associated with strong antitumor activity. The cellular effect of short-chain fatty acids is mediated through G-protein-coupled receptors GPR41/FFAR3 (Samuel et al. 2008) and GPR43/FFAR2 affecting inflammation (Maslowski et al. 2009). It also helps in the regulation of antidiabetic hormone GLP-1 secretion (Tolhurst et al. 2012). SCFAs are anti-inflammatory, regulate glucose homeostasis, promote integrity of intestinal epithelium and gut homeostasis, and regulate the number and function of the regulatory T-cells in the colon (Thorburn et al. 2015). Two predominant phyla *Bacteroidetes* and *Firmicutes* are main producers of SCFAs in human gut. Acetate and propionate are mainly produced by several members of *Bacteroidetes*, whereas butyrate is produced in high amounts by the *Firmicutes* (Macfarlane and Macfarlane 2003). Human milk is rich in milk oligosaccharides (MOS), most of which are indigestible glycans and possess prebiotic property (Stark and Lee 1982). MOS, phosphate, and milk proteins are known to encourage the growth of *Bifidobacteria* spp. and *Lactobacillus* spp. (Yoshioka et al. 1983). Breast-fed babies harbor more *Bifidobacteria* and *Lactobacillus* sp. but less of *Bacteroides* spp., *Enterobacteria*, and *Clostridium* spp. in comparison to formula-fed babies (Harmsen et al. 2000; Fallani et al. 2011).

The fermentation of different types of cellulose in human colon is influenced by the presence or absence of methanogens which directs the configuration of cellulolytic microbial consortium in human gut. The cellulose fermenters belong mainly to *Bacteroidetes* in non-methane excreting individuals, whereas members of *Firmicutes* degrade cellulose in methane excreting individuals. This diversity of cellulose degraders enables the fermentation of different kinds of cellulose (Chassard et al. 2010). Dietary modifications can influence host-microbe interaction, the immune system, and subsequently inflammation in different tissues. Experiments on mice have shown that mice fed on high-fiber diet had higher concentration of circulating SCFAs and were protected from lung allergy in comparison to mice fed on low-fiber diet. Influence of diets in the well-being of individuals is likely to be through microbes, microbial metabolites, and hematopoiesis, though we still do not understand detailed pathways linking diet to disease processes (Trompette et al. 2014).

Arabinogalactan and inulin have the potential to promote a gut bacterial profile found in a lean individual. Propionate and acetate are the two SCFAs predominantly generated by arabinogalactan fermentation. Fermentation of arabinogalactan by microbiota of an obese individual showed a higher production of propionate in comparison to *n*-butyrate. The fermentation of both substrates by lean microbiota produces more *n*-butyrate in comparison to that in obese microbiota (Aguirre et al. 2016). Arabinogalactan fermentation has been shown to be associated with a statistically significant increase in the population of bacteria particularly *Bacteroidetes* and *Faecalibacterium prausnitzii* with a simultaneous decrease in pathogenic bacteria *Clostridium perfringens* (Terpend et al. 2013). The effect of dietary glycans on gut microbiota and overall metabolic functions may vary depending on initial gut microbial configuration (Walker et al. 2011).

21.4.2 Effect of Noncaloric/Nonnutritive Sweeteners on Health and Gut Microbiota

Noncaloric/nonnutritive sweeteners (synthetic and natural) are regularly being used as additives in food and beverages across the globe in order to control negative impact of high-caloric sugar on human health that includes weight gain and obesity, among other diseases (Swithers 2013; Suez et al. 2014). Aspartame (L-aspartyl-L-phenylalanine methyl ester), sucralose (trichlorogalactosucrose), saccharin (1,1-dioxo-1,2-benzothiazol-3-one), acesulfame-K (5,6-dimethyl-1,2,3-oxathiazine-4(3H)-one 2,2-dioxide), stevia (steviol glycosides, rebaudioside A, stevioside), neotame (*N*-[*N*-(3,3-dimethylbutyl)-L-α-aspartyl]-L-phenylalanine-1-methyl ester), and luo han guo extract (cucurbitane glycosides, mogrosides II, III, IV, V, VI) are the artificial sweeteners approved by FDA, USA (Fitch and Keim 2012). All of these noncaloric/nonnutritive sweeteners have also been approved by FSSAI, India, for use as an additive in food and beverages except luo han guo extract. Nonnutritive sweeteners are generally considered healthy alternative to caloric sugars, but recent animal studies suggest that replacing caloric sugars with synthetic sweeteners leads to metabolic dysregulation. Saccharin and aspartame induce greater weight gain than sucrose in rats with similar caloric consumption (Feijo et al. 2013). Artificial sweeteners such as acesulfame-K and saccharin and natural nonnutritive sweeteners such as stevioside (sweeteners extracted from the leaves of the *Stevia rebaudiana* plant) both increase food intake, weight gain, and accumulation of body fat in Wistar rats in comparison to glucose (Swithers and Davidson 2008; Swithers et al. 2010; Swithers 2013). Chronic dietary consumption of sucralose induces reversible chronic sweet/energy imbalance and results in activation of neuronal starvation response leading to increased food intake, perturbed glucose homeostasis, insomnia, and hyperactivity in fruit flies and mammals (Wang et al. 2016). As most of the noncaloric sweeteners are not digested in human intestine, chronic ingestion of some of the noncaloric sweeteners, such as saccharin sucralose and aspartame, disturbs glucose homeostasis and gut flora in C57B1/6 mice. Metagenomic analysis of gut flora of saccharin-consuming mice showed alteration in more than 40 OTU with relative enrichment of several taxa belonging to *Bacteroides* genus and *Clostridiales* order. Saccharin-consuming mice produced high level of SCFAs (acetate and propionate) than glucose-consuming mice, and glycan-degrading pathways are overrepresented in their GM. Saccharin induced changes in GM, and consequently glucose intolerance was also observed in humans (Suez et al. 2014).

21.4.3 Fat

Fat is an important constituent of our diet. It enhances the flavor and palatability of food and is an important component of cooking across the world. Dietary fat provides energy, helps in absorption of fat-soluble vitamins, and is considered as the main source of essential fatty acids such as ω-6 linoleic acid and ω-3 linolenic acid (Schmid 2011). Fat is an essential structural component of cell wall, a substrate for hormones and participates as cell-signaling molecule. Fat plays a very crucial role in the development of neurological and brain function in young children (Milner and Allison 1999). It is synthesized in our body and only a small amount of fat is needed for our daily needs. Dietary fat mainly contains three types of fats, viz., saturated, unsaturated, and trans fat. Most of the food items contain mixed types of fat. Unsaturated fats contain mono- and polyunsaturated fatty acids mostly generated from plant products such as seeds, nuts, and vegetable oils. On the other hand, saturated fat is generated mainly from animal-derived food, such as meat, fish, and dairy products. However, few plant products such as palm oil and coconut oil are also sources of saturated fat. The main source of trans fats is hydrogenated vegetable oil, but it is also present in small amount in beef and dairy fat. The quality and quantity of dietary fat affect human health, and excess consumption of fat has been associated with obesity and cardiovascular diseases. The nature of dietary fat consumed can induce significant changes in microbiota and gut barrier function as shown in mice (Lam et al. 2015). Diets enriched in saturated lipids have been found to be linked with white adipose tissue (WAT) inflammation, obesity, and insulin resistance in mice. When two groups of mice were fed with diet differing only in fat composition (lard or fish oil), the mice on saturated fat diet (lard) gained more weight, consumed more food, and had higher level of fasting glucose and insulin level compared to mice fed on fish oil. After 11 weeks, mice on lard diet showed increased TLR activation, high WAT inflammation, and decreased insulin sensitivity than that of fish oil-fed mice (Caesar et al. 2015). On analyzing the gut flora, it was observed that fish oil promotes the growth of *Lactobacillus* which has earlier been shown to reduce inflammation in IBD models in mice (Guarner et al. 2005). Fish oil also promotes the enrichment of *Akkermansia mucinophila* in gut flora, and this is thought to be associated with decrease in weight gain, infiltration of WAT by macrophages, and improvement of gut barrier function and glucose metabolism (Everard et al. 2013). Intake of high-fat diet leads to more secretion of bile acids (Reddy 1981). Higher concentration of bile acids creates conducive environment for growth of a number of bile-resistant gastrointestinal pathogens such as *B. wadsworthia*, *Helicobacter hepaticus*, and *Listeria monocytogenes* and protozoan parasites such as giardia, microsporidia, and cryptosporidia. Bacterial antigens and metabolic products of these bacteria such as H_2S or secondary bile acids damage the gut mucosal barrier resulting in increased immune cell infiltration and damaged tissues. High levels of bile acids change microbial assemblage by promoting the growth of *Bilophila wadsworthia*, a sulfite-reducing pathobiont, known for exacerbating colitis by inducing Th-1 immune response in genetically susceptible mice $IL10^{-/-}$ (Devkota et al. 2012). The quantity and quality of dietary fat affect the composition of gut flora. The consumption of low-fat diet promotes the growth of *Firmicutes* and reduces the abundance of others. Intake of polyunsaturated (safflower oil) fat (PUFA) and saturated (milk-derived) fat diets (MF) increases the population of *Bacteroidetes* and decreases abundance of *Firmicutes*. MF uniquely promotes the growth of *B. wadsworthia* otherwise hard to detect in healthy individuals. The abundance of *B. wadsworthia* has been associated with appendicitis and other intestinal inflammatory disorders. This kind of alteration in gut flora may shift the balanced immune state toward chronic diseases in genetically susceptible hosts. High-fat and high-sugar Western diets had been linked with obesity, imbalanced glucose homeostasis, IBD, and other chronic inflammatory diseases in genetically susceptible hosts.

Cheese, seafood, eggs, and meat are considered to be rich sources of phospholipid, phosphatidylcholine, choline, and carnitine. Choline is an important constituent of cell membrane and is considered as a substrate for synthesis of acetylcholine and provides a pool of methyl groups during metabolism of amino acids methionine and homocysteine (Zeisel 2006). Choline is involved in the formation of very-low-density lipoprotein (VLDL) and lipid metabolism in the liver. Though choline is partially synthesized by humans, insufficient dietary consumption may lead to disturbed gut flora and nonalcoholic fatty liver disease (NAFLD) in mice (Henao-Mejia et al. 2012). Microbial biotransformation of choline into TMA in human gut decreases bioavailability of choline for the host. Choline deficiency has been linked with a number of liver diseases such as NAFLD, steatosis, and hepatocarcinomas. Hepatic steatosis phenotype is influenced by GM and can be characterized using metabolomic profiling (Corbin and Zeisel 2012). Changes in the relative levels of *Gammaproteobacteria* and *Erysipelotrichia* have been reported to be associated with NAFLD in humans (Spencer et al. 2011). In excessive phosphatidylcholine-rich diet, choline gets biotransformed at a higher rate because of bacterial TMA lyase activity into TMA in gut which further gets converted into TMAO in the liver by hepatic flavin monooxygenase 3 (FMNO3), thereby increasing the risk of atherosclerosis and cardiovascular diseases (Tang et al. 2013). Recently, it has been shown that DMB (3,3-dimethyl-1-butanol), a structural analog of choline, can act as an inhibitor of bacterial TMA lyase activity and thus can control the biotransformation of choline into TMA without any side effects, thereby reducing the risk of atherosclerosis (Wang et al. 2015).

21.4.4 Protein-Rich Diet

Proteins are a major constituent of a balanced diet. Our system is unable to synthesize a number of amino acids and thus need to be taken from dietary sources for sustenance. Absorption of fat and protein normally occurs in the small intestine, while gut bacteria are mainly located in the large intestine/colon. As undigested carbohydrates reach the large intestine, some proteins also reach into the colon where fermentation of protein by the gut microflora takes place. Some of main dominant proteolytic bacteria observed in stool samples are *Bacteroides* spp., *Propionibacterium* spp., *Streptococcus*, *Clostridium*, *Bacillus*, and *Staphylococcus* (Macfarlane et al. 1986). Association of *Bacteroides* enterotype has been reported in individuals consuming diet high in animal protein (De Filippo et al. 2010). Residual undigested proteins that enter the large intestine are fermented by *Bacteroides* and *Clostridium* into varied by-products based on the amino acid composition of the proteins (Macfarlane et al. 1992).

Microbial fermentation of proteins leads to formation of different gases as well as other metabolites increasing the nitrogenous substrate for the gut microbiota and consequently putrefactive fermentation products (Silvester and Cummings 1995). This process largely takes place in the colon region. Putrefactive fermentation has been identified as one of the main causes of common bowel diseases such as CRC and IBD. Among the beneficial properties of the gut bacteria, colonic health is the major outcome. When the diet consists of fermentable carbohydrate substrates, colonic bacteria grow optimally and contribute amino acids and proteins. It has been estimated that a substantial amount (1 and 20%) of amino acids, especially lysine and threonine derived from the gut flora, have been observed in circulating plasma of adults (Laparra and Sanz 2010; Compare et al. 2012). When the gut bacteria undergo proteolytic fermentation, it produces some beneficial compounds such as polyphenols that induce anti-inflammatory, anti-oxidative, and antiaging effects and promotes generation of a reasonable concentration of SCFAs. However, anaerobic metabolism of proteins known as putrefaction leads to the production of toxic substances including ammonia, hydrogen sulfide, amines, phenols, thiols, and indoles. These compounds are recognized as cytotoxins, genotoxins, and carcinogens (Hughes et al. 2000) and are considered to be detrimental for the host's health (Cummings et al. 1979; Macfarlane et al. 1986; Compare et al. 2010; Lopez-Legarrea et al. 2014; Conlon and Bird 2015). Putrefactive fermentation is suggested to play a critical role in the development and progression of diseases like CRC and IBD in the distal colon (Macfarlane and Macfarlane 2012) (Toden et al. 2005). Western diets rich in red meat, milk, and eggs deliver higher amounts of sulfur compounds to the colon and hence favor sulfidogenic hydrogen disposal by GM (Conlon and Bird 2015). High level of sulfides in fecal samples is positively associated with high dietary protein consumption by humans. It has been shown in a mouse model that higher intake of red meat is associated with damage to colonic mucosa during shortage of fermentable dietary carbohydrates (Toden et al. 2007). Red meat is a rich source of amino acid L-carnitine. Recently it was shown that GM metabolizes dietary L-carnitine into TMA (trimethylamine) that later gets converted to trimethylamine-*N*-oxide (TMAO) by flavin monooxygenases in the liver. Increased TMAO levels have been associated with increase in atherosclerosis in mice and risk of cardiovascular diseases in humans. Microbiota of omnivorous human was found to be more efficient in producing TMAO from dietary carnitine than that of vegans or vegetarians (Koeth et al. 2013). Dietary intake of proteins may also affect drug bioavailability. Bioavailability of digoxin, a glycoside widely used as a cardiac drug in humans, is reduced in some patients due to excretion of inactive digoxin metabolite, dihydrodigoxin. It has been observed that the colonization by different strains of *Eggerthella lenta* and host diet acts together and affect bioavailability of digoxins. Pharmacokinetic studies in gnotobiotic mice have shown that dietary protein reduces in vivo microbial metabolism of digoxin by *E. lenta*, resulting in significant changes in the drug concentration both in the serum and urine (Haiser et al. 2013).

Large epidemiological studies have indicated an association of colorectal cancer (CRC) with the high consumption of red and processed meat (Norat et al. 2005). In animal models, significant changes were observed in the composition of the GMt with increase in age and high-fat-containing diet. However, the effect can be nullified if the diet is supplemented with high protein to sucrose ratio (Kiilerich et al. 2016). Similar findings were observed in another study where in rodents fed with high-fat diet, obesity was prevented with a supplement diet high in protein to sucrose ratio (Pichon et al. 2006; Freudenberg et al. 2013). It was also observed that a diet high in protein and low in carbohydrates affected the GMt and fatty acid profiles in obese individuals. After consumption of high-protein diet for about 4 weeks, there was an increase in branched-chain fatty acids with a simultaneous decrease in butyrate concentration. This was further confirmed by a decrease in *Roseburia/Eubacterium* numbers. Further, it was also noted that due to high intake of proteins and low carbohydrate, concentration of fiber-derived antioxidant phenolic acids decreased (Russell et al. 2011). The decrease in *Roseburia/E. rectale* population decreases the butyrate level in fecal SCFAs (Lopez-Legarrea et al. 2014). Controlled dietary studies in obese men on high-protein/low-carbohydrate diet for 3–4 weeks have shown reduction in population of *Collinsella aerofaciens*, *E. rectale*, *Roseburia*, and *Bifidobacterium* spp. (Duncan et al. 2007; Russell et al. 2011).

Zonulin is a marker of intestinal permeability. The plasma level of zonulin is reported to be higher in celiac disease and type 1 diabetes. The plasma level of zonulin has been reported to be inversely proportional to the amount of protein taken up through diet and is associated with diet composition (Zak-Golab et al. 2013). Bacteria and gluten cause zonulin levels to rise in all individuals, suggesting that gluten-free diet may prevent an individual from leaky gut condition.

Gluten proteins are mainly found in barley, wheat, and rye, and consumption of these causes celiac disease in genetically predisposed subjects. Gluten-degrading microorganisms in the GI tract have been identified as different species of *Rothia* possessing gliadin enzyme. This enzyme helps in the digestion of dietary gluten. Since this organism and its enzyme is capable of neutralizing the harmful effects of gluten in celiac disease patients, therefore, development of a novel therapy based on different species of *Rothia* is being considered (Zamakhchari et al. 2011). A gluten-free diet (GFD) reduces beneficial gut bacteria populations (*Bifidobacterium*, *Lactobacillus*, and *B. longum*) and the ability to stimulate host's immunity in celiac disease patients. So, the impact of GFD on gut health needs to be taken under consideration in treatment of celiac disease patients (De Palma et al. 2009).

21.5 Effect of Different Dietary Patterns on the Microbiota

21.5.1 Vegetarian vs. Nonvegetarian Diet

The consumption of plant-based food increases taxonomic and bacterial gene diversity, leads to higher levels of short-chain fatty acid, and is linked with higher *Prevotella/Bacteroides* ratio. Refined foods from plant sources such as cereals and potatoes are not considered healthy food choice and are often linked with increased risk of metabolic and cardiovascular diseases (Fung et al. 2001; Willett et al. 2002).

Adventist Health Studies (AHS-2), conducted by researchers at the Loma Linda University School of Public Health, provides an opportunity to examine the profile of microbiota of individuals consuming vegan, vegetarian, or nonvegetarian diets and also those who are omnivorous. Vegetarian diets in general were found to be more beneficial compared to omnivore diets; however, vegan diets confer an additional advantage in bringing down the odds ratio for developing type 2 diabetes (Tonstad et al. 2013). *Bacteroidetes* phylum was more abundant in vegans and vegetarians compared with omnivores ($p < 0.05$), and higher *Firmicutes*/*Bacteroidetes* ratio was predominantly observed in omnivores (De Filippis et al. 2015). *Lachnospira* and *Prevotella* were significantly associated to plant-based diets, whereas L-Ruminococcus (*Ruminococcus* genus assigned to *Lachnospiraceae* family) was positively associated to omnivore diets.

Further, individuals consuming vegan diets exhibited reduced inflammations and were less prone to female-specific cancer risks. However, therapeutic effect of vegan diet that can be prescribed for long-term health benefits needs to be established (Glick-Bauer and Yeh 2014). Examination of fecal samples of lacto-vegetarian and omnivore South Indian population showed enrichment of clostridium and butyrate-producing taxa in omnivores (Kabeerdoss et al. 2012). Consumption of animal-based diet is linked to higher deoxycholic acid (DCA) formation in gut. DCA, a secondary bile acid produced by the microbial action on bile in gut, promotes liver cancer. DCA is reported to decrease the growth of members of the *Bacteroidetes* and *Firmicutes* phyla while promoting the growth of a bile-resistant *B. wadsworthia*, a sulfite-reducing bacterium known to cause gut inflammation in mouse models (Yoshimoto et al. 2013).

21.5.2 Western Diet

Classical Western diet generally contains high amount of fat, animal proteins (red meat), and refined sugars/carbohydrates but low amount of fresh fruit, vegetables, and whole grain cereals (Hou et al. 2011). The intake of high-fat and high-sugar diet leads to reduction in microbial richness, taxonomic diversity, and metabolic activity of gut bacterial flora in comparison to low-fat and plant-based diet of agrarian societies (De Filippo et al. 2010). Such diet-induced changes in GMt due to consumption of Western diet along with sedentary lifestyle are thought to be responsible for high incidence of inflammatory diseases such as IBD and other metabolic diseases such as obesity, heart disease, and diabetes (Law 2000). Intake of dietary saturated fats is associated with progressive increase in population of pro-inflammatory gut microbes such as *B. wadsworthia*, *Staphylococcus*, *E. coli*, and members of *Enterobacteriaceae*. Recovery of healthy diverse GM may require long-term dietary intervention in individuals on Western diet.

21.5.3 Mediterranean Diet

The Mediterranean-style diet (MD) is an ensemble of traditional dietary habits of people living in different countries of the Mediterranean region (Cyprus, Croatia,

Spain, Greece, Italy, Morocco, Portugal on the coast of the Mediterranean Sea) (Sofi et al. 2010). It is one of the healthiest dietary styles and has been listed in Intangible Cultural Heritage of Humanity of UNESCO (http://www.unesco.org/culture/ich/index.php?lg5en&pg50001). According to the European Food Safety Authority,

> "Mediterranean diet is generally characterized by high consumption of vegetables, fruits, cereals (unprocessed, whole grain), legumes, nuts, and seeds; moderate consumption of dairy products (mostly in the form of cheese or yogurt), fish, poultry, eggs, and unsaturated fats, such as olive oil as the primary source of monounsaturated fat for cooking and dressing; low to moderate intake of wine during meals; and little intake of red, processed meats and saturated fats" (Trichopoulou et al. 2003).

Metagenomic studies conducted on individuals adhering to the MD pattern of diet exhibited increased number of genes associated with polysaccharide degradation and SCFA metabolism (De Filippis et al. 2015). A relationship between MD diet and TMAO level was also established in this study providing a valuable insight for a possible modulation of TMAO levels through diet. A study using Greek population showed that the Mediterranean diet significantly reduced overall mortality due to coronary heart disease and cancer, even after taking care of confounders like age, sex, body mass index, and physical activity level (Trichopoulou et al. 2003; Martinez-Gonzalez et al. 2012).

21.6 Longitudinal Effect of Diet: Effect of Short-Term and Long-Term Dietary Intervention

Short-term dietary interventions have been found to affect gut microbial composition and gene expression within 24 hours. The changes in gut microbial composition and gene expression were correlated with the changes in diet composition (plant- or animal-based diet). This dynamic and quick response of gut flora to diet is probably due to survival pressure on GMt because of day-to-day variation in food intake of individuals. However, it is also probable that there may be other reasons as short-term dietary intervention (10 days) failed to achieve any changes in enterotype (David et al. 2014). Despite rapid response of gut flora to the consumed diet, it is the long-term consumption that shapes microbial configuration and functionality of gut flora in an individual. Long-term consumption of diet strongly correlates with formation of gut microbial enterotypes, and some of the human health problems may be a consequence of variations in diets (Wu et al. 2011). The GMt of patients with inflammatory bowel disease (IBD), obese individuals, and elderly patients with inflammation is less diverse in comparison to that of healthy people (Le Chatelier et al. 2013). Though short-term dietary intervention improves gut microbial gene richness and clinical parameters in low gene content (LGC) individuals, complete recovery of gene richness is not achieved. However, permanent changes in gut microbial consortium can be achieved by long-term dietary intervention with appropriate diet (Wu et al. 2011; Cotillard et al. 2013). It is important to note here that due to individualized nature of GM, there is substantial variation in the gut flora in response to a particular diet and people can be categorized as responders and nonresponders. Responders show better metabolic functions than non-responders. So, any therapy based on dietary intervention designed to improve gut flora should identify the potential beneficiary first for optimum benefit (Kovatcheva-Datchary et al. 2015). It appears that human genotype may have a role in the type of GM and the non-responsiveness to dietary intervention phenotype seen in some individuals.

21.7 Tools being Used to Elucidate the Microbial Metabolites

Our quest to study various metabolic syndromes has significantly advanced, as more and more metagenomic studies have suggested that microbial metabolites significantly vary during disease conditions. LC-MS and nuclear magnetic resonance (NMR) are two important and complimentary molecular techniques that are being used to assess the potential microbial metabolites involved in health and disease (Wikoff et al. 2009). These techniques are suitable for analysis of complex samples due to their reproducibility and their wide dynamic range. Untargeted MS-based metabolomics can be employed in animal models to successfully assess the impact of GM on blood chemistry.

Tools such as CASINO (Community And Systems-level Interactive Optimization) are now available to quantify the metabolic changes due to alterations in the diet in the human gut. For example, with the help of CASINO, one can predict the responses of specific metabolites such as SCFAs to individual food. This is a comprehensive computational platform for understanding and analyzing GM through metabolic modeling and gives a precise information regarding intolerance of a person to a specific dietary component. This information can subsequently be used to design functional foods that can increase colonic levels of SCFAs. This model can also be used to assess changes in the level of amino acids in stool and serum in response to dietary intervention (Shoaie et al. 2015).

21.8 Limitation and Opportunities of Gut Flora Research

Despite of enormous progress made in gut flora research gaps exist in our understanding of the GM and its interaction with diet. Here we present some of the issues that have emerged from this review and need to be addressed. Advancement made in the detection and analysis of hundreds of microorganisms from the human gut using metagenomic analysis has enabled us to enumerate bacterial population and extrapolate functional features including metabolism being carried out in the gut. However, to get a clear insight into their role in disease conditions, we need to understand the role of the bacteria that are poorly represented in healthy condition but can amplify in a disease state leading to major changes in the harmful metabolic products released by them.

It is very important to understand the impact of diet in the maintenance of fine dynamic equilibrium between different phylotypes. It has been observed that diet low in carbohydrate but high in protein tilts the balance in favor of *Bacteroidetes*, whereas a diet rich in fiber but low in protein shows more abundance of *Prevotella*. Metagenomic analysis has shown that this leads to low gene content (LGC) in the former case with low diversity of microbiota, whereas high gene content (HGC) has been recorded in the latter case showing higher diversity of microbiota (Cotillard et al. 2013). This study needs to be further extended in order to understand the fluctuations in some useful bacteria, such as butyrate-producing bacteria.

The consumption of low-fiber diet increases the expression of host glycan (mucin) utilizing genes and the number of glycan utilizing generalists such as *Bacteroides thetaiotaomicron* (Sonnenburg et al. 2005). It also causes slow bowel movement, low production of SCFAs, and more calorie intake from other diet components such as fat, protein, and refined sugar and selects a microbial consortium enriched in mucus-utilizing gut bacteria. The subsequent metabolic changes have been associated with several diseases of Western world (Sonnenburg and Sonnenburg 2014). The long-term impact of low-fiber diet on the integrity of the gut mucosa, inflammation, erosion of mucus layer, and pathobiology of associated diseases calls for further investigation by detailed studies for better management of diseases through diet.

The gut mucosal integrity is of vital importance to human health; a better understanding of the role of GM in shaping our immune system is required. It is also known that dietary components help in the maintenance of barrier integrity by lowering the concentrations of some microbes that secrete various microbial products including toxins. These compounds function as stimulant causing damage to the tissue and result in inflammation. Further knowledge in this area would help to tailor specific diets that can reduce access of these toxins to the tissue.

Studies integrating the factors like environmental, host genetics, and diet in the development of microbial profile are limited. Research should focus on the impact of diet and the environmental factors on children with different genetic background so as to achieve an optimal diet for maintaining a healthy gut profile at later period of life. Variations in macro and micronutrients of diet have demonstrated high potential in the pathogenesis of a disease; however, their role in modulating gut flora is not yet well understood. Studies on the impact of macronutrients like carbohydrate, protein, and lipids on the GM have been initiated, and preliminary results have just started emerging on the sources of these macronutrients (animal or plant source) and their role modulating the gut flora. Information is lacking on the impact of micronutrients, such as specific vitamins in changing the gut profile.

So far most of the studies concentrated on profiling of gut flora of the large intestine due to more abundance of bacteria in this region. However, detail profiling of the bacterial flora of the small intestine is crucial so as to understand the mechanisms that allow digestion process in the small intestine. Higher rate of digestion in the small intestine will pass on the lesser level of undigested products to the large intestine increasing chances of complete digestion with minimal generation of harmful by-products. There is a need to conduct case-controlled studies supplemented with metabolic approaches to assess the physiological impact of identified microbial metabolites and decipher its specific mechanism of action.

Recently the role of gut-brain axis (GBA) has been proposed to have an important role in health and disease. Evidence of GBA-microbiota relationship has come to the surface since dysbiosis in GMt has been linked with diseases such as autism, multiple sclerosis, and IBS. Understanding the specific microbial products that reach the brain and how dietary manipulations of microbiota can impact our health is yet not convincingly established.

So far the knowledge in this field has clearly demonstrated that the predominant bacteria are crucial for maintaining a healthy gut. Now it is important to culture those bacteria and maintain a library of these isolates that can be delivered to diseased individuals in the form of prebiotics/probiotics as and when required for stimulating the growth of beneficial flora lost due to pathogenesis. This exercise also calls for a tailored diet in the form of prebiotics to maintain the healthy gut. High-throughput techniques must be employed for identification of a minimal set of biomarkers screened from microbial metabolites that are associated with the disease state.

Conclusion

Due to influence of diet, host genetics, and various environmental factors, healthy microbiota of an individual remains undefined. Changes in gut flora affect the development of several diseases such as obesity, diabetes, IBD, atherosclerosis, airway allergy, and multiple sclerosis as evident from animal studies; however, any interventions should be designed with a pinch of salt as alteration reported in animal studies may not be equally strong and reproducible in human subjects. However, it is noteworthy that several of these associations have shown promising results in IBD patients after fecal transplant. The human genome is relatively static, whereas gut microbiome is dynamic. This plasticity of genomic asymmetry between the human genome and gut microbiome gives an opportunity to take advantage of the plasticity of our gut residents for pragmatic and logical manipulation of gut flora to fortify human health and better management or prevention of various diseases. Remarkable progress in our understanding on diet-microbiota-host interactions has opened avenues to explore therapeutic approaches to manipulate the microbiota selectively. This can produce beneficial metabolite to promote health and prevent disease condition. However, whether changes in lifestyle, diet, and genetic predisposition all together or individually affect microbial profile in a disease state needs to be further investigated.

Acknowledgments The authors wish to acknowledge the financial support received from the DBT-European Union Grant (BT/IN/EU/Infect-Era/01/AB/2015). AB acknowledges support from the Department of Science & Technology for J.C. Bose fellowship.

References

Aguirre M, Bussolo de Souza C, Venema K (2016) The gut microbiota from lean and obese subjects contribute differently to the fermentation of arabinogalactan and inulin. PLoS One 11. doi:10.1371/journal.pone.0159236

Arumugam M, Raes J, Pelletier E, Le Paslier D, Yamada T, Mende DR, Fernandes GR, Tap J, Bruls T, Batto JM, Bertalan M, Borruel N, Casellas F, Fernandez L, Gautier L, Hansen T, Hattori M, Hayashi T, Kleerebezem M, Kurokawa K, Leclerc M, Levenez F, Manichanh

C, Nielsen HB, Nielsen T, Pons N, Poulain J, Qin J, Sicheritz-Ponten T, Tims S, Torrents D, Ugarte E, Zoetendal EG, Wang J, Guarner F, Pedersen O, de Vos WM, Brunak S, Doré J, MetaHIT Consortium, Antolín M, Artiguenave F, Blottiere HM, Almeida M, Brechot C, Cara C, Chervaux C, Cultrone A, Delorme C, Denariaz G, Dervyn R, Foerstner KU, Friss C, van de Guchte M, Guedon E, Haimet F, Huber W, van Hylckama-Vlieg J, Jamet A, Juste C, Kaci G, Knol J, Lakhdari O, Layec S, Le Roux K, Maguin E, Mérieux A, Melo Minardi R, M'rini C, Muller J, Oozeer R, Parkhill J, Renault P, Rescigno M, Sanchez N, Sunagawa S, Torrejon A, Turner K, Vandemeulebrouck G, Varela E, Winogradsky Y, Zeller G, Weissenbach J, Ehrlich SD, Bork P (2011) Enterotypes of the human gut microbiome. Nature 473:174–180. doi:10.1038/nature09944

Bergogne-Berezin E (2000) Treatment and prevention of antibiotic associated diarrhea. Int J Antimicrob Agents 16:521–526. doi:10.1016/S0924-8579(00)00293-4

Burgess SL, Buonomo E, Carey M, Cowardin C, Naylor C, Noor Z, Wills-Karp M, Petri WA Jr (2014) Bone marrow dendritic cells from mice with an altered microbiota provide interleukin 17A-dependent protection against Entamoeba histolytica colitis. MBio 5:e01817. doi:10.1128/mBio.01817-14

Caesar R, Tremaroli V, Kovatcheva-Datchary P, Cani PD, Backhed F (2015) Crosstalk between gut microbiota and dietary lipids aggravates WAT inflammation through TLR signaling. Cell Metab 22:658–668. doi:10.1016/j.cmet.2015.07.026

Cantarel BL, Lombard V, Henrissat B (2012) Complex carbohydrate utilization by the healthy human microbiome. PLoS One 7:e28742. doi:10.1371/journal.pone.0028742

Carmody RN, Gerber GK, Luevano JM Jr, Gatti DM, Somes L, Svenson KL, Turnbaugh PJ (2015) Diet dominates host genotype in shaping the murine gut microbiota. Cell Host Microbe 17:72–84. doi:10.1016/j.chom.2014.11.010

Chassard C, Delmas E, Robert C, Bernalier-Donadille A (2010) The cellulose-degrading microbial community of the human gut varies according to the presence or absence of methanogens. FEMS Microbiol Ecol 74:205–213. doi:10.1111/j.1574-6941.2010.00941.x

Claesson MJ, Jeffery IB, Conde S, Power SE, O'Connor EM, Cusack S, Harris HM, Coakley M, Lakshminarayanan B, O'Sullivan O, Fitzgerald GF, Deane J, O'Connor M, Harnedy N, O'Connor K, O'Mahony D, van Sinderen D, Wallace M, Brennan L, Stanton C, Marchesi JR, Fitzgerald AP, Shanahan F, Hill C, Ross RP, O'Toole PW (2012) Gut microbiota composition correlates with diet and health in the elderly. Nature 488:178–184. doi:10.1038/nature11319

Clemente JC, Pehrsson EC, Blaser MJ, Sandhu K, Gao Z, Wang B, Magris M, Hidalgo G, Contreras M, Noya-Alarcon O, Lander O, McDonald J, Cox M, Walter J, Oh PL, Ruiz JF, Rodriguez S, Shen N, Song SJ, Metcalf J, Knight R, Dantas G, Dominguez-Bello MG (2015) The microbiome of uncontacted Amerindians. Sci Adv 1:e1500183. doi:10.1126/sciadv.1500183

Compare D, Rocco A, Nardone G (2010) Risk factors in gastric cancer. Eur Rev Med Pharmacol Sci 14:302–308

Compare D, Coccoli P, Rocco A, Nardone OM, De Maria S, Carteni M, Nardone G (2012) Gut-liver axis: the impact of gut microbiota on non alcoholic fatty liver disease. Nutr Metab Cardiovasc Dis 22:471–476. doi:10.1016/j.numecd.2012.02.007

Conlon MA, Bird AR (2015) The impact of diet and lifestyle on gut microbiota and human health. Nutrients 7:17–44. doi:10.3390/nu7010017

Corbin KD, Zeisel SH (2012) Choline metabolism provides novel insights into nonalcoholic fatty liver disease and its progression. Curr Opin Gastroenterol 28:159–165. doi:10.1097/MOG.0b013e32834e7b4b

Cotillard A, Kennedy SP, Kong LC, Prifti E, Pons N, Le Chatelier E, Almeida M, Quinquis B, Levenez F, Galleron N, Gougis S, Rizkalla S, Batto JM, Renault P, ANR MicroObes Consortium, Doré J, Zucker JD, Clément K, Ehrlich SD (2013) Dietary intervention impact on gut microbial gene richness. Nature 500:585–588. doi:10.1038/nature12480

Cummings JH, Hill MJ, Bone ES, Branch WJ, Jenkins DJ (1979) The effect of meat protein and dietary fiber on colonic function and metabolism. II. Bacterial metabolites in feces and urine. Am J Clin Nutr 32:2094–2101

David LA, Maurice CF, Carmody RN, Gootenberg DB, Button JE, Wolfe BE, Ling AV, Devlin AS, Varma Y, Fischbach MA, Biddinger SB, Dutton RJ, Turnbaugh PJ (2014) Diet rapidly and reproducibly alters the human gut microbiome. Nature 505:559–563. doi:10.1038/nature12820

Davie JR (2003) Inhibition of histone deacetylase activity by butyrate. J Nutr 133:2485s–2493s

De Filippis F, Pellegrini N, Vannini L, Jeffery IB, La Storia A, Laghi L, Serrazanetti DI, Di Cagno R, Ferrocino I, Lazzi C, Turroni S, Cocolin L, Brigidi P, Neviani E, Gobbetti M, O'Toole PW, Ercolini D (2015) High-level adherence to a Mediterranean diet beneficially impacts the gut microbiota and associated metabolome. Gut. doi:10.1136/gutjnl-2015-309957

De Filippo C, Cavalieri D, Di Paola M, Ramazzotti M, Poullet JB, Massart S, Collini S, Pieraccini G, Lionetti P (2010) Impact of diet in shaping gut microbiota revealed by a comparative study in children from Europe and rural Africa. Proc Natl Acad Sci U S A 107:14691–14696. doi:10.1073/pnas.1005963107

De Palma G, Nadal I, Collado MC, Sanz Y (2009) Effects of a gluten-free diet on gut microbiota and immune function in healthy adult human subjects. Br J Nutr 102:1154–1160. doi:10.1017/S0007114509371767

De Vadder F, Kovatcheva-Datchary P, Goncalves D, Vinera J, Zitoun C, Duchampt A, Backhed F, Mithieux G (2014) Microbiota-generated metabolites promote metabolic benefits via gut-brain neural circuits. Cell 156:84–96. doi:10.1016/j.cell.2013.12.016

Dethlefsen L, Relman DA (2011) Incomplete recovery and individualized responses of the human distal gut microbiota to repeated antibiotic perturbation. Proc Natl Acad Sci U S A 108:4554–4561. doi:10.1073/pnas.1000087107

Devkota S, Wang Y, Musch MW, Leone V, Fehlner-Peach H, Nadimpalli A, Antonopoulos DA, Jabri B, Chang EB (2012) Dietary-fat-induced taurocholic acid promotes pathobiont expansion and colitis in Il10-/- mice. Nature 487:104–108. doi:10.1038/nature11225

Dominguez-Bello MG, Costello EK, Contreras M, Magris M, Hidalgo G, Fierer N, Knight R (2010) Delivery mode shapes the acquisition and structure of the initial microbiota across multiple body habitats in newborns. Proc Natl Acad Sci U S A 107:11971–11975. doi:10.1073/pnas.1002601107

Duncan SH, Holtrop G, Lobley GE, Calder AG, Stewart CS, Flint HJ (2004) Contribution of acetate to butyrate formation by human faecal bacteria. Br J Nutr 91:915–923. doi:10.1079/BJN20041150

Duncan SH, Belenguer A, Holtrop G, Johnstone AM, Flint HJ, Lobley GE (2007) Reduced dietary intake of carbohydrates by obese subjects results in decreased concentrations of butyrate and butyrate-producing bacteria in feces. Appl Environ Microbiol 73:1073–1078. doi:10.1128/AEM.02340-06

Eckburg PB, Bik EM, Bernstein CN, Purdom E, Dethlefsen L, Sargent M, Gill SR, Nelson KE, Relman DA (2005) Diversity of the human intestinal microbial flora. Science 308:1635–1638. doi:10.1126/science.1110591

El Kaoutari A, Armougom F, Gordon JI, Raoult D, Henrissat B (2013) The abundance and variety of carbohydrate-active enzymes in the human gut microbiota. Nat Rev Microbiol 11:497–504. doi:10.1038/nrmicro3050

Enattah NS, Sahi T, Savilahti E, Terwilliger JD, Peltonen L, Jarvela I (2002) Identification of a variant associated with adult-type hypolactasia. Nat Genet 30:233–237. doi:10.1038/ng826

Everard A, Belzer C, Geurts L, Ouwerkerk JP, Druart C, Bindels LB, Guiot Y, Derrien M, Muccioli GG, Delzenne NM, de Vos WM, Cani PD (2013) Cross-talk between Akkermansia muciniphila and intestinal epithelium controls diet-induced obesity. Proc Natl Acad Sci U S A 110:9066–9071. doi:10.1073/pnas.1219451110

Fallani M, Amarri S, Uusijarvi A, Adam R, Khanna S, Aguilera M, Gil A, Vieites JM, Norin E, Young D, Scott JA, Doré J, Edwards CA, INFABIO Team (2011) Determinants of the human infant intestinal microbiota after the introduction of first complementary foods in infant samples from five European centres. Microbiology 157:1385–1392. doi:10.1099/mic.0.042143-0

Feijo FD, Ballard CR, Foletto KC, Batista BAM, Neves AM, Ribeiro MFM, Bertoluci MC (2013) Saccharin and aspartame, compared with sucrose, induce greater weight gain in adult Wistar rats, at similar total caloric intake levels. Appetite 60:203–207. doi:10.1016/j.appet.2012.10.009

Fitch C, Keim KS (2012) Position of the academy of nutrition and dietetics: use of nutritive and nonnutritive sweeteners. J Acad Nutr Diet 112:739–758. doi:10.1016/j.jand.2012.03.009

Flint HJ, Scott KP, Louis P, Duncan SH (2012) The role of the gut microbiota in nutrition and health. Nat Rev Gastroenterol Hepatol 9:577–589. doi:10.1038/nrgastro.2012.156

Folseraas T, Melum E, Rausch P, Juran BD, Ellinghaus E, Shiryaev A, Laerdahl JK, Ellinghaus D, Schramm C, Weismuller TJ, Gotthardt DN, Hov JR, Clausen OP, Weersma RK, Janse M, Boberg KM, Björnsson E, Marschall HU, Cleynen I, Rosenstiel P, Holm K, Teufel A, Rust C, Gieger C, Wichmann HE, Bergquist A, Ryu E, Ponsioen CY, Runz H, Sterneck M, Vermeire S, Beuers U, Wijmenga C, Schrumpf E, Manns MP, Lazaridis KN, Schreiber S, Baines JF, Franke A, Karlsen TH (2012) Extended analysis of a genome-wide association study in primary sclerosing cholangitis detects multiple novel risk loci. J Hepatol 57:366–375. doi:10.1016/j.jhep.2012.03.031

Freudenberg A, Petzke KJ, Klaus S (2013) Dietary L-leucine and L-alanine supplementation have similar acute effects in the prevention of high-fat diet-induced obesity. Amino Acids 44:519–528. doi:10.1007/s00726-012-1363-2

Fukuda S, Toh H, Hase K, Oshima K, Nakanishi Y, Yoshimura K, Tobe T, Clarke JM, Topping DL, Suzuki T, Taylor TD, Itoh K, Kikuchi J, Morita H, Hattori M, Ohno H (2011) Bifidobacteria can protect from enteropathogenic infection through production of acetate. Nature 469:543–U791. doi:10.1038/nature09646

Fung TT, Willett WC, Stampfer MJ, Manson JE, Hu FB (2001) Dietary patterns and the risk of coronary heart disease in women. Arch Intern Med 161:1857–1862. doi:10.1001/archinte.161.15.1857

Gilchrist CA, Petri SE, Schneider BN, Reichman DJ, Jiang N, Begum S, Watanabe K, Jansen CS, Elliott KP, Burgess SL, Ma JZ, Alam M, Kabir M, Haque R, Petri WA Jr (2016) Role of the gut microbiota of children in diarrhea due to the protozoan parasite Entamoeba histolytica. J Infect Dis 213:1579–1585. doi:10.1093/infdis/jiv772

Glick-Bauer M, Yeh MC (2014) The health advantage of a vegan diet: exploring the gut microbiota connection. Nutrients 6:4822–4838. doi:10.3390/nu6114822

Goodrich JK, Waters JL, Poole AC, Sutter JL, Koren O, Blekhman R, Beaumont M, Van Treuren W, Knight R, Bell JT, Spector TD, Clark AG, Ley RE (2014) Human genetics shape the gut microbiome. Cell 159:789–799. doi:10.1016/j.cell.2014.09.053

Graf D, Di Cagno R, Fak F, Flint HJ, Nyman M, Saarela M, Watzl B (2015) Contribution of diet to the composition of the human gut microbiota. Microb Ecol Health Dis 26:26164. doi:10.3402/mehd.v26.26164

Guarner F, Perdigon G, Corthier G, Salminen S, Koletzko B, Morelli L (2005) Should yoghurt cultures be considered probiotic? Br J Nutr 93:783–786. doi:10.1079/BJN20051428

Haiser HJ, Gootenberg DB, Chatman K, Sirasani G, Balskus EP, Turnbaugh PJ (2013) Predicting and manipulating cardiac drug inactivation by the human gut bacterium Eggerthella lenta. Science 341:295–298. doi:10.1126/science.1235872

Hamer HM, Jonkers D, Venema K, Vanhoutvin S, Troost FJ, Brummer RJ (2008) Review article: the role of butyrate on colonic function. Aliment Pharmacol Ther 27:104–119. doi:10.1111/j.1365-2036.2007.03562.x

Harmsen HJ, Wildeboer-Veloo AC, Raangs GC, Wagendorp AA, Klijn N, Bindels JG, Welling GW (2000) Analysis of intestinal flora development in breast-fed and formula-fed infants by using molecular identification and detection methods. J Pediatr Gastroenterol Nutr 30:61–67

Hehemann JH, Correc G, Barbeyron T, Helbert W, Czjzek M, Michel G (2010) Transfer of carbohydrate-active enzymes from marine bacteria to Japanese gut microbiota. Nature 464:908–U123. doi:10.1038/nature08937

Henao-Mejia J, Elinav E, Jin C, Hao L, Mehal WZ, Strowig T, Thaiss CA, Kau AL, Eisenbarth SC, Jurczak MJ, Camporez JP, Shulman GI, Gordon JI, Hoffman HM, Flavell RA (2012) Inflammasome-mediated dysbiosis regulates progression of NAFLD and obesity. Nature 482:179–185. doi:10.1038/nature10809

Hogenauer C, Hammer HF, Krejs GJ, Reisinger EC (1998) Mechanisms and management of antibiotic-associated diarrhea. Clin Infect Dis 27:702–710. doi:10.1086/514958

Hou JK, Abraham B, El-Serag H (2011) Dietary intake and risk of developing inflammatory bowel disease: a systematic review of the literature. Am J Gastroenterol 106:563–573. doi:10.1038/ajg.2011.44

Hughes R, Magee EA, Bingham S (2000) Protein degradation in the large intestine: relevance to colorectal cancer. Curr Issues Intest Microbiol 1:51–58

Jeffery IB, Lynch DB, O'Toole PW (2016) Composition and temporal stability of the gut microbiota in older persons. ISME J 10:170–182. doi:10.1038/ismej.2015.88

Kabeerdoss J, Devi RS, Mary RR, Ramakrishna BS (2012) Faecal microbiota composition in vegetarians: comparison with omnivores in a cohort of young women in southern India. Br J Nutr 108:953–957. doi:10.1017/S0007114511006362

Kashyap PC, Marcobal A, Ursell LK, Smits SA, Sonnenburg ED, Costello EK, Higginbottom SK, Domino SE, Holmes SP, Relman DA, Knight R, Gordon JI, Sonnenburg JL (2013) Genetically dictated change in host mucus carbohydrate landscape exerts a diet-dependent effect on the gut microbiota. Proc Natl Acad Sci U S A 110:17059–17064. doi:10.1073/pnas.1306070110

Kiilerich P, Myrmel LS, Fjaere E, Hao Q, Hugenholtz F, Sonne SB, Derrien M, Pedersen LM, Petersen RK, Mortensen A, Licht TR, Rømer MU, Vogel UB, Waagbø LJ, Giallourou N, Feng Q, Xiao L, Liu C, Liaset B, Kleerebezem M, Wang J, Madsen L, Kristiansen K (2016) Effect of a long-term high-protein diet on survival, obesity development, and gut microbiota in mice. Am J Physiol Endocrinol Metab 310:E886–E899. doi:10.1152/ajpendo.00363.2015

Koeth RA, Wang Z, Levison BS, Buffa JA, Org E, Sheehy BT, Britt EB, Fu X, Wu Y, Li L, Smith JD, DiDonato JA, Chen J, Li H, Wu GD, Lewis JD, Warrier M, Brown JM, Krauss RM, Tang WH, Bushman FD, Lusis AJ, Hazen SL (2013) Intestinal microbiota metabolism of L-carnitine, a nutrient in red meat, promotes atherosclerosis. Nat Med 19:576–585. doi:10.1038/nm.3145

Kovatcheva-Datchary P, Nilsson A, Akrami R, Lee YS, De Vadder F, Arora T, Hallen A, Martens E, Bjorck I, Backhed F (2015) Dietary fiber-induced improvement in glucose metabolism is associated with increased abundance of Prevotella. Cell Metab 22:971–982. doi:10.1016/j.cmet.2015.10.001

Lam YY, Ha CWY, Hoffmann JMA, Oscarsson J, Dinudom A, Mather TJ, Cook DI, Hunt NH, Caterson ID, Holmes AJ, Storlien LH (2015) Effects of dietary fat profile on gut permeability and microbiota and their relationships with metabolic changes in mice. Obesity 23:1429–1439. doi:10.1002/oby.21122

Laparra JM, Sanz Y (2010) Interactions of gut microbiota with functional food components and nutraceuticals. Pharmacol Res 61:219–225. doi:10.1016/j.phrs.2009.11.001

Law M (2000) Dietary fat and adult diseases and the implications for childhood nutrition: an epidemiologic approach. Am J Clin Nutr 72:1291s–1296s

Le Chatelier E, Nielsen T, Qin J, Prifti E, Hildebrand F, Falony G, Almeida M, Arumugam M, Batto JM, Kennedy S, Leonard P, Li J, Burgdorf K, Grarup N, Jørgensen T, Brandslund I, Nielsen HB, Juncker AS, Bertalan M, Levenez F, Pons N, Rasmussen S, Sunagawa S, Tap J, Tims S, Zoetendal EG, Brunak S, Clément K, Doré J, Kleerebezem M, Kristiansen K, Renault P, Sicheritz-Ponten T, de Vos WM, Zucker JD, Raes J, Hansen T, MetaHIT Consortium, Bork P, Wang J, Ehrlich SD, Pedersen O (2013) Richness of human gut microbiome correlates with metabolic markers. Nature 500:541–546. doi:10.1038/nature12506

Ley RE, Peterson DA, Gordon JI (2006) Ecological and evolutionary forces shaping microbial diversity in the human intestine. Cell 124:837–848. doi:10.1016/j.cell.2006.02.017

Livanos AE, Greiner TU, Vangay P, Pathmasiri W, Stewart D, McRitchie S, Li H, Chung J, Sohn J, Kim S, Gao Z, Barber C, Kim J, Ng S, Rogers AB, Sumner S, Zhang XS, Cadwell K, Knights D, Alekseyenko A, Bäckhed F, Blaser MJ (2016) Antibiotic-mediated gut microbiome perturbation accelerates development of type 1 diabetes in mice. Nat Microbiol 1:16140. doi:10.1038/nmicrobiol.2016.140

Lopez-Legarrea P, Fuller NR, Zulet MA, Martinez JA, Caterson ID (2014) The influence of Mediterranean, carbohydrate and high protein diets on gut microbiota composition in the

treatment of obesity and associated inflammatory state. Asia Pac J Clin Nutr 23:360–368. doi:10.6133/apjcn.2014.23.3.16
Lozupone CA, Stombaugh JI, Gordon JI, Jansson JK, Knight R (2012) Diversity, stability and resilience of the human gut microbiota. Nature 489:220–230. doi:10.1038/nature11550
Macfarlane S, Macfarlane GT (2003) Regulation of short-chain fatty acid production. Proc Nutr Soc 62:67–72. doi:10.1079/PNS2002207
Macfarlane GT, Macfarlane S (2012) Bacteria, colonic fermentation, and gastrointestinal health. J AOAC Int 95:50–60. doi:10.5740/jaoacint.SGE_Macfarlane
Macfarlane GT, Cummings JH, Allison C (1986) Protein degradation by human intestinal bacteria. J Gen Microbiol 132:1647–1656. doi:10.1099/00221287-132-6-1647
Macfarlane GT, Gibson GR, Cummings JH (1992) Comparison of fermentation reactions in different regions of the human colon. J Appl Bacteriol 72:57–64. doi:10.1111/j.1365-2672.1992.tb04882.x
Mahana D, Trent CM, Kurtz ZD, Bokulich NA, Battaglia T, Chung J, Muller CL, Li H, Bonneau RA, Blaser MJ (2016) Antibiotic perturbation of the murine gut microbiome enhances the adiposity, insulin resistance, and liver disease associated with high-fat diet. Genome Med 8:48. doi:10.1186/s13073-016-0297-9
Martinez I, Stegen JC, Maldonado-Gomez MX, Eren AM, Siba PM, Greenhill AR, Walter J (2015) The gut microbiota of rural papua new guineans: composition, diversity patterns, and ecological processes. Cell Rep 11:527–538. doi:10.1016/j.celrep.2015.03.049
Martinez-Gonzalez MA, Guillen-Grima F, De Irala J, Ruiz-Canela M, Bes-Rastrollo M, Beunza JJ, del Burgo CL, Toledo E, Carlos S, Sanchez-Villegas A (2012) The Mediterranean diet is associated with a reduction in premature mortality among middle-aged adults. J Nutr 142:1672–1678. doi:10.3945/jn.112.162891
Maslowski KM, Vieira AT, Ng A, Kranich J, Sierro F, Yu D, Schilter HC, Rolph MS, Mackay F, Artis D, Xavier RJ, Teixeira MM, Mackay CR (2009) Regulation of inflammatory responses by gut microbiota and chemoattractant receptor GPR43. Nature 461:1282–U1119. doi:10.1038/nature08530
McGovern DPB, Jones MR, Taylor KD, Marciante K, Yan XF, Dubinsky M, Ippoliti A, Vasiliauskas E, Berel D, Derkowski C, Dutridge D, Fleshner P, Shih DQ, Melmed G, Mengesha E, King L, Pressman S, Haritunians T, Guo X, Targan SR, Rotter JI, International IBD Genetics Consortium (2010) Fucosyltransferase 2 (FUT2) non-secretor status is associated with Crohn's disease. Hum Mol Genet 19:3468–3476. doi:10.1093/hmg/ddq248
Milner JA, Allison RG (1999) The role of dietary fat in child nutrition and development: summary of an ASNS workshop. American Society for Nutritional Sciences. J Nutr 129:2094–2105
Modi SR, Collins JJ, Relman DA (2014) Antibiotics and the gut microbiota. J Clin Invest 124:4212–4218. doi:10.1172/JCI72333
Morton ER, Lynch J, Froment A, Lafosse S, Heyer E, Przeworski M, Blekhman R, Segurel L (2015) Variation in rural African gut microbiota is strongly correlated with colonization by Entamoeba and subsistence. PLoS Genet 11:e1005658. doi:10.1371/journal.pgen.1005658
Moschen AR, Wieser V, Tilg H (2012) Dietary factors: major regulators of the Gut's microbiota. Gut Liver 6:411–416. doi:10.5009/gnl.2012.6.4.411
Norat T, Bingham S, Ferrari P, Slimani N, Jenab M, Mazuir M, Overvad K, Olsen A, Tjonneland A, Clavel F, Boutron-Ruault MC, Kesse E, Boeing H, Bergmann MM, Nieters A, Linseisen J, Trichopoulou A, Trichopoulos D, Tountas Y, Berrino F, Palli D, Panico S, Tumino R, Vineis P, Bueno-de-Mesquita HB, Peeters PH, Engeset D, Lund E, Skeie G, Ardanaz E, González C, Navarro C, Quirós JR, Sanchez MJ, Berglund G, Mattisson I, Hallmans G, Palmqvist R, Day NE, Khaw KT, Key TJ, San Joaquin M, Hémon B, Saracci R, Kaaks R, Riboli E (2005) Meat, fish, and colorectal cancer risk: the European Prospective Investigation into cancer and nutrition. J Natl Cancer Inst 97:906–916. doi:10.1093/jnci/dji164
Pichon L, Huneau JF, Fromentin G, Tome D (2006) A high-protein, high-fat, carbohydrate-free diet reduces energy intake, hepatic lipogenesis, and adiposity in rats. J Nutr 136:1256–1260

Qin J, Li R, Raes J, Arumugam M, Burgdorf KS, Manichanh C, Nielsen T, Pons N, Levenez F, Yamada T, Mende DR, Li J, Xu J, Li S, Li D, Cao J, Wang B, Liang H, Zheng H, Xie Y, Tap J, Lepage P, Bertalan M, Batto JM, Hansen T, Le Paslier D, Linneberg A, Nielsen HB, Pelletier E, Renault P, Sicheritz-Ponten T, Turner K, Zhu H, Yu C, Li S, Jian M, Zhou Y, Li Y, Zhang X, Li S, Qin N, Yang H, Wang J, Brunak S, Doré J, Guarner F, Kristiansen K, Pedersen O, Parkhill J, Weissenbach J, MetaHIT Consortium, Bork P, Ehrlich SD, Wang J (2010) A human gut microbial gene catalogue established by metagenomic sequencing. Nature 464:59–65. doi:10.1038/nature08821

Reddy BS (1981) Diet and excretion of bile acids. Cancer Res 41:3766–3768

Rombeau JL, Kripke SA (1990) Metabolic and intestinal effects of short-chain fatty-acids. JPEN J Parenter Enteral Nutr 14:S181–S185. doi:10.1177/014860719001400507

Russell WR, Gratz SW, Duncan SH, Holtrop G, Ince J, Scobbie L, Duncan G, Johnstone AM, Lobley GE, Wallace RJ, Duthie GG, Flint HJ (2011) High-protein, reduced-carbohydrate weight-loss diets promote metabolite profiles likely to be detrimental to colonic health. Am J Clin Nutr 93:1062–1072. doi:10.3945/ajcn.110.002188

Samuel BS, Shaito A, Motoike T, Rey FE, Backhed F, Manchester JK, Hammer RE, Williams SC, Crowley J, Yanagisawa M, Gordon JI (2008) Effects of the gut microbiota on host adiposity are modulated by the short-chain fatty-acid binding G protein-coupled receptor, Gpr41. Proc Natl Acad Sci U S A 105:16767–16772. doi:10.1073/pnas.0808567105

Schmid A (2011) The role of meat fat in the human diet. Crit Rev Food Sci Nutr 51:50–66. doi:10.1080/10408390903044636

Schnorr SL, Candela M, Rampelli S, Centanni M, Consolandi C, Basaglia G, Turroni S, Biagi E, Peano C, Severgnini M, Fiori J, Gotti R, De Bellis G, Luiselli D, Brigidi P, Mabulla A, Marlowe F, Henry AG, Crittenden AN (2014) Gut microbiome of the Hadza hunter-gatherers. Nat Commun 5:3654. doi:10.1038/ncomms4654

Shoaie S, Ghaffari P, Kovatcheva-Datchary P, Mardinoglu A, Sen P, Pujos-Guillot E, de Wouters T, Juste C, Rizkalla S, Chilloux J, Hoyles L, Nicholson JK, MICRO-Obes Consortium, Dore J, Dumas ME, Clement K, Bäckhed F, Nielsen J (2015) Quantifying diet-induced metabolic changes of the human gut microbiome. Cell Metab 22:320–331. doi:10.1016/j.cmet.2015.07.001

Silvester KR, Cummings JH (1995) Does digestibility of meat protein help explain large-bowel cancer risk. Nutr Cancer 24:279–288. doi:10.1080/01635589509514417

Sofi F, Abbate R, Gensini GF, Casini A (2010) Accruing evidence on benefits of adherence to the Mediterranean diet on health: an updated systematic review and meta-analysis. Am J Clin Nutr 92:1189–1196. doi:10.3945/ajcn.2010.29673

Sonnenburg ED, Sonnenburg JL (2014) Starving our microbial self: the deleterious consequences of a diet deficient in microbiota-accessible carbohydrates. Cell Metab 20:779–786. doi:10.1016/j.cmet.2014.07.003

Sonnenburg JL, Xu J, Leip DD, Chen CH, Westover BP, Weatherford J, Buhler JD, Gordon JI (2005) Glycan foraging in vivo by an intestine-adapted bacterial symbiont. Science 307:1955–1959. doi:10.1126/science.1109051

Spencer MD, Hamp TJ, Reid RW, Fischer LM, Zeisel SH, Fodor AA (2011) Association between composition of the human gastrointestinal microbiome and development of fatty liver with choline deficiency. Gastroenterology 140:976–986. doi:10.1053/j.gastro.2010.11.049

Stark PL, Lee A (1982) The microbial ecology of the large bowel of breast-fed and formula-fed infants during the first year of life. J Med Microbiol 15:189–203. doi:10.1099/00222615-15-2-189

Suez J, Korem T, Zeevi D, Zilberman-Schapira G, Thaiss CA, Maza O, Israeli D, Zmora N, Gilad S, Weinberger A, Kuperman Y, Harmelin A, Kolodkin-Gal I, Shapiro H, Halpern Z, Segal E, Elinav E (2014) Artificial sweeteners induce glucose intolerance by altering the gut microbiota. Nature 514:181–186. doi:10.1038/nature13793

Swithers SE (2013) Artificial sweeteners produce the counterintuitive effect of inducing metabolic derangements. Trends Endocrinol Metab 24:431–441. doi:10.1016/j.tem.2013.05.005

Swithers SE, Davidson TL (2008) A role for sweet taste: calorie predictive relations in energy regulation by rats. Behav Neurosci 122:161–173. doi:10.1037/0735-7044.122.1.161

Swithers SE, Martin AA, Davidson TL (2010) High-intensity sweeteners and energy balance. Physiol Behav 100:55–62. doi:10.1016/j.physbeh.2009.12.021

Tang WHW, Wang ZE, Levison BS, Koeth RA, Britt EB, Fu XM, Wu YP, Hazen SL (2013) Intestinal microbial metabolism of phosphatidylcholine and cardiovascular risk. N Engl J Med 368:1575–1584. doi:10.1056/NEJMoa1109400

Terpend K, Possemiers S, Daguet D, Marzorati M (2013) Arabinogalactan and fructo-oligosaccharides have a different fermentation profile in the Simulator of the Human Intestinal Microbial Ecosystem (SHIME (R)). Environ Microbiol Rep 5:595–603. doi:10.1111/1758-2229.12056

Thorburn AN, McKenzie CI, Shen S, Stanley D, Macia L, Mason LJ, Roberts LK, Wong CHY, Shim R, Robert R, Chevalier N, Tan JK, Mariño E, Moore RJ, Wong L, McConville MJ, Tull DL, Wood LG, Murphy VE, Mattes J, Gibson PG, Mackay CR (2015) Evidence that asthma is a developmental origin disease influenced by maternal diet and bacterial metabolites. Nat Commun 6:7320. doi:10.1038/ncomms8320

Toden S, Bird AR, Topping DL, Conlon MA (2005) Resistant starch attenuates colonic DNA damage induced by higher dietary protein in rats. Nutr Cancer 51:45–51. doi:10.1207/s15327914nc5101_7

Toden S, Bird AR, Topping DL, Conlon MA (2007) High red meat diets induce greater numbers of colonic DNA double-strand breaks than white meat in rats: attenuation by high-amylose maize starch. Carcinogenesis 28:2355–2362. doi:10.1093/carcin/bgm216

Tolhurst G, Heffron H, Lam YS, Parker HE, Habib AM, Diakogiannaki E, Cameron J, Grosse J, Reimann F, Gribble FM (2012) Short-chain fatty acids stimulate glucagon-like peptide-1 secretion via the G-protein-coupled receptor FFAR2. Diabetes 61:364–371. doi:10.2337/db11-1019

Tonstad S, Stewart K, Oda K, Batech M, Herring RP, Fraser GE (2013) Vegetarian diets and incidence of diabetes in the Adventist Health Study-2. Nutr Metab Cardiovasc Dis 23:292–299. doi:10.3945/ajcn.113.071233

Torres MF, Uetanabaro AP, Costa AF, Alves CA, Farias LM, Bambirra EA, Penna FJ, Vieira EC, Nicoli JR (2000) Influence of bacteria from the duodenal microbiota of patients with symptomatic giardiasis on the pathogenicity of Giardia duodenalis in gnotoxenic mice. J Med Microbiol 49:209–215. doi:10.1099/0022-1317-49-3-209

Tremaroli V, Backhed F (2012) Functional interactions between the gut microbiota and host metabolism. Nature 489:242–249. doi:10.1038/nature11552

Trichopoulou A, Costacou T, Bamia C, Trichopoulos D (2003) Adherence to a Mediterranean diet and survival in a Greek population. N Engl J Med 348:2599–2608. doi:10.1056/NEJMoa025039

Trompette A, Gollwitzer ES, Yadava K, Sichelstiel AK, Sprenger N, Ngom-Bru C, Blanchard C, Junt T, Nicod LP, Harris NL, Marsland BJ (2014) Gut microbiota metabolism of dietary fiber influences allergic airway disease and hematopoiesis. Nat Med 20:159–166. doi:10.1038/nm.3444

Verma AK, Verma R, Ahuja V, Paul J (2012) Real-time analysis of gut flora in Entamoeba histolytica infected patients of Northern India. BMC Microbiol 12:183. doi:10.1186/1471-2180-12-183

Walker AW, Ince J, Duncan SH, Webster LM, Holtrop G, Ze XL, Brown D, Stares MD, Scott P, Bergerat A, Louis P, McIntosh F, Johnstone AM, Lobley GE, Parkhill J, Flint HJ (2011) Dominant and diet-responsive groups of bacteria within the human colonic microbiota. ISME J 5:220–230. doi:10.1038/ismej.2010.118

Wang Z, Roberts AB, Buffa JA, Levison BS, Zhu W, Org E, Gu X, Huang Y, Zamanian-Daryoush M, Culley MK, DiDonato AJ, Fu X, Hazen JE, Krajcik D, DiDonato JA, Lusis AJ, Hazen SL (2015) Non-lethal inhibition of gut microbial trimethylamine production for the treatment of atherosclerosis. Cell 163:1585–1595. doi:10.1016/j.cell.2015.11.055

Wang QP, Lin YQ, Zhang L, Wilson YA, Oyston LJ, Cotterell J, Qi Y, Khuong TM, Bakhshi N, Planchenault Y, Browman DT, Lau MT, Cole TA, Wong AC, Simpson SJ, Cole AR, Penninger JM, Herzog H, Neely GG (2016) Sucralose promotes food intake through NPY and a neuronal fasting response. Cell Metab 24:75–90. doi:10.1016/j.cmet.2016.06.010

Whitman WB, Coleman DC, Wiebe WJ (1998) Prokaryotes: the unseen majority. Proc Natl Acad Sci U S A 95:6578–6583

Wikoff WR, Anfora AT, Liu J, Schultz PG, Lesley SA, Peters EC, Siuzdak G (2009) Metabolomics analysis reveals large effects of gut microflora on mammalian blood metabolites. Proc Natl Acad Sci U S A 106:3698–3703. doi:10.1073/pnas.0812874106

Willett W, Manson J, Liu SM (2002) Glycemic index, glycemic load, and risk of type 2 diabetes. Am J Clin Nutr 76:274s–280s

Wold AE (1998) The hygiene hypothesis revised: is the rising frequency of allergy due to changes in the intestinal flora? Allergy 53:20–25

Wu GD, Chen J, Hoffmann C, Bittinger K, Chen YY, Keilbaugh SA, Bewtra M, Knights D, Walters WA, Knight R, Sinha R, Gilroy E, Gupta K, Baldassano R, Nessel L, Li H, Bushman FD, Lewis JD (2011) Linking long-term dietary patterns with gut microbial enterotypes. Science 334:105–108. doi:10.1126/science.1208344

Yatsunenko T, Rey FE, Manary MJ, Trehan I, Dominguez-Bello MG, Contreras M, Magris M, Hidalgo G, Baldassano RN, Anokhin AP, Heath AC, Warner B, Reeder J, Kuczynski J, Caporaso JG, Lozupone CA, Lauber C, Clemente JC, Knights D, Knight R, Gordon JI (2012) Human gut microbiome viewed across age and geography. Nature 486:222–227. doi:10.1038/nature11053

Yoshimoto S, Loo TM, Atarashi K, Kanda H, Sato S, Oyadomari S, Iwakura Y, Oshima K, Morita H, Hattori M, Honda K, Ishikawa Y, Hara E, Ohtani N (2013) Obesity-induced gut microbial metabolite promotes liver cancer through senescence secretome. Nature 499:97–101. doi:10.1038/nature12347

Yoshioka H, Iseki K, Fujita K (1983) Development and differences of intestinal flora in the neonatal period in breast-fed and bottle-fed infants. Pediatrics 72:317–321

Zak-Golab A, Kocelak P, Aptekorz M, Zientara M, Juszczyk L, Martirosian G, Chudek J, Olszanecka-Glinianowicz M (2013) Gut microbiota, microinflammation, metabolic profile, and zonulin concentration in obese and normal weight subjects. Int J Endocrinol 2013:674106. doi:10.1155/2013/674106

Zamakhchari M, Wei GX, Dewhirst F, Lee J, Schuppan D, Oppenheim FG, Helmerhorst EJ (2011) Identification of Rothia bacteria as gluten-degrading natural colonizers of the upper gastrointestinal tract. PLoS One 6:e24455. doi:10.1371/journal.pone.0024455

Zeisel SH (2006) Choline: critical role during fetal development and dietary requirements in adults. Annu Rev Nutr 26:229–250. doi:10.1146/annurev.nutr.26.061505.111156

Zhou D, Zhang H, Bai Z, Zhang A, Bai F, Luo X, Hou Y, Ding X, Sun B, Sun X, Ma N, Wang C, Dai X, Lu Z (2016) Exposure to soil, house dust and decaying plants increases gut microbial diversity and decreases serum immunoglobulin E levels in BALB/c mice. Environ Microbiol 18:1326–1337. doi:10.1111/1462-2920.12895

Zoetendal EG, Akkermans ADL, Akkermans-van Vliet WM, de Visser JAGM, de Vos WM (2001) The host genotype affects the bacterial community in the human gastrointestinal tract. Microb Ecol Health Dis 13:129–134

Microfungi for the Removal of Toxic Triphenylmethane Dyes

22

Si Hui Chen and Adeline Su Yien Ting

Abstract

Triphenylmethane (TPM) dyes are a group of aromatic, synthetic dyes used widely in industrial processes. The discharge of these dyes into the environment demands strict monitoring and treatment due to their toxicity and cancer-inducing possibilities. The inexpensive, environmental-friendly biological treatment of TPM dyes using microfungi is an attractive remediation approach compared to the conventional physico-chemical methods. A diverse population of microfungi (comprising of members with microscopic fruiting bodies), such as white-rot and non-white-rot fungi, have demonstrated potential in removing TPM dyes via biosorption and biodegradation. Enzymes involved in dye decolourization include laccase, lignin peroxidase, manganese peroxidase and reductases. The biosorption and biodegradation activities of microfungi are influenced by nutrients, pH, temperature, initial dye concentration and biomass concentration. This chapter discusses the various strains of microfungi with TPM dye-decolourizing potential, as well as their mechanisms, optimum conditions and some current technological applications for these useful microfungi to remove TPM dyes.

Keywords

Biodegradation • Biosorption • Decolourization • Microfungi • Triphenylmethane dyes

S.H. Chen • A.S.Y. Ting (✉)
School of Science, Monash University Malaysia,
Jalan Lagoon Selatan, 47500 Bandar Sunway, Selangor, Malaysia
e-mail: adeline.ting@monash.edu; adelsuyien@yahoo.com

V.C. Kalia et al. (eds.), *Mining of Microbial Wealth and MetaGenomics*,
DOI 10.1007/978-981-10-5708-3_22

Abbreviations

Lac	Laccase
LiP	Lignin peroxidase
MnP	Manganese peroxidase
MS	Mass spectroscopy
TPM	Triphenylmethane

22.1 Introduction

Triphenylmethane (TPM) dyes belong to a class of inexpensive, water-soluble synthetic colourants (red, blue, green, violet) with examples such as Basic Fuchsin, Brilliant Green, Bromophenol Blue, Coomassie Brilliant Blue, Cresol Red, Ethyl Violet, Crystal Violet, Cotton Blue, Malachite Green, Methyl Violet, New Fuchsin and Pararosaniline (Bumpus and Brock 1988; Shedbalkar et al. 2008; Casas et al. 2009; Kumar et al. 2012; Chen and Ting 2015a). These dyes are toxic to both living organisms and the environment, with Crystal Violet identified as a mitotic poisoning agent that causes tumour growth and cancer in bacteria, mammalian cells, fishes and rodents; Methyl Violet showing genotoxic interactions with calf thymus DNA; and Malachite Green with cytotoxicity towards mammalian cells and an irritant to the gastrointestinal tract, skin and eyes (Aidoo et al. 1990; Bose et al. 2006; Chi et al. 2009; Khataee et al. 2011). TPM dyes are persistent in the environment as they are resistant to degradation. These dyes have complex aromatic structures, characterized by a chromophoric centre comprising of three phenyl groups surrounding a central carbon atom (Shedbalkar et al. 2008; Jasińska et al. 2012). This structural nature of TPM dyes contributes to their resistance to degradation by microorganisms as well as abiotic oxidizing agents (e.g. soap, sunlight).

Various TPM dyes are used in industries producing textiles, paper, leather, plastic, food and pharmaceuticals (Bumpus and Brock 1988; Malik et al. 2007). However, ineffective textile dyeing processes have led to the release of approximately 10–15% of dyes used (approximately 280,000 tonnes) into the environment via effluents (Maas and Chaudhari 2005). The impact of dye pollution in the environment is devastating, as dye concentrations as low as 1 mg/L could trigger significant reduction of sunlight permeability, leading to reduced photosynthesis and dissolved oxygen and an imbalance ecosystem (Jadhav and Govindwar 2006). Dyes then remain in the environment due to their resistance to degradation, causing toxicity to the living organisms. Hence, treatment of dyes in industrial effluents is pertinent prior to release into the environment. The removal of TPM dyes from industrial effluents is challenging as the effluents may contain different types (and concentrations) of various dyes, inorganic salts, lubricants, solvents, insecticides, bleaches and heavy metals (Haroun and Idris 2009; Halimoon and Goh 2010). These combinations of varying pollutants complicate dye removal, as the effluent itself is toxic and limits the growth of microorganisms responsible for the aerobic biological treatment in sewage treatment systems (Kabra et al. 2012). Therefore, conventional approaches to treating dye effluents are primarily based on physical and chemical treatments such as coagulation-flocculation, ozonation, adsorption and membrane filtration (Azizian et al. 2009; Rodrigues et al. 2013; Zhou et al. 2013).

Coagulation-flocculation is a chemical treatment that is extensively used to remove synthetic dyes (Abu Hassan et al. 2009). Coagulants such as aluminium sulphate, ferric sulphate, ferric chloride and calcium chloride are used to form flocks with dye, which are subsequently separated from water via sedimentation or filtration and finally disposed as sludge (Grinevicius et al. 2009; Rodrigues et al. 2013). This method is however hampered by the generation of toxic sludge and the high cost for sludge disposal. On the contrary, ozonation is a much "cleaner" approach. Ozonation is an advanced oxidation process that utilizes ozone (O_3, oxidizing agent) and chemical catalysts (manganese (Mn) ions, titanium dioxide (TiO_2)) to generate hydroxyl radicals to rapidly oxidize (complete mineralization) TPM dyes into water and carbon dioxide (Zhou et al. 2013). This treatment involves simple procedures without producing toxic sludge or metabolites. Nevertheless, this treatment is limited by the short half-life of ozone (20 min) and has been found to be effective towards only certain types of dyes. The oxidizing efficiency was also discovered to be strongly influenced by concentrations of pollutants and environmental conditions (Turhan et al. 2012).

Dyes are also removed via adsorption, a physical process via binding onto a suitable interface (Malik et al. 2007). This process is effective in removing various types of dyes, even at low dye concentrations (Malik et al. 2007; Azizian et al. 2009). Adsorbents used for this purpose usually have high affinity and capacity for dyes as well as regeneration capability. Activated carbon is one of the most commonly used adsorbents for dye removal (Azizian et al. 2009). Unfortunately, they are expensive with losses incurred during the regeneration process. As a result, cheaper adsorbents are sought as alternatives, particularly waste materials generated from industrial processes. Several waste materials have been reported as useful adsorbents. They include bottom ash from coal combustion at power generation plants (Gupta et al. 2006), de-oiled soya from soybean oil industries (Mittal et al. 2008), hen feathers (Mittal 2006), Pu-erh tea residual powder (Li et al. 2010), walnut shell (Dahri et al. 2014) and chemically modified sphagnum peat moss (Hemmati et al. 2016).

In addition, membrane filtration is another useful approach to remove dye. This technology encompasses the use of various filtration techniques, which include micro-, ultra-, nanofiltration as well as reverse osmosis (Vergili et al. 2012). Both microfiltration and ultrafiltration have been used for removal of TPM dyes. The membranes are typically made of polysulfones, polyamides or cellulose acetate. Several studies have established the success of membrane filtration technology, with Brilliant Green (3.0×10^{-6} M) successfully removed using polyurethane membrane from dye solution containing either 1 M NaCl (63%) or 0.01 M HCl (100%) (Rzeszutek and Chow 2001). Another useful membrane is the polysulfone asymmetric hollow fibre ultrafiltration membrane with anionic poly(sodium-4-styrenesulfonate) polymer, which could result in 98, 100 and 97% removal of

Malachite Green, Brilliant Green and New Fuchsin solutions (5 mg/L) (Tan et al. 2006). Membrane filtration is fast and efficient in removing dyes, but the usage is limited by high costs incurred, risks of membrane fouling, production of toxic sludge and small volume of wastewater treated (Vergili et al. 2012). Hence, biological approaches, specifically the use of biological organisms, are garnering interest as potential alternatives to dye removal.

Biological organisms, particularly microorganisms such as fungi, bacteria, algae and yeast, have the potential to degrade or transform toxic dyes into simpler, less hazardous molecules (Pokharia and Ahluwalia 2013). In comparison to physico-chemical approaches, biodegradation incurs lower cost, produces less toxic sludge and is more environmental-friendly (especially when complete mineralization of dyes occurs) (Shedbalkar et al. 2008; Jasińska et al. 2015). Of the many types of microorganisms, microfungi are favoured as they have been reported to be able to degrade a broad range of pollutants, such as synthetic dyes, pesticides, polycyclic aromatic hydrocarbons, chlorophenols and explosives (Harms et al. 2011). To date, microfungi have been reported to remove azo dyes, anthraquinone dyes as well as TPM dyes (Zhang et al. 2007; Yang et al. 2009b; Torres et al. 2011; Chen and Ting 2015a, b). Members of microfungi consist of a diverse group of eukaryotic organisms that are ubiquitous in nature and have a distinct characteristic of harbouring microscopic fruiting bodies (Martin and Rygiewicz 2005). Microfungi can adapt well to changing environmental conditions and are also amenable to genetic and morphological modification to enhance elimination of environmental pollutants (Zhuo et al. 2011; Tamayo-Ramos et al. 2012). The biodegradation potential of dyes by microfungi is attributed to their ability to produce non-specific, extracellular enzymes which are responsible in degrading dyes (pollutants) (dos Santos Bazanella et al. 2013). Microfungi can also be used as biosorbents, removing dyes via surface binding of dye molecules to functional groups in the cell wall (Shedbalkar and Jadhav 2011; Chaudhry et al. 2014). The high capacity of microfungi for binding with dyes is attributed to the growth and large cell-to-surface ratio of fungal mycelia. In addition, fungal biomass is also easily produced in large quantities at reasonably low cost, especially as a by-product from industrial fermentation processes (from synthesis of enzymes and antibiotics) (Svecova et al. 2006). This chapter will discuss the use of microfungi in removing TPM dyes, which includes the diversity of microfungi known as efficient dye removers, their mechanisms of dye removal, conditions optimum for dye removal and some of the current technological applications of dye removal by microfungi.

22.2 Diversity of Microfungi for Dye Removal

The most extensively studied microfungi for dye removal is the white-rot fungus *Phanerochaete chrysosporium*. White-rot fungi are Basidiomycetes that form fruiting caps. Nevertheless, they are also considered as microfungi as their fruiting bodies (basidiospores) are characteristically microscopic. *P. chrysosporium* is extremely well-studied as this species is capable of degrading various toxic dyes (Levin et al.

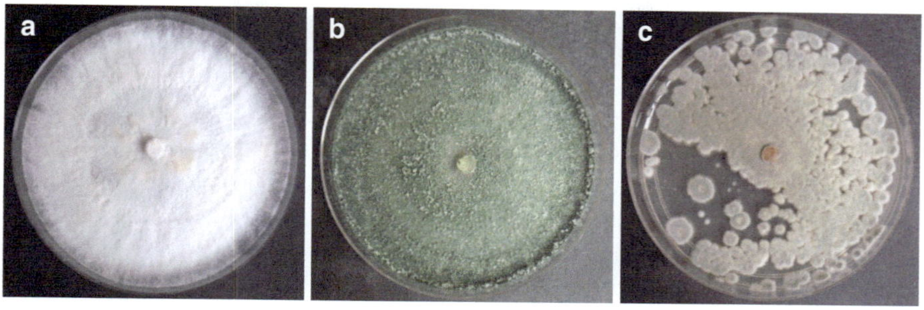

Fig. 22.1 Colony morphologies of isolates (**a**) *Coriolopsis* sp. (1c3), (**b**) *Trichoderma asperellum* (11) and (**c**) *Penicillium simplicissimum* (10) on Potato Dextrose Agar. These are microfungi with TPM dye-decolourizing potentials

2004; Sedighi et al. 2009). The isolate was first described in 1983 for degradation of polymeric dyes (e.g. Poly B-411, Poly R-481, Poly Y-606) (Glenn and Gold 1983). It has also shown biodegradation activities on Bromophenol Blue, Cresol Red, Crystal Violet, Ethyl Violet, Methyl Violet, Methyl Green, Malachite Green, Brilliant Green, Pararosaniline and Victoria Blue (Bumpus and Brock 1988; Papinutti and Forchiassin 2004; Radha et al. 2005; Gomaa et al. 2008). In recent years, other species of white-rot fungi with similar potential to degrade TPM dyes were discovered. They include species of *Trametes* (Liu et al. 2004), *Coriolopsis* (Chen and Ting 2015a), *Pycnoporus* (Pointing and Vrijmoed 2000), *Coriolus* (Levin et al. 2004) and *Irpex* (Novotný et al. 2004) (Fig. 22.1). Each species varies in their decolourization efficiency attributed to the species-dye interaction. These may range from 79 to 98% by *P. chrysosporium* for Methyl Violet (Radha et al. 2005); 92 and 79% by *Trametes versicolor* for Coomassie Brilliant Blue and Cresol Red, respectively (Liu et al. 2004); 94, 97, 91 and 52% by *Coriolopsis* sp. for Crystal Violet, Methyl Violet, Cotton Blue and Malachite Green (Chen and Ting 2015a); 100, 98 and 20% by *Pycnoporus sanguineus* for Malachite Green, Bromophenol Blue and Crystal Violet (Pointing and Vrijmoed 2000); as well as 96% by *Irpex lacteus* for Bromophenol Blue (Novotný et al. 2004).

Several non-white-rot fungi have also been explored for removal of TPM dyes. Species of *Aspergillus* (Kumar et al. 2011), *Penicillium* (Shedbalkar et al. 2008; Chen and Ting 2015b), *Trichoderma* (Chew and Ting 2015) and *Fusarium* (Abedin 2008) have been found to show good potentials for dye decolourization (Fig. 22.1). Methyl Violet and Brilliant Green (10 mg/L) were completely decolourized (>95%) by *Aspergillus* sp. CB-TKL-1 via biodegradation within 1 and 3 days, respectively (Kumar et al. 2011, 2012). Aerobic *Fusarium solani* (Martius) Saccardo cultures adsorbed and degraded Crystal Violet and Malachite Green (3.5 mg/L) at relatively high decolourization efficiencies of 84–91% (by day 4) and 93–94% (by day 2), respectively (Abedin 2008). In contrast, the complete degradation of Cotton Blue (50 mg/L) by *Penicillium ochrochloron* occurred much faster (within 2.5 h) under static incubation (Shedbalkar et al. 2008). Dried biomass of *Trichoderma asperellum* immobilized on alginate polymers demonstrated high efficiency in removing

100 mg/L Crystal Violet (60.64 mg/g), Methyl Violet (50.29 mg/g), Malachite Green (16.61 mg/g) and 50 mg/L Cotton Blue (49.91 mg/g) through biosorption (Chew and Ting 2015).

Microfungi with dye removal potential can be found either as indigenous species to the polluted sites or are isolated from non-polluted sources and introduced to the contaminated sites (Kumar et al. 2011; Chew and Ting 2015; Chen and Ting 2015a, b). Polluted sites are often a good source for useful isolates as the prolonged exposure to the pollutant allows tolerance and adaptation of isolate towards the pollutant. In this case, industrial effluents, rivers or soil in contact with dyes, are excellent sources. Several studies documented this successfully. *Trichoderma asperellum* SPF1 isolated from river sediments of the Penchala River (Selangor, Malaysia) showed decolourization activities towards Crystal Violet, Malachite Green, Methyl Violet and Cotton Blue (Chew and Ting 2015). Effective decolourization of Crystal Violet and Malachite Green was also reported for *Fusarium solani* (Martius) Saccardo isolated from dye effluents in Egypt (Abedin 2008). Jasińska et al. (2012) isolated *Penicillium pinophilum* and *Myrothecium roridum* from soil surrounding a textile dyeing factory (Poland), which decolourized 10 mg/L Malachite Green (in liquid cultures) with 87 and 97% decolourization efficacy, respectively. Microfungi isolated from soil acquired from textile dye factories in India (*Mucor mucedo*, *Lenzites betulina*, *Polyporus elegans* and *Trametes versicolor*) demonstrated 26–78% decolourization of Crystal Violet and Malachite Green (20 mg/L) (Moturi and Singara Charya 2009).

Microfungi from non-polluted sites have also showed similar potential as isolates from polluted sources. Lesser-known white-rot fungi such as *Irpex* spp., *Pycnoporus sanguineus* (Fr) Murill and *Trametes elegans* isolated from dead forest trees of Chirinda and Chimanimani (Zimbabwe) showed decolourization activities for 0.02% Cresol Red, Crystal Violet and Bromophenol Blue on solid media (Tekere et al. 2001). Another isolate, *Trametes* sp. SQ01 isolated from decayed woods of a temperate forest (China), was capable of removing 100% of Bromophenol Blue but only 30–70% of Malachite Green and Coomassie Brilliant Blue G250 in liquid cultures (Yang et al. 2009b). Other microfungi such as *Penicillium* sp. and *Cladosporium* sp. isolated from marine sediments, seawater and live seagrasses (Philippines) removed 87–91% of 0.01% Crystal Violet (Torres et al. 2011).

In the early years, diversity studies were performed by identifying fungal isolates based solely on the morphological and cultural characteristics (Denoyes and Baudry 1995) (Fig. 22.2). Careful examinations on the colour, texture and size of the colonies, formation of exudates on the aerial part of actively growing mycelia and presence of characteristic microscopic structures aided in species identification (Tekere et al. 2001; Hutwimmer et al. 2010; Chen and Ting 2015a). This is a long process, complicated by each taxonomic fungal group having specific features and terminology, which at times are difficult to distinguish between species. In recent years, fungal identification is easier and faster with technological advancements in the fields of DNA sequencing, PCR amplification and bioinformatics (White et al. 1990; Kõljalg et al. 2005; Tsui et al. 2011) (Fig. 22.2). These approaches involve sequencing the 18S, 5.8S and 28S ribosomal RNA genes, especially the internal

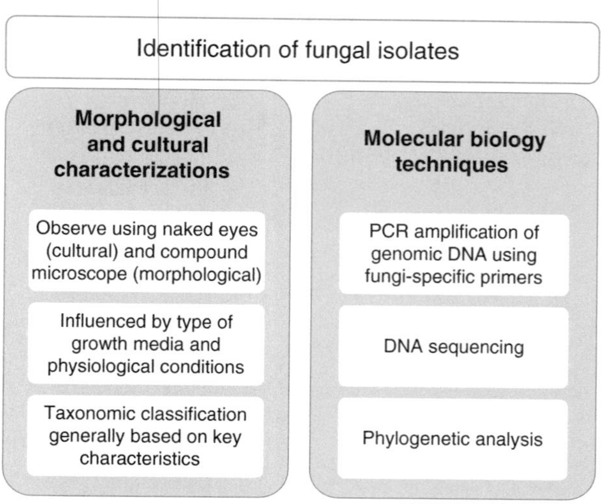

Fig. 22.2 General scheme for identifying fungal isolates

transcribed spacer (ITS) gene region, to identify fungal isolates at genus and species levels (Yang et al. 2009a, b; Jasińska et al. 2012). The identification depends on the rate of mutations accumulated in these gene regions, which correlates to the rate of speciation (Nilsson et al. 2008). The sequences obtained are subsequently compared to accession sequences available at public genetic sequence databases, such as GenBank, the European Molecular Biology Laboratory (EMBL) as well as other relevant databases such as DNA Data Bank of Japan (DDBJ) and User-friendly Nordic ITS Ectomycorrhiza Database (UNITE), to deduce the identity of the isolate (Kõljalg et al. 2005; Tedersoo et al. 2010; Jasińska et al. 2015).

22.3 Mechanisms of TPM Dye Removal by Microfungi

22.3.1 Biosorption of Dyes

The treatment of wastewater containing TPM dyes may involve whole microbial cells, immobilized cells or purified extracellular enzymes extracted from microorganisms (Liu et al. 2004; dos Santos Bazanella et al. 2013; Chew and Ting 2015). The removal of TPM dyes by microfungi can occur via two mechanisms: biosorption and/or bioaccumulation (Jasińska et al. 2012; Chen and Ting 2015a). Biosorption involves a combination of active and passive transport mechanism (e.g. adsorption, chelation, complexation, electrostatic interaction and ion exchange) for dye molecules to bind onto the microbial biomass (Asgher 2012). This process occurs for both live and dead cells, attributed to the rich lipid and heteropolysaccharide components (glucan, chitin and chitosan) in the cell wall of fungal species. These components have functional groups such as carboxyl, hydroxyl, phosphate and amino groups, which attract and bind dye molecules (Saeed et al. 2009; Chew and Ting 2015). The use of dead cells for biosorption is

Table 22.1 Dead cells of microfungi investigated for removal of TPM dyes via biosorption

Fungal strain	TPM dye(s)	References
Aspergillus fumigatus	Acid Violet 49	Chaudhry et al. (2014)
	Bromophenol Blue	Zeroual et al. (2006a)
Aspergillus niger	Methyl Violet	Bhole et al. (2004)
	Basic Fuchsin	
Fusarium solani (Martius) Saccardo	Crystal Violet	Abedin (2008)
	Malachite Green	
Fusarium sp.	Bromophenol Blue	Zeroual et al. (2006a)
Geotrichum sp.	Bromophenol Blue	Zeroual et al. (2006a)
Rhizopus stolonifer	Bromophenol Blue	Zeroual et al. (2006a)
Trichoderma asperellum	Crystal Violet	Chew and Ting (2015)
	Cotton Blue	
	Malachite Green	
	Methyl Violet	

often preferred to live cells as the former is unaffected by the toxic and harsh environmental conditions (Zeroual et al. 2006a). In comparison to live cells, dead cells also do not require nutrient input to sustain growth, can be stored for longer period of time (dried biomass for storage) and are more amenable to regeneration and reuse for dye removal (Svecova et al. 2006; Abedin 2008). Nevertheless, usage of dead cells is relatively limited compared to using live cells, which generates more desired outcomes. When dead cells are used, dyes are entrapped in fungal biomass without fragmentation, creating toxic biomass residues with disposal problems. The lack of structural difference of dye molecules after fungal biosorption was indicated by a proportional decrease in absorption peaks of UV-visible spectrum of the supernatant (Gomaa et al. 2008; Almeida and Corso 2014). In addition, complete removal of dyes from solutions via fungal biosorption does not always occur. For instance, dead biomass of *Aspergillus* sp. CB-TKL-1 adsorbed only 10% of Brilliant Green after 15 days of incubation as compared to live cells, which showed a 99% decolourization (Kumar et al. 2012) (Tables 22.1 and 22.2). The biosorption of Methyl Violet by live cells of *Phanerochaete chrysosporium* was documented at only 39%, with 100% removal achieved upon dye degradation by fungal enzymes (Radha et al. 2005). Similar observations were reported by Jasińska et al. (2012) for the biosorption of Malachite Green by live *Penicillium pinophilum* cells. The complete degradation of dyes is more desirable as the degraded products are often less harmful than the original dye molecules. This was proven to be so through toxicity assays on bacteria and plant seeds. For instance, high growth intensities of bacterial strains (e.g. *Escherichia coli*, *Staphylococcus aureus* and *Pseudomonas aeruginosa*) were observed when exposed to the degradation products of Malachite Green by *Myrothecium roridum*, but not for the original dye compound (Jasińska et al. 2015). These indicated the non-toxic nature of the former. The by-products of Cotton Blue degradation by *Penicillium ochrochloron* were also found to have no

Table 22.2 Live cells of microfungi investigated for removal of TPM dyes via biodegradation

Fungal strain	TPM dye(s)	References
Aspergillus niger	Bromocresol Purple	Tamayo-Ramos et al. (2012)
	Malachite Green	
Aspergillus ochraceus	Malachite Green	Gomashe et al. (2011)
Aspergillus sp.	Methyl Violet	Kumar et al. (2011)
	Brilliant Green	Kumar et al. (2012)
Cladosporium sp.	Crystal Violet	Torres et al. (2011)
Coriolopsis sp.	Crystal Violet	Chen and Ting (2015a)
	Cotton Blue	
	Malachite Green	
	Methyl Violet	
Coriolus versicolor f. *antarcticus*	Malachite Green	Levin et al. (2004)
Cunninghamella elegans	Malachite Green	Cha et al. (2001)
Ganoderma sp.	Bromophenol Blue	Zhuo et al. (2011)
	Crystal Violet	
	Malachite Green	
Geobacillus thermoglucosidasius	Malachite Green	Gao et al. (2015)
Irpex lacteus	Bromophenol Blue	Novotný et al. (2004)
Lenzites betulina	Crystal Violet	Moturi and Singara Charya (2009)
	Malachite Green	
Mucor mucedo	Crystal Violet	Moturi and Singara Charya (2009)
	Malachite Green	
Myrothecium roridum	Malachite Green	Jasińska et al. (2015)
Penicillium ochrochloron	Cotton Blue	Shedbalkar et al. (2008)
	Malachite Green	Shedbalkar and Jadhav (2011)
Penicillium pinophilum	Malachite Green	Jasińska et al. (2012)
Penicillium simplicissimum	Crystal Violet	Chen and Ting (2015b)
	Cotton Blue	
	Malachite Green	
	Methyl Violet	
Penicillium sp.	Crystal Violet	Torres et al. (2011)
	Malachite Green	Yang et al. (2011b)
Phanerochaete chrysosporium	Crystal Violet	Bumpus and Brock (1988); Sani et al. (1998)
	Methyl Violet	Radha et al. (2005)
Polyporus elegans	Crystal Violet	Moturi and Singara Charya (2009)
	Malachite Green	
Pycnoporus sanguineus	Crystal Violet	Lu et al. (2009)
	Bromophenol Blue	Pointing and Vrijmoed (2000)
	Crystal Violet	
	Malachite Green	

(continued)

Table 22.2 (continued)

Fungal strain	TPM dye(s)	References
Trametes elegans	Bromophenol Blue	Tekere et al. (2001)
	Cresol Red	
	Crystal Violet	
Trametes sp.	Bromophenol Blue	Yang et al. (2009b)
	Coomassie Brilliant Blue G250 Malachite Green	
Trametes trogii Berk	Gentian Violet	Yan et al. (2014)
	Malachite Green	
Trametes versicolor	Crystal Violet	Moturi and Singara Charya (2009)
	Malachite Green	
	Coomassie Brilliant Blue	Liu et al. (2004)
	Cresol Red	

toxic effects on the seeds of *Triticum aestivum* and *Ervum lens Linn*, as well as bacteria *Azotobacter vinelandii* (Shedbalkar et al. 2008). Thus, the biodegradation process by microfungi also plays an important role in removing TPM dyes in addition to dye biosorption.

22.3.2 Biodegradation of Dyes

Biodegradation is a process that is energy-dependent and utilizes non-specific, ligninolytic enzymes to catalyse the breakdown of dyes into smaller and simpler compounds (Casas et al. 2009). This process is exclusive to live cells. Dyes are decolourized when the chromophoric centres are cleaved via enzymatic reactions (Abedin 2008). Complete breakdown (mineralization) of dyes may result in less toxic compounds such as carbon dioxide, water and inorganic salts. The ligninolytic enzymes involved in such process are mainly laccase, lignin peroxidase, manganese peroxidase and/or NADH-DCIP reductase (Yang et al. 2011; Jasińska et al. 2012). These enzymes are secreted during secondary metabolism where nutrient availability (mostly nitrogen or carbon) is limited and is not affected by pollutants (Levin et al. 2010). Briefly, laccase (Lac) (EC 1.10.3.2), a multicopper oxidase, oxidizes dye molecules aerobically to generate free radicals for dye degradation (Zhuo et al. 2011). Two heme-containing enzymes, lignin peroxidase (LiP) (EC 1.11.1.14) and manganese peroxidase (MnP), perform degradation of non-phenolic aromatic compounds via catalysis with hydrogen peroxide, converting the aromatic compounds into less harmful water compounds (Yang et al. 2011; dos Santos Bazanella et al. 2013). NADH-2,6-dichloroindophenol (DCIP) reductase (EC 1.6.99.3) is a flavoprotein where NADH acts as an electron donor to reduce DCIP (Nishiya and Yamamoto 2007). Certain microfungi (e.g. *Coriolopsis* sp. *Penicillium ochrochloron*) produce several of these degradative enzymes to remove dyes, while others only produce one or two of them (e.g. *Penicillium simplicissimum*, *Myrothecium roridum*) (Shedbalkar et al. 2008; Jasińska et al. 2012; Chen and Ting 2015a, b) (Table 22.2).

A typical pathway for biodegradation of dye by microfungi is as follows: secretion of extracellular enzymes to catalyse dye degradation and/or adsorption of dyes onto fungal biomass, followed by accumulation within cells for degradation catalysed by intracellular enzymes (Jasińska et al. 2012). The secretion of fungal extracellular enzymes contributes primarily to degradation of TPM dyes, where the type of enzymes produced depend on fungal species and type of dyes. This was discovered by Chen and Ting (2015a, b) where *Coriolopsis* sp. produced Lac, LiP and NADH-DCIP reductase to aid in the degradation of Crystal Violet, Methyl Violet, Malachite Green and Cotton Blue, while *Penicillium simplicissimum* secreted only LiP and NADH-DCIP for the degradation of the same dyes. The occurrence of biodegradation was confirmed by the shift in major absorption peaks and appearance of new peaks (by-products of biodegradation) using UV-visible spectroscopy analyses (Fig. 22.3). The decolourization of Malachite Green has also been attributed to extracellular Lac activities of *Myrothecium roridum* (2.90 mM/min/L) (Jasińska et al. 2012) and *Coriolus versicolor* f. *antarcticus* (0.13 U/mL) (Levin et al. 2004), as well as extracellular LiP and NADH-DCIP reductase activities of *Penicillium* sp. YW01 (Yang et al. 2011). The degradation of Cotton Blue by *Penicillium ochrochloron* MTCC 517 involved a wide range of enzymes, including LiP

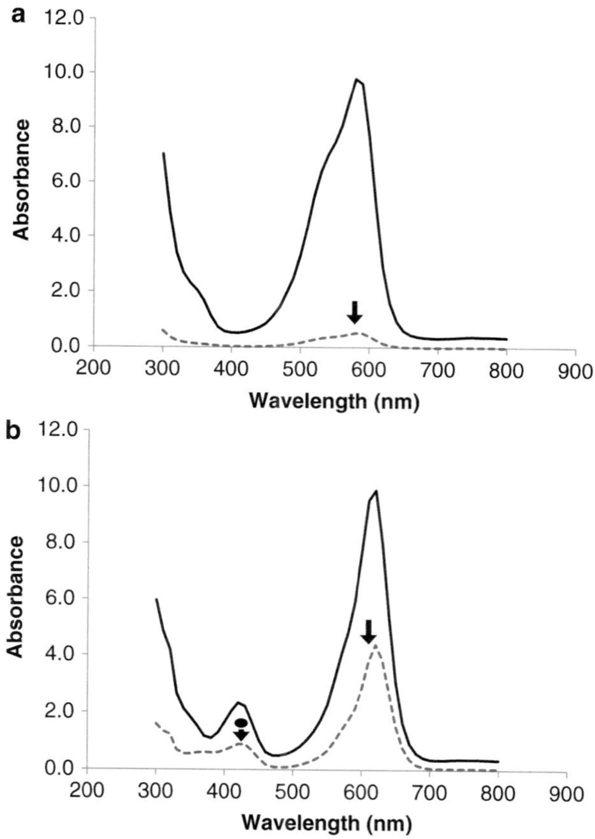

Fig. 22.3 UV-visible spectra of (**a**) Methyl Violet and (**b**) Malachite Green before (*solid line*) and after decolourization (*dashed line*) by *Penicillium simplicissimum* (iso 10), with (*down arrow*) indicating disappearance of original peak and (*circle with down arrow*) as shift in wavelengths of peaks

(0.025 U/min/mL), Lac (0.045 U/min/mL), tyrosinase (0.23 U/min/mL), NADH-DCIP reductase (483 μg DCIP reduced/min/mL), MG reductase (12.49 μg MG reduced/min/mL) and aminopyrine *N*-demethylase (0.22 *n* mol formaldehyde liberated/min/mg protein) (Shedbalkar et al. 2008). The involvement of membrane-associated oxidoreductive enzymes such as cytochrome P450 monooxygenases has also been reported for Malachite Green removal by *Cunninghamella elegans* (strain ATCC 36112) (Cha et al. 2001).

The pathways of enzymatic degradation for certain TPM dyes by microfungi have been elucidated in several studies, primarily through the analysis of products from the degradation process. Although UV-visible spectra offered an insight into this, validation is typically achieved through high-performance liquid chromatography (HPLC) analysis. The HPLC analysis detects the simplified compounds or products from the biodegradation process, which reflects the sequential breakdown of TPM dyes. In one of the earlier studies by Bumpus and Brock (1988), sequential *N*-demethylation of Crystal Violet by ligninolytic cultures of *Phanerochaete chrysosporium* was revealed through the detection of three metabolites: the first a N,N,N′,N′,N″-penta-, secondly a N,N,N′,N″-tetra- and the third compound, a N,N′,N″-trimethylpararosaniline (Bumpus and Brock 1988). Similarly, Jasińska et al. (2012) documented the reduction of Malachite Green to a compound known as leucomalachite green (LMG), via conversion by *N*-demethylation to generate *N*-demethyl-leucomalachite green (desmethyl-LMG) and *N*-demethyl-malachite green (desmethyl-MG) by *Penicillium pinophilum* and *Myrothecium roridum.* This was deduced based on the use of the liquid chromatography-tandem mass spectrometry (LC/MS/MS). Shedbalkar et al. (2008) included FTIR spectroscopy and gas chromatography coupled with mass spectrometer (GC-MS) analysis, as evidence of biodegradation of Cotton Blue by LiP of *Penicillium ochrochloron* via detection of sulphonamide and triphenylmethane as the by-products of the biodegradation process.

The removal of TPM dyes by biodegradation is promising as complete mineralization of dyes may occur such that no toxic products are released into the environment. However, the process is complicated by factors such as toxic effect of dye on live cells, nitrogen-limiting conditions, low pH for optimum enzymatic activities, long growth cycle of fungi and long hydraulic retention times for complete dye decolourization (Abedin 2008; Kumar et al. 2012; Chen and Ting 2015a). Thus, there is a need for optimization to promote dye removal by microfungi.

22.4 Optimum Conditions for TPM Dye Removal by Microfungi

The efficiency of microfungi in decolourizing TPM dyes depends on various parameters such as pH, agitation, temperature, type of dye, initial dye concentration, biomass concentration and media composition (nitrogen and carbon sources) (Radha et al. 2005; Abedin 2008; Chen and Ting 2015a). These factors generally can be classified as factors affecting the characteristics of dye solutions and fungal growth conditions. The degrees in which these factors influence decolourization via biosorption or biodegradation vary, as the former takes place mainly in dead fungal cells, while the latter involves live cells.

The pH value of the dye solution is crucial in dye biosorption as it influences the type of surface charge found on the biosorbent, which subsequently influences their adsorption efficacy to oppositely charged dye groups (Zeroual et al. 2006b). In addition, pH influences the ionization status of dye molecules in water as well. The biosorption capacity of microfungi for acidic TPM dyes (Bromophenol Blue) has been reported to increase at lower pH values, while basic TPM dyes (Basic Fuchsin) occur at higher pH (Bhole et al. 2004; Zeroual et al. 2006b). These could be due to the electrostatic interactions occurring between the biomass and dye compounds. At low pH, the protonation of weak base groups on the biomass is increased, leading to the biomass acquiring a net positive charge that allows binding with anionic groups of the acidic dye. On the contrary, at high pH, deprotonation of functional groups occurs resulting in negatively charged biomass having strong interactions with positively charged dye molecules. The pH factor also influences the biodegradation activities of live fungal cells as fungal growth and ligninolytic enzyme activities are generally at their maximum at lower pH (4.5–7) (Radha et al. 2005; Lu et al. 2009; Kumar et al. 2012). The pH range of 4.0–5.5 was optimum for the decolourization of basic TPM dyes Methyl Violet (>90%) and Brilliant Green (92–99%) by *Phanerochaete chrysosporium* and *Aspergillus* sp. CB-TKL-1, respectively (Radha et al. 2005; Kumar et al. 2012). Similarly, *Myrothecium roridum* removed more Malachite Green at pH 4 (56–94%) compared to pH 7 (24–42%) (Jasińska et al. 2012). Nevertheless, some microfungi showed better removal efficacy at higher pH, such as in the case of *Fusarium solani* which demonstrated complete decolourization of Crystal Violet and Malachite Green at pH 8–9 (Abedin 2008). In some rare cases, microfungi may not be influenced by pH conditions, as for *Penicillium ochrochloron* which removed Malachite Green most effectively at pH 7 (93%), followed by pH 5 (88%) and pH 9 (64%) (Shedbalkar and Jadhav 2011). Decolourization of the same TPM dye by *Penicillium pinophilum* showed less than 10% difference in removal efficiency across pH values of 4–7 (Jasińska et al. 2012). This is beneficial as wastewater effluents from textile industries may vary from pH values of 3.85 to 12.99 (Ogunlaja and Aemere 2009; Lim et al. 2010).

Few studies have conducted investigations on the influence of temperature on TPM dye biosorption by microfungi, although temperature is known to affect the surface activity and kinetic energy of various dye molecules. The optimum temperature for TPM dye decolourization by majority of microfungi is within the range of 30–35 °C (Shedbalkar and Jadhav 2011; Kumar et al. 2012). Enhanced decolourization of Malachite Green was demonstrated by live *Penicillium ochrochloron* cells at 30 °C (93%), instead of 20 °C (68%) and 40 °C (70%) (Shedbalkar and Jadhav 2011). Temperatures of 35 °C were found optimal for *Phanerochaete chrysosporium* and *Aspergillus* sp. CB-TKL-1 to degrade Methyl Violet and Brilliant Green (Kumar et al. 2012; Radha et al. 2005). The few literatures available concluded that at low temperatures (20–25 °C), dye molecules have low kinetic energy such that there is less interaction between the molecules and biomass for biosorption and biodegradation to occur and vice versa. At optimal temperatures, fungal respiration and substrate metabolism were postulated to be enhanced. The decrease in degradation capabilities at very high temperatures may be due to the inability of fungi to produce enzymes for dye decolourization or denaturation of enzymes. Interestingly, certain fungal strains showed better decolourization efficiencies at higher temperatures of 45–70 °C. The biosorption capacity of pretreated *Rhizopus stolonifer* for Bromophenol Blue remained stable at 45–55 °C (716–717 mg/g) (Zeroual et al. 2006b), while crude Lac extract of *Trametes trogii* Berk S0301 decolourized Malachite Green (79%) and Gentian Violet (81%) maximally at 50–70 °C (Yan et al. 2014). Microfungi with the ability to decolourize dyes at high temperatures are gaining attention as actual dye wastewater are at such temperatures (Ogunlaja and Aemere 2009; Lim et al. 2010).

Another relatively important factor of dye decolourization by fungi is the initial dye concentration, especially at higher concentrations. Low initial dye concentrations did not affect the adsorption of dyes onto the fungal biomass as the ratio of dye molecules is too low at such concentrations. At 1.5–3.0 mg/L of Crystal Violet and Malachite Green, similar decolourization efficiencies of dead *Fusarium solani* biomass for the TPM dyes (approximately 80–94%) were demonstrated (Abedin 2008). Decolourization of Brilliant Green by *Aspergillus* sp. CB-TKL-1 was similarly high (97–100%) at 10 and 70 mg/L (Kumar et al. 2012). On the contrary, increasing the initial dye concentrations to much higher values generally resulted in enhanced decolourization by microfungi. The availability of more dye molecules could have provided a stronger driving force in overcoming the mass transfer resistance between the dye molecules and cells. However, high decolourization efficiencies were only observed for up to certain initial dye concentrations as any further increase in concentrations leads to the saturation of available binding sites on fungal biomass. The biosorption capacity of dead *Rhizopus stolonifer* for Bromophenol Blue increased with the increase in initial dye concentrations, as observed in 200 mg/L (194 mg/g) to 1000 mg/L (750 mg/g) but at the expanse of percentage of dye removed (97–75%) (Zeroual et al. 2006b). The decolourization efficiencies of live cells of *Coriolopsis* sp. and *Penicillium simplicissimum* decolourized Crystal Violet, Methyl Violet and Malachite Green were most effectively at 50 mg/L (64–90%) and decreased with increasing initial dye concentrations of 100 (40–82%) and 200 mg/L (7–79%) (Chen and Ting 2015a, b). Radha et al. (2005) also reported similar observations for Methyl Violet degradation by *Phanerochaete chrysosporium*. In addition to saturating binding sites, high initial dye concentrations may also exert toxic effects on the metabolic activities of live fungal cells, which in turn reduce secretion of enzymes for dye decolourization (Chen and Ting 2015a). Thus, the decolourization efficiencies of microfungi can be enhanced with increasing initial dye concentrations but only up to a certain range of dye concentrations.

The concentration of biomass influences the biosorption and biodegradation of dyes by affecting the active sites available for binding to dye molecules (Radha et al. 2005). Increasing the biomass concentration increases the surface area of the biomass, which in turn provides more active sites so that dye molecules can bind onto for removal from the solution. Crystal Violet was decolourized at a higher percentage (89%) when 5.0 and 7.5 g/L of dead *Fusarium solani* biomass were used instead of 2.5 g/L (about 45%) (Abedin 2008). High decolourization efficiencies were also reported for live cells of *Coriolopsis* sp., *Penicillium simplicissimum* and *Phanerochaete chrysosporium* for the degradation of TPM dyes (Radha et al. 2005; Chen and Ting 2015a, b). The presence of more cells may have also contributed to the secretion of higher amounts of enzymes responsible for biodegradation, as well as cell viability and metabolic functions in the presence of toxic dyes. Nevertheless, a plateau (stationary) stage of decolourization was reached where any further increase in biomass concentration did not yield higher decolourization efficiency. The percentage of Crystal Violet and Methyl Violet degraded by 2 g of fresh *Coriolopsis* sp. and 2 mL of *P. chrysosporium* (inoculum size) was insignificantly different than when higher biomass concentrations were used at 4–8 g and 3–4 mL, respectively (approximately 80–90%) (Radha et al. 2005; Chen and Ting 2015a). This was because at high biomass concentrations, the decolourization process is limited by the number of dye molecules that bound to the active sites as well as physical blockage of binding sites due to interference from many cells interacting with one another. On that account, using an optimally low biomass concentration may be sufficient to achieve maximum decolourization efficiency, simultaneously reducing biomass resources.

The decolourization process of TPM dyes is also influenced by agitation, with agitated cultures generally more effective in removing dyes compared to static conditions. This could be due to increased oxygen contact by fungal cells and oxidative enzymes such as Lac for oxidative metabolism (Kumar et al. 2012). Agitation of *Aspergillus ochraceus* at 150 rpm was reported to yield higher decolourization of Malachite Green (83%) than at static conditions (55%) within 1 day (Gomashe et al. 2011). Decolourization of Crystal Violet by shaking *P. chrysosporium* cultures reached maximum (approximately 80%) within 13 days, while under static conditions, low dye removal was observed (about 15%) even after 13 days (Sani et al. 1998). Most biosorption studies of microfungi are conducted with agitation to achieve maximum biosorption capacities for TPM dyes (Zeroual et al. 2006a; Abedin 2008; Chew and Ting 2015). In some cases of biodegradation, static incubation conditions have been reported as better for decolourization of TPM dyes by certain fungal strains. Cotton Blue together with Crystal Violet and Methyl Violet was decolourized better by *Coriolopsis* sp. with agitation (82–86%), while Malachite Green favoured anaerobic, static conditions (48%) (Chen and Ting 2015a). The four dyes were removed by 46, 74, 66 and 40% at inferior conditions. In contrast, only Cotton Blue was decolourized effectively by *Penicillium simplicissimum* with agitation (64%), whereas Crystal Violet, Methyl Violet and Malachite Green were highly decolourized under static conditions (73–82%) (Chen and Ting 2015b). Better dye decolourization under static conditions may be attributed to the induction of reductive enzymes, i.e. NADH-DCIP reductase, for dye degradation (Chen and Ting 2015b). Based on these observations, it may be advantageous to use mixed fungal cultures (consortium) instead of pure cultures as each strain could interact with the dye molecules at different positions or further decompose the degradation products generated by another strain, depending on oxygen availability (Yang et al. 2009a). Effective decolourization of TPM dyes may also occur through sequential aerobic/anaerobic treatment involving microbial consortium, though majority of the studies were conducted on bacteria treatment of azo dyes (Elisangela et al. 2009).

The type of media and nutrient inputs (nitrogen and carbon) also play important roles in the biodegradation of TPM dyes as they determine the rate of fungal growth and metabolism. Fungal biomass is often established in pure nutrient media in the absence of dyes (e.g. Potato Dextrose Broth, Czapek Dox Broth, Kirk's liquid media) (Shedbalkar et al. 2008; Yang et al. 2011; Chen and Ting 2015a). Studies have shown fungi capable of decolourizing TPM dyes in media with relatively simple compositions. The biodegradation of the dye Malachite Green by *Aspergillus ochraceus* NCIM-1146 was demonstrated higher in either potassium phosphate buffer (pH 7.4) or purely distilled water, compared to Potato Dextrose Broth (contains 20 g/L potato infusion and 20 g/L dextrose) and distilled water containing 0.1 g/L yeast extract (Gomashe et al. 2011). The requirement of nitrogen and glucose inputs for dye degradation by fungi is dependent on factors such as fungal strain and type of dyes. Earlier studies on macrofungi *Cyathus bulleri* biomass and ligninolytic cultures of *P. chrysosporium* demonstrated effective degradation of Crystal Violet under nitrogen-limited conditions (Bumpus and Brock 1988; Vasdev et al. 1995). However, recent studies showed that the addition of nitrogen and carbon sources may aid in the dye decolourization process of microfungi. The decolourization efficiency of *Aspergillus* sp. CB-TKL-1 for Brilliant Green was higher in the presence of sodium nitrate (nitrogen source) (about 99%) than without any nitrogen input (51%) (Kumar et al. 2012). Abedin (2008) documented similar beneficial effects of sodium nitrate on the decolourization of Crystal Violet and Malachite Green by live *Fusarium solani* (96–99%). The supply of nitrogen could have enhanced fungal growth, which in turn led to more biomass and enzyme secretions to decolourize toxic dyes. However, the concentration of nitrogen supplied requires careful monitoring. Ammonium chloride at 0.05 g/L enabled higher degradation of Methyl Violet by *P. chrysosporium*, while 0.1–0.2 g/L drastically lowered the decolourization rate (Radha et al. 2005). This was postulated to be due to reduction reactions involving nitrogen in both the medium and dyes. The presence of carbon sources also has similar effect as nitrogen sources on dye decolourization (Radha et al. 2005). This may be explained by the usage of glucose for fungal growth instead of decolourizing dyes. The most effective carbon source commonly reported is glucose (Radha et al. 2005; Abedin 2008; Kumar et al. 2012). Unfortunately, glucose is costly, which led to the search for inexpensive carbon sources such as starch, sucrose, xylose, arabinose and fructose. Hence, it is possible to enhance the decolourization potentials of microfungi by controlling the influencing factors (e.g. addition of nitrogen and carbon sources, pH, temperature, agitation, initial dye concentration and biomass concentration) to suit the type of dyes, fungal strains and other environment conditions present in wastewater.

22.5 Current Technological Applications for TPM Dye Removal by Microfungi

Studies over the years have discovered several microfungi species with potential to decolourize TPM dyes under laboratory conditions. However, these decolourizing potentials may diminish when applied to industrial wastewater containing dyes as some microfungi are not naturally found in such waterbodies. Industrial wastewaters also contain other interfering chemicals such as inorganic salts, solvents and heavy metals, and these pollutants may overwhelm the biosorption/biodegradation potential of fungi, demanding that more fungal biomass is generated for sufficient dye decolourization (Haroun and Idris 2009; Halimoon and Goh 2010). Consequently, several technological applications are being researched to improve the potential of microfungi as biosorbents or biodegraders of TPM dyes. These innovations involved improving/modifying the surface of the fungal biomass (through pretreatments, immobilization and desorption) to enhance biosorption activities and usability for several decolourization cycles, as well as improving the properties, purity and production of dye-degrading enzymes (through genetic modifications and solid-state fermentation) (Fig. 22.4).

The binding sites of fungal biomass can be modified via physical and chemical pretreatments such as exposure to heat (autoclave), sodium bicarbonate ($NaHCO_3$) and sodium hydroxide (NaOH), to enhance adsorption of TPM dyes. Dried biomass of *Rhizopus stolonifer* pretreated with these heat and chemical approaches showed biosorption capacity for Bromophenol Blue (317–385 mg/g) compared to

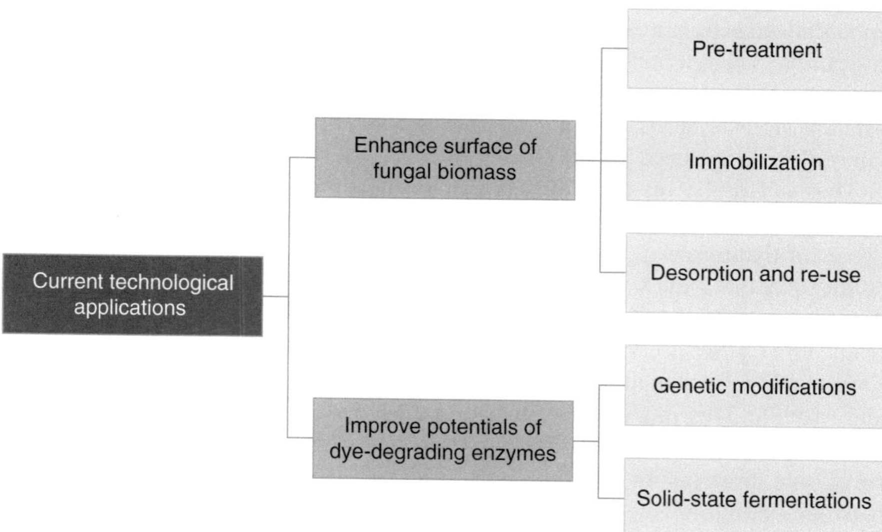

Fig. 22.4 Various technological applications currently research on to improve removal of TPM dyes by microfungi

their native form (280 mg/g) (Zeroual et al. 2006b). Similar results were described by Patel and Suresh (2008) for non-TPM Reactive Black 5 removal by *Aspergillus foetidus* pretreated with either autoclaving or 0.1 M sodium hydroxide. Heating of the fungal biomass through the autoclave process was postulated to cause rupturing of cells, resulting in increased cell surface area and revealing certain binding sites for dye adsorption (Zeroual et al. 2006b). On the other hand, the presence of bicarbonate ions from exposure to $NaHCO_3$ could have contributed to the protonation of fungal cell surface and enhanced electrostatic interactions with dye molecules. Pretreatment of the biomass with NaOH would have removed the proteins and glucans from fungal cell wall, which in turn increases the percentage of chitin/chitosan available for binding to dye molecules. Other chemicals such as hydrochloric acid (HCl), sulphuric acid (H_2SO_4), calcium chloride ($CaCl_2$), sodium carbonate (Na_2CO_3) and sodium chloride (NaCl) have also been studied for pretreatments of fungal biomass, with varying degrees of success reported for increasing the biosorption capacity for non-TPM dyes (Fu and Viraraghavan 2002; Arica and Bayramoğlu 2007).

Another form of improvement to microfungi as biosorbents is the use of immobilization. This is a process of entrapping or attaching fungal biomass to an inert matrix such as sodium alginate. The immobilization of fungal biomass mitigates clogging problems in water systems caused by small fungal particles and facilitates the separation and removal of biosorbents from the wastewater (Saeed et al. 2009). Such process was also demonstrated to enhance the biosorption capacities of microfungi for TPM dyes. Free cells of dead *Trichoderma asperellum* demonstrated low biosorption capacities for Crystal Violet (12.97 mg/g), Methyl Violet (12.54 mg/g), Cotton Blue (14.34 mg/g) and Malachite Green (11.44 mg/g), which was further enhanced when immobilized on alginate beads (60.64, 50.29, 49.91 and 16.61 mg/g, respectively) (Chew and Ting 2015). Similar observations were described by Radha et al. (2005) for decolourization of Methyl Violet by *P. chrysosporium* entrapped in calcium alginate beads. The alginate beads were also found to contribute to dye removal by adsorption.

Dyes adsorbed on the biomass can be desorbed using chemicals so that regeneration of biomass can occur, and biomass could be reused for following cycles of dye removal. Various chemicals have been studied for dye desorption studies. HCl (0.1 M) was found to be more effective than ethanol (50%) for the desorption of TPM dye Basic Fuchsin, Methyl Violet or mixture of both dyes from *Aspergillus niger* BU10 biomass (Bhole et al. 2004). HCl treatment of biomass was postulated to increase electrostatic repulsion between the biomass and cationic dyes (e.g. Basic Fuchsin and Methyl Violet). Similarly, the treatment of fungal biomass with ethanol may have caused the lipid content to dissolve and proteins to be dehydrated, which weaken the interactive bonds between the biding sites on fungal surface with the dye molecules, resulting in the release of dyes (Chakraborty et al. 2013). Treatment of *Rhizopus stolonifer* with NaOH (0.1 M) resulted in 90–100% desorption of Bromophenol Blue, where regenerated biomass showed reusability for five sorption-desorption cycles with stable dye biosorption capacities (Zeroual et al. 2006b). Similar observations were reported for desorption of TPM dye Acid Violet 49 from dead *Aspergillus fumigatus* biomass (Chaudhry et al. 2014). Treating the biomass with NaOH would have increased deprotonation, resulting in negatively charged binding sites on the surface of the biomass that desorb anionic dye molecules (Bromophenol Blue) via electrostatic repulsion (Zeroual et al. 2006b). Dye desorption is affected by the concentration of NaOH used, where higher strengths of the chemical (>0.1 M) reduced the desorption efficiency of the dye (Patel and Suresh 2008).

Molecular biology techniques have also been found useful in enhancing bioremediation of TPM dyes by microfungi, particularly the enzymes. These techniques are often employed to determine genes conferring dye degradation capabilities and expression of such gene in organism with no prior dye decolourization or tolerance abilities (Tamayo-Ramos et al. 2012). A commonly used vector for enzyme expression is the methylotrophic yeast *Pichia pastoris.* The genetic composition of this yeast is well known. The enzymes can be extracted via simple purification procedures for recombinant enzymes and perform post-translational modifications, and the genes are amenable to high levels of gene expression to enable secretion of the extracellular proteins (enzymes) (Macauley-Patrick et al. 2005). Zhuo et al. (2011) successfully cloned laccase gene *lac-En3-1* from TPM dye degrader *Ganoderma* sp. En3, which was expressed in *Pichia pastoris.* The purified recombinant Lac exhibited better decolourization efficiency for Bromophenol Blue (52%), Crystal Violet (80%) and Malachite Green (91%) without mediators compared to *Ganoderma* sp. En3 fungus. Recombinant Lac with moderate decolourization efficiency for Crystal Violet (52%) was also expressed and purified from *Pichia pastoris* containing laccase-encoding cDNA from another TPM dye degrader, the white-rot fungus *Pycnoporus sanguineus* (Lu et al. 2009). Purified laccase-like multicopper oxidase McOB expressed from microfungus *Aspergillus niger* not only decolourized TPM dyes Bromocresol Purple (41%) and Malachite Green (83%) but also withstand high temperatures (50–75 °C), metal salts (such as $CaCl_2$, $MgCl_2$, NaCl and $CuSO_4$) and inhibitors (such as SDS and EDTA) commonly found in wastewater (Tamayo-Ramos et al. 2012). Another commonly used expression vector was *Escherichia coli*, which was used by Gao et al. (2015) to express the putative triphenylmethane reductase-like protein produced by the microfungus *Geobacillus thermoglucosidasius* C56-Y593. The pure recombinant protein, which was stable in pH (4.5–10.5), temperature (of below 65 °C) and tolerated presence of metal ions and organic solvents, interacted better with azo dye Methyl Red compared to TPM dye Malachite Green.

Amenability to culturing on solid substrates (solid-state fermentation) is another advantage. Microfungi on solid matrix are enriched with nutrients (Tychanowicz et al. 2004). This technique mimics the natural environment in which fungi grow and is advantageous as it stimulates the production of large amounts of enzymes prior to contact with toxic dyes. Most of such studies have been conducted on macrofungi, where solid-state cultures of *Pleurotus pulmonarius* on wheat bran, pineapple peel and orange bagasse exhibited enhanced productions of Lac (2100–2860 U/L) and MnP (up to 2200 U/L) (dos Santos Bazanella et al. 2013). The filtrates containing enzymes effectively decolourized TPM dyes Ethyl Violet to a residual dye percentage of 22–40%. In addition to enhancing enzyme productions, solid-state fermentation was also reported to aid dye removal by concentrating dyes via adsorption prior to biological treatment by fungi. Other forms of solid substrates that induced production of enzymes with good degradation activities for TPM dyes include glucose-ammonium tartrate-corncob for *Pleurotus pulmonarius* (Tychanowicz et al. 2004), rice straw for *Pleurotus ostreatus* BP (Yan et al. 2009), banana skin for *Trametes pubescens* (Osma et al. 2007) as well as chestnut shell and barley bran for *Coriolopsis rigida* (Gómez et al. 2005). These materials are often selected as substrates due to their low cost and large availability. Solid-state fermentations are also employed for fungal strains that were genetically modified to express dye-degrading enzymes such as Lac and peroxidase. *Aspergillus niger* expressing Lac gene from *Trametes versicolor* produced the enzyme with high activities (362–592 U/L) when cultivated on polyurethane foam (Téllez-Jurado et al. 2006). Peroxidase production by *Aspergillus oryzae* (carrying peroxidase gene DyP from *Geotrichum candidum*) showed a 4.1-fold increase in solid-state cultures on wheat bran compared to submerge cultured in liquid media (Sugano et al. 2001).

22.6 Conclusion and Future Prospects

Triphenylmethane (TPM) dyes comprise of various types of dyes with extensive usage in industrial processes. The release of such toxic dyes into the environment should be controlled and treated prior to release as effluents. The use of microfungi for biological treatment (biosorption, biodegradation) of TPM dyes has been gaining attention due to their dye decolourization efficiencies and availability in large quantities at low cost. Diverse species of microfungi, ranging from white-rot to non-white-rot fungi and isolated from various sources (polluted and non-polluted sites), have demonstrated this potential, particularly in biodegrading the dyes using enzymes. Whether these enzymes work collectively or sequentially to degrade TPM dyes is undetermined as more studies are required to further elucidate enzymatic roles in dye degradation pathways. The biological process of removing TPM dyes is influenced by several parameters that affected fungal growth and dye characteristics in wastewater, such as growth media composition, pH, temperature, agitation, initial dye concentration and biomass concentration. These factors require further optimization for successful upscaling for application on wastewater. Current technological advances in this field of dye bioremediation involve approaches to improve the bioremediation capabilities of microfungi as well as increase expression and production of the functional enzymes with excellent dye-degrading properties.

Acknowledgements The authors express gratitude to the Malaysian Ministry of Higher Education for providing research funding under the FRGS grant scheme (FRGS/2/2013/STWN01/MUSM/02/2) and Monash University Malaysia for the funds and research facilities.

References

Abedin RMA (2008) Decolourization and Biodegradation of crystal violet and malachite green by *Fusarium solani* (Martius) Saccardo. A comparative study on biosorption of dyes by the dead fungal biomass. Am Eurasian J Bot 1:17–31

Abu Hassan MA, Tan PL, Noor ZZ (2009) Coagulation and flocculation treatment of wastewater in textile industry using chitosan. J Chem Nat Resour Eng 4:43–53

Aidoo A, Gao N, Neft RE, Schol HM, Hass BS, Minor TY, Heflich RH (1990) Evaluation of the genotoxicity of gentian violet in bacterial and mammalian cell systems. Teratog Carcinog Mutagen 10:449–462. doi:10.1002/tcm.1770100604

Almeida EJ, Corso CR (2014) Comparative study of toxicity of azo dye Procion Red MX-5B following biosorption and biodegradation treatments with the fungi *Aspergillus niger* and *Aspergillus terreus*. Chemosphere 112:317–322. doi:10.1016/j.chemosphere.2014.04.060

Arica MY, Bayramoğlu G (2007) Biosorption of Reactive Red-120 dye from aqueous solution by native and modified fungus biomass preparations of *Lentinus sajor-caju*. J Hazard Mater 149(2):499–507. doi:10.1016/j.jhazmat.2007.04.021

Asgher M (2012) Biosorption of reactive dyes: a review. Water Air Soil Pollut 223:2417–2435. doi:10.1007/s11270-011-1034-z

Azizian S, Haerifar M, Bashiri H (2009) Adsorption of methyl violet onto granular activated carbon: equilibrium, kinetics and modeling. Chem Eng J 146:36–41. doi:10.1016/j.cej.2008.05.024

Bhole BD, Ganguly B, Madhuram A, Deshpande D, Joshi J (2004) Biosorption of methyl violet, basic fuchsin and their mixture using dead fungal biomass. Curr Sci 86:1641–1645

Bose B, Gour RR, Motiwale L, Rao KV (2006) Decreased phosphoactive ERKs and JNKs in Malachite-green-transformed Syrian hamster embryo fibroblasts are associated with increased phosphoactive p38 kinase: possible therapeutic importance. Chemotherapy 52:210–214. doi:10.1159/000093038

Bumpus JA, Brock BJ (1988) Biodegradation of crystal violet by the white rot fungus *Phanerochaete chrysosporium*. Appl Environ Microbiol 54(5):1143–1150

Casas N, Parella T, Vicent T, Caminal G, Sarrà M (2009) Metabolites from the biodegradation of triphenylmethane dyes by *Trametes versicolor* or laccase. Chemosphere 75(10):1344–1349. doi:10.1016/j.chemosphere.2009.02.029

Cha C-J, Doerge DR, Cerniglia CE (2001) Biotransformation of malachite green by the fungus *Cunninghamella elegans*. Appl Environ Microbiol 67(9):4358–4360. doi:10.1128/aem.67.9.4358-4360.2001

Chakraborty S, Basak B, Dutta S, Bhunia B, Dey A (2013) Decolorization and biodegradation of congo red dye by a novel white rot fungus *Alternaria alternata* CMERI F6. Bioresour Technol 147:662–666. doi:10.1016/j.biortech.2013.08.117

Chaudhry MT, Zohaib M, Rauf N, Tahir SS, Parvez S (2014) Biosorption characteristics of *Aspergillus fumigatus* for the decolorization of triphenylmethane dye acid violet 49. Appl Microbiol Biotechnol 98:3133–3141. doi:10.1007/s00253-013-5306-y

Chen SH, Ting ASY (2015a) Biodecolorization and biodegradation potential of recalcitrant triphenylmethane dyes by *Coriolopsis* sp. isolated from compost. J Environ Manag 150:274–280. doi:10.1016/j.jenvman.2014.09.014

Chen SH, Ting ASY (2015b) Biosorption and biodegradation potential of triphenylmethane dyes by newly discovered *Penicillium simplicissimum* isolated from indoor wastewater sample. Int Biodeter Biodegr 103:1–7. doi:10.1016/j.ibiod.2015.04.004

Chew SY, Ting ASY (2015) Common filamentous *Trichoderma asperellum* for effective removal of triphenylmethane dyes. Desalin Water Treat 57. doi:10.1080/19443994.2015.1060173

Chi Z, Liu R, Zhao X, Sun Y, Yang B, Gao C (2009) Study on the genotoxic interaction of methyl violet with calf thymus DNA. Appl Spectrosc 63:1331–1335. doi:10.1366/000370209790109085

Dahri MK, Kooh MRR, Lim LBL (2014) Water remediation using low cost adsorbent walnut shell for removal of malachite green: equilibrium, kinetics, thermodynamic and regeneration studies. J Environ Chem Eng 2:1434–1444. doi:10.1016/j.jece.2014.07.008

Denoyes B, Baudry A (1995) Species identification and pathogenicity study of French *Colletotrichum* strains isolated from strawberry using morphological and cultural characteristics. Phytopathology 85:53–57. doi:10.1094/phyto-85-53

dos Santos Bazanella GC, de Souza DF, Castoldi R, Oliveira RF, Bracht A, Peralta RM (2013) Production of laccase and manganese peroxidase by *Pleurotus pulmonarius* in solid-state cultures and application in dye decolorization. Folia Microbiol (Praha) 58:641–647. doi:10.1007/s12223-013-0253-7

Elisangela F, Andrea Z, Fabio DG, de Menezes Cristiano R, Regina DL, Artur C-P (2009) Biodegradation of textile azo dyes by a facultative *Staphylococcus arlettae* strain VN-11 using a sequential microaerophilic/aerobic process. Int Biodeter Biodegr 63(3):280–288. doi:10.1016/j.ibiod.2008.10.003

Fu Y, Viraraghavan T (2002) Removal of Congo Red from an aqueous solution by fungus *Aspergillus niger*. Adv Environ Res 7:239–247. doi:10.1016/s1093-0191(01)00123-x

Gao F, Ding H, Shao L, Xu X, Zhao Y (2015) Molecular characterization of a novel thermal stable reductase capable of decoloration of both azo and triphenylmethane dyes. Appl Microbiol Biotechnol 99:255–267. doi:10.1007/s00253-014-5896-z

Glenn JK, Gold MH (1983) Decolorization of several polymeric dyes by the lignin-degrading basidiomycete *Phanerochaete chrysosporium*. Appl Environ Microbiol 45(6):1741–1747

Gomaa OM, Linz JE, Reddy CA (2008) Decolorization of Victoria blue by the white rot fungus, *Phanerochaete chrysosporium*. World J Microbiol Biotechnol 24:2349–2356. doi:10.1007/s11274-008-9750-2

Gomashe A, Suriya SS, Dharmik PG (2011) Decolorisation of textile dyes by *Aspergillus ochraceus* (NCIM-1146). Ecoscan 5(3–4):169–171

Gómez J, Pazos M, Couto SR, Sanromán MA (2005) Chestnut shell and barley bran as potential substrates for laccase production by *Coriolopsis rigida* under solid-state conditions. J Food Eng 68:315–319. doi:10.1016/j.jfoodeng.2004.06.005

Grinevicius VM, Geremias R, Laus R, Bettega KF, Laranjeiras MC, Fávere VT, Wilhelm Filho D, Pedrosa RC (2009) Textile effluents induce biomarkers of acute toxicity, oxidative stress, and genotoxicity. Arch Environ Contam Toxicol 57:307–314. doi:10.1007/s00244-008-9263-x

Gupta VK, Mittal A, Krishnan L, Mittal J (2006) Adsorption treatment and recovery of the hazardous dye, Brilliant Blue FCF, over bottom ash and de-oiled soya. J Colloid Interface Sci 293:16–26. doi:10.1016/j.jcis.2005.06.021

Halimoon N, Goh RSY (2010) Removal of heavy metals from textile wastewater using Zeolite. Environ Asia 3:124–130

Harms H, Schlosser D, Wick LY (2011) Untapped potential: exploiting fungi in bioremediation of hazardous chemicals. Nat Rev Microbiol 9:177–192. doi:10.1038/nrmicro2519

Haroun M, Idris A (2009) Treatment of textile wastewater with an anaerobic fluidized bed reactor. Desalination 237(1–3):357–366. doi:10.1016/j.desal.2008.01.027

Hemmati F, Norouzbeigi R, Sarbisheh F, Shayesteh H (2016) Malachite green removal using modified sphagnum peat moss as a low-cost biosorbent: kinetic, equilibrium and thermodynamic studies. J Taiwan Inst Chem Eng 58:482–489. doi:10.1016/j.jtice.2015.07.004

Hutwimmer S, Wang H, Strasser H, Burgstaller W (2010) Formation of exudate droplets by *Metarhizium anisopliae* and the presence of destruxins. Mycologia 102:1–10. doi:10.3852/09-079

Jadhav JP, Govindwar SP (2006) Biotransformation of malachite green by *Saccharomyces cerevisiae* MTCC 463. Yeast 23:315–323. doi:10.1002/yea.1356

Jasińska A, Różalska S, Bernat P, Paraszkiewicz K, Długoński J (2012) Malachite green decolorization by non-basidiomycete filamentous fungi of *Penicillium pinophilum* and *Myrothecium roridum*. Int Biodeter Biodegr 73:33–40. doi:10.1016/j.ibiod.2012.06.025

Jasińska A, Paraszkiewicz K, Sip A, Długoński J (2015) Malachite green decolorization by the filamentous fungus *Myrothecium roridum*—mechanistic study and process optimization. Bioresour Technol 194:43–48. doi:10.1016/j.biortech.2015.07.008

Kabra AN, Khandare RV, Waghmode TR, Govindwar SP (2012) Phytoremediation of textile effluent and mixture of structurally different dyes by *Glandularia pulchella* (Sweet) Tronc. Chemosphere 87:265–272. doi:10.1016/j.chemosphere.2011.12.052

Khataee AR, Dehghan G, Zarei M, Ebadi E, Pourhassan M (2011) Neural network modeling of biotreatment of triphenylmethane dye solution by a green macroalgae. Chem Eng Res Des 89:172–178. doi:10.1016/j.cherd.2010.05.009

Kõljalg U, Larsson KH, Abarenkov K, Nilsson RH, Alexander IJ, Eberhardt U, Erland S, Høiland K, Kjøller R, Larsson E, Pennanen T, Sen R, Taylor AF, Tedersoo L, Vrålstad T, Ursing BM (2005) UNITE: a database providing web-based methods for the molecular identification of ectomycorrhizal fungi. New Phytol 166:1063–1068. doi:10.1111/j.1469-8137.2005.01376.x

Kumar CG, Mongolla P, Basha A, Joseph J, Sarma VUM, Kamal A (2011) Decolorization and biotransformation of triphenylmethane dye, methyl violet, by *Aspergillus* sp. isolated from Ladakh, India. J Microbiol Biotechnol 21:267–273

Kumar CG, Mongolla P, Joseph J, Sarma VUM (2012) Decolorization and biodegradation of triphenylmethane dye, brilliant green, by *Aspergillus* sp. isolated from Ladakh, India. Process Biochem 47:1388–1394. doi:10.1016/j.procbio.2012.05.015

Levin L, Papinutti L, Forchiassin F (2004) Evaluation of Argentinean white rot fungi for their ability to produce lignin-modifying enzymes and decolorize industrial dyes. Bioresour Technol 94:169–176. doi:10.1016/j.biortech.2003.12.002

Levin L, Melignani E, Ramos AM (2010) Effect of nitrogen sources and vitamins on ligninolytic enzyme production by some white-rot fungi. Dye decolorization by selected culture filtrates. Bioresour Technol 101:4554–4563. doi:10.1016/j.biortech.2010.01.102

Li P, Su YJ, Wang Y, Liu B, Sun LM (2010) Bioadsorption of methyl violet from aqueous solution onto Pu-erh tea powder. J Hazard Mater 179:43–48. doi:10.1016/j.jhazmat.2010.02.054

Lim SL, Chu WL, Phang SM (2010) Use of *Chlorella vulgaris* for bioremediation of textile wastewater. Bioresour Technol 101:7314–7322. doi:10.1016/j.biortech.2010.04.092

Liu WX, Chao YP, Yang XQ, Bao HB, Qian SJ (2004) Biodecolorization of azo, anthraquinonic and triphenylmethane dyes by white-rot fungi and a laccase-secreting engineered strain. J Ind Microbiol Biotechnol 31:127–132. doi:10.1007/s10295-004-0123-z

Lu L, Zhao M, Liang SC, Zhao LY, Li DB, Zhang BB (2009) Production and synthetic dyes decolourization capacity of a recombinant laccase from *Pichia pastoris*. J Appl Microbiol 107:1149–1156. doi:10.1111/j.1365-2672.2009.04291.x

Maas R, Chaudhari S (2005) Adsorption and biological decolourization of azo dye Reactive Red 2 in semicontinuous anaerobic reactors. Process Biochem 40:699–705. doi:10.1016/j.procbio.2004.01.038

Macauley-Patrick S, Fazenda ML, McNeil B, Harvey LM (2005) Heterologous protein production using the *Pichia pastoris* expression system. Yeast 22:249–270. doi:10.1002/yea.1208

Malik R, Ramteke DS, Wate SR (2007) Adsorption of malachite green on groundnut shell waste based powdered activated carbon. Waste Manag 27:1129–1138. doi:10.1016/j.wasman.2006.06.009

Martin KJ, Rygiewicz PT (2005) Fungal-specific PCR primers developed for analysis of the ITS region of environmental DNA extracts. BMC Microbiol 5:28. doi:10.1186/1471-2180-5-28

Mittal A (2006) Adsorption kinetics of removal of a toxic dye, Malachite Green, from wastewater by using hen feathers. J Hazard Mater 133:196–202. doi:10.1016/j.jhazmat.2005.10.017

Mittal A, Gajbe V, Mittal J (2008) Removal and recovery of hazardous triphenylmethane dye, Methyl Violet through adsorption over granulated waste materials. J Hazard Mater 150:364–375. doi:10.1016/j.jhazmat.2007.04.117

Moturi B, Singara Charya MA (2009) Decolourisation of crystal violet and malachite green by fungi. Sci World J 4:28–33. doi:10.4314/swj.v4i4.53585

Nilsson RH, Kristiansson E, Ryberg M, Hallenberg N, Larsson K-H (2008) Intraspecific *ITS* variability in the kingdom *Fungi* as expressed in the international sequence databases and its implications for molecular species identification. Evol Bioinform Online 4:193–201. doi:10.4137/EBO.S653

Nishiya Y, Yamamoto Y (2007) Characterization of a NADH:dichlorophenol oxidoreductase from *Bacillus subtilis*. Biosci Biotechnol Biochem 71:611–614. doi:10.1271/bbb.60548

Novotný Č, Svobodová K, Kasinath A, Erbanová P (2004) Biodegradation of synthetic dyes by *Irpex lacteus* under various growth conditions. Int Biodeter Biodegr 54:215–223. doi:10.1016/j.ibiod.2004.06.003

Ogunlaja OO, Aemere O (2009) Evaluating the efficiency of a textile wastewater treatment plant located in Oshodi, Lagos. Afr J Pure Appl Chem 3:189–196

Osma JF, Herrera JLT, Couto SR (2007) Banana skin: a novel waste for laccase production by *Trametes pubescens* under solid-state conditions. Application to synthetic dye decolouration. Dyes Pigments 75:32–37. doi:10.1016/j.dyepig.2006.05.021

Papinutti VL, Forchiassin F (2004) Modification of malachite green by *Fomes sclerodermeus* and reduction of toxicity to *Phanerochaete chrysosporium*. FEMS Microbiol Lett 231:205–209. doi:10.1016/s0378-1097(03)00957-1

Patel R, Suresh S (2008) Kinetic and equilibrium studies on the biosorption of reactive black 5 dye by *Aspergillus foetidus*. Bioresour Technol 99:51–58. doi:10.1016/j.biortech.2006.12.003

Pointing SB, Vrijmoed LLP (2000) Decolorization of azo and triphenylmethane dyes by *Pycnoporus sanguineus* producing laccase as the sole phenoloxidase. World J Microbiol Biotechnol 16:317–318. doi:10.1023/A:1008959600680

Pokharia A, Ahluwalia SS (2013) Isolation and screening of dye decolorizing bacterial isolates from contaminated sites. Text Light Ind Sci Technol 2:54–61. doi:10.12691/jaem-2-5-7

Radha KV, Regupathi I, Arunagiri A, Murugesan T (2005) Decolorization studies of synthetic dyes using *Phanerochaete chrysosporium* and their kinetics. Process Biochem 40(10):3337–3345. doi:10.1016/j.procbio.2005.03.033

Rodrigues CSD, Madeira LM, Boaventura RAR (2013) Treatment of textile dye wastewaters using ferrous sulphate in a chemical coagulation/flocculation process. Environ Technol 34:719–729. doi:10.1080/09593330.2012.715679

Rzeszutek K, Chow A (2001) Extraction of metal–dye ion-association complexes by thin ether-type polyurethane membranes. J Membr Sci 181:265–277. doi:10.1016/s0376-7388(00)00545-7

Saeed A, Iqbal M, Zafar SI (2009) Immobilization of *Trichoderma viride* for enhanced methylene blue biosorption: batch and column studies. J Hazard Mater 168:406–415. doi:10.1016/j.jhazmat.2009.02.058

Sani RK, Azmi W, Banerjee UC (1998) Comparison of static and shake culture in the decolorization of textile dyes and dye effluents by *Phanerochœte chrysosporium*. Folia Microbiol 43:85–88. doi:10.1007/bf02815550

Sedighi M, Karimi A, Vahabzadeh F (2009) Involvement of ligninolytic enzymes of *Phanerochaete chrysosporium* in treating the textile effluent containing Astrazon Red FBL in a packed-bed bioreactor. J Hazard Mater 169:88–93. doi:10.1016/j.jhazmat.2009.03.070

Shedbalkar U, Jadhav JP (2011) Detoxification of malachite green and textile industrial effluent by *Penicillium ochrochloron*. Biotechnol Bioprocess Eng 16:196–204. doi:10.1007/s12257-010-0069-0

Shedbalkar U, Dhanve R, Jadhav J (2008) Biodegradation of triphenylmethane dye cotton blue by *Penicillium ochrochloron* MTCC 517. J Hazard Mater 157:472–479. doi:10.1016/j.jhazmat.2008.01.023

Sugano Y, Matsuo C, Shoda M (2001) Efficient production of a heterologous peroxidase, DyP from *Geotrichum candidum* Dec 1, on solid-state culture of Aspergillus oryzae RD005. J Biosci Bioeng 92:594–597. doi:10.1016/S1389-1723(01)80323-6

Svecova L, Spanelova M, Kubal M, Guibak E (2006) Cadmium, lead and mercury biosorption on waste fungal biomass issued from fermentation industry. I. Equilibrium studies. Sep Purif Technol 52:142–153. doi:10.1016/j.seppur.2006.03.024

Tamayo-Ramos JA, van Berkel WJH, de Graaff LH (2012) Biocatalytic potential of laccase-like multicopper oxidases from *Aspergillus niger*. Microb Cell Fact 11:165. doi:10.1186/1475-2859-11-165

Tan XY, Kyaw NN, Teo WK, Li K (2006) Decolorization of dye-containing aqueous solutions by the polyelectrolyte-enhanced ultrafiltration (PEUF) process using a hollow fiber membrane module. Sep Purif Technol 52:110–116. doi:10.1016/j.seppur.2006.03.028

Tedersoo L, Nilsson RH, Abarenkov K, Jairus T, Sadam A, Saar I, Bahram M, Bechem E, Chuyong G, Kõljalg U (2010) 454 Pyrosequencing and Sanger sequencing of tropical mycorrhizal fungi provide similar results but reveal substantial methodological biases. New Phytol 188:291–301. doi:10.1111/j.1469-8137.2010.03373.x

Tekere M, Mswaka AY, Zvauya R, Read JS (2001) Growth, dye degradation and ligninolytic activity studies on Zimbabwean white rot fungi. Enzym Microb Technol 28:420–426. doi:10.1016/s0141-0229(00)00343-4

Téllez-Jurado A, Arana-Cuenca A, González Becerra AE, Viniegra-González G, Loera O (2006) Expression of a heterologous laccase by *Aspergillus niger* cultured by solid-state and submerged fermentations. Enzyme Microb Technol 38:665–669. doi:10.1016/j.enzmictec.2005.07.021

Torres JMO, Cardenas CV, Moron LS, Guzman APA, Dela Cruz TEE (2011) Dye colorization activities of marine-derived fungi isolated from Manila Bay and Calatagan Bay, Philippines. Philipp J Sci 140:133–143

Tsui CKM, Woodhall J, Chen W, Lévesque CA, Lau A, Schoen CD, Baschien C, Najafzadeh MJ, de Hoog GS (2011) Molecular techniques for pathogen identification and fungus detection in the environment. IMA Fungus 2:177–189. doi:10.5598/imafungus.2011.02.02.09

Turhan K, Durukan I, Ozturkcan SA, Turgut Z (2012) Decolorization of textile basic dye in aqueous solution by ozone. Dyes Pigments 92(3):897–901. doi:10.1016/j.dyepig.2011.07.012

Tychanowicz GK, Zilly A, de Souza CGM, Peralta RM (2004) Decolourisation of industrial dyes by solid-state cultures of *Pleurotus pulmonarius*. Process Biochem 39:855–859. doi:10.1016/s0032-9592(03)00194-8

Vasdev K, Kuhad RC, Saxena RK (1995) Decolorization of triphenylmethane dyes by the bird's nest fungus *Cyathus bulleri*. Curr Microbiol 30:269–272. doi:10.1007/bf00295500

Vergili I, Kaya Y, Sen U, Göndera ZB, Aydiner C (2012) Techno-economic analysis of textile dye bath wastewater treatment by integrated membrane processes under the zero liquid discharge approach. Resour Conserv Recycl 58:25–35. doi:10.1016/j.resconrec.2011.10.005

White TJ, Bruns T, Lee S, Taylor J (1990) Amplification and direct sequencing of fungal ribosomal RNA genes for phylogenetics. In: Innis MA, Gelfand DH, Sninsky JJ, White TJ (eds) PCR protocols: a guide to methods and applications. Academic, San Diego, pp 315–322. doi:10.1016/b978-0-12-372180-8.50042-1

Yan K, Wang H, Zhang X, Yu H (2009) Bioprocess of triphenylmethane dyes decolorization by *Pleurotus ostreatus* BP under solid-state cultivation. J Microbiol Biotechnol 19:1421–1430. doi:10.4014/jmb.0901.0033

Yan JP, Niu JZ, Chen DD, Chen YH, Irbis C (2014) Screening of *Trametes* strains for efficient decolorization of malachite green at high temperatures and ionic concentrations. Int Biodeter Biodegr 87:109–115. doi:10.1016/j.ibiod.2013.11.009

Yang Q, Li C, Li H, Li Y, Yu N (2009a) Degradation of synthetic reactive azo dyes and treatment of textile wastewater by a fungi consortium reactor. Biochem Eng J 43:225–230. doi:10.1016/j.bej.2008.10.002

Yang X, Zhao X, Liu C, Zheng Y, Qian S (2009b) Decolorization of azo, triphenylmethane and anthraquinone dyes by a newly isolated *Trametes* sp. SQ01 and its laccase. Process Biochem 14:1185–1189. doi:10.1016/j.procbio.2009.06.015

Yang YY, Wang G, Wang G, Du LN, Jia XM, Zhao YH (2011) Decolorization of malachite green by a newly isolated *Penicillium* sp. YW 01 and optimization of decolorization parameters. Environ Eng Sci 28:555–562. doi:10.1089/ees.2010.0172

Zeroual Y, Kim BS, Kim CS, Blaghen M, Lee KM (2006a) A comparative study on biosorption characteristics of certain fungi for bromophenol blue dye. Appl Biochem Biotechnol 134:51–60. doi:10.1385/abab:134:1:51

Zeroual Y, Kim BS, Kim CS, Blaghen M, Lee KM (2006b) Biosorption of Bromophenol Blue from aqueous solutions by *Rhizopus Stolonifer* biomass. Water Air Soil Pollut 177:135–146. doi:10.1007/s11270-006-9112-3

Zhang X, Liu Y, Yan K, Wu H (2007) Decolorization of anthraquinone-type dye by bilirubin oxidase-producing nonligninolytic fungus *Myrothecium* sp. IMER1. J Biosci Bioeng 104(2):104–110. doi:10.1263/jbb.104.104

Zhou XJ, Guo WQ, Yang SS, Zheng HS, Ren NQ (2013) Ultrasonic-assisted ozone oxidation process of triphenylmethane dye degradation: evidence for the promotion effects of ultrasonic on malachite green decolorization and degradation mechanism. Bioresour Technol 128:827–830. doi:10.1016/j.biortech.2012.10.086

Zhuo R, Ma L, Fan F, Gong Y, Wan X, Jiang M, Zhang X, Yang Y (2011) Decolorization of different dyes by a newly isolated white-rot fungi strain *Ganoderma* sp.En3 and cloning and functional analysis of its laccase gene. J Hazard Mater 192:855–873. doi:10.1016/j.jhazmat.2011.05.106

Applied Aspect of Microalgae in Monitoring of Heavy Metals

23

Santlal Jaiswar and Prakram Singh Chauhan

Abstract

Microalgae which forms the primary energy in any aquatic ecosystem plays great role in the food chain. Heavy metal enters into the aquatic system by the way of various anthropogenic, natural weathering, mining, smelting, and industrial activities. The transport of heavy metals into algae comprises two phases: metabolism-independent phase (enter via surface binding of physico-chemical nature into cell) and a metabolism-dependent phase (metal ions are transported from one side to the other of the cell membrane to the cell). Metal sorption by algal cells may be influenced by many environmental factors such as pH, temperature, conductivity, dissolved oxygen, etc. Heavy metal affects the growth and physiology of algae either by attaching to sulfhydryl group proteins or the distraction of protein structure. Modifications in cell size or morphology are general symptoms of heavy metal toxicity in microalgae. However luxurious growth and bioconcentration of heavy metals indicate the use of microalgae in bioremediation. With the benefits of large accumulation capacity and no other secondary pollution, growth of unknown species in natural habitat, algae are very promising for monitoring the wastewater containing heavy metals.

Keywords

Microalgae • Biomonitoring • Metal • Pollution

S. Jaiswar (✉) • P.S. Chauhan
Marine Biotechnology and Ecology Division, CSIR-Central Salt and Marine Chemicals Research Institute, Bhavnagar 364 002, Gujarat, India
e-mail: santlal@csmcri.org

V.C. Kalia et al. (eds.), *Mining of Microbial Wealth and MetaGenomics*,
DOI 10.1007/978-981-10-5708-3_23

23.1 Introduction

Heavy metals which are cytotoxic, mutagenic, as well as carcinogenic in nature are the main environmental pollutant studied by ecotoxicologists nowadays (Muley et al. 2000; Narula et al. 2015). Some metals are required for algal metabolism and physiology as macroelements (K, Mg) or microelements (Co, Cu, Fe, Mn, Mo, Ni, V, Zn) and must be obtained from the external environment. However, the essential elements, such as Cu, Mn, Ni, and Zn, can be toxic at high concentrations (Machado et al. 2015). Other metals, such as Cd, Hb, and Pb, have no known biological function and are always toxic. Some of the essential and nonessential heavy metals for the organism are summarized in Table 23.1. Due to the waste discharge from industries of the different origin like tanning, marble, etc., domestic sewage, drainages from mines, spills of oil from petroleum or crude oil refineries, extensively using of modern and latest techniques of fishing, etc. are some of the major causes of degradation of water quality of the coastal areas, estuaries, rivers, lakes, and similar other water bodies around the world (Zutshi and Raghuprasad 2008). Monitoring and remediation of heavy metal from aquatic body are the burning topics for the researcher (Priyadarshani et al. 2011; Prabha et al. 2016).

Algae are phototrophic organisms and vary differently in size and morphology. Most of them are aquatic and inhabiting both fresh and saline waters. Algae commonly occurred at the surface layers in water bodies exposed to light. Myxophyceae and Bacillariophyceae have been shown to be correlated with intensive pollution. Based on occurrence and diversity patterns, algae are used as the indicator of pollution in a particular water body (Jaiswar et al. 2015; Jena et al. 2005). The freshwater biodiversity is greatly and severely affected and threatened by pollutants like heavy metals exhausted by the industries, smelting, and mines. Productivity of crop, quality of water in atmosphere, health, and life of all living organisms (animals and plants) were directly or indirectly affected by the heavy metal pollutants as these metals have already been entered in our food chain which in turn entered in our food web (Javed 2006). Results of heavy metal pollutants including metals like Cu, Zn, Ni, Pb, Cd, Co, Hg, Cr, and Ar are commonly found in the surrounding environment, due to human activities, and are long term and irreversible (Volland et al. 2014; Karman et al. 2015).

Table 23.1 Some of the essential and nonessential heavy metals for the organism present in the environment

Metal	Essential or not essential	Role
Ni	Essential	Structure of ribosomes, nucleic acid stabilization
Cr	Essential	Glucose metabolism in mammals
Cu	Essential	Phenolic structure metabolism; polyphenol oxidase and ascorbic acid; metalloenzyme cytochrome oxidase
Cd	Not essential	NA
Pb	Not essential	NA

23.2 Sources of Heavy Metal Accumulation

There are several factors which promote the heavy metal pollution. The heavy metals enter into the water by way of natural weathering, mining, smelting, and other industrial activities (Chaoyang et al. 2009). However metals do occur in the environment both as a result of natural processes and pollutants from anthropogenic activities (Franca et al. 2005). The major anthropogenic activities noticed were disposal of dead animal, cattle wading, bathing, open defecation in open places, washing of clothes, dumping of the wastes, and ashes (Verma and Saksena 2010).

An enormous amount of metal pollutants is present in the surface water and soil and remains retained in them as their leaching is not an easy and possible task. Accumulation of these metal pollutants may cause the threat of surrounding biota; therefore important step should have been taken for protecting the environment from these contaminants (metal pollutant) with immediate effect, present in the soil, sediment, and water bodies in excess amount (Lasat 2002; Cheng 2003). Polluted sediment is another significant source of water pollution. The suspended material in a water course as well as the riverbed surface can absorb chemicals and metal from water. When the suspended material settled down, then the toxic material present forms a sink; the extent to which it can cause harm to aquatic life depends upon the link between chemical and particles. However total metal concentrations in sediment do not necessarily reflect concentration available in biota (Dadrasnia et al. 2015).

Localized point is the source of pollutants in which pollutants existed from single and identifiable pipes or ditches, e.g., discharge from a sewage treatment plant. On the contrary, collective or new effect of small amount of diffused contamination which does not originate from a single discrete source and gathered from a large area is referred to as nonpoint source of pollutants, for example, coal-fired power generation (source of heavy metal pollution). Deposition of heavy metal like mercury takes place in water bodies which are emitted from the boiler flues. It is a great challenge to protect the water resources, from getting, infected by these metals pollutants. Large amount of ashes released by the industries and factories contains many types of heavy metal pollutants (Mustapha and Halimoon 2015).

23.3 Factors Affecting Heavy Metal Accumulation in Algae

The most important physicochemical properties of contaminant affecting bioconcentration appear to be the water solubility. It is very difficult to predict metal speciation aquatic system due to the chemical complexity of the aquatic environment (Kallqvist 2009). Heavy metal sorption by algal cells may be influenced by many environmental variables, and pH is one of the most significant factors because of its effect on the charge of functional groups and chemical speciation of metals. Low external pH decreased algal sorption of cadmium adsorption (Topcuoğlu et al. 2010). However, sorption of anionic metal complexes such as TiO increased with the decreasing external pH. Abiotic environmental factors including oxygen, water hardness, pH, temperature, etc. affect the aquatic environment by increasing the level of metal toxicity, pH, and temperature (Ye et al. 2015).

Rainfall, photosynthetic activity, and discharge of industrial and domestic effluents affect the pH of the water body. Reducing the pH has been found to increase the amount of dissolved metal in the aquatic system (Vannini et al. 2009). In aquatic ecosystems it is also important to understand that added metal ions may be entering the water body due to metals dissolved in acidic rainwater as it washes over the surrounding topography; the increase in dissolved metals may not just be from sediment mobilization. Generally, toxicity of such metals as Cu and Zn toward algal cells decreased as the pH of surrounding solution decreased (Karyn et al. 2006; Báscik-Remisiewicz et al. 2009; Soto et al. 2011). Cobalt, copper, and nickel were less toxic to the green alga *Chlamydomonas reinhardtii* at pH 5 than at pH 7 (Macfie et al. 1994; Branzini et al. 2012). Also the decrease of copper and uranium toxicity at pH 5.7 compared to those at pH 6.5 was reported for the freshwater alga *Chlorella* sp. (Franklin et al. 2000).

The toxic effect of metal, a different type of interaction competitively held between aquatic organisms, is affected and influenced by many cations and anions along with the hydrogen ions. The organisms are protected by the cations like Ca^{+2} and Mn^{+2} from uptaking of the metal as well as from expressing their toxic effects. At lower pH values, there is increased competition between protons and metal ions for negative binding sites on particulates (Kovacik et al. 2010). This results in less sites for metal adsorption and more of the metal stays in solution. The ligands with which metals can form complexes, such as OH^- and CO_3^-, can also become protonated, thus reducing their metal binding capacity. pH changes may not be a factor of concern in well-buffered aquatic systems.

Temperature effect on the surface sorption of heavy metals is relatively small. Higher energy demand causes higher respiration rate in organism which in turn increases the toxicity along with the increase in temperature. Chemical composition of water could get affected by the temperature which in turn might affect the toxicity of heavy metal (Fritioff et al. 2005). With the increase in temperature, the reason behind the decrease in the toxicity is not cleared yet. TDS in the water is mainly contributed by a large number of salt dissolved such as carbonate bicarbonates, sulfate, phosphate, chloride, nitrate of Ca, Mg, Na, etc. present in natural water. The high content of dissolved solid denotes the density of the water. Conductivity represents the capacity of water to conduct the electric current reflecting dissolved ions. The conductivity indirectly reveals the concentration of ionic species in water. The higher conductivity is observed due to the high sewages and wastewater disposal. There is currently no official guideline as to what is considered a safe level for conductivity (Karikari et al. 2007). Dissolved oxygen is the most important parameter in water quality studies. The increased rate of dissolved oxygen may be due to high photosynthetic rate by algae that cause release of more oxygen. Good quality water is saturated with dissolved oxygen, while polluted water tends to lower the dissolved oxygen content which proves to be harmful to the organisms that survived. The DO is essential to maintain the biotic community in water and to minimize the effect of the waste discharge into the water body. The lowest DO is occurring due to the disposal of domestic and vegetable waste; sometimes high organic waste is responsible for the oxygen depletion. For the degradation of the organic matter, oxygen in the water may be used, and it leads to anoxic condition. Like any other green plant, phototrophic algae require oxygen for respiration to maintain the metabolic process during the dark phase. To get some basic information about the oxygen concentration in the pond and river, the amount of O_2 transferred from air to the pond and river is important to determine the sufficient O_2 available for the algae to respire during nighttime.

23.4 The Mechanism of Heavy Metal Uptake by Algae

Metals must be in a certain chemically available form to make easy uptake by organisms, but the simple guidelines are based on total metal concentrations and do not consider metal chemistry. Therefore, it is important to know the metal chemistry before predicting the impacts of metal concentrations in aquatic ecosystem. Metals can take many forms: free ions, dissolved complexes, precipitates, complexes with organic or inorganic compounds, and adsorbed to particulates (Jamers et al. 2013).

The mechanism of heavy metal uptake by algal cells comprises two phases: a fast, metabolic independent surface binding of physicochemical nature and a metabolism-dependent phase, in which the metal ions are transported through the cell membrane to the cells (Volland et al. 2011; Wang et al. 2013). From the measurable point of view, in the case of planktonic green and blue-green algae, the surface absorption may be the major amount of the total metal uptake (about 80–98%), as reported for Cd, Cu, Zn, and Pb (Volland et al. 2012, 2013). However, in the case of very low concentrations of essential metals, up to 80% of the total cellular copper in *Scenedesmus subspicatus* was located intracellularly, and only 20% was located on the cell surface. With the increasing external free Zn concentrations from 10^{-9} to 10^{-5}, the amount of superficially adsorbed Zn reduced from 5 to 80% of the total cellular Zn concentration in this alga.

The huge surface area of microalgae may extremely affect the metal proportion in the surrounding water and, in result, the metal obtainability to other organisms. This concerns both living and dead algal biomass; dead cells can often adsorb metal ions in larger quantities than living biomass. A systematic diagram of mechanism of metal uptake in algae is presented in Fig. 23.1.

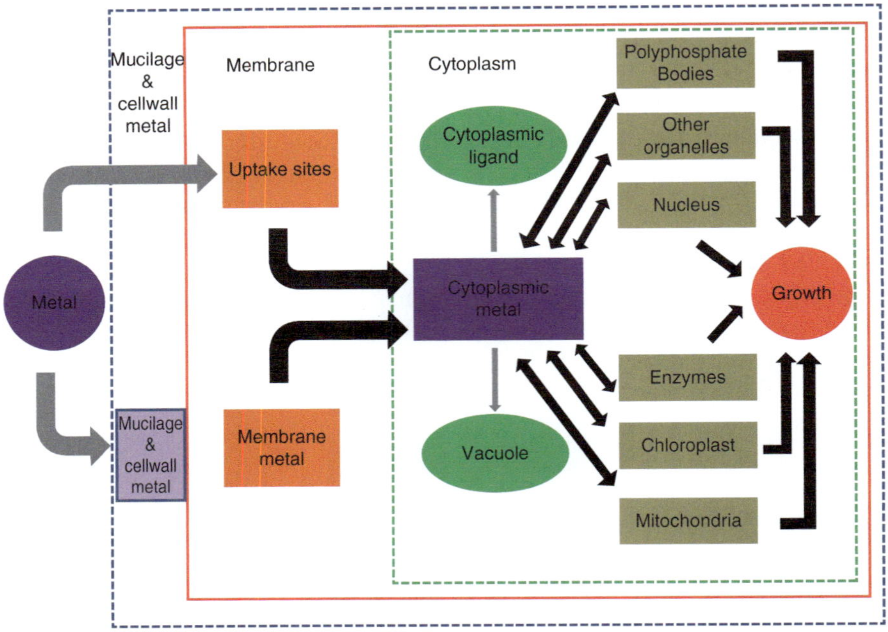

Fig. 23.1 Systematic diagram of mechanism of metal uptake in algae

23.5 Biosensing of Heavy Metal by Algae

To analyze the biological effects of heavy metals in the environment, it is very important to know both the total and the biological availability of heavy metals. The total heavy metal is mostly determined by atomic absorption or mass spectroscopy through acid digestion of the samples. Biologically available heavy metals in the environment indirectly depend on the total metals present in an organism. However, this type of bioassay is sometimes not very useful because of the different behavioral aspect of the organism to the available heavy metals. Therefore an easy, effective, and less harmful approach to monitor bioavailability of heavy metals as indicator species has great importance. These organisms are very sensitive to heavy metals and have visible effects (Machado and Soares 2013; Volland et al. 2014).

Rajamani (2006) established a fluorescence resonance energy transfer (FRET) mechanism based on the heavy metal biosensor, and its expression has been confirmed in *Chlamydomonas*. FRET depends upon distance phenomenon which involved a donor and an acceptor fluorescent molecule, which occurred naturally in the modified form of the jellyfish green fluorescent protein (GFP). FRET occurs when donor and acceptor pair space is <10 nm. Under favorable conditions, inferior energy acceptor molecule received the excited donor molecule; as a result there will be decreased donor fluorescence and increased acceptor fluorescence (Machado and Soares 2014; Machado and Soares 2015). FRET has been proved to be very useful in monitoring distance changes in biological macromolecular systems and found to

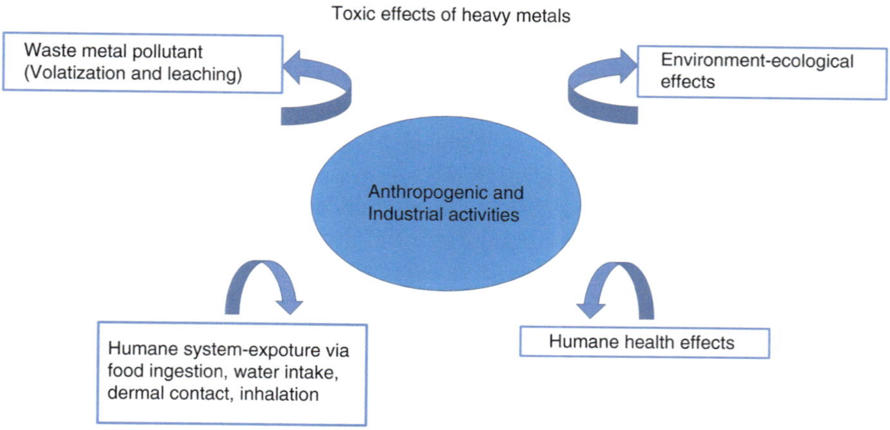

Fig. 23.2 Toxic effects of heavy metals available through anthropogenic and industrial activities in the environment

have very tremendous applications in structural biology and biochemistry. Due to the advancement of GFP and its variant FRET biomonitoring, both in vitro and in vivo studies have great importance (Machado et al. 2015). Calcium biosensor by tandem fusion of cyan fluorescent protein (CFP)-calmodulin-calmodulin binding protein and M13-yellow fluorescent protein (YFP) for monitoring calcium signaling was proved by them. From this onward GFP variant, particularly CFP-YFP FRET pairs, has been started in intramolecular and intermolecular FRET studies (Zaccolo 2004).

More recently, several authors have agreed on the concept of the whole-cell biosensor (WCB), which is a very useful alternative to classical biosensors (Gutierrez et al. 2015). A WCB uses the intact prokaryotic or eukaryotic cell as a single reporter, which incorporates both bioreceptor and transducer elements into the same cell. Organisms which are used for WCB are generally genetically modified by incorporating transducer capacity or by increasing its sensitivity. WCB is generally divided into two types of assay, *turn-off* and *turn-on* assays (Belkin 2003). *Turn-off* assays are general microbial toxicological bioassays. Toxicity is generally assayed in the cell by the degree of inhibition of a cellular activity (e.g., growth inhibition, respiration, motility depletion, etc.) or a particular reporter gene expression. In *turn-on* assays, a particular gene promoter is fused to a quantifiable molecular, by activating chemical or environmental pollutant in the form of heavy metals. The toxicity of the sample is proportional to the gene expression of the reporter molecule (Fig. 23.2).

23.6 Toxic Effects of Metals on Algae

The heavy metals affect the growth and physiology of algae. Toxicity in algae mainly occurs due to attachment of metal to protein containing sulfhydryl groups or else the distraction of protein arrangement (Affenzeller et al. 2009; Andosch et al. 2012). Metals have the capacity to degrade the permanency of the microalgae, encouraging substance that inhibits oxidation, i.e., superoxide dismutase, glutathione peroxidase, and ascorbate peroxidase which led to various biomolecules of algal cells that enter into stress conditions (Okamoto et al. 1996; Szivak et al. 2009). Severe injury occurs inside the microalgal cell either due to rise in amount of reactive oxygen species (Pandey et al. 2009; Macfarlane and Burchett 2001; Bielen et al. 2013) or by dropping the repair capacity of the cell against these stresses (Taskila et al. 2015).

Simple structure of algae is helpful for the study of the ultrastructural effects of heavy metal toxicity. A significant number of ultrastructural effects for single and complex metals have been described in many different algal species. Besides, a direct discrepancy between severe and persistent effects has been only addressed time to time (Qian et al. 2009). There are many reports which described the change of metals on algae; those studies involve only small structural and ultrastructural differences. Variations in size of the cell are obvious effects of heavy metal toxicity in microalgae, although the type and power of the effects vary from species to species, type of metal, and its concentration (Table 23.2). Carfagna et al. (2013) studied

Table 23.2 Toxic effects of different heavy metals on growth and morphology of microalgae

Strains	Metals	Effects	References
Micrasterias denticulata	Cu, Zn, Al, Cd, Cr, Fe, Pb	Growth, morphogenesis, photosynthetic and respiratory activity ultrastructure, and function of organelles	Volland et al. (2014)
Tetraselmis gracilis	Cd	Oxidized proteins and lipids in the algal cell shows the stress	Okamoto et al. (1996)
Chlorella vulgaris	Cu, Cd	Ultrastructural effects for different metals individually or in mixture	Qian et al. (2009)
Chlorella sorokiniana	Pb, Cd	Algal cell ultrastructure and its physiological features	Carfagna et al. (2013)
Oscillatoria mougeotii, *Scenedesmus quadricauda*	Zn, Cd, Ni, Cu, Cr	Physiological and morphological characteristic	Shehata et al. (1999)
Scenedesmus obliquus, *Desmodesmus pleiomorphus*	Cd, Zn	Growth inhibition	Monteiro et al. (2011)
Chlorella and *Dunaliella*	Pb	Growth inhibitor	Muhaemin (2004)
Chlorella marina, *Isochrysis galbana*, *Tetraselmis* sp., *Nannochloropsis* sp., *Dunaliella salina*	Zn	Major reduction in cell density	Kumar et al. (2014)
Phormidium ambiguum, *Pseudochlorococcum typicum*, *Scenedesmus quadricauda*	Hg, Pb, Cd	Changes at ultrastructural level	Shanab et al. (2012)
Chlorella vulgaris	Co, Cu, Zn	Stimulatory and inhibitory effects	Afkar et al. (2010)
Pseudokirchneriella subcapitata	Cd, Cr, Cu, Zn	Loss of membrane integrity, inhibition of esterase activity, reduction of chlorophyll a, and inhibition of esterase activity	Machado et al. (2015)

the effect of Pb or Cd in terms of cellular and morphological responses to *Chlorella sorokiniana* and showed that these metals change the algal cell ultrastructure and its features. After 24 h of treatment with 250 μM lead and cadmium, carbohydrate synthesis is repressed to a drastic level. The toxic effects of multi-metal mixture on some physiological and morphological characteristic of freshwater algae were reported by Shehata et al. (1999). Some of the general toxic effects of heavy metals available anthropogenically and industrially are accessible in Fig. 23.2.

Conclusion

With the advantage of large accumulation ability and no secondary pollution, algae are very promising for indicators of wastewater containing different heavy metals. The accumulation of algae had been studied widely for biomonitoring or bioremediation purpose. By observing the condition of organisms alive in a metal contaminated area may tell more about its pollution status than measuring sediment and water concentrations. Recently the immobilization of whole cells attracted much attention due to their potential in industrial application. This kind of work will lead to valuable use of algae and subsequently speed up the development of more effective biosorbent.

References

Affenzeller MJ, Darehshouri A, Andosch A, Lütz C, Lütz-Meindl U (2009) Salt stress-induced cell death in the unicellular green alga *Micrasterias denticulata*. J Exp Bot 60:939–954. doi:10.1093/jxb/ern348

Afkar E, Ababna H and Fathi AA (2010) Toxicological Response of the Green Alga Chlorella vulgaris, to Some Heavy Metals. Am J Environ Sci 6:230–237. doi:10.3844/ajessp.2010.230.237

Andosch A, Affenzeller MJ, Lütz C, Lütz-Meindl U (2012) A freshwater green alga under cadmium stress: ameliorating calcium effects on ultrastructure and photosynthesis in the unicellular model *Micrasterias*. J Plant Physiol 169:1489–1500. doi:10.1016/j.jplph.2012.06.002

Báscik-Remisiewicz A, Tomaszewska E, Labuda K, Tukaj Z (2009) The effect of Zn and Mn on the toxicity of Cd to the green microalga *Desmodesmus armatus* cultured at ambient and elevated (2%) CO_2 concentrations. Pol J Environ Stud 18:775–780

Belkin S (2003) Microbial whole-cell sensing systems of environmental pollutants. Curr Opin Microbiol 6:206–212. doi:10.1016/S1369-5274(03)00059-6

Bielen A, Remans T, Vangronsveld J, Cuypers A (2013) The influence of metal stress on the availability and redox state of ascorbate, and possible interference with its cellular functions. Int J Mol Sci 14:6382–6413. doi:10.3390/ijms14036382

Branzini A, Gonzalez RS, Zubillaga M (2012) Absorption and translocation of copper, zinc and chromium by *Sesbania virgata*. J Environ Manag 102:50–54. doi:10.1016/j.jenvman.2012.01.033

Carfagna S, Lanza N, Salbitani G, Basile A, Sorbo S, Vona V (2013) Physiological and morphological responses of Lead or Cadmium exposed *Chlorella sorokiniana* 211-8K (Chlorophyceae). Springerplus 2:147. doi:10.1186/2193-1801-2-147

Chaoyang W, Cheng W, Linsheng Y (2009) Characterizing spatial distribution and sources of heavy metals in the soils from mining-smelting activities in Shuikoushan, Hunan Province, China. J Environ Sci 21:1230–1236. doi:10.1016/S1001-0742(08)62409-2

Cheng S (2003) Heavy metals in plants and phytoremediation. Environ Sci Pollut Res 10:335–340. doi:10.1065/espr2002.11.141.3

Dadrasnia A, Wei KSC, Shahsavari N, Azirun MS, Ismail S (2015) Biosorption potential of *Bacillus salmalaya* strain 139SI for removal of Cr(VI) from aqueous solution. Int J Environ Res Public Health 12:15321–15338. doi:10.3390/ijerph121214985

Franca S, Vinagre C, Cacador I, Cabral HN (2005) Heavy metal concentrations in sediment, benthic invertebrates and fish in three salt marsh areas subjected to different pollution loads in the Tagus Estuary (Portugal). Mar Pollut Bull 50:998–1003. doi:10.1016/j.marpolbul.2005.06.040

Franklin NM, Stauber JL, Markich SJ, Lim RP (2000) pH dependent toxicity of copper and uranium to a tropical freshwater alga (*Chlorella* sp.) Aquat Toxicol 48:275–289. doi:10.1016/s0166-445x(99)00042-9

Fritioff A, Kautsky L, Greger M (2005) Influence of temperature and salinity on heavy metal uptake by submersed plants. Environ Pollut 133:265–274. doi:10.1016/j.envpol.2004.05.036

Gutierrez JC, Amaro F, Gonzalez AM (2015) Heavy metal whole-cell biosensors using eukaryotic microorganisms: an updated critical review. Front Microbiol 6:48. doi:10.3389/fmicb.2015.00048

Jaiswar S, Kazi MA, Mehta S (2015) Biomonitoring of heavy metals using freshwater algal species of Bhavnagar, Gujarat. J Environ Biol 36:1361–1366

Jamers A, Blust R, De Coen W, Griffin JL, Jones OAH (2013) An omics based assessment of cadmium toxicity in the green alga *Chlamydomonas reinhardtii*. Aquat Toxicol 126:355–364. doi:10.1016/j.aquatox.2012.09.007

Javed M (2006) Studies on metal contamination levels in plankton and their role as biological indicator of water pollution in the River Ravi. Pak J Biol Sci 9:313–317. doi:10.3923/pjbs.2006.313.317

Jena M, Ratha SK, Adhikary SP (2005) Algal diversity changes in Kathajodi River after receiving sewage of Cuttack and its ecological implications. Indian Hydrobiol 8:67–74

Kallqvist T (2009) Effect of water hardness on the toxicity of cadmium to the green alga *Pseudokirchneriella subcapitata* in an artificial growth medium and nutrient-spiked natural lake waters. J Toxicol Environ Health A 72:277–283. doi:10.1080/15287390802539368

Karikari AY, Bernasko JK, Bosque-Hamilton EKA (2007) An assessment of water quality of Angaw River in Southeastern Coastal Plains of Ghana. West Afr J Appl Ecol 11:77–87. doi:10.4314/wajae.v11i1.45728

Karman SB, Diah ZM, Gebeshuber IC (2015) Raw materials synthesis from heavy metal industry effluents with bioremediation and phytomining: a biomimetic resource management approach. doi:10.1155/2015/185071

Karyn LW, Jennifer LS, Scott JM, Natasha MF, Paul LB (2006) The effect of pH on the uptake and toxicity of copper and zinc in a tropical freshwater alga (*Chlorella* sp.) Arch Environ Contam Toxicol 51:174–185. doi:10.1007/s00244-004-0256-0

Kovacik J, Klejdus B, Hedbavny J, Backor M (2010) Effect of copper and salicylic acid on phenolic metabolites and free amino acids in *Scenedesmus quadricauda* (Chlorophyceae). Plant Sci 178(3):307–311. doi:10.1016/j.plantsci.2010.01.009

Kumar SD, Santhanam P, Ananth S, Devi SA, Nandakumar R, Prasath BB, Jeyanthi S, Jayalakshmi T, Ananthi P (2014) Effect of different dosages of zinc on the growth and biomass in five marine microalgae. Int J Fish Aquac 6:1–8. doi:10.5897/IJFA2013.0393

Lasat MM (2002) Reviews and analyses: phytoextraction of toxic metals—a review of biological mechanisms. J Environ Qual 31:109–120. doi:10.1111/j.1365-3040.2011.02338.x

Macfarlane GR, Burchett MD (2001) Photosynthetic pigments and peroxidase activity as indicators of heavy metal stress in the Grey mangrove, *Avicennia marina* (Forsk.) Vierh. Mar Pollut Bull 42:233–240. doi:10.1016/s0025-326x(00)00147-8

Macfie SM, Tarmohamed Y, Welbourn PM (1994) Effects of cadmium, cobalt, copper and nickel on growth of the green alga *Chlamydomonas reinhardtii*: the influence of cell wall and pH. Arch Environ Contam Toxicol 27:454–458. doi:10.1007/bf00214835

Machado MD, Soares EV (2013) Optimization of a microplate-based assay to assess esterase activity in the alga *Pseudokirchneriella subcapitata*. Water Air Soil Pollut 224:1–11. doi:10.1007/s11270-012-1358-3

Machado MD, Soares EV (2014) Modification of cell volume and proliferative capacity of *Pseudokirchneriella subcapitata* cells exposed to metal stress. Aquat Toxicol 147:1–6. doi:10.1016/j.aquatox.2013.11.017

Machado MD, Soares EV (2015) Use of a fluorescence-based approach to assess short-term responses of the alga *Pseudokirchneriella subcapitata* to metal stress. J Appl Phycol 27:805–813. doi:10.1007/s10811-014-0351-1

Machado MD, Lopes AR, Soares EV (2015) Responses of the alga *Pseudokirchneriella subcapitata* to long-term exposure to metal stress. J Hazard Mater 296:82–92. doi:10.1016/j.jhazmat.2015.04.022

Monteiro CM, Fonseca SC, Castro PML, Malcata FX (2011) Toxicity of cadmium and zinc on two microalgae, *Scenedesmus obliquus* and *Desmodesmus pleiomorphus*, from Northern Portugal. J Appl Phycol 23:97–103. doi:10.1007/s10811-010-9542-6

Muhaemin M (2004) Toxicity and bioaccumulation of lead in *Chlorella* and *Dunaliella*. J Coast Dev 8:27–33

Muley DV, Kamble GB, Bhilave MP (2000) Effect of heavy metals on nucleic acids in *Cyprinus carpio*. J Environ Biol 21:367–370

Mustapha MU, Halimoon N (2015) Microorganisms and biosorption of heavy metals in the environment: a review paper. Microb Biochem Technol 7:5. doi:10.4172/1948-5948.1000219

Narula P, Mahajan A, Gurnani C, Kumar V, Mukhija S (2015) Microalgae as an indispensable tool against heavy metals toxicity to plants: a review. Int J Pharm Sci Rev Res 31:86–93

Okamoto OK, Asano CS, Aidar E, Colepicolo P (1996) Effects of cadmium on growth and superoxide dismutase activity of the marine microalga *Tetraselmis gracilis*. J Phycol 32:74–79. doi:10.1111/j.0022-3646.1996.00074.x

Pandey V, Dixit V, Shyam R (2009) Chromium effect on ROS generation and detoxification in pea (*Pisum sativum*) leaf chloroplasts. Protoplasma 236:85–95. doi:10.1007/s00709-009-0061-8

Prabha Y, Soni SK, Gupta S, Sonal (2016) Potential of algae in bioremediation of wastewater: current research. Int J Curr Microbiol Appl Sci 5(2):693–700. doi:10.20546/ijcmas.2016.502.076

Priyadarshani I, Sahu D, Rath B (2011) Microalgal bioremediation: current practices and perspectives. J Biochem Technol 3(3):299–304

Qian H, Li J, Sun L, Chen W, Sheng DG, Liu W, Fu Z (2009) Combined effect of copper and cadmium on *Chlorella vulgaris* growth and photosynthesis-related gene transcription. Aquat Toxicol 94:56–61. doi:10.1016/j.aquatox.2009.05.014

Rajamani S (2006) Small molecule signaling and detection systems in protists and bacteria. PhD Thesis, The Ohio State University

Shanab S, Essa A, Shalaby E (2012) Bioremoval capacity of three heavy metals by some microalgae species (Egyptian Isolates). Plant Signal Behav 7:1–8. doi:10.4161/psb.19173

Shehata SA, Lasheen MR, Kobbia IA, Ali GH (1999) Toxic effect of certain metals mixture on some physiological and morphological characteristics of freshwater algae. Water Air Soil Pollut 110:119–135. doi:10.1023/a:1005058607002

Soto P, Gaete H, Hidalgo ME (2011) Assessment of catalase activity, lipid peroxidation, chlorophyll-a, and growth rate in the freshwater green algae *Pseudokirchneriella subcapitata* exposed to copper and zinc. Lat Am J Aquat Res 39:280–285. doi:10.3856/vol39-issue2-fulltext-9

Szivak I, Behra R, Sigg L (2009) Metal-induced reactive oxygen species production in *Chlamydomonas Reinhardtii* (Chlorophyceae). J Phycol 45:427–435. doi:10.1111/j.1529-8817.2009.00663.x

Taskila S, Leiviskä T, Haapalainen OP, Tanskanen J (2015) Utilization of industrial microbe side streams for biosorption of heavy metals from wastewaters. J Bioremed Biodegr 6:285. doi:10.4172/2155-6199.1000285

Topcuoğlu S, Kılıç O, Belivermiş M, Ergül HA, Kalayci G (2010) Use of marine algae as biological indicator of heavy metal pollution in Turkish marine environment. J Black Sea/Mediterr Environ 16:43–52

Vannini C, Marsoni M, Domingo G, Antognoni F, Biondi S, Bracale M (2009) Proteomic analysis of chromate-induced modifications in *Pseudokirchneriella subcapitata*. Chemosphere 76:1372–1379. doi:10.1016/j.chemosphere.2009.06.022

Verma AK, Saksena DN (2010) Influence of pollutants on bottom sediment of sewage collecting Kalpi (Morar) River, Gwalior, Madhya Pradesh (MP). J Toxicol Environ Health 2:38–49

Volland S, Andosch A, Milla M, Stöger B, Lütz C, Lütz-Meindl U (2011) Intracellular metal compartmentalization in the green algal model system *Micrasterias denticulata* (Streptophyta) measured by transmission electron microscopy-coupled electron energy loss spectroscopy. J Phycol 47(3):565–579. doi:10.1111/j.1529-8817.2011.00988.x

Volland S, Lütz C, Michalke B, Lütz-Meindl U (2012) Intracellular chromium localization and cell physiological response in the unicellular alga *Micrasterias*. Aquat Toxicol 109:59–69. doi:10.1016/j.aquatox.2011.11.013

Volland S, Schaumlöffel D, Wesenberg D, Krauss G-J, Lütz-Meindl U (2013) Identification of phytochelatins in the cadmium-stressed conjugating green alga *Micrasterias denticulata*. Chemosphere 91:448–454. doi:10.1016/j.chemosphere.2012.11.064

Volland S, Bayer E, Baumgartner V, Andosch A, Lütz C, Sima E, Lütz-Meindl U (2014) Rescue of heavy metal effects on cell physiology of the algal model system Micrasterias by divalent ions. J Plant Physiol 171:154–163. doi:10.1016/j.jplph.2013.10.002

Wang SZ, Zhang DY, Pan XL (2013) Effects of cadmium on the activities of photosystems of *Chlorella pyrenoidosa* and the protective role of cyclic electron flow. Chemosphere 93:230–237. doi:10.1016/j.chemosphere.2013.04.070

Ye J, Xiao H, Xiao B, Xu W, Gao L, Lin G (2015) Bioremediation of heavy metal contaminated aqueous solution by using red algae *Porphyra leucosticta*. Water Sci Technol 72(9):16626. doi:10.2166/wst.2015.386

Zaccolo M (2004) Use of chimeric fluorescent proteins and fluorescence resonance energy transfer to monitor cellular responses. Circ Res 94:866–873. doi:10.1161/01.res.0000123825.83803.cd

Zutshi B, Raghuprasad SG (2008) Impact of pollution on fresh and marine water resources. Pollut Res 27:461–466

Mining Bacterial Diversity for Biosurfactants

24

Tayebeh Soltanighias, Athoiba Elangbam Singh, and Praveen Rahi

Abstract

A wide range of compounds (biosurfactants) is synthesized by microorganisms especially bacteria. Microbial surfactants are amphiphilic compounds that exhibit surface activity. There is increasing interest in the production and utilization of biosurfactants for several reasons such as environmentally friendly because they are low toxic and biodegradable. Apart from that, biosurfactants carry potential advantages over chemical surfactants due to their unique structures. They have shown great potential applications in various industries, such as medicine, cosmetics, pharmaceutical, food processing and oil enhanced recovery, agriculture, and environmental protection. There are different groups of biosurfactant produced by diverse bacteria. The main focus of this chapter is to give an overview on structure and function of different groups of bacterial surfactants and their application in bioremediation and petroleum oil degradation as well as their role in biomedical as therapeutic agents.

T. Soltanighias
Department of Environmental Biotechnology, Rajiv Gandhi Institute of IT and Biotechnology, Bharati Vidyapeeth Deemed University, Katraj, Pune 411 046, India

National Centre for Microbial Resource, National Centre for Cell Science, Pune, Maharashtra 411 007, India

A.E. Singh
Department of Environmental Biotechnology, Rajiv Gandhi Institute of IT and Biotechnology, Bharati Vidyapeeth Deemed University, Katraj, Pune 411 046, India

P. Rahi (✉)
National Centre for Microbial Resource, National Centre for Cell Science, Pune, Maharashtra 411 007, India
e-mail: praveen@nccs.res.in

V.C. Kalia et al. (eds.), *Mining of Microbial Wealth and MetaGenomics*,
DOI 10.1007/978-981-10-5708-3_24

Keywords
Antimicrobial activity • Bacterial surfactant • Bioremediation • Emulsification • Polyaromatic hydrocarbon (PAH)

24.1 Introduction

Surfactants are amphiphilic compounds that carry both hydrophobic and hydrophilic moieties that exhibit surface activity (Cooper et al. 1981; Ron and Rosenberg 2002; Mulligan 2005; Das et al. 2008a; Moldes et al. 2013; Saha and Orvig 2010). The term "surfactant" is derived from a surface-active agent and is called so because of its ability to solubilize oil and other such immiscible substances. The curiosity and interest for bio-based material production in various industries have extremely developed in recent decades. Bacteria are known for producing a large number of bioactive secondary metabolites that biosurfactants are one of such compounds. Chemically synthesized surfactants are mostly noxious and nonbiodegradable and can stand for an additive source of contamination (Oliveira et al. 2013; Panjiar et al. 2013). The persistent toxicity of three chemically synthesized surfactants (PSE-61, Triton x-100, and Corexit 9500) and three microbially synthesized surfactants (emulsan, rhamnolipid, and biological cleanser PES-51) was studied, which are normally applied in oil spills remediation. However, the result pointed that PSE-61 and emulsan showed the lowest toxicity, while Triton x-100 had the most toxicity (Edward et al. 2003). Hence, the potential use of bacterial cells for production of biosurfactants has been recognized and considered by many industries (Makkar et al. 2011). Bacterial surfactants may be the most prospecting substitute to chemical surfactants (Henkel et al. 2012). These surface-active agents carry numerous advantages when they are compared to chemically synthesized counterparts, their biodegradability, low toxicity, structural diversity, enhanced environmental compatibility, high specificity, and satisfactory production cost from renewable waste substrates (Dastgheib et al. 2008; Angelini et al. 2009; Chen et al. 2015). These surface-active agents are used as an antimicrobial, coagulating, anti-adhesive thickening, gelling, wetting, dispersion, moisturizing, foaming, emulsifiers, corrosion inhibition, and detergency agents, in almost every aspect of various industries such as medical, food processing, oil enhanced recovery, and bioremediation of organic pollutants (Banat et al. 2000; Oliveira et al. 2013; Uzoigwe et al. 2015). There are two main categories of biosurfactants, according to their physiochemical properties and molecular weight (Banat et al. 2010). High-molecular-weight molecules are referred to as biopolymers or bioemulsifier, and low-molecular-weight molecules are called biosurfactants. The low-molecular-weight biosurfactants (normally 500–1500 Da) have a lower surface and interfacial tension and include groups of macromolecules such as proteins, lipopeptides, and glycolipids. For instance, surfactin is a low-molecular-weight biosurfactant with 802 Da molecular mass (Souza et al. 2014). High-molecular-weight biosurfactant is mostly polyanionic heteropolysaccharides that consist of proteins and polysaccharides (Perfumo et al. 2010; Uzoigwe et al. 2015). The great example of high-molecular-weight biosurfactant is a lipopolysaccharide biosurfactant produced by *Acinetobacter radioresistens* KA53 and *Acinetobacter calcoaceticus* with

1000 kDa and 1 MDa molecular mass (Uzoigwe et al. 2015). Bioemulsifier and biosurfactant have often been explained as surface-active biomolecules. Moreover, they differ from each other according to their particular physicochemical features and physiological performances (Rahman and Randhawa 2015). Since biosurfactants and bioemulsifiers are both amphiphile compounds, the chemical composition of them differs, and they play specific roles in the biotechnological applications as well as in nature (Uzoigwe et al. 2015; Rahman and Randhawa 2015). Therefore, bioemulsifiers are not same as biosurfactants. They both possess emulsification activity, but in comparison, bioemulsifiers are not as effective in surface tension reduction and they may lack. However, bioemulsifiers are considered as emulsion-stabilizing agent (Das et al. 2014). Bioemulsifiers can successfully form an emulsion between two immiscible phases like hydrophobic molecules and hydrocarbons but not exhibit any surface activity (Satpute et al. 2010; Mnif et al. 2015). *Inquilinus limosus* strain KB3 produced lipopeptide biosurfactant that formed a high stable emulsion with a hydrocarbon (Saimmai et al. 2013). Bioemulsifiers are useful agents in the food industry as stabilizing agents (Fracchia et al. 2012). RAG-1 emulsan was a polysaccharide bioemulsifier produced by *Acinetobacter calcoaceticus* and is commercially available (Suthar et al. 2008). The main focus of this chapter is to give an overview of the structure and function of bacterial surfactants and their applications in bioremediation and petroleum oil degradation as well as their applications in biomedicine as therapeutic agents.

24.2 Surfactants Production by Bacteria

Biosurfactants have turned into interesting hydrocarbon-degrading molecules in the late 1960s for the first time, and they have been extensively used in the recent decades (Cameotra and Makkar 2004). These surface-active molecules produced by bacteria may be either released out of the cells (extracellular) or may be attached to the surface of the cell membrane (Mulligan 2005; Banat et al. 2014). These specific biomolecules are produced by diverse bacteria and may be glycolipids, phospholipids, lipopeptides, neutral lipids, fatty acids, and polymeric biosurfactants (Mulligan 2005; Banat et al. 2010; Cameotra et al. 2010; Franzetti et al. 2010; Reis et al. 2010).

In order to synthesize the biosurfactant, the vital ingredient is the availability if a carbon source, since hydrocarbon-degrading bacteria in oil-contaminated sites is the most dominant biosurfactant producer. Many researchers used various wastes (rich in carbon source) as raw material to produce biosurfactant (Kim et al. 2007; Guo et al. 2007; Wadekar et al. 2012; Xia et al. 2012). Oil-contaminated sites (terrestrial and aquatic ecosystems) possess a rich microbial community of biosurfactant-producing bacteria. The petroleum hydrocarbons present at sites are solubilized and utilized by microorganisms as a carbon source; therefore, the probability of isolation of novel biosurfactant-producing strains is encouraged at these sites. According to previous research, diverse bacteria have been isolated from oil-contaminated sites for biosurfactant production. *Pseudomonas* sp. have been reported in many studies for production of glycolipid biosurfactant. Due to its abundance in soil, easy growth condition, and high degradation capacity for crude oil, it is extensively used as biosurfactant producer in various industries (Maier and Soberon-Chavez 2000). The

Table 24.1 Biosurfactant production by diverse bacteria from different sources

Bacterium	Isolation source	Biosurfactant type	References
Acinetobacter sp.	Marine	Polymeric	Patil and Chopade (2003)
Aneurinibacillus aneurinilyticus SBP-11	Sea water	Exopolysaccharide	Balan et al. (2017)
Antarctobacter sp. TG22	Marine	Glycolipid	Gutiérrez et al. (2007b)
Arthrobacter sp. YIM CS25	Soil	Glycolipid	Hu et al. (2016)
Azotobacter sp.	Soil	Lipopeptide	Fauzi and Suryatmana (2016)
Bacillus licheniformis W16	Oil well soil	Lipopeptide	Joshi et al. (2016)
Bacillus megaterium sp. CG MCC 7086	Marine	Lipopeptide	Ma et al. (2016)
Bacillus subtilis M 2–7	Soil	Lipopeptide	Plaza et al. (2014)
Bacillus thuringiensis CMB26	Soil	Lipopeptide	Nguyen and Sabatini (2011)
Corynebacterium sp. ATCC 49955	Soil	Glycolipid	Erdem and Cutright (2016)
Enterobacter cloacae	Marine sediment	Exopolysaccharide	Ramasamy et al. (2017)
Halomonas sp. TG39 TGG7	Marine	Glycolipid	Gutiérrez et al. (2007a)
Pseudomonas aeruginosa	Sea water	Glycolipid	Thavasi et al. (2011)
Rhodococcus sp.HL-6	Oil-contaminated soil	Glycolipid	Tian et al. (2016)
Vibrio alginolyticus	Marine	Exopolysaccharide	Jayaraman and Seetharaman (2003)

biosurfactants released by maturing bacteria are responsible for both the decrease in the critical micelle concentration (CMC) and the surface and interfacial tension between two immiscible phases (Banat et al. 2010; Sivapathasekaran and Sen 2017). Some strains of *Bacillus* sp. were capable of lowering the surface tension from 72 to 27 mN/m in laboratory condition (Wei et al. 2004). Many researchers have isolated various biosurfactant-producing bacteria from diverse environments (Table 24.1).

24.3 Physicochemical Properties of Bacterial Surfactants

Surfactants are amphipathic compounds that carry both moieties (hydrophobic and hydrophilic) in the same molecules (Banat et al. 2010). The polar moiety of biosurfactant carries a carbohydrate, an amino acid, and/or a phosphate group. The nonpolar domain contains long-chain fatty acids (Usman et al. 2016). Lichenysin, fengycin, surfactin, bacillomycin, and iturin have been reported as major sources of lipopeptides produced by different strains of *Bacillus* sp (Romera et al. 2007; Deleu et al. 2008; Nguyen and Sabatini 2011; Perez-Ameneiro et al. 2015). Surfactin is the first lipopeptide derived from *Bacillus* sp. that contains heptapeptide, β-hydroxy fatty acid, and an acyl chain with 12–18 carbons. (Singh et al. 2007; Sen 2010). Various *Bacillus* sp. produce lipopeptide biosurfactants through the naturally conserved multimodular enzyme complexes called as non-ribosomal peptide synthetases (NRPs) (Peypoux et al. 1999; Thaniyavarn et al. 2003; Das et al. 2010; Sen 2010). Biosurfactants also can be categorized according to their polar head group, a nonionic group (no charged group in its head) such as glycolipid and gylcolipopeptide, an ionic group (negative charge), a cationic group (positive charge) such as lipopeptide, and zwitter ionic (contain both positive and negative charges in its head) (Ron and Rosenberg 2002; Sen and Swaminathan 2005; Sen 2010; Bustamante et al. 2012). The position and size of the hydrophobic and hydrophilic operative groups orientation determine the specific application of surfactant in diverse industries. Fengycins are one of the well-known lipopeptide biosurfactants. Their structure contains fatty acids linked to a peptide domain. The isoforms of fengycins differ in constituents of the peptide moiety, the length of the fatty acid chain as well as a connection between them (Sun et al. 2006). The ionic backbone and low critical micelle concentration define the physiochemical properties of biosurfactants. Rhamnolipid of *Pseudomonas* sp. (with glycosyl head moiety and alkanoic acid) is one of the glycolipid biosurfactants on which intense research work is performed. Physicochemical features have been studied and reported for a number of other such biosurfactant-producing bacteria (Table 24.2).

24.4 Prerequisite for Biosurfactants Production

To produce biosurfactant, carbon and nitrogen are the fundamental requirements (Fonseca et al. 2007; Brzonkalik et al. 2012; Putra et al. 2015). Higher cell concentrations and an improvement in the biosurfactant yield showed an increment of 23% with adding an external source of carbon (Sen 1997; Raza et al. 2007). Lang (2000) used nitrogen and C/N ratio as nutrient sources to produce rhamnolipid from *Pseudomonas aeruginosa*. A broad range of C/N ratio were studied by using frying oil and urea, carbon, and nitrogen sources, respectively, and finally it was observed that the ratio 30:1 produced more biomass and biosurfactant (2.80 g l^{-1}) (Elazzazy et al. 2015). Hydrocarbon substrates and agro-industrial wastes with high carbon content can be used as economical raw materials as a carbon source for production of biosurfactant (Das et al. 2008b; Banat et al. 2014; Vecino et al. 2015). Similarly, a number of studies have been reported in the significance of biosurfactant stability at different conditions such as temperature, pH, and salinity (Silva et al. 2010; Thavasi et al. 2011). These factors will determine the surface activity and solubility of biosurfactants for biodegradation purpose (Abouseoud et al. 2010; Bello et al. 2012). For instance, biosurfactant produced by *Bacillus subtilis* shows a great solubility at 100 °C, pH ranging from 3.0–11.0, and 1.0 M NaCl (Yeh et al. 2005; Sen 2008). Similarly, a large group of bacteria are capable to uptake hydrocarbons as a source of carbon. Oil wastes, olive oil, vegetable oils, and carbohydrates are used as carbon sources for biosurfactant production. The major limitation of the biosurfactant is the high production cost, and hence substitute production techniques should

Table 24.2 Physicochemical properties of bacterial surfactants and their applications

Biosurfactant	Group	Bacterium	Physicochemical properties	Applications	References
Low molecular weight					
Glycolipid	Rhamnolipid	*Pseudomonas aeruginosa*	Rhamnose sugars linked to one or two hydroxydecanoic acid molecules	Bioremediation and oil degradation	Muthusamy et al. (2008), Thavasi et al. (2011)
Lipopeptides	Surfactin	*Bacillus subtilis*	Seven amino acids joined to a hydroxyl(OH-) and carboxyl (COO-) groups of C14 acids	Enhanced oil recovery and antimicrobial activity	Rosenberg et al. (2016), Sana et al. (2017)
		Bacillus stratosphericus A15	Protein linked with β-hydroxy fatty acid moiety containing 14 carbons in normal, iso, or anteiso forms	Medical application	
	Fengycin	*Bacillus subtilis*	β-hydroxy fatty acid chain linked with small peptide containing ten amino acid	Antifungal	Liu et al. (2008), Deleu et al. (2008)
High molecular weight					
Bioemulsifiers	Emulsan	*Acinetobacter* sp. RAG-1 (ATCC 31012)	Lipopolysaccharide. Lipid domain (unsaturated fatty acid of C10–18) polysaccharide domain (D-galactosamine, D-galactosaminuronic acid, di-amino-6-deoxy-D-glucose)	Bioavailability of substrates with increasing surface area	Ron and Rosenberg (2002)
	Uronic acid bioemulsifier	*Halomonas* sp. *Klebsiella* sp.	Polysaccharides-proteins-uronic acids	Detoxification of PHAs and emulsification activity	Martínez-Checa et al. (2002), Jain et al. (2013)

be developed. Hence, the production of biosurfactant can be more economical and productive with using low-cost renewable wastes as substrate. Application of vegetable and waste oils in order to enhance the rate of biosurfactant production has been employed by many researchers (Nitschke et al. 2005; Xiong et al. 2008; Guo et al. 2007; Scherr et al. 2012; Xia et al. 2012; Wadekar et al. 2012).

Elazzazy et al. (2015) reused old frying oil of sunflower for the production of rhamnolipid and sophorolipids. In addition, biosynthesis of glycolipid was achieved in the presence of cashew apple juice as a nutritional source (Rocha et al. 2007). Zhang et al. (2012) used waste frying oil as a cheap carbon source for production of biosurfactant by *Pseudomonas* sp. Radzuan et al. (2017) utilized palm oil agricultural refinery waste as a source of carbon for rhamnolipid biosurfactant biosynthesis by *Pseudomonas aeruginosa* PAO1. Their results demonstrated that palm oil could be used as a low-cost substrate with its transformation into a valuable biosurfactant product.

24.5 Growth Kinetic of Biosurfactant Production

The sustainability of further development in fermentation or industrial application is defined by various kinetic models. The profile of important products can be predicted by these kinetic models. Monod kinetics, modified Gompertz equation, and Luedeking Piret (L-p) are the available tools to develop and optimize the production processes of the desired product (Zajsek and Gorsek 2010; Farias et al. 2014; Putra et al. 2015). Biosurfactants can be produced in various growth kinetic such as growth associate, pseudo growth associated, nongrowth associated, and mix growth associated (Desai and Banat 1997; Karanth et al. 1999; Nitschke and Pastore 2004; Nitschke and Pastore 2006).In growth-associated production, the rate of produced biosurfactant is directly proportional to the growth rate. Most of the biosurfactants are produced in this manner. *Bacillus subtilis* LB5a and *Bacillus licheniformis* JF-2 are excellent examples of growth-associated biosurfactant production (Lin et al. 1993). In mixed growth-associated production, biosurfactant is produced during the exponential and stationary phase. Pseudo growth-associated product formation occurs in the growth phase of bacteria such as surfactin produced by *Bacillus subtilis* (Sen and Swaminathan 1997). In nongrowth-associated growth, the production of biosurfactant takes place in stationary phase with zero growth rate. Biosurfactants from *Pseudomonas* sp. and many glycolipids from other bacteria exhibit the nongrowth-associated production (Davila et al. 1992; Babu et al. 1996; Persson et al. 1998).

24.6 Potential Applications of Bacterial Surfactants

Biosurfactants are valuable biotechnological products along with their broad scope of applications as detergents, emulsifier, foaming, wetting and dispersing, antimicrobial agents, and also useful for the removal of toxic environmental pollutants.

Distinct industries such as food, textile, cosmetics, pharmaceutical, oil recovery, and metal mining are inclined to the application of these bioactive products. Surfactants with bacterial origin may be valuable elements in bioremediation of oil spills, heavy metals, plastics, microbial oil enhanced recovery (MEOR), mining, and food and beverage industries according to previous research (Perfumo et al. 2010; Vedaraman and Venkatesh 2011). The detoxification and biodegradation of industrial effluents and waste management may be other potential uses of biosurfactants (Joutey et al. 2013). In spite of the fact, a large number of chemical surfactants are available, biosurfactants present unique and novel properties that are limited or absent in chemical surfactants (Cameotra and Makkar 2004). Biosurfactants application has also been increased in nanotechnology. Biosurfactants help in forming nanoparticles (Martinez et al. 2014; Plaza et al. 2014). They have the potential to conjugate with nano-Sio2 that helps to enhance the surface activity with reducing interfacial tension in the solution (Sivapathasekaran and Sen 2017). The lipopeptide surfactin and fengycin produced by *Bacillus subtilis* are found to be a highly effective and nontoxic mediator for a carbon nanotube and show immense compatibility. Das et al. (2016) found that biosurfactant isolated from *Pseudomonas aeruginosa* MKVIT3 can be used for the synthesis of microemulsion silver nanoparticles. They also show the efficiency of biosurfactant for antimicrobial and cytotoxic activity. A large number of diverse applications have been reported for biosurfactants/bioemulsifiers.

24.6.1 Environmental Application

24.6.1.1 Biodegradation of Petroleum Products

As an outcome of increasing population and fast-growing industrialization, the global demands for energy are achieved by utilizing different natural sources especially hydrocarbons. Consequently, huge amount of hydrocarbons and their products are released into the environment and caused a serious concern for health and sustainability of human and other living organisms in both aquatic and terrestrial ecosystems and indirectly affect the economic losses especially in developing countries (Samanta et al. 2002; Ismail et al. 2013; Mnif et al. 2015; Montagnolli et al. 2015; Joy et al. 2017). Petroleum hydrocarbons carry a big share of environmental pollutants around the world especially in oil-producing countries (Mnif et al. 2015). Environmental oil contamination occurs during exploration, transportations, storage, and refinery of oil (Patowary et al. 2014). Apart from that, accidents of oil tankers and leakage of oil pipelines are other major sources of oil pollution (Montagnolli et al. 2015). For instance, Exxon Valdez oil spill in the USA, 1989 (Atlas and Hazen 2011), and Korean oil spill (yellow sea) in China, 2007, may be counted as worst oil spills during the history. Polyaromatic hydrocarbons (PAHs) are omnipresent in nature. The environment pollution caused due to petroleum products are a serious threat today, due to their toxicity, resistance, carcinogenicity, and mutagenicity (Clemente et al. 2001; Lau et al. 2014; Zhang et al. 2012). The physicochemical and biological features of soil and water can be excessively changed by oil spills, leaks, and drilling waste. However, several reports have proved that biodegradation, biotransformation, and mineralization of petroleum-derived products using bacteria mediated remediation. Bioremediation of oil spills and oil-contaminated sites are effective technology and can be recognized as a powerful substitute to conventional remedy to solve environmental problems (Karhu et al. 2009; Ismail et al. 2013). Biosurfactant properties make them suitable candidates for a wide range of oil industries, such as oil recovery, oil spill cleanup, cleaning the oil contaminate tankers, emulsification, viscosity control, and removal of crude oil from sludge (Makkar et al. 2011; Oliveira et al. 2013; Mnif et al. 2015). Large groups of diverse bacteria have been isolated from contaminated sites and are reported to effectively remove oil pollutants. As highlighted earlier, aspects like temperature, pH, salinity, and application of nutrients (phosphorus and nitrogen) have a great influence on hydrocarbon-degrading bacteria growth (Desai and Banat 1997). Montagnolli et al. (2015) showed the successful biodegradation of crude oil and kerosene with biosurfactant produced by *Bacillus subtilis*. Vargas et al. (2017) reported the effectiveness of bioaugmentation by nitrogen-fixing bacteria that were able to produce biosurfactant. According to their results, the removal of total hydrocarbon (TH) reached 80% in oil spill impacted area. A number of bacteria genera isolated from oil-polluted sites have a significant performance during biodegradation, and they are highly potent to degrade the polycyclic aromatic hydrocarbons (PAHs) into less or nontoxic compounds such as *Pseudomonas*, *Acinetobacter*, *Bacillus*, *Mycobacterium*, and *Arthrobacter* sp. (Cameotra and Makkar 1998). In spite of broad variety of hydrocarbon-degrading bacteria, *Bacillus* and *Pseudomonas* sp. have higher capability and potential for degradation (Al-Bahry et al. 2013; Oliveira et al. 2013; Al-Wahaibi et al. 2014; Parthipan et al. 2017). Nie et al. (2010) found that *Pseudomonas aeruginosa* NY3 could degrade a mixture of an equal volume of phenanthrene, fluorence, pyrene, anthracene, and fluoranthene at different rates. Enzymes synthesized by biosurfactant-producing bacteria show critical performance in hydrocarbon biodegradation (Yong and Zhong 2010). Parthipan et al. (2017) reported the capability of biosurfactant-producing *Bacillus subtilis* A1 to produce degradative enzymes such as alkane hydroxylase and alcohol dehydrogenase for biodegradation of crude oil. Their results indicated that compounds with low molecular weight (C10–C14) were totally degraded, while degradation of hydrocarbons with a length of C15–C19 was up to 97% in 7 days. The oxygenation of terminal methyl group is an important mechanism for alkane removal. Alkane-degrading bacteria carry a number of genes that are responsible for synthesis of alkane hydroxylases, which can degrade the broad range of alkanes conveniently. Yuliani et al. (2012) showed the capability of *Bacillus pumilus* C15 for biodegradation of phenanthrene and pyrene due to the presence of the dioxyrenese nid A and Nan AC genes that are responsible for PAH degradation. Pi et al. (2017) compared the contribution of GM-2 (chemical dispersant) and biosurfactant produced by *Pseudomonas* sp. LSH-7 on the degradation of crude oil. They reported that the GM-2 did not increase biodegradation, while rhamnolipid produced by *Pseudomonas* had no toxic effect and enhanced oil degradation. Mnif et al. (2017) used the microbial consortium cultured with *Bacillus subtilis* and *Acinetobacter radioresistance*

RI7 strains to enhance biodegradation of diesel oil. Their result suggested the potential application of selected consortium for bioremediation of diesel-contaminated soil and water.

24.6.1.2 Mode of Action of Bacterial Surfactants on Contaminants

Biosurfactant-producing bacteria carry diverse enzymatic pathways to degrade many organic and inorganic contaminants. There are considerable challenges to illustrate how bacteria perform an ecological function in the presence of contaminants in the environment (Kataria and Ruhal 2014). Microbial transformation and degradation of organic compounds generally occur due to the capability of biosurfactant-producing bacteria for utilizing the contaminants as their growth. Organic pollutants play an important role in degradation by bacteria in two ways; they provide electron and carbon source that both support biodegradation process. During the biodegradation of pollutants by bacteria, the energy-producing chemical reactions are catalyzed. These reactions are associated with transferring electron and breaking chemical bonds from contaminants. This type of chemical reaction is known as an oxidation-reduction reaction. The organic compounds (petroleum products) play as an electron donor and are oxidized, and some electron acceptors such as O_2, SO_4, and NO_3 are reduced. Degradation of organic contaminants is carried out in a basic metabolic pathway (Fig. 24.1). Apart from that, carbon present in contaminants is oxidized to CO_2, and the other part of carbon contents are used by bacteria for growth and reproduction of new cell mass. The final product of these reactions will be the transformation and degradation of pollutants into less or non-toxic compounds. The process of biodegradation can be either aerobic or anaerobic respiration. Biosurfactant-producing bacteria degrade the organic contaminant with reducing the surface and interfacial tension between oil aqueous phases and consequently reduce the mobility of organic compounds. Similarly, with forming aggregates (micelles) solubilize the pollutants. Application of biosurfactants enhances the

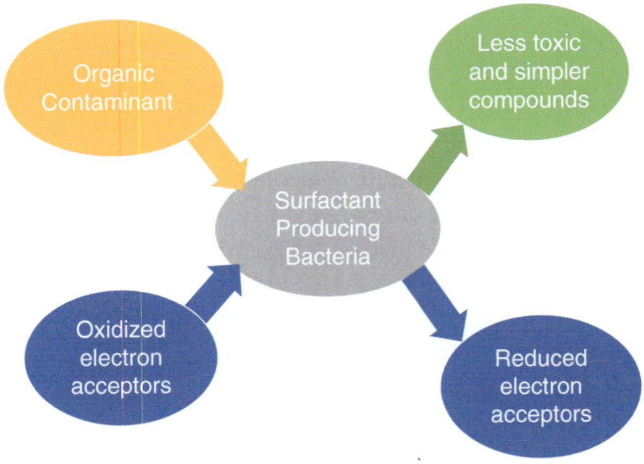

Fig. 24.1 Oxidative biodegradation of organic contaminants by biosurfactant-producing bacteria

bioremediation rates with emulsification, solubilization, and mobilization of organic compounds (Whang et al. 2008; Uzoigwe et al. 2015).

24.7 Potential Applications of Biosurfactants in Medicine

The potential applications of biosurfactants in medicine have been rapidly developed with therapeutic properties like antimicrobial, antifungal, antiviral, and antitumor. Currently, the drug resistance against some pathogenic bacteria is a serious threat; therefore, high demand is needed for discovery of new antimicrobial agents (Mnif et al. 2015). Microbial surfactants may be promising candidates for a new drug in chemotherapy and can employ as significant agents in biopharmaceutical industries (Desai and Banat 1997; Rodrigues et al. 2006). Biosurfactants act as antimicrobial agents because they possess membrane permeabilization properties. Similarly, they are able to form stoma and ion channels in lipid bilayer membrane. Consequently, biosurfactants act on the bacterial membrane and prevent their growth or even their dead (Mnif et al. 2015). Lipopeptide has become a valuable biosurfactant in medical, due to its antibacterial, anticancer, antifungal activities, and anti-mycoplasma agents (Silva et al. 2014). Similarly, rhamnolipid, which is powerful biosurfactant in biodegradation of contaminants, shows anticancer activity (Sivapathasekaran and Sen 2017). Fengycin, surface-active lipopeptide, produced by *Bacillus* sp. carry antifungal activity. The novel lipopeptide (Aneurinifactin) was isolated from marine *Aneurinibacillus aneurinilyticus* SBP-11 that showed promising antimicrobial activity; therefore, it could be a good candidate for biosurfactant production in the various medical fields (Balan et al. 2017). Zouari et al. (2017) evaluated the effect of *Bacillus subtilis* SPB1 on hyperglycemia and kidney function in rat fed on hypercaloric (high-fat, high-fructose) diet. Their funding clearly showed that the peak of blood glucose concentration was decreased (34%) by administration of *Bacillus subtilis* SPB1 biosurfactant in 1 h significantly. The treatment with *Bacillus subtilis* SPB1 lipopeptide reduced the urea and creatinine level (near normal level) notably. Subsequently, kidney function was improved. The excisional wound healing activity of biosurfactant produced by *Bacillus stratosphericus* A15 was investigated. The wound closure was enhanced (97%) by an ointment of test lipopeptide biosurfactant in 10 days, compared to the untreated control group. Interestingly, biosurfactant also showed antioxidant and antimicrobial activity against *Staphylococcus aureus* ATCC 25923 and *Escherichia coli* ATCC 8739 that proved its additional influence in wound protection (Sana et al. 2017). Perez et al. (2017) showed the potential application of two bacterial strains (*Bacillus* sp. P5 and *Bacillus* sp. P3) for biosurfactant production. Apart from the antifungal activity of *Bacillus* sp. P5, it showed significant antimicrobial action against pathogens such as *Listeria monocytogenes* and *Bacillus cereus*. The genes encoding (sboA and ituD) for the production of the antimicrobial substances (iturin A, surfactin, bacteriocin subtilosin A, and the antimicrobial lipopeptides) were identified in *Bacillus* sp. P5. Multiple-step enzymatic processes are responsible for the production of these important substances (Stein 2008; Abriouel et al. 2011). Velho et al. (2013) detected genes (sboA and spaS) responsible for the production of antimicrobial compounds in novel *Bacillus* sp.

Conclusions

Biosurfactants are amphiphilic surface-active compounds that are synthesized by a particular group of bacteria. Application of bacterial surfactants appears advantageous over the chemical surfactants. The biosurfactants are less toxic, easily degradable, highly efficient, and compatible in the environment. They have the potential applications in various industries including food processing, pharmaceutical, petroleum biotechnology, bioremediation of environmental pollutions, and medicine. A lot of research has already been done in laboratory scale on removal and degradation of hydrocarbons, but their applications in the field are not very consistent. The major reason for this inconsistency appears to be optimum condition for bacterial growth. More efforts are required to select appropriate nutrient sources, which can support the bacterial growth in field conditions to achieve the high rate of biosurfactant production, which will lead to efficient degradation and removal of hydrocarbons. Application of consortia comprising two or more than two bacterial strain has shown better results in degradation processes. However, designing highly efficient consortia of biosurfactant-producing bacteria is quite complex, as it requires compatible bacterial partners. The majority of bacteria reported for biosurfactant production belong to the genera *Pseudomonas* and *Bacillus*, which are highly dominant in most of the environments. It is expected that screening of newer strains belonging to the less dominant or rare groups of bacteria for surfactant activity will lead to the discovery of novel biosurfactants.

Acknowledgments This work was supported by the Department of Biotechnology, Government of India (BT/Coord.II/01/03/2016).

References

Abouseoud M, Yataghene A, Amrane A, Maachi R (2010) Effect of pH and salinity on the emulsifying capacity and naphthalene solubility of a biosurfactant produced by *Pseudomonas fluorescens*. J Hazard Mater 180:131–136. doi:10.1016/j.jhazmat2010.04.003

Abriouel H, Franz CM, Ben Omar N, Gálvez A (2011) Diversity and applications of *Bacillus* bacteriocins. FEMS Microbiol Rev 35:201–232. doi:10.1111/j.1574-6976.2010.00244.x

Al-Bahry SN, Al-Wahaibi YM, Elshafie AE, Al-Bemani AS, Joshi SJ, Al-Makhmari HS, Al-Sulaimani HS (2013) Biosurfactant production by *Bacillus subtilis* B20 using date molasses and its possible application in enhanced oil recovery. Int Biodeter Biodegr 81:141–146. doi:10.1016/j.ibiod.2012.01.006

Al-Wahaibi Y, Joshi S, Al-Bahry S, Elshafie A, Al-Bemani A, Shibulal B (2014) Biosurfactant production by *Bacillus subtilis* B30 and its application in enhancing oil recovery. Colloids Surf B Biointerfaces 114:324–333. doi:10.1016/j.colsurfb.2013.09.022

Angelini TE, Roper M, Kolter R, Weitz DA, Brenner MP (2009) *Bacillus subtilis* spreads by surfing on the waves of surfactant. Proc Natl Acad Sci U S A 106:18109–18113. doi:10.1073/pnas.0905890106

Atlas RM, Hazen TC (2011) Oil biodegradation and bioremediation: a tale of the two worst spills in U.S. history. Environ Sci Technol 45:6709–6715. doi:10.1021/es2013227

Babu PS, Vaidya AN, Bal AS, Kapur R, Juwarkar A, Khanna P (1996) Kinetics of biosurfactants production by *Pseudomonas aeruginosa* strain BS2 from industrial wastes. Biotechnol Lett 18:263–268. doi:10.1007/BF00142942

Balan SS, Kumar CG, Jayalakshmi S (2017) Aneurinifactin, a new lipopeptide biosurfactant produced by a marine *Aneurinibacillus aneurinilyticus* SBP-11 isolated from Gulf of Mannar: purification, characterization and its biological evaluation. Microbiol Res 194:1–9. doi:10.1016/j.micres.2016.10.005

Banat IM, Makkar RS, Cameotra SS (2000) Potential commercial applications of microbial surfactants. Appl Microbiol Biotechnol 53:495–508. doi:10.1007/s002530051648

Banat IM, Franzetti A, Gandolfi I, Bestetti G, Martinotti MG, Fracchia L, Smyth TJ, Marchant R (2010) Microbial biosurfactants production, applications and future potential. Appl Microbiol Biotechnol 87:427–444. doi:10.1007/s00253-010-2589-0

Banat IM, Satpute SK, Cameotra SS, Patil R, Nyayanit NV (2014) Cost effective, technologies and renewable substrates for biosurfactant production. Front Microbiol 5:697. doi:10.3389/fmicb.2014.00697

Bello XV, Rey RD, Cruz JM, Moldes AB (2012) Study of the synergistic effect of salinity, pH and temperature on the surface-active properties of bio surfactants produced by *Lactobacillus pentosus*. J Agric Food Chem 60:1258–1265. doi:10.1021/jf205095d

Brzonkalik K, Hümmer D, Syldatk C, Neumann A (2012) Influence of pH and carbon to nitrogen ratio on mycotoxin production by *Alternaria alternata* in submerged cultivation. AMB Express 2:28. doi:10.1186/2191-0855-2-28

Bustamante M, Duran N, Diez MC (2012) Biosurfactants are useful tools for the bioremediation of contaminated soil: a review. Soil Sci Plant Nutr 12:667–687. doi:10.4067/S0718-95162012005000024

Cameotra SS, Makkar RS (1998) Synthesis of biosurfactants in extreme conditions. Appl Microbiol Biotechnol 50:520–529. doi:10.1007/s002530051329

Cameotra SS, Makkar RS (2004) Recent applications of biosurfactants as biological immunological molecules. Curr Opin Microbiol 7:262–266. doi:10.1016/j. mib.2004.04.006

Cameotra SS, Makkar RS, Kaur J, Mehta S (2010) Synthesis of biosurfactants and their advantages to microorganisms and mankind. Adv Exp Med Biol 672:261–280. doi:10.1007/978-1-4419-5979-9_20

Chen WC, Juang RS, Wei YH (2015) Applications of a lipopeptide biosurfactant, surfactin, produced by microorganisms. Biochem Eng 103:158–169. doi:10.1016/j.bej.2015.07.009

Clemente AR, Anazawa TA, Durrant LR (2001) Biodegradation of polycyclic aromatic hydrocarbons by soil fungi. Braz J Microbiol 32:255–260. doi:10.1590/S1517-83822001000400001

Cooper DG, Macdonald CR, Duff JBS, Kosaric N (1981) Enhanced production of surfactin from *Bacillus subtilis* by continuous product removal and metal cation additions. Appl Environ Microbiol 42:408–412. doi:0099-2240/81/090408-05$02.00/0

Das P, Mukherjee S, Sen R (2008a) Genetic regulation of the biosynthesis of microbial surfactants: an overview. Biotechnol Genet Eng Rev 25:165–185. doi:10.5661/bger-25-165

Das P, Mukherjee S, Sen R (2008b) Improved bioavailability and biodegradation of a model polyaromatic hydrocarbon by a biosurfactant producing bacterium of marine origin. Chemosphere 72:1229–1243. doi:10.1016/j.chemosphere.2008.05.015

Das P, Mukherjee S, Sivapathasekaran C, Sen R (2010) Microbial surfactants of marine origin: potential and prospects. In: Sen R (ed) Biosurfactants. Springer, New York. ISBN: 978-1-4419-5978-2, pp 88–101. doi:10.1007/978-1-4419-5979-9_7

Das P, Yang X-P, Ma LZ (2014) Analysis of biosurfactants from industrially viable Pseudomonas strain isolated from crude oil suggests how rhamnolipids congeners affect emulsification property and antimicrobial activity. Front Microbiol 5:696. doi:10.3389/fmicb.2014.00696

Das M, Patowary K, Vidya R, Malipeddi H (2016) Microemulsion synthesis of silver nanoparticles using biosurfactant extracted from *Pseudomonas aeruginosa* MKVIT3 strain and com-

parison of their antimicrobial and cytotoxic activities. IET Nanobiotechnol 10(6):411–418. doi:10.1049/iet-nbt.2015.0119
Dastgheib SM, Amoozegar MA, Elahi E, Asad S, Banat IM (2008) Bioemulsifier production by a halothermophilic Bacillus strain with potential applications in microbially enhanced oil recovery. Biotechnol Lett 30:63–720. doi:10.1007/s10529-007-9530-3
Davila AM, Marchal R, Vandescasteele JP (1992) Kinetics and balance of fermentation free from product inhibition: sophorose lipid production by *Candida bombicola*. Appl Microbiol Biotechnol 38:6–11. doi:10.1007/BF00169410
Deleu M, Paquot M, Nylander T (2008) Effect of fengycin, a lipopeptide produced by *Bacillus subtilis,* on model biomembranes. Biophys J 94:2667–2679. doi:10.1529/biophysj.107.114090
Desai J, Banat IM (1997) Microbial production of surfactants and their commercial potential. Microbiol Mol Biol Rev 61:47–64. doi:0146-0749/97/$04.0010
Edward KP, Lepo JE, Lewis MA (2003) Toxicity comparison of biosurfactants and synthetic surfactants used in oil spill remediation to two estuarine species. Mar Pollut Bull 46:1309–1316. doi:10.1016/S0025-326X(03)00238-8
Elazzazy AM, Abdelmoneim TS, Almaghrabi OA (2015) Isolation and characterization of biosurfactant production under extreme environmental conditions by alkali-halo-thermophilic bacteria from Saudi Arabia. Saudi J Biol Sci 22:466–475. doi:10.1016/j.sjbs.2014.11.018
Erdem Z, Cutright TJ (2016) Biotransformation of 1,1,1-trichloro-2,2-bis(p-chlorophenyl) ethane (4,4′-DDT) on a sandy loam soil using aerobic bacterium *Corynebacterium* sp. Environ Earth Sci 75:1267. doi:10.1007/s12665-016-6057-8
Farias D, de Andrade RR, Maugeri-Filho F (2014) Kinetic modeling of ethanol production by *Scheffersomyces stipitis* from xylose. Appl Microbiol Biotechnol 172:361–379. doi:10.1007/s12010-013-0546-y
Fauzi M, Suryatmana P (2016) Bioremediation of crude oil waste contaminated soil using petrophilic consortium and *Azotobacter* sp. J Degraded Mining Lands Managment 3:521–526. doi:10.15243/jdmlm.2016.032.521
Fonseca RR, Silva AJ, França FPD, Cardoso VL, Sérvulo EF (2007) Optimizing carbon/nitrogen ratio for biosurfactant production by a *Bacillus subtilis* strain. Appl Biochem Biotechnol 137:471–486. doi:10.1007/978-1-60327-181-3_40
Fracchia L, Cavallo M, Martinotti MG, Banat IM (2012) Biosurfactants and bioemulsifiers biomedical and related applications present status and future potentials. In: Ghista DN (ed) Biomedical science, engineering and technology. InTech, Rijeka. ISBN: 978-953-307-471-9, pp 325–370. doi:10.5772/23821
Franzetti A, Tamburini E, Banat IM (2010) Application of biological surface active compounds in remediation technologies. In: Sen R (ed) Biosurfactants. Springer, New York. ISBN: 978-1-4419-5978-2, pp 121–134. doi:10.1007/978-1-4419-5979-9_7
Guo Z, Xing R, Liu S, Zhong Z, Ji X, Wang L, Li P (2007) Antifungal properties of Schiff bases of chitosan, N-substituted chitosan and quaternized chitosan. Carbohydr Res 342:1329–1332. doi:10.1016/j.carres.2007.04.006
Gutiérrez T, Mulloy B, Black K, Green DH (2007a) Glycoprotein emulsifiers from two marine Halomonas species: chemical and physical characterization. J Appl Microbiol 103:1716–1727. doi:10.1111/j.1365-2672.2007.03407.x
Gutiérrez T, Mulloy B, Bavington C, Black K, Green DH (2007b) Partial purification and chemical characterization of a glycoprotein (putative hydrocolloid) emulsifier produced by a marine bacterium *Antarctobacter*. Appl Microbiol Biotechnol 76:1017–1026. doi:10.1007/s00253-007-1091-9
Henkel M, Muller MM, Kugler JH, Lovaglio RB, Contiero J, Syldatk C, Hausmann R (2012) Rhamnolipids as biosurfactants from renewable resources: concepts for next-generation rhamnolipid production. Process Biochem 47:1207–1226. doi:10.1016/j.procbio.2012.04.018
Hu QW, Chu X, Xiao M, Li CT, Yan ZF, Hozzein WN, Kim CJ, Zhi XY, Li WJ (2016) *Arthrobacter deserti* sp. nov. isolated from a desert soil sample. Int J Syst Evol Microbiol 66:2035–2040. doi:10.1099/ijsem.0.000986

Ismail W, Al-Rowaihi IS, Al-Humam AA, Hamza RY, El Nayal AM, Bououdina M (2013) Characterization of a lipopeptide biosurfactant produced by a crude-oil-emulsifying *Bacillus* sp. I-15. Int Biodeter Biodegr 84:168–178. doi:10.1016/j.ibiod.2012.04.017

Jain RM, Mody K, Joshi N, Mishra A, Jha B (2013) Production and structural characterization of biosurfactant produced by an alkaliphilic bacterium, *Klebsiella* sp: evaluation of different carbon sources. Colloids Surf B Biointerfaces 108:199–204. doi:10.1016/j.colsurfb.2013.03.002

Jayaraman M, Seetharaman J (2003) Phsicochemical analysis of the exopolysaccharides produced by a marine biofouling bacterium, *Vibrio alginolyticus*. Process Biochem 38:841–847. doi:10.1016/S0032-9592(02)00021-3

Joshi SJ, Al-Wahaibi YM, Al-Bahry SN, Elshafie AE, Al-Bemani AS, Al-Bahri A, Al-Mandhari MS (2016) Production, characterization, and application of *Bacillus licheniformis* W16 biosurfactant in enhancing oil recovery. Front Microbiol 7:1853. doi:10.3389/fmicb.2016.01853

Joutey NT, Bahafid W, Sayel H, ElGhachtouli N (2013) Biodegradation: involved microorganisms and genetically engineered microorganisms. In: Chamy R, Rosenkranz F (eds) Biodegradation life of science. Intech, Rijeka. ISBN: 978-953-51-1154-2, pp 289–320. doi:10.5772/56194

Joy S, Butalia T, Sharma S, Rahman PKSM (2017) Biosurfactant producing bacteria from hydrocarbon contaminated environment. In: Heimann K, Karthikeyan OP, Muthu SS (eds) Biodegradation and bioconversion of hydrocarbons. Springer, Singapore. ISBN: 978-981-10-0199-4, pp 259–305. doi:10.1007/978-981-10-020-4_8

Karanth NGT, Deo G, Veenanadig NK (1999) Microbial production of biosurfactants and their importance. Curr Sci 77:116–126

Karhu M, Kaakinen J, Kuokkanen T, Rämö J (2009) Biodegradation of light fuel oils in water and soil as determined by the manometric respirometric method. Water Air Soil Pollut 197:3–14. doi:10.1007/s11270-008-9752-6

Kataria R, Ruhal R (2014) Microbiological metabolism under chemical stress. In: Das S (ed) Microbial biodegradation and bioremediation. Elsevier, Amsterdam. ISBN: 9780128004821, pp 499–511. doi:10.1016/B978-0-12-800021-2.00021-2

Kim S, Kim JY, Kim SH, Bae HJ, Yi H, Yoon SH, Koo BS, Kwon M, Cho JY, Lee CH, Hong S (2007) Surfactin from *Bacillus subtilis* displays antiproliferative effect via apoptosis induction, cell cycle arrest and survival signaling suppression. FEBS Lett 581:865–871. doi:10.1016/j.febslet.2007.01.059

Lang S (2000) Biological amphiphiles (microbial biosurfactants). Curr Opin Colloid Interface Sci 7:12–20. doi:10.1016/S1359-0294(02)00007-9

Lau EV, Gan S, Ng HK, Poh PE (2014) Extraction agents for the removal of polycyclic aromatic hydrocarbons (PAHs) from soil in soil washing technologies. Environ Pollut 184:640–649. doi:10.1016/j.envpol.2013.09.010

Lin SC, Sharma MM, Georgiou G (1993) Production and deactivation of biosurfactant by *Bacillus licheniformis* JF-2. Biotechnol Prog 9:138–145. doi:10.1021/bp0002 0a004

Liu XH, Yang SZ, Mu BZ (2008) Isolation and characterization of a C12-lipo peptide produced by *Bacillus subtilis* HSO 121. J Pept Sci 14:864–875. doi:10.1002/psc.1017

Ma Y, Kong Q, Qin C, Chen Y, Chen Y, Lv R, Zhou G (2016) Identification of lipopeptides in *Bacillus megaterium* by two-step ultrafiltration and LC-ESI-MS/MS. AMB Express 6:79. doi:10.1186/s13568-016-0252-6

Maier RM, Soberon-Chavez G (2000) *Pseudomonas aeruginosa* rhamnolipid: biosynthesis and potential applications. Appl Microbiol Biotechnol 54:625–633. doi:10.1007/s002530000443

Makkar SR, Swaranjit SC, Ibrahim MB (2011) Advances in utilization of renewable substrates for biosurfactant production. AMB Express 1:5. doi:10.1186/2191-0855-1-5

Martinez DS, Faria AF, Berni E, Filho AGS, Almeida G, Oliveira ACGMJ, Durrant LR, Umbuzeiro GA, Alves OL (2014) Exploring the use of biosurfactants from *Bacillus subtilis* in bionanotechnology: a potential dispersing agent for carbon nanotube ecotoxicological studies. Process Biochem 49:1162–1168. doi:10.1016/j.procbio.2014.04.006

Martínez-Checa F, Toledo FL, Vilchez R, Quesada E, Calvo C (2002) Yield production, chemical composition and functional properties of emulsifier H28 synthesized by *Halomonas eurihalina* strain H-28 in media containing various hydrocarbons. Appl Microbial Biotechnol 58:358–363. doi:10.1007/s00253-001-0903-6

Mnif I, Mnif S, Sahnoun R, Maktouf S, Ellouze-Chaabouni YAS, Ghribi D (2015) Biodegradation of diesel oil by a novel microbial consortium: comparison between co-inoculation with biosurfactant-producing strain and exogenously added biosurfactants. Environ Sci Pollut Res Int 22:14852–14861. doi:10.1007/s11356-015-4488-5

Mnif I, Sahnoun R, Ellouz-Chaabouni S, Ghribi D (2017) Application of bacterial biosurfactants for enhanced removal and biodegradation of diesel oil in soil using a newly isolated consortium. Process Saf Environ Prot 109:72–81. doi:10.1016/j.psep.2017.02.002

Moldes AB, Paradelo R, Vecino X, Cruz JM, Gudina E, Rodrigues L, Teixeria JA, Domínguez JM, Barral MT (2013) Partial characterization of biosurfactant from *Lactobacillus pentosus* and comparison with sodium dodecyl sulphate for the bioremediation of hydrocarbon contaminated soil. Biomed Res Int 2013:1–26. doi:10.1155/2013/961842

Montagnolli RN, Lopes PRM, Bidoia ED (2015) Assessing *Bacillus subtilis* biosurfactant effects on the biodegradation of petroleum products. Environ Monit Assess 187:4116. doi:10.1007/s10661-014-4116-8

Mulligan CN (2005) Environmental applications for biosurfactants. Environ Pollut J 133:183–198. doi:10.1016/j.envpol.2004.06.009

Muthusamy K, Gopalakrishnan S, Ravi TK, Sivachidambaram P (2008) Biosurfactants: properties, commercial production and application. Curr Sci 94:736–747

Nguyen TT, Sabatini DA (2011) Characterization and emulsification properties of rhamnolipid and sophorolipid biosurfactants and their applications. Int Mol Sci 12:1232–1244. doi:10.3390/ijms12021232

Nie M, Yin X, Ren C, Wang Y, Xu F, Shen Q (2010) Novel rhamnolipid biosurfactants produced by polycyclic aromatic hydrocarbon degrading bacterium *Pseudomonas aeruginosa* strain NY3. Biotechnol Adv 28:635–643. doi:10.1016/j.biotechadv.2010.05.013

Nitschke M, Pastore GM (2004) Biosurfactant production by *Bacillus subtilis* using cassava-processing effluent. Appl Biochem Biotechnol 112:163–172. doi:10.1385/ABAB:112:3:163

Nitschke M, Pastore GM (2006) Production and properties of a surfactant obtained from *Bacillus subtilis* grown on cassava wastewater. Bioresour Technol 97:336–341. doi:10.1016/j.biortech.2005.02.044

Nitschke M, Costa SG, Haddad R, Gonçalves LAG, Eberlin MN, Contiero J (2005) Oil wastes as unconventional substrates for rhamnolipid biosurfactant production by *Pseudomonas aeruginosa* LBI. Biotechnol Prog 21:1562–1566. doi:10.1021/bp050198x

Oliveira DWF, Franca IWL, Felix AKN, Martins JJL, Giro MEA, Melo VMM (2013) Kinetic study of biosurfactants production by *Bacillus subtilis* LAMI005 grown in clarified cashew apple juice. Colloids Sur Biointerfaces 101:34–43. doi:10.1016/j.colsurfb.2012.06.011

Panjiar N, Gabrani R, Sarethy IP (2013) Diversity of biosurfactant producing *Streptomyces* sp. isolates from hydrocarbon-contaminated soil. Int Pharma Biol Sci 4:524–535

Parthipan P, Preetham E, Machuca LL, Pattanathu KSMR, Kadarkarai M, Aruliah R (2017) Biosurfactant and degradative enzymes mediated crude oil degradation by bacterium *Bacillus subtilis* A1. Front Microbiol 8:193. doi:10.3389/fmicb.2017.00193

Patil J R, Chopade B A (2003) Bioemulsifier production by *Acinetobacter* strains isolated from healthy human skin. US Patent No. US 2004/0138429 A1

Patowary K, Saikia RR, Kalita MC, Deka S (2014) Degradation of polyaromatic hydrocarbons employing biosurfactant producing *Bacillus pumilus* KS2. Ann Microbiol 65:225–234. doi:10.1007/s13213-014-0854-7

Perez KJ, Viana JD, Lopes FC, Pereira JQ, Dos Santos DM, Oliveira JS, Velho RV, Crispim SM, Nicoli JR, Brandelli A, Nardi RM (2017) *Bacillus* sp. isolated from Puba as a source of biosurfactants and antimicrobial lipopeptides. Front Microbiol 8:61. doi:10.3389/fmicb.2017.00061

Perez-Ameneiro M, Vecino X, Cruz JM, Moldes AB (2015) Wastewater treatment enhancement by applying a lipopeptide biosurfactant to a lignocellulosic biocomposite. Carbohydr Polym 131:186–196. doi:10.1016/j.carbpol.2015.05.075

Perfumo A, Smyth TJP, Marchant R, Banat IM (2010) Production and roles of biosurfactants and bioemulsifiers in accessing hydrophobic substrates. In: Timmis KN (ed) Handbook of hydrocarbon and lipid microbiology. Springer, Berlin. ISBN: 978-3-540-77584-3, pp 1501–1512. doi:10.1007/978-3-540-77587-4_103

Persson A, Osterberg E, Dostalek M (1998) Biosurfactant production from *Pseudomonas fluorescens* 378: growth and product characteristic. Appl Microbiol Biotechnol 29:1–4. doi:10.1007/BF00258342

Peypoux F, Bonmatin JM, Wallach J (1999) Recent trends in the biochemistry of surfactin. Appl Microbiol Biotechnol 54:553–563. doi:10.1007/s0025300514

Pi Y, Bao M, Liu Y, Lu T, He R (2017) The contribution of chemical dispersants and biosurfactants on crude oil biodegradation by *Pseudomonas* sp. LSH-7. J Clean Prod 153:74–82. doi:10.1016/j.jclepro.2017.03.120

Plaza GA, Chojniak J, Banat IM (2014) Biosurfactnat mediated biosynthesis of selected metallic nanoparticles. Inter Mol Sci 15:1372–13737. doi:10.3390/ijms150813720

Putra MD, Abasaeed AE, Atiyeh HK, Al-Zahrani SM, Gaily MH, Sulieman AK (2015) Kinetic modeling and enhanced production of fructose and ethanol from date fruit extract. Chem Eng Commun 202:1618–1627. doi:10.1080/00986445.2014.968711

Radzuan MN, Banat IM, Winterburn J (2017) Production and characterization of rhamnolipid using palm oil agricultural refinery waste. Bioresour Technol 225:99–105. doi:10.1016/j.biortech.2016.11.052

Rahman KSM, Randhawa KKS (2015) Editorial: microbiotechnology based surfactants and their applications. Front Microbiol 6:1344. doi:10.3389/fmicb.2015.01344

Ramasamy S, Arumugam A, Chandran P (2017) Optimization of *Enterobacter cloacae* (KU923381) for diesel oil degradation using response surface methodology (RSM). J Microbiol 55:104. doi:10.1007/s12275-017-6265-2

Raza ZA, Khan MS, Khalida ZM (2007) Evaluation of distant carbon sources in biosurfactant production by a gamma ray-induced *Pseudomonas putida* mutant. Process Biochem 42:686–692. doi:10.1016/j.procbio.2006.10.001

Reis RS, da Rocha SLG, Chapeaurouge DA, Domont GB, Santa Anna LMM, Freire DMG, Perales J (2010) Effects of carbon and nitrogen sources on the proteome of *Pseudomonas aeruginosa* PA1 during rhamnolipid production. Process Biochem 45:1504–1510. doi:10.1016/j.procbio.2010.05.032

Rocha MV, Souza MCM, Benedicto SC, Bezerra MS, Macedo GR, Pinto GAS, Goncalves LRB (2007) Production of biosurfactant by *Pseudomonas aeruginosa* grown on cashew apple juice. Appl Biochem Biotechnol 137-140:185–194

Rodrigues L, Banat IM, Teixeria J, Oliveira R (2006) Biosurfactants: potential applications in medicine. J Antimicrob Chemother 57:609–618. doi:10.1093/jac/dkl024

Romera D, Vincente AD, Rakotoaly RH, Dufour SE, Veening JW, Arre Bola E, Cazorla FM, Kuipers OP, Paquot M, Garcia AP (2007) The Iturin and Fengycin family of lipopeptides are key factor in antagonism of *Bacillus subtilis* towards *Podosphaera fusca*. Mol Plant-Microbe Interact 20:430–440. doi:10.1094/MPMI-20-4-0430

Ron EZ, Rosenberg E (2002) Biosurfactant and oil remediation. Curr Opin Biotechnol 13:249–252. doi:10.1016/S0958-1669(02)00316-6

Rosenberg G, Steinberg N, Oppenheimer-Shaanan Y, Olender T, Doron S, Ben-Ari J, Sirota-Madi A, Bloom-Ackermann Z, Kolodkin-Gal I (2016) Not so simple, not so subtle: the interspecies competition between *Bacillus simplex* and *Bacillus subtilis* and its impact on the evolution of biofilms. Biofilms Microbiomes:1–11. doi:10.1038/npjbiofilms.2015.27

Saha B, Orvig C (2010) Biosorbents for hexavalent chromium elimination from industrial and municipal effluents. Coord Chem Rev 254:2959–2972. doi:10.1016/j.ccr.2010.06.005

Saimmai A, Udomsilp S, Maneerat S (2013) Production and characterization of biosurfactant from marine bacterium *Inquilinus limosus* KB3 grown on low-cost raw materials. Ann Microbiol 63:1327–1339. doi:10.1007/s13213-012-0592-7

Samanta SK, Singh OV, Jain RK (2002) Polycyclic aromatic hydrocarbons: environmental pollution and bioremediation. Trends Biotechnol 20:243–248. doi:10.1016/S0167-7799(02)01943-1

Sana S, Mazumder A, Datta S, Biswas D (2017) Towards the development of an effective in vivo wound healing agent from *Bacillus* sp. derived biosurfactant using *Catla catla* fish fat. RSC Adv 7:13668–13677. doi:10.1039/c6ra26904d

Satpute SK, Banpurkar AG, Dhakephalkar PK, Banat IM, Balu A (2010) Methods for investigating biosurfactants and bioemulsifiers: a review. Crit Rev Biotechnol 30:127–144. doi:10.3109/07388550903427280

Scherr KE, Hasinger M, Loibner AP (2012) Increasing PAH (bio) accessibility in historically PAH-contaminated soils: waste versus food-grade oils. J Chem Technol Biotechnol 87:1229–1236. doi:10.1002/jctb.3823

Sen R (1997) Response surface optimization of the critical media components for the production of surfactin. Chem Technol Biotechnol 68:263–270. doi:10.1002/(SICI)1097-4660(199703)68-3<263

Sen R (2008) Biotechnology in petroleum recovery: the microbial EOR. Prog Energy Combust Sci 34:714–724. doi:10.1016/j.pecs.2008.05.001

Sen R (2010) Surfactin biosynthesis, genetics and potential applications. Adv Exp Med Biol 672:315–319. doi:10.1007/978-1-4419-5979-9

Sen R, Swaminathan T (1997) Application of response surface methodology to evaluate the optimum environmental conditions for the enhanced production of surfactin. Appl Microbiol Biotechnol 47:358–363. doi:10.1007/s002530050940

Sen R, Swaminathan T (2005) Characterization of concentration and purification parameters and operating conditions for the small-scale recovery of surfactin. Process Biochem 40:2953–2958. doi:10.1016/j.procbio.2005.01.014

Silva SN, Farias CBB, Rufino RD, Luna JM, Sarubbo LA (2010) Glycerol as substrate for the production of biosurfactant by *Pseudomonas aeruginosa* UCP0992. Colloids Surf B Biointerfaces 79:174–183. doi:10.1016/j.colsurfb.2010.03.050

Silva DS, Castro CC, Silva FS, Sant AV, Vargas GD, Lima M (2014) Antiviral activity of a *Bacillus* sp. P34 peptide against pathogenic viruses of domestic animals. Braz J Microbiol 45:1089–1094. doi:10.1590/S1517-83822014000300043

Singh A, Hamme JDV, Ward OP (2007) Surfactants in microbiology and biotechnology: part 2 application aspects. Biotechnol Adv 25:99–121. doi:10.1016/j.biotechadv.2006.10.004

Sivapathasekaran C, Sen R (2017) Origin, properties, production and purification of microbial surfactants as molecules with immense commercial potential. Tenside Surfactant Deterg 54:92–107. doi:10.3139/113.110482

Souza EC, Vessoni-Penna TC, de Souza ORP (2014) Biosurfactant-enhanced hydrocarbon bioremediation: an overview. Int Biodeter Biodegr 89:88–94. doi:10.1016/j.ibiod.2014.01.007

Stein T (2008) Whole-cell matrix-assisted laser desorption/ionization mass spectrometry for rapid identification of bacteriocin/lantibiotic-producing bacteria. Rapid Commun Mass Spectrom 22:1146–1152. doi:10.1002/rcm.3481

Sun L, Lu Z, Bie X, Lu F, Yang S (2006) Isolation and characterization of a co-producer of fengycins and surfactins, endophytic *Bacillus amyloliquefaciens* ES-2, from *Scutellaria baicalensis* Georgi. World J Microbiol Biotechnol 22:1259–1266. doi:10.1007/s11274-006-9170-0

Suthar H, Hingurao K, Desai A, Nerurkar A (2008) Evaluation of bioemulsifier mediated microbial enhanced oil recovery using sand pack column. J Microbiol Methods 75:225–230. doi:10.1016/j.mimet.2008.06.007

Thaniyavarn J, Roonsawang N, Kameyama T, Haruki M, Imanaka TMM, Kanaya S (2003) Production and characterization of biosurfactants from *Bacillus licheniformis* F2.2. Biosci Biotechnol Biochem 67:1239–1244. doi:10.1271/bbb.67.1239

Thavasi R, Subramanyam NVRM, Jayalakshmi S, Balasubramanian T, Banat IM (2011) Biosurfactant production by *Pseudomonas aeruginosa* from renewable resources. Indian J Microbiol 51:30–36. doi:10.1007/s12088-011-0076-7

Tian ZT, Chen LY, Li DH, Pang HY, Wu S, Jin-Biaoliu HL (2016) Characterization of a biosurfactant-producing strain *Rhodococcus* sp. HL-6. Romanian Biotechnol Lett 6:976. doi:10.3389/fmicb.2015.00976

Usman MM, Dadrasnia A, Lim KT, Mahmud AF, Ismail S (2016) Application of biosurfactants in environmental biotechnology; remediation of oil and heavy metal. AIMS Bioeng 3:289–304. doi:10.3934/bioeng.2016.3.289

Uzoigwe C, Burgess JG, Ennis CJ, Rahman PKSM (2015) Bioemulsifiers are not biosurfactants and require different screening approaches. Front Microbiol 6:245. doi:10.3389/fmicb.2015.00245

Vargas JP, Carmona SEV, Moreno EZ, Casado NAR, Calva GC (2017) Bioremediation of soils from oil spill impacted sites using bioaugmentation with biosurfactants producing, native, free-living nitrogen fixing bacteria. Rev Int Contam Ambie 33:105–114. doi:10.20937/RICA.2017.33.esp01.09

Vecino X, Rodríguez-López L, Cruz JM, Moldes AB (2015) Sewage sludge polycyclic aromatic hydrocarbon (PAH) decontamination technique based on the utilization of a lipopeptide biosurfactant extracted from corn steep liquor. Agric Food Chem 63:7143–7150. doi:10.1021/acs.jafc.5b02346

Vedaraman N, Venkatesh N (2011) Production of surfactin by *Bacillus subtilis* MTCC 2423 from waste frying oils. Braz J Chem Eng 28:175–180. doi:10.1590/S0104-66322011000200001

Velho RV, Basso AP, Segalin J, Costa-Medina LF, Brandelli A (2013) The presence of sboA and spaS genes and antimicrobial peptides subtilosin a and subtilin among *Bacillus* strains of the Amazon basin. Genet Mol Biol 36:101–104. doi:10.1590/S1415-47572013000100014

Wadekar S, Kale S, Lali A, Bhowmick D, Pratap AP (2012) Microbial synthesis of rhamnolipids by *Pseudomonas aeruginosa* (ATCC 10145) on waste frying oil as low cost carbon source. Prep Biochem Biotechnol 42:249–266. doi:10.1080/10826068.2011.603000

Wei YH, Wang LF, Chang JS (2004) Optimizing iron supplement strategies for enhanced surfactin production with *Bacillus subtilis*. Biotechnol Prog 20:979–983. doi:10.1021/bp030051a

Whang L-M, Liu P-WG, Ma CC, Cheng SS (2008) Application of biosurfactants, rhamnolipid, and surfactin, for enhanced biodegradation of diesel-contaminated water and soil. J Hazard Mater 151:155–163. doi:10.1016/j.jhazmat.2007.05.063

Xia WJ, Luo Z, Dong HP, Yu L, Cui QF, Bi YQ (2012) Synthesis, characterization and oil recovery application of biosurfactant produced by indigenous *Pseudomonas aeruginosa* WJ-1 using waste vegetable oils. Appl Biochem Biotechnol 166:1148–1166. doi:10.1007/s12010-011-9501-y

Xiong Z, Jiang Y, Qi D, Lu H, Yang F, Yang J, Chen L, Sun L, Xu X, Xue Y, Zhu Y, Jin Q (2008) Complete genome sequence of the extremophilic *Bacillus cereus* strain Q1 with industrial applications. J Bacteriol 191:1120–1121. doi:10.1128/JB.01629-08

Yeh MS, Wei YH, Chang JS (2005) Enhanced production of surfactin from *Bacillus subtilis* by addition of solid carriers. Biotechnol Prog 21:1329–1334. doi:10.1021/bp050040c

Yong YC, Zhong JJ (2010) Recent advances in biodegradation in China: new microorganisms and pathways, biodegradation engineering, and bioenergy from pollutant biodegradation. Process Biochem 45:1937–1943. doi:10.1016/j.procbio.2010.04.009

Yuliani H, Sahlan M, Hermansyah H, Wijanarko A (2012) Selection and identification of polyaromatic hydrocarbon degrading bacteria. World Appl Sci J 20:1133–1138. doi:10.5829/idosi.wasj.2012.20.08.2123

Zajsek K, Gorsek A (2010) Modelling of batch kefir fermentation kinetics for ethanol production by mixed natural microflora. Food Bioprod Process 88:55–60. doi:10.1016/j.fbp.2009.09.002

Zhang X, Xu D, Zhu C, Lundaa T, Scherr KE (2012) Isolation and identification biosurfactant producing and crude oil degrading *Pseudomonas aeruginosa* strains. Chem Eng J 209:138–146. doi:10.1016/j.marpolbul.2015.09.053

Zouari R, Hamden K, El Feki A, Chaabouni K, Makni-Ayadi F, Sallemi F (2017) Evaluation of *Bacillus subtilis* SPB1 biosurfactant effects on hyperglycemia, angiotensin I-converting enzyme (ACE) activity and kidney function in rats fed on high-fat-high-fructose diet. Arch Physiol Biochem 123:112–120. doi:10.1080/13813455.2016.1261902

FSC
www.fsc.org
MIX
Papier aus verantwortungsvollen Quellen
Paper from responsible sources
FSC® C105338

If you have any concerns about our products,
you can contact us on
ProductSafety@springernature.com

In case Publisher is established outside the EU,
the EU authorized representative is:
Springer Nature Customer Service Center GmbH
Europaplatz 3, 69115 Heidelberg, Germany

Printed by Libri Plureos GmbH
in Hamburg, Germany